LINEAR SYSTEMS
with Applications
and Discrete Analysis

LINEAR SYSTEMS

with Applications
and Discrete Analysis

CARTER M. GLASS
United States Air Force Academy,
Colorado

WEST PUBLISHING CO.
St. Paul • New York • Boston
Los Angeles • San Francisco

Copyright © 1976 By WEST PUBLISHING CO.
All rights reserved
Printed in the United States of America

Library of Congress Cataloging in Publication Data

Glass, Carter M
    Continuous systems with theory, applications and discrete analysis.
    Bibliography: p.
    Includes index.
    1. Electric engineering—Mathematics. 2. System analysis. 3. Electric engineering—Data processing. I. Title.
TK153.G58   620'.0042'2   75-40282
ISBN 0-8299-0081-0

For Gayle, Julie and Carter

Preface

The current emphasis on analyzing engineering systems by discrete methods reflects the proliferation of small, inexpensive digital computers. Increasingly, colleges and universities have introduced into their engineering curricula courses in digital signal processing, real-time computation, and discrete analysis. The effect of this trend on the teaching of linear systems has been somewhat unexpected. Rather than replacing continuous topics with discrete ones, engineering educators have found it advantageous to move the continuous studies to an earlier point in the curriculum. This is because many of the discrete analysis techniques can be viewed as extensions or reformulations of classical continuous systems topics: digital filtering theory, for example, is, to a large degree, simply an extension of continuous filtering theory. We have thus come to view continuous systems analysis in part as a necessary prerequisite to the study of discrete systems and signals. And, of course, continuous mathematical models still conform more closely to "reality" in many applications, a fact we cannot disregard just because the most convenient analysis tool happens to lend itself to discrete mathematics.

The current revolution in electronics caused by the advent of integrated circuits has had a profound impact on the teaching of dynamic systems. Just as we now have special purpose digital computers or parts of digital processors on integrated-circuit chips, so we also have parts of analog computers--operational amplifiers, multipliers, summers, etc.--on chips. Thus at a time when the general purpose analog computer has begun to disappear, we find that the principles of analog devices have become more important than ever. For example, active filters implemented with integrated-circuit operational amplifiers are nothing more than special-purpose analog computers. Whereas courses in general purpose analog computation have begun to vanish from university curricula, the operational amplifier, which is the basic building block of the computer, has assumed the role of a fundamental circuit component. As a result, most circuit-theory, electronics and instrumentation courses devote considerable time to studying operational amplifiers.

In addition to presenting the standard linear systems topics, this book was written with a view to exploiting three trends in engineering education: (1) the need to move the study of continuous systems to an earlier point in the curriculum, (2) the need to introduce the operational amplifier as a fundamental building block, and (3) most important, to show how to use the digital computer as a computational tool for analyzing continuous systems. The book evolved over a period of years and is a synthesis of material used in a number of courses taught by the author at the United States Air Force

Preface

Academy. It is intended to support a first systems course for engineering students. The mathematical prerequisites are limited to a course in calculus and analytic geometry.

The book divides itself naturally into two parts: Chapters 1 through 10 show how to derive and analyze the mathematical models needed to analyze continuous engineering systems. Chapters 11, 12, and 13 then show how continuous system models can be discretized and subsequently analyzed by digital computers. Computer programming, however, is not discussed, but the material is written so that the student who has had no computer programming is not at a disadvantage.

Chapters 1 and 2 introduce such topics as signals, differential equation terminology, notation, dimensions and units. Chapter 3, a "mini-course" in electrical circuit analysis, includes a thorough introduction to the operational amplifier.

Chapter 4 deals with solving the differential equations which constitute the mathematical models of linear systems. Emphasis is on first- and second-order linear equations. Chapter 5 follows with an extensive introduction to the Laplace transform, the all-important frequency-domain language of linear systems analysis. Chapter 6 discusses the s-plane, a geometrical interpretation of the Laplace transform that lets the analyst learn useful information about a system without solving its describing differential equations.

Chapter 7, an introduction to transfer functions and feedback theory, is followed in Chapter 8 by techniques for analyzing linear systems in the sinusoidal steady state. Power and energy relationships and phasors are also treated in Chapter 8.

The continuous portion of the book concludes with Chapter 9 on the frequency response of linear systems (including Bode plots) and Chapter 10 which deals with the important modern topic of state-variable analysis. The matrix theory needed for understanding state variables is included in this chapter.

Chapter 11 begins the discrete portion of the book by introducing the idea of a discrete system, discussing the

Preface

sampling theorem (which establishes the criterion by which one can sample a continuous variable with assurance that no information is lost) and concludes with a discussion of practical methods for converting analog signals to digital form.

Chapter 12 is an in-depth presentation of the z-transform, a language of discrete systems which plays a role analagous to that of the Laplace transform with respect to continuous systems. The z-transform is applied both to solving discretized versions of differential equations and to representing discrete systems by z-domain transfer functions and block diagrams. Chapter 13 addresses the numerical solution of differential equations. The treatment includes both classical and z-transform methods.

The example problems used throughout the book to illustrate the principles of linear systems analysis are taken mostly from electrical engineering. For those who need a book with more of an interdisciplinary flavor, I have included a rather long appendix, Appendix H, which introduces mechanical systems, both rotational and translational, and also includes some material on linearizing nonlinear differential equations. Appendix H contains numerous worked examples and exercise problems.

The first 10 chapters of the book can be used for a one-quarter or even a one-semester course in linear systems analysis with an electrical engineering flavor. If Chapters 11, 12 and 13 are included, the student is equipped not only with the techniques for modeling and analyzing continuous dynamic systems, but also with the methods needed to analyze the systems on the digital computer. The end-of-chapter problems (with answers for the odd problems) and numerous illustrative examples make the book easily "teachable" whether the instructor uses the lecture-discussion method or one of the popular self-paced or self-managed teaching schemes.

In conclusion, I would like to acknowledge my indebtedness to my colleagues and students at the United States Air

Preface

Force Academy for providing, over the years, the inspiration and opportunity to develop the material which is included in this book. My thanks too to Barbara Flynn who typed the manuscript (several times). And without the patience and understanding of my children, Julie and Carter, and the encouragement and proofreading of my wife Gayle, the project would never have been completed.

CARTER M. GLASS

Contents

1 INTRODUCTION 1
Preliminaries 1
Models for Dynamic Systems 3
The Systems Approach 4
Computers in Continuous Systems Analysis 5
References 8

2 SIGNALS, FUNCTIONS AND DIFFERENTIAL EQUATIONS 11
Variables, Parameters and Constants 12
Linearity 13
Differential Equations 15
Types of Differential Equations 18
Signal Classification 21
Standard Forcing Functions 22
Notation, Dimensions and Units 43
References 52

3 INTRODUCTORY NETWORK ANALYSIS 55

Fundamental Laws and Quantities 56
Electrical Components 64
Loop Analysis 71
Nodal Analysis 79
Other Analysis Techniques 85
Transient Models and Initial Conditions 99
Operation Amplifiers 114
References 140

4 TRANSIENT RESPONSE OF DYNAMIC SYSTEMS 143

The Natural Response 144
Forced Response 163
Second-order Transients 176
Coupled Systems 180
Nonlinear and Time-varying Systems 181
References 189

5 THE LAPLACE TRANSFORM

The Laplace Transform Defined 193
Laplace Transforms of Useful Functions 196
Properties of the Laplace Transform 201
The Inverse Transform: Partial Fractions 211
s-Domain Models of Electrical Components 226
s-Domain Network Analysis 232
Coupled Differential Equations 242
The Impulse Response 245
Convolution 247
References 258

6 THE COMPLEX FREQUENCY PLANE 261

The s-Plane 262
Evaluating Residues in the s-Plane 274
Stability 283
References 295

7 SYSTEMS ANALYSIS   297

Preliminaries 297
Transfer Functions 299
Block Diagrams 301
Feedback Systems 317
Transfer Function Simulation 324
References 333

8 THE SINUSOIDAL STEADY STATE   335

*Transfer Functions in the
  Sinusoidal Steady State 336*
Phasors 342
Resonance 354
Power in the Sinusoidal Steady State 357
References 388

9 FREQUENCY RESPONSE OF LINEAR SYSTEMS   391

Frequency Response of First-Order Systems 392
Frequency Response of Second-Order Systems 398
Magnitude and Phase from the Pole-Zero Diagram 404
Distortion in the Sinusoidal Steady State 406
Introduction to Bode Plots 409
Experimental Determination of Transfer Functions 423
References 428

10 STATE SPACE ANALYSIS   431

Matrices 432
State Variables 443
Equivalence Transformations 453
Solving State Equations 457
References 468

11 INTRODUCTION TO DISCRETE ANALYSIS   471

Issues in Discrete Analysis   472
The Sampling Process   476
Signal Conversion   492
Summary   501
References   507

12 SYSTEM ANALYSIS BY z-TRANSFORM   511

The z-Transform Defined   512
The Inverse z-Transform   523
Theorems of the z-Transform   526
Discrete Systems   530
Convolution   537
Discrete Transfer Functions   541
Poles, Zeros and Transients   544
Synthesis of Transfer Functions   554
References   567

13 NUMERICAL SOLUTION OF DIFFERENTIAL EQUATIONS   569

Discretization in the Time Domain   570
Rectangular Integration   578
Trapezoidal Integration   583
State-Variable Systems   588
Runge-Kutta Integration   597
References   603

APPENDICES   606

A. Complex Algebra   607
B. Electrical Units   613
C. Unit Prefixes   614
D. Laplace Transform Pairs   615
E. Properties of the Laplace Transform   616
F. z-Transform Pairs   617
G. Bode Plots for Common Factors   618
H. Mechanical Systems   619
I. Translational Mechanical Units   653
J. Rotational Mechanical Units   654

Contents

ANSWERS TO SELECTED PROBLEMS   655

INDEX   667

CHAPTER 1

Introduction

1.1 PRELIMINARIES

A continuous system is a system whose variables change smoothly or continuously. Such systems are said to be <u>dynamic</u> if the changes are with respect to the independent variable time. A dynamic system may be electrical, mechanical, biological, aerodynamic, or even financial. Dynamic systems are all around us; some are man made, others are natural phenomena. An automobile or a space ship is clearly a dynamic system, but so is a TV set, the stock market, or the human body. Any person who intends to work in the modern technological community must have more than a passing acquaintance with the language and techniques of dynamic systems analysis.[1]*

What specifically does the engineer need to know about continuous dynamic systems? As a partial answer to this question, we observe that the ultimate job of an engineer is to <u>design</u>, that is, to create systems or devices whose behavior is predictable. The designer is given a set of specifications and must construct a system whose response conforms to those specifications. Of course not all engineers are designers, just as all musicians are not composers. But before we can design we must first be able to <u>analyze</u>: to determine the response of a dynamic system once we know its inputs and internal construction. This is because much real-world design work is nothing more than repeated analyses of various proposed system configurations. This brings us to question whether there are common features which allow us to examine dynamic systems from a variety of engineering disciplines in a unified manner.

The answer is an emphatic yes. Although the physical laws which govern the dynamic behavior of ocean waves differ from those which control a transistor amplifier, the processes by which we analyze them are quite similar. The physical laws are distilled into mathematical models that conform to formulas, theorems, empirical relationships, and equations. Once the engineer has mastered the mathematical models which relate to his discipline, the physics receeds into the background, and dynamic analysis becomes a proposition of manipulating a standard set of models using a standard set of mathematical tools. While the models common to one engineering discipline may differ from those needed for another, the mathematical tools have a surprisingly wide scope of utility. Thus the engineer competent in calculus, analytic geometry,

---

*Numbers in superscript parenthesis refer to end-of-chapter references.

Introduction

complex variables, differential equations, and Laplace transforms is well equipped to handle dynamic systems from aeronautics, electrical engineering, mechanics, or from a wide spectrum of other areas.

1.2 MODELS FOR DYNAMIC SYSTEMS

A fair question, especially for those who are not mathematically inclined, is why not study dynamic systems by the more direct method of building physical rather than mathematical models--prototypes, as they are called in the aircraft industry. Why not build an experimental aircraft, for example, and let a test pilot fly it under controlled conditions? The answer is that this procedure is expensive and often dangerous; in most instances we need ways to verify proposed designs short of building prototypes.

Well then, how about an inexpensive miniature working model of a proposed system? The idea of building a scale model of a system is appealing, provided we know how to "scale up" the test results so they apply to the full-sized system. This approach has proved successful in designing ocean-going ships, where scale models are tested in tow tanks to verify hull design. But what would we mean by a scale model of an electronic amplifier? In most cases scale models are also impractical, and we must turn instead to some kind of mathematical approach.

A mathematical model for a dynamic system is a set of differential equations which predict its behavior under typical inputs. The difficulties of mathematical modeling are two fold: how to develop a model which truly represents the system under study, and once the model is at hand, how to solve the equations which comprise it. Although this book is concerned somewhat with both of these problems, it concentrates more heavily on the second--how to analyze a system given its mathematical model.

1.3 THE SYSTEMS APPROACH [2]

A system is an interconnection of components unified by the need to perform some identifiable function. The components or subsystems derive their identities from the overall function of the parent system. Viewing systems in this manner, the engineer is concerned with the behavior of component parts only insofar as they **affect** the overall system objectives. The advantage of the system viewpoint is that it lets us take an axiomatic approach toward component modeling. We can, for instance, postulate the existence of a number of ideal mechanical and electrical components with rigid mathematical descriptions. Actual components always differ from the ideal ones, but if our goal is to understand fundamental concepts, we can ignore model imperfections. The reader interested in a more detailed discussion of the physical laws which underlie the electrical and mechanical models presented in subsequent chapters is referred to any of a number of standard physics texts.[3,4]

This book treats dynamic analysis from a generalized, systems viewpoint. To illustrate the fundamental principles, however, we need specific examples, and these are drawn primarily from electrical engineering. Electrical and electronic examples are used because the book is aimed partly at the electrical engineering student and because many non-electrical systems and processes are instrumented electronically. On the other hand, many of the principles of linear systems are more easily understood if they are applied to mechanical systems. This is because we have more of a feel for mechanical variables such as force and displacement than we do for voltage and current. And of course many of the users of linear systems analysis are neither electrical nor instrumentation engineers. Appendix H at the end of the book introduces translational and rotational mechanical systems (including gear trains). The reader who starts with Appendix H will be able to apply the theoretical material in the mainstream of the book to mechanical systems as he progresses.

Introduction

1.4 COMPUTERS IN CONTINUOUS SYSTEMS ANALYSIS

Analyzing system models requires a variety of intellectual skills. Some of these are purely mathematical, like differential-equation theory or Laplace transforms; others are specialized system concepts such as block-diagram algebra. Moreover, modern engineering systems have become so complex that pencil-and-paper methods of analysis are inadequate for all but the simplest of systems. For this reason, any worthwhile treatment of continuous systems must not only discuss the appropriate theory, but also introduce the computational tools needed to manipulate that theory.

Digital Computation[5]

The digital computer has become the dominant computational tool of the engineering community. Virtually all universities, corporations, and government laboratories have digital computer centers (plus numbers of smaller minicomputers scattered here and there). Any organization which deals with large engineering systems is sure to have suitable digital computing equipment with which to analyze those systems. Below are listed some of the advantages and disadvantages of the digital computer as a tool for analyzing continuous systems:

Advantages

•High-level digital computer languages are easily learned and do not depend on the particular computer being used. The engineer who knows FORTRAN, BASIC, ALGOL, or one of the other higher-order programming languages will find himself at home in most computer centers.

•Digital computer solutions are extremely accurate and are generated ultra-rapidly compared to hand calculation.

•Computational costs, including programming manhours, computer operating costs and other such factors are low compared to those for analog or hybrid methods, and, more important, are continuing to decrease (relatively).

- Most computer centers have program libraries which contain standard computing packages. These can be used "off the shelf" by the analyst so that he does not have to do all of his own programming.

Disadvantages

- Digital computers cannot directly handle the smoothly changing variables found in continuous dynamic systems. The continuous variables must first be <u>discretized</u>, a process which is both time consuming and, if not done properly, can introduce errors which offset the inherent accuracy of the computer.
- Digital output is numerical. The numerical solution of a differential equation is, for example, given as a list of points. Expensive peripheral equipment is needed to convert tabulated output data to continuous graphs.
- In off-line operation of large computers the user has little control over the solution as it progresses. This is a disadvantage compared to analog computation where the computer operator can vary parameters quickly between runs.
- Computation is done in series rather than in parallel, making the digital computer slower than its analog competitor for certain classes of problems.
- A digital computer program is not a true analogy of the system under study; hence, some of the internal variables are usually missing.

Analog Computation

The general purpose analog computer is a natural tool for analyzing continuous dynamic systems.[6,7] If properly programmed the analog computer is itself a dynamic system with voltages that vary continuously with time. Programming the computer consists of establishing analogies between computer voltages and problem variables (hence the name analog computer). All of the internal variables of the system preserve their identities, and the output is graphical, making for efficient interpretation of results. The operator has intimate interaction and control over the solution as it progresses: he can alter parameters at will to try different system configurations; and he can stop the solution at a given instant to examine internal variables. Computation is done in parallel, often giving the analog computer a significant speed advantage over its digital counterpart. Moreover, time scaling lets us match the running time of the problem to the inherent speed of the computer components. Nonlinear or time-varying systems present no special problems.[8]

The advent of integrated-circuit operational amplifiers has aroused new interest in analog computation. We now have the "distributed" computer--analog computer components and subsystems incorporated into larger systems. Active filters, for example, are special-purpose analog computers.[9]

On the debit side, analog computers have limited accuracy, and programming is often more of an art than a science. The computer operator must understand a great deal about the internal workings of his machine. Finally, the analog computer can be used only to simulate dynamic systems: it has no alternate uses and for that reason may not be worth the investment at some installations. These disadvantages have precipitated a gradual disappearance of the general purpose analog computer and along with it courses in pure analog computation at most universities.

Hybrid Computation

The hybrid computer attempts to combine in a single machine the advantages of both digital and analog computers.[10,11] Hybrid computers have a patch panel with standard analog components such as integrators, potentiometers, and summing amplifiers. They also have logic elements--AND gates, OR gates, registers, counters, and flip-flops--and hybrid elements which interface between the analog and logic components. Hybrid elements include comparators, electronic switches, relays, analog-to-digital converters, and digital-to-analog converters.[12] Finally, a sophisticated hybrid system will also contain a complete general-purpose digital computer which acts as a function generator, as an arithmetic unit, and as a control unit for the analog machine.

Hybrid computers are not only expensive and difficult to program, but they differ so greatly from one installation to the next that efforts to develop general hybrid-computer programming languages have largely met with failure. As a result, hybrids have failed to gain wide acceptance in the engineering community. All-digital equipment has improved so rapidly in the past ten years that hybrid computers have become lost in the stampede to develop real-time mini- and microcomputers. Nevertheless, for certain specialized applications in the aerospace industry, hybrid computers still outshine their digital competitors. In addition, a new breed of desk-top hybrids with a few analog, hybrid, and logic elements has been developed. These versatile yet economical computers are ideal for use in teaching dynamic analysis. Even these, however, will doubtless give way to the even smaller and cheaper digital microcomputers.

REFERENCES FOR CHAPTER 1

(1) Cannon, R. H. Jr.: <u>Dynamics of Physical Systems</u>, McGraw-Hill Book Co., Inc., New York, 1967.

(2) Goode, H. H. and R. E. Machol: <u>System Engineering</u>, McGraw-Hill Book Co., Inc., New York, 1957.

(3) Borowitz, S. and A. Beiser: <u>Essentials of Physics</u>, Addison-Wesley Publishing Co., Inc., Reading, Massachusetts, 1966.

(4) Halliday, D. and R. Resnick: <u>Physics for Students of Science and Engineering</u>, John Wiley and Sons, New York, 1962.

(5) Desmonde, W. H.: <u>Computers and Their Uses</u>, Prentice-Hall Inc., Englewood Cliffs, New Jersey, 1972.

(6) James, M. L., Smith, G. M. and J. C. Wolford: *Analog Computer Simulation for Scientists and Engineers*, Intext Educational Publishers, Inc., Scranton, Pennsylvania, 1971.

(7) Jackson, A. S.: *Analog Computation*, McGraw-Hill Book Co., Inc., New York, 1960.

(8) Johnson, C. L.: *Analog Computer Techniques*, McGraw-Hill Book Co., Inc., New York, 1956.

(9) Mitra, S. K. ed.: *Active Inductorless Filters*, IEEE Press, New York, 1971.

(10) Bekey, G. A. and W. J. Karplus: *Hybrid Computation*, John Wiley and Sons, New York, 1971.

(11) Bennett, A. W.: *Introduction to Computer Simulation*, West Publishing Co., Inc., St. Paul, Minnesota, 1974.

(12) Hausner, A.: *Analog and Analog/Hybrid Computer Programming*, Prentice-Hall, Inc., Englewood Cliffs, New Jersey, 1971.

CHAPTER 2

Signals, Functions and Differential Equations

If mathematics is the language of engineering, then surely differential equations are a large part of the vocabulary of the dynamic systems analyst. For this reason our study commences with a discussion of differential equations and their associated terminology. The treatment differs from that found in traditional mathematics texts in that we are more concerned with using differential equations to represent engineering systems. Finding solutions is of secondary importance.

2.1 VARIABLES, PARAMETERS AND CONSTANTS

Dynamic systems are classified according to the kinds of differential equations required to describe their behavior. The classification of differential equations depends in turn on the kinds of variables and parameters used in their development. In general, a variable is a quantity which changes during an analysis while parameters and constants do not. More specific definitions follow:

Independent Variables

Independent variables are quantities which change, but over which neither we nor the system exercise control. The most common independent variables are time (t) and the spatial coordinates (xyz). The derivatives in differential equations are usually taken with respect to independent variables.

Dependent Variables

Dependent variables identify the magnitudes of system quantities--the voltage across a resistor or the angular velocity of a rotating shaft, for example. Dependent variables, as the name implies, depend on the values of other quantities. More specifically, they are functions of the independent variables. Throughout the text we use lower-case letters near the end of the alphabet to represent dependent variables. The explicit dependence on an independent variable is shown by parentheses; thus y(t), read "y of t", is a dependent variable which depends on time, t. At times we omit the explicit dependence of the independent variable when the context makes the situation unambiguous. We might, for example, use either v(t) or v for a voltage variable.

A _signal_ is a special kind of dependent variable, one which depends only on time. In general we make no distinction among the terms signal, function, and dependent variable when discussing continuous dynamic systems.

Signals, Functions and Differential Equations 13

Parameters

Parameters are fixed quantities associated with systems. The mass of a wheel or the value of a resistor are examples of system parameters. Strictly speaking, however, parameters may also be functions: the density of a nonhomogeneous material may vary with the space coordinates, or the mass of a rocket might change with time as it consumes fuel. Parameters appear in differential equations as coefficients by which the derivatives or other terms are multiplied.

Constants

Constants are fixed quantities which differ from parameters in that they are not associated with specific systems. The conversion factor from degrees to radians is a constant.

Variables, parameters, signals, and constants all may have both <u>dimensions</u> and <u>units</u> which relate them to the physical world. Section 2.7 discusses the dimensions and units of quantities used in the following chapters. Depending on the context, we may occasionally omit units to avoid obscuring important results with details.

2.2 LINEARITY

Dynamic systems are classed as either <u>linear</u> or <u>nonlinear</u>. Linear systems are mathematically tractable and have spawned a great body of theory (much of it a restatement of the superposition principle). Nonlinear systems, on the other hand, are difficult to investigate using hand calculation, and little general theory is available. Each nonlinear system must be treated as a special case.

A linear system is one which satisfies the principle of <u>superposition</u>.[1] Figure 2-1 shows a single-input-single-output system with input $x(t)$ and output $y(t)$.

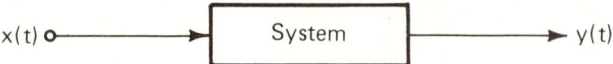

Figure 2-1. A system operates on an input variable to produce an output variable.

Let $y_1(t)$ be the response of the system to the input $x_1(t)$ and $y_2(t)$ be the response to $x_2(t)$. The system is <u>linear</u> if the input

$$x_3(t) = ax_1(t) + bx_2(t) \qquad (2-1)$$

produces the output

$$y_3(t) = ay_1(t) + by_2(t) \qquad (2-2)$$

where a and b are constants. This is the principle of superposition.

The term linear is applicable to systems, differential equations, and functional operators. A linear system satisfies superposition of inputs as just described. A linear differential equation satisfies superposition of its solutions. To understand linearity of mathematical operators or functions consider the operation of squaring. Because $y_1 = x_1^2$, if we form the combined input $x_3 = ax_1 + bx_2$, then the output y_3 is

$$y_3 = x_3^2 = (ax_1 + bx_2)^2 \qquad (2-3)$$

and

$$y_3 = a^2 y_1 + 2abx_1 x_2 + b^2 y_2 \qquad (2-4)$$

If the squaring operation were linear, it would produce the output $y_3 = ay_1 + by_2$. As the two expressions for y_3 differ, we conclude that squaring is a nonlinear operation. Integration with respect to time, on the other hand, is linear because

$$y_3 = \int_0^t x_3 \, dt = \int_0^t (ax_1 + bx_2) \, dt \qquad (2-5)$$

and $y_3 = ay_1 + by_2$ as required.

Example 2-1. Investigate the linearity of the differentiating system shown in Fig. 2-2.

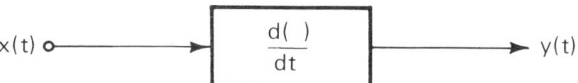

Figure 2-2. The operation of differentiation can be interpreted as a system.

Solution. Let $y_1 = dx_1/dt$ and $y_2 = dx_2/dt$. If we form the combined input $x_3 = ax_1 + bx_2$, then the output y_3 is $y_3 = dx_3/dt$ or

$$y_3 = \frac{d}{dt}\{ax_1 + bx_2\} = a\left\{\frac{dx_1}{dt}\right\} + b\left\{\frac{dx_2}{dt}\right\} \qquad (2\text{-}6)$$

and $y_3 = ay_1 + by_2$ as required for superposition; hence, differentiation is a linear operation.

2.3 DIFFERENTIAL EQUATIONS [2]

A differential equation is defined as an equation which contains one or more differentials or derivatives. The following are examples of differential equations:

$$dy = 0 \qquad (2\text{-}7)$$

$$\frac{dy}{dx} + ax = f(x) \qquad (2\text{-}8)$$

$$y'' + 5y' + y = 0 \qquad (2\text{-}9)$$

$$\ddot{y} + 2\zeta\omega_n \dot{y} + \omega_n^2 y = 0 \qquad (2\text{-}10)$$

$$\dot{y} + 3y - x = 0 \qquad \dot{x} + 2x - y = 5 \qquad (2\text{-}11)$$

Equation (2-7) is simple but, as it contains a differential, it meets the requirements. The prime notation used in Eq. (2-9) means that $y' = dy/dx$ and $y'' = d^2y/dx^2$. Similarly, Eq. (2-10) contains the widely used "dot" notation for derivatives with respect to time—$\dot{y} = dy/dt$, $\ddot{y} = d^2y/dt^2$, etc. Finally, Eqns. (2-11) are <u>simultaneous differential equations</u>. These are needed to describe so-called "coupled" systems which may have several dependent variables but only one independent variable. In Eqns. (2-11) both x(t) and y(t) are dependent variables, but both depend solely on the independent variable time.

The order of a differential equation is the order of the highest-order derivative contained in the equation. Equations (2-7) and (2-8) are of the first order. Equations (2-9) and (2-10) are second-order equations. The order of a system is always less than or equal to the sum of the orders of the differential equations needed to model the system mathematically. The coupled first-order equations, Eqns. (2-11), describe a second-order system.

Initial Conditions

The example differential equations, Eqns. (2-7) through (2-11), are not complete. Before it can be solved, a differential equation must have initial or boundary conditions. For an ordinary differential equation, the number of initial conditions equals the order of the equation. An ordinary second-order equation, for example, requires two initial conditions, one for the dependent variable and one for its first derivative.

Physically, initial conditions arise from energy stored in system components and released as the motion takes place. A charged capacitor contains stored electrical energy, and the differential equation of a network containing a charged capacitor requires one or more initial conditions to account for this stored energy.

Solutions

A function is a solution of a differential equation if, (a) when substituted into the equation, it reduces the differential equation to an identity and (b) it satisfies the initial conditions. The function y(t)=exp(-at), for example, is a solution of the differential equation ẏ(t)+ay(t)=0 with

Signals, Functions and Differential Equations

y(0)=1. This is true because y(0)=exp(0)=1 (the initial contion is satisfied) and

$$\frac{d}{dt}e^{-at} + ae^{-at} = -ae^{-at} + ae^{-at} \equiv 0 \qquad (2\text{-}12)$$

In other words, substituting y(t)=exp(-at) into the differential equation has reduced it to the identity 0≡0.

Do all differential equations have solutions? It depends. If by solution we mean finite sums and products of the so-called <u>elementary functions</u>--sinusoids, polynomials, and exponentials--the answer is an emphatic no. Few differential equations have "closed-form" solutions in terms of elementary functions. Many differential equations have solutions which must be written either as infinite series or in terms of special constructions such as Bessel functions or Legendre polynomials. Fortunately, the linear, constant-coefficient differential equations needed to model most continuous systems have simple solutions. Even more fortunately, we do not need to find the solutions too often--just talk about them. Solutions are discussed in several later chapters.

Forcing Functions

The standard form for a differential equation is with all terms containing the dependent variable or its derivatives collected on the left-hand side of the equation, and with the coefficient of the highest-order derivative equal to unity. All other terms are on the right-hand side. When the equation is written in this form, the sum of the terms on the right-hand side is called the <u>forcing function</u>. The equation

$$\dot{y} + 5y = t + 8\cos(3t) \qquad (2\text{-}13)$$

has the forcing function f(t)=t+8cos(3t). The forcing function is the input to a dynamic system which, as the name implies, forces it to move. In general, dynamic systems are dynamic for one of two reasons: (a) they are forced to move by the application of forcing functions, or (b) they "relax" by dissipating stored energy. This means that if the right-hand side of a differential equation is zero (no forcing function) and all initial conditions are zero (no stored energy in the system), then the solution is identically zero for all time--nothing moves. As an illustration, the functions $y(t)=0$ and $y(t)=\exp(-at)$ are both potential solutions of $\dot{y}+ay=0$ depending on the initial conditions. Section 2.6 introduces some of the forcing functions encountered in the remainder of the book.

2.4 TYPES OF DIFFERENTIAL EQUATIONS

The general ordinary differential equation has the form

$$a_n \frac{d^n t}{dt^n} + a_{n-1}\frac{d^{n-1}y}{dt^{n-1}} + \ldots a_1\frac{dy}{dt} + a_0 y = f(t) \tag{2-14}$$

where y is a function only of t. The parameters a_i may be constants, functions of t, functions of y, or functions of both t and y. We can, with no loss of generality, divide the equation by a_n to achieve a leading coefficient of 1. Equation (2-14) is accompanied by a set of n initial conditions, $y(0), \dot{y}(0), \ldots y(0)^{(n-1)}$.

Constant-Coefficient Equations

If the a_i are all constants, then Eq. (2-14) is a linear, constant-coefficient differential equation. To show linearity we must show that if $y_1(t)$ is a solution for input $f_1(t)$, and $y_2(t)$ is a solution for $f_2(t)$, then $y_3(t)=Ay_1(t)+By_2(t)$ is a solution for the input $f_3(t)=Af_1(t)+Bf_2(t)$. Thus, for the second-order equation

$$\ddot{y} + a_1\dot{y} + a_0 y = f(t) \tag{2-15}$$

if $y_1(t)$ is a solution for $f_1(t)$, we have

$$\ddot{y}_1 + a_1\dot{y}_1 + a_0 y_1 \equiv f_1(t) \tag{2-16}$$

Signals, Functions and Differential Equations

and similarly for $y_2(t)$. If the input $f_3 = Af_1 + Bf_2$ has $y_3 = Ay_1 + By_2$ as its solution, then for linearity the identity

$$\frac{d^2}{dt^2}\{Ay_1+By_2\} + a_1\frac{d}{dt}\{Ay_1+By_2\} + a_0\{Ay_1+By_2\} \equiv Af_1+Bf_2 \tag{2-17}$$

must hold. Using the properties of derivatives and regrouping, we can rewrite Eq. (2-17) as

$$A(\ddot{y}_1+a_1\dot{y}_1+a_0y_1) + B(\ddot{y}_2+a_1\dot{y}_2+a_0y_2) \equiv Af_1 + Bf_2 \tag{2-18}$$

and, finally,

$$A(\ddot{y}_1+a_1\dot{y}_1+a_0y_1-f_1) + B(\ddot{y}_2+a_1\dot{y}_2+a_0y_2-f_2) \equiv 0 \tag{2-19}$$

From Eq. (2-16) and its equivalent for y_2, the bracketed terms are zero and the identity holds. This proves linearity. Similar reasoning holds for the nth-order case. Most of the material in the remainder of the book deals with linear, constant-coefficient differential equations.

Time-Varying Systems

A system may be linear but still be time varying.[3] By this we mean that one or more parameters, though invariant with respect to the space variables, changes with time. A rocket vehicle in flight consumes its fuel so rapidly that its mass must be treated as a time-varying parameter. Such systems are represented mathematically by time-varying differential equations. This in turn means that at least one of the a_i in Eq. (2-14) is a function of time.

Nonlinear Systems

If at least one of the a_i in Eq. (2-14) is a function of the dependent variable or its derivatives, say $a_i = a_i(y)$, then the system described by the equation is <u>nonlinear</u>.[4,5]

Higher-order nonlinear differential equations seldom have closed-form solutions in terms of elementary functions. The following are typical nonlinear differential equations:

$$\dot{y} + y^2 = f(t) \tag{2-20}$$

$$\dot{y} + A\sin(y) = 0 \tag{2-21}$$

$$\ddot{y} + \dot{y}|y| + Ky = f(t) \tag{2-22}$$

As compared with linear systems, a limited amount of theory has been developed for nonlinear systems, and they must usually be analyzed on a case-study basis. Fortunately for the analyst, nonlinear effects can often be treated as minor distortions of a basically linear model, permitting us to achieve good results with linear theory. The following chapters deal mainly with linear systems.

While both nonlinear and time-varying systems are difficult to analyze using hand calculation, they present no special problems to a digital or an analog computer. Even an equation which has no closed-form solution can be solved on the computer and its solution displayed graphically or tabulated point by point.

Finally, we can have differential equations which are both nonlinear **and** time-varying. This occurs when at least one of the a_i in Eq. (2-14) is a function of both the dependent and the independent variable, say $a_i = a_i(t,y)$. The equation

$$\dot{y} + yt^2 \ln(y) = f(t) \tag{2-23}$$

is nonlinear and time-varying because the coefficient of y is $t^2 \ln(y)$. While no general methods exist for solving such equations, they can be handled using analog or digital computers. Happily, such monstrosities are seldom needed as mathematical models for real-world dynamic systems.

2.5 SIGNAL CLASSIFICATION

The advent of the digital computer as a tool for systems analysis has led to increased interest in discrete mathematics. Classical numerical analysis as a mathematical sub-discipline has been with us literally for centuries, but the ability of the digital computer to execute numerical algorithms with great speed has brought such methods from the realm of the pure mathematician into the everyday world of the systems analyst. Unfortunately, the continuous variables needed to describe most dynamic systems cannot be handled by the computer; such variables must first be discretized.

Dynamic systems are classed as either <u>continuous</u> or <u>discrete</u> depending on whether the associated signals are continuous or discrete functions of time.[6] (The terms signal and variable are often used interchangeably, although as noted above a signal is a function whose independent variable is always time, whereas a variable may depend on quantities other than time.) Signals may be classed as follows (see Fig. 2-3):

- A <u>continuous signal</u> exists for all values of time in a given range.
- A <u>discrete signal</u> is defined only for discrete values of time.
- An <u>analog signal</u> is a continuous signal whose magnitude is also a continuous function.
- A <u>sampled-data signal</u> is a discrete signal whose amplitude is continuous.
- A <u>digital signal</u> is a discrete signal whose amplitude is also discrete.
- A <u>binary signal</u> is a signal, either continuous or discrete, whose amplitude assumes only two values.

The signal in Fig. 2-3(a) is continuous and analog while the one in Fig. 2-3(b) is continuous, but, due to the discontinuities is not analog. Figure 2-3(c) shows a discrete signal which may be either a sampled-data or a digital signal, depending on whether or not the amplitude is also quantized.

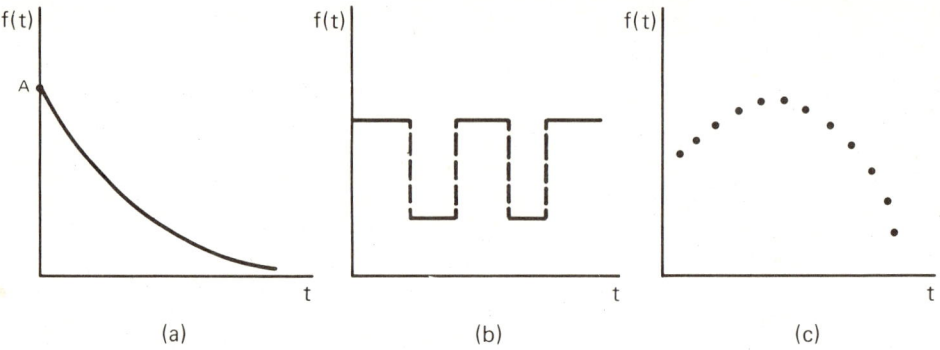

Figure 2-3. Signal classification depends on the nature (continuous or discrete) of the associated variables. (a) A continuous analog signal; (b) A continuous non-analog signal; (c) A discrete signal.

2.6 STANDARD FORCING FUNCTIONS

The function $f(t)$ in Eq. (2-14) is the forcing function of the system modeled by the differential equation. Forcing functions are the system inputs. As such, they are either inherent in the environment of the system, or they are constructed by the analyst as test inputs. We have little control over inherent forcing functions, but we are free within wide limits to choose test inputs. The term "test inputs" does not imply that we actually operate the system: the test may be an analysis by hand calculation or a computer simulation.

The standard forcing functions are the step, the polynomial, the sinusoid, the exponential, and the impulse.[7,8] These five functions can be manipulated (added, multiplied, divided, integrated, or differentiated) to produce virtually all functions needed in the analysis or design of engineering systems.

The Step Function

The step function corresponds to a constant value applied suddenly to the system and maintained for a relatively long time. A large body of knowledge concerning the step response of linear systems has been accumulated. The unit step function, denoted hereafter by u(t), is shown in Fig. 2-4(a) and defined by

$$u(t) = 1 \text{ for } t \geq 0 \qquad (2-24)$$

and u(t)=0 for t<0. Mathematically, the step function has a jump discontinuity at t=0.

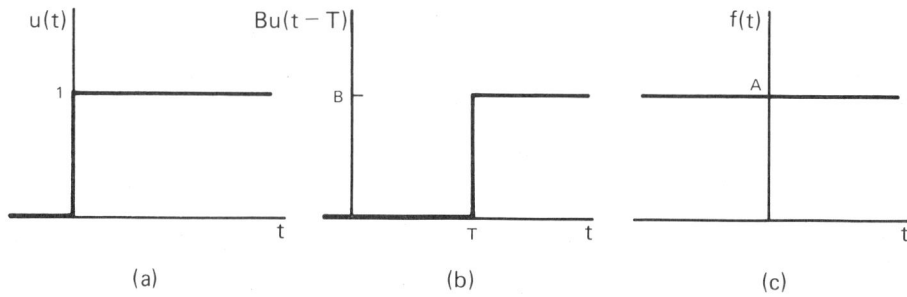

Figure 2-4. The step function rises abruptly from zero. It differes from the constant function which never changes. (a) A unit step function; (b) A delayed step function; (c) An eternal constant.

A more complete definition includes the possibility that the step function has a magnitude other than 1 and is "turned on" at some time other than t=0. The delayed step function of Fig. 2-4(b) satisfies the relationship

$$Bu(t-T) = B \text{ for } t \geq T \qquad (2-25)$$

and Bu(t-T)=0 for t<T. The constant B is the amplitude and T is the time when the step rises. In general, a unit step function has magnitude 1 when its argument is positive or zero and magnitude 0 when its argument is negative. In many applications it is important to distinguish between an "eternal" constant, Fig. 2-4(c), and one which suddenly

appears at t=0. The use of step functions always clarifies the issue.

It is impossible to generate a true step function in the real world because no physical variable can undergo a jump change in its value due to inertia. We can, however, approximate step functions closely; all we require is that the duration of the constant value be long compared to the time needed for the function to rise to that value. The rise time of the step function must also be short compared to time constants of the system to which it is applied (about which more later).

Although the step function may seem to be an almost trivial functional building block, we can use it to construct a variety of more complex waveforms. (We use the term <u>waveform</u> to describe the pattern of variations of a signal.)

Example 2-2. Express the pulse of Fig. 2-5(a) in terms of step functions.

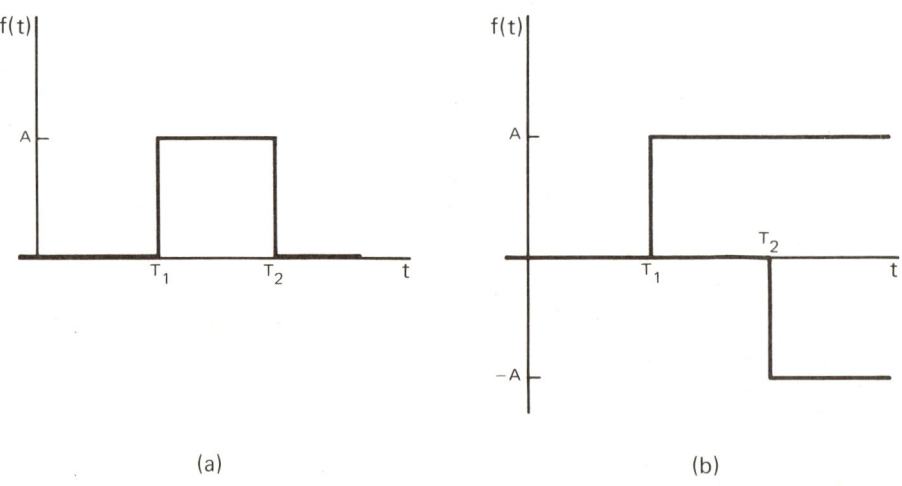

Figure 2-5. A rectangular pulse can be written as the sum of two step functions. (a) A rectangular pulse; (b) Step functions which can be summed to give the rectangular pulse.

Signals, Functions and Differential Equations

Solution. The pulse is the sum of two delayed step functions,

$$f(t) = A\{u(t-T_1) - u(t-T_2)\} \qquad (2-26)$$

as shown in Fig. 2-5(b). When $t \geq T_2$ the negative and positive step functions cancel to give the required value zero.

Example 2-3. Express the pulse sequence of Fig. 2-6 in terms of step functions.

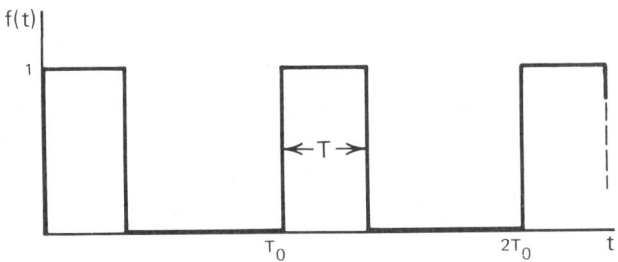

Figure 2-6. This pulse sequence can be written as an infinite sum of suitably delayed step functions according to Eq. (2-27).

Solution. Extend the result of Example 2-2 and write the infinite summation

$$f(t) = \sum_{i=0}^{\infty} \{u(t-iT_0) - u(t-iT_0-T)\} \qquad (2-27)$$

Both the derivative and the integral of the step function give rise to other useful functions, the impulse and the ramp. These are discussed in later paragraphs.

The Sinusoid

Sinusoidal functions arise in continuous systems analysis because (1) many dynamic systems have natural oscillations which are inherently **sinusoidal**, (2) a variety of non-sinusoidal functions can be expressed as sums of sinusoids using Fourier series, and (3) sinusoidal functions have mathematical properties which make them useful test inputs for dynamic systems.

A sinusoidal function is defined by its amplitude A, angular frequency ω_o, and phase $\emptyset$. The defining equation is

$$y(t) = A\cos(\omega_o t - \emptyset) \qquad (2\text{-}28)$$

The quantity $\emptyset/\omega_o$ measures the distance between the origin and the first positive peak of the function (Fig. 2-7). The argument of a sinusoidal function must be dimensionless. This implies that if t is time in seconds, then the radian frequency ω_o has units of $(\sec)^{-1}$ and the phase angle $\emptyset$ is dimensionless (or in radians as the radian is a dimensionless unit of measure). In practice, ω_o has the units of radians per second. If $\emptyset$ is given in degrees it must be converted to radians before it is combined with the dimensionless term $\omega_o t$. The amplitude A has the same dimensions as y(t) and is the maximum value attainable by y(t). The peak-to-peak value of y(t) is 2A.

The sinusoid of Fig. 2-7 is an eternal function: it extends to infinity in both directions. In addition, it is a periodic function, endlessly repeating the same waveform. A periodic function is one for which

$$y(t) = y(t-T_o) \qquad (2\text{-}29)$$

If T_o is the smallest constant for which Eq. (2-29) holds, then T_o is the period of the function. The period is thus the minimum time interval between successive identical functional values. For a sinusoid the period is related to the radian frequency ω_o by the expression $T_o = 2\pi/\omega_o$. The conventional frequency f_o satisfies $f_o = \omega_o/2\pi = 1/T_o$ and has the units of hertz (Hz). These relationships lead to the alternate sinusoidal forms

$$y(t) = A\cos(2\pi f_o t - \emptyset) \qquad (2\text{-}30)$$

and

$$y(t) = A\cos(2\pi t/T_o - \emptyset) \qquad (2\text{-}31)$$

Signals, Functions and Differential Equations

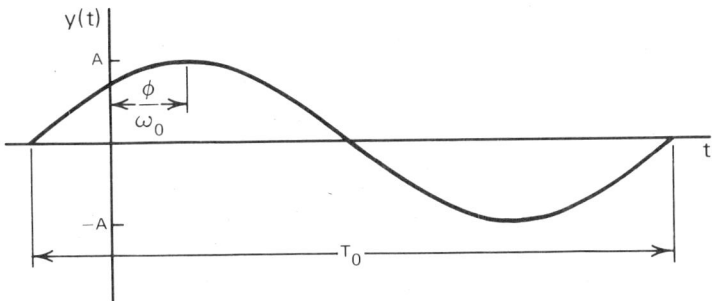

Figure 2-7. The general sinusoid $y(t) = A\cos(\omega_0 t - \phi)$ is completely specified by its amplitude A, phase ϕ, and radian frequency ω_0.

By a proper choice of ϕ in Eq. (2-28) we can achieve either a pure (zero phase angle) sine or a pure cosine function. The general form is equivalent to the sum of a sine and a cosine. To see this, expand Eq. (2-28) using the trigonometric identity for the cosine of the difference of two angles,

$$\cos(x-y) = \cos(x)\cos(y) + \sin(x)\sin(y) \qquad (2\text{-}32)$$

The result is

$$y(t) = A\cos(\omega_0 t)\cos(\phi) + A\sin(\omega_0 t)\sin(\phi) \qquad (2\text{-}33)$$

or

$$y(t) = C_1 \cos(\omega_0 t) + C_2 \sin(\omega_0 t) \qquad (2\text{-}34)$$

where $C_1 = A\cos(\phi)$ and $C_2 = A\sin(\phi)$. If $\phi = \pi/2$, $C_1 = 0$ and a pure sine wave results; if $\phi = 0$, $C_2 = 0$ and we have a pure cosine wave. In many derivations throughout the remainder of the book we assume cosine inputs to dynamic systems. This restriction does not limit our generality as $\sin(x) = \cos(\pi/2 - x)$. Finally, the expressions $C_1 = A\cos(\phi)$ and $C_2 = A\sin(\phi)$ can be "inverted". Thus,

$$C_1^2 + C_2^2 = A^2\{\cos^2(\phi) + \sin^2(\phi)\} \qquad (2\text{-}35)$$

or

$$A = \sqrt{c_1^2 + c_2^2} \qquad (2-36)$$

and

$$\frac{C_2}{C_1} = \frac{A\sin(\phi)}{A\cos(\phi)} \qquad (2-37)$$

or

$$\phi = \tan^{-1}\left\{\frac{C_2}{C_1}\right\} \qquad (2-38)$$

When a sinusoid is plotted, the abcissa can be labeled either in time, periods, radians, or degrees. Figure 2-8 illustrates this idea for the function $y(t) = A\sin(\omega_o t)$.

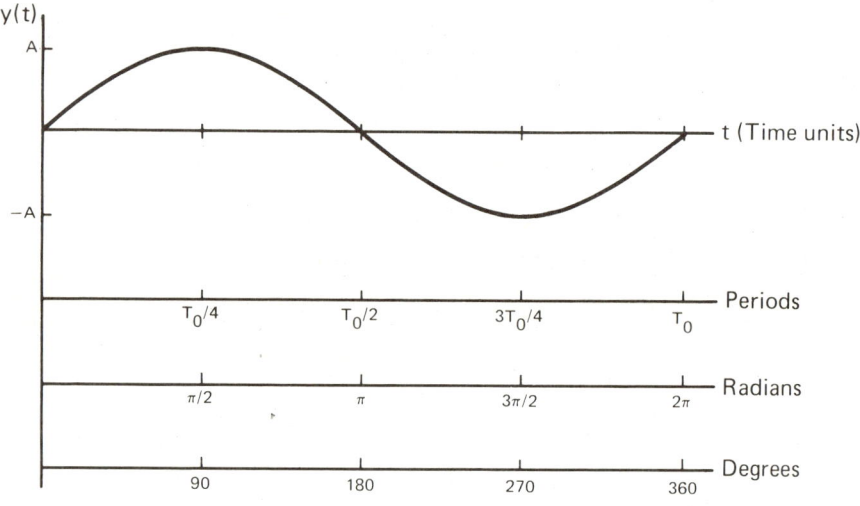

Figure 2-8. A sinusoid, in this case $y(t) = A\sin(\omega_o t)$, can be plotted with the abcissa labelled in time, periods, radians, or degrees.

<u>The derivative property</u>. The sinusoidal function maintains its waveshape when differentiated or integrated. This is seen from

$$\frac{d}{dt} A\cos(\omega_o t) = -\omega_o A\sin(\omega_o t) \qquad (2-39)$$

$$= -\omega_o A\cos(\omega_o t - 90°) \qquad (2-40)$$

$$= \omega_o A\cos(\omega_o t + 90°) \qquad (2-41)$$

and

$$\int A\cos(\omega_o t)dt = \frac{A}{\omega_o}\sin(\omega_o t) + C \qquad (2\text{-}42)$$

$$= \frac{A}{\omega_o}\cos(\omega_o t - 90°) + C \qquad (2\text{-}43)$$

The sinusoid which results from differentiating or integrating a sinusoid has the same frequency as the parent function but a different amplitude and phase angle.

The Exponential Function

The exponential function, though seldom used as a test input, is important in continuous systems analysis. This is because the exponential function is the basic solution form for linear, constant-coefficient differential equations. As a result, if the output of one system is fed to the input of another, the second system is often forced by some combination of exponential functions.

The exponential function is defined by

$$f(t) = u(t)Ae^{-t/T_c} \qquad (2\text{-}44)$$

The constant A is the initial value of the function and also the maximum value if $f(t)$ decreases. The unit step function multiplier, which insures that the function is 0 for $t<0$, is often omitted in problems which "start" at $t=0$. The quantity T_c, the <u>time constant</u> of the exponential function, has the dimensions of time because the argument must be dimensionless. (The reason for this is that the exponential is defined by the infinite series

$$e^x = 1 + x + \frac{x^2}{2!} + \frac{x^3}{3!} + \ldots \qquad (2\text{-}45)$$

If the argument x were not dimensionless, the right-hand side of Eq. (2-45) could not possibly be correct.) If the time constant is negative, the exponential grows rather than decays with time. Figure 2-9 shows both the decay and growth cases.

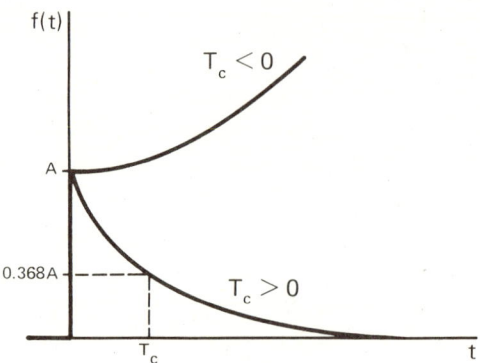

Figure 2-9. The exponential function $f(t) = u(t)A\exp(-t/T_c)$ decays for positive T_c and grows for negative T_c.

The time constant determines the decay rate. For $t=T_c$, Eq. (2-44) becomes $f(T_c)=A\exp(-1)$, which is approximately 0.368A. This means that the exponential function decays to approximately 37% of its initial value during the first time constant. After five time constants, $f(5T_c)=A\exp(-5)$, which is approximately 0.00674A. The exponential thus decays to less than 1% of its initial value after a time span of five time constants. Although theoretically the exponential never decays to 0, for practical purposes we take the signal duration as five time constants. Table 2-1 gives the values of the exponential function for the first five time constants.

<u>Properties of the exponential</u>. A useful property of the exponential is that its form is invariant under the operations of differentiation and integration; thus,

$$\frac{de^{-at}}{dt} = -ae^{-at} \qquad (2\text{-}46)$$

and

$$\int e^{-at} dt = -\frac{1}{a}e^{-at} + C \qquad (2\text{-}47)$$

Equation (2-46) leads to the <u>slope property</u>. If $f(t) = \exp(-t/T_c)$* is plotted, then the slope of the curve is $df(t)/dt$, where

---

*We use this alternate notation for the exponential when the function is written in the text rather than displayed.

Signals, Functions and Differential Equations 31

Table 2-1.

Some Values of the Exponential Function

Time(t)	Ae^{-t/T_c}
0	A
T_c	0.368A
$2T_c$	0.135A
$3T_c$	0.498A → .0498A
$4T_c$	0.183A → .0183A
$5T_c$	0.00674A

$$\frac{df(t)}{dt} = -\frac{1}{T_c}\exp(-t/T_c) = -\frac{1}{T_c}f(t) \qquad (2\text{-}48)$$

Equation (2-48) shows that the slope at any point is proportional to the value of the function--large slopes for small t and small slopes for large t. More important, Eq. (2-48) reveals that small time constants lead to large slopes and rapid decay while large time constants imply small slopes and slow decay.

Example 2-4. Show that the magnitude of an exponential changes by an equal percent in equal time intervals, regardless of the location of the time interval. This is the so-called <u>decrement property</u>.

Solution. Recall the rules for exponents, $\exp(a)\exp(b) = \exp(a+b)$, and $\{\exp(a)\}^b = \exp(ab)$. The value of the exponential Eq. (2-44) at some time $t_1 > 0$ is

$$f(t_1) = Ae^{-t_1/T_c} \qquad (2\text{-}49)$$

and at some time later, say $t_1 + T$, the value is

$$f(t_1 + T) = Ae^{-(t_1+T)/T_c} \qquad (2\text{-}50)$$

The ratio of the two values is

$$\frac{f(t_1+T)}{f(t_1)} = \frac{Ae^{-(t_1+T)/T_c}}{Ae^{-(t_1/T_c)}} = e^{-T/T_c} \qquad (2\text{-}51)$$

Thus, the **percent** change in interval T does not depend on t_1. If, for example, an exponential decreases by 10% during the first second, it decreases by 10% during every one-second interval thereafter.

   Complex exponentials. In much of our subsequent work we will make extensive use of the intimate relationship which exists between complex exponential functions and real sines and cosines. If the reciprocal of the time constant T_c in Eq. (2-44) is a complex number, say $1/T_c = (\sigma+j\omega)$, where σ and ω are positive constants and $j=\sqrt{-1}$, then f(t) is a complex exponential. Equation (2-44) then becomes

$$f(t) = Ae^{-(\sigma+j\omega)t} = Ae^{-\sigma t}e^{-j\omega t} \qquad (2\text{-}52)$$

where we have omitted the unit step with the understanding that we are referring to times after t=0. Using series expansions for the sine and cosine functions, it can be shown that

$$e^{-j\omega t} = \cos(\omega t) - j\sin(\omega t) \qquad (2\text{-}53)$$

An alternate interpretation of Eq. **(2-53)** is that the <u>real part</u> of exp($-j\omega t$) is cos(ωt), or

$$\text{Re}\{e^{-j\omega t}\} = \cos(\omega t) \qquad (2\text{-}54)$$

and the <u>imaginary part</u> is $-\sin(\omega t)$, or

$$\text{Im}\{e^{-j\omega t}\} = -\sin(\omega t) \qquad (2\text{-}55)$$

Substituting Eq. (2-53) into Eq. (2-52) yields

$$f(t) = Ae^{-\sigma t}\{\cos(\omega t) - j\sin(\omega t)\} \qquad (2\text{-}56)$$

The complex time function f(t) is seen to consist of a real sinusoid plus an imaginary sinusoid, each with the <u>envelope</u> Aexp($-\sigma t$).

Signals, Functions and Differential Equations 33

Equation (2-53) can be solved for the trigonometric functions in terms of exponentials, with the results

$$\cos(x) = \frac{1}{2}(e^{jx} + e^{-jx}) \quad (2\text{-}57)$$

and

$$\sin(x) = \frac{1}{2j}(e^{jx} - e^{-jx}) \quad (2\text{-}58)$$

These are called Euler's equations for the sine and cosine functions. We will find much use for them in later work.

Example 2-5. Show that if A and B are complex conjugates the expression

$$f(t) = Ae^{-(\sigma - j\omega_d)t} + Be^{-(\sigma + j\omega_d)t} \quad (2\text{-}59)$$

can be written as the product of a real exponential and a sinusoid.

Solution. If A and B are complex conjugates we can write A=a+jb and B=a-jb. (See Appendix A for a review of the algebra of complex numbers.) Putting these expressions in magnitude-angle form yields A=Cexp(jØ) and B=Cexp(-jØ) where

$$C = \sqrt{a^2 + b^2} \quad (2\text{-}60)$$

and Ø=tan^{-1}(b/a). Substituting for A and B in Eq. (2-59) yields

$$f(t) = Ce^{j\emptyset}e^{-(\sigma - j\omega_d)t} + Ce^{-j\emptyset}e^{-(\sigma + j\omega_d)t} \quad (2\text{-}61)$$

$$= Ce^{-\sigma t}\left\{e^{j(\omega_d t + \emptyset)} + e^{-j(\omega_d t + \emptyset)}\right\} \quad (2\text{-}62)$$

and, using Eq. (2-57),

$$f(t) = 2Ce^{-\sigma t}\cos(\omega_d t + \emptyset) \quad (2\text{-}63)$$

as required. Equation (2-63) is one form of the <u>damped sinusoid</u> encountered so often in linear systems work.

The Impulse Function[9]

Impulse functions are needed when there is an almost instantaneous transfer of energy (e.g. hitting a nail with a hammer). The unit impulse function, $\delta(t-a)$, is defined by the expressions

$$\delta(t-a) = 0 \quad t \neq a \tag{2-64}$$

and

$$\int_{-\infty}^{\infty} \delta(t-a)dt = 1 \tag{2-65}$$

The impulse is undefined at t=a. Geometrically, the impulse can be developed by starting with a rectangle of height 1/A and width A. In the limit as A approaches 0, the area of the rectangle remains A(1/A)=1 even through the height becomes infinite and the width approaches 0 (Fig. 2-10(a)).

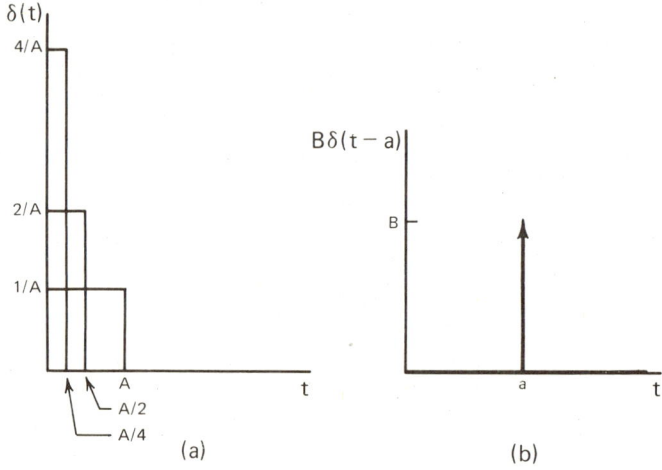

Figure 2-10. Graphical representations of the impulse function. (a) The impulse function can be interpreted as a rectangle the height of which has grown without bound as the width approaches zero; (b) The impulse is represented graphically as an arrow with height equal to the strength of the impulse.

Strictly speaking, it is impossible to graph an impulse function as its height is infinite. To circumvent this

Signals, Functions and Differential Equations

problem we represent the impulse by an arrow whose height is the **strength** or area of the impulse. Figure 2-10(b) illustrates this idea.

The sampling property. True impulses do not occur in nature but they are **approximated by many phenomena.** Moreover, impulses have properties which make them useful in proofs and derivations. An example is the so-called sampling property. Let f(t) be an arbitrary function. Then

$$\int_{-\infty}^{\infty} f(t)\delta(t-a)dt = f(a) \qquad (2\text{-}66)$$

To prove Eq. (2-66) we write

$$\int_{-\infty}^{\infty} f(t)\delta(t-a)dt = \int_{-\infty}^{a^-} f(t)\delta(t-a)dt + \int_{a^-}^{a^+} f(t)\delta(t-a)dt +$$

$$\int_{a^+}^{\infty} f(t)\delta(t-a)dt \qquad (2\text{-}67)$$

where a^- and a^+ are points infinitesimally to the left and right of the point t=a. Because the impulse rises at t=a, the first and third integrals on the right-hand side of Eq. (2-67) cover regions where the impulse is 0 and are themselves 0. Further, the interval covered by the middle integral is so short that, practically speaking, f(t) has no chance to vary and we can replace f(t) by f(a). This leads to

$$\int_{-\infty}^{\infty} f(t)\delta(t-a)dt = \int_{a^-}^{a^+} f(a)\delta(t-a)dt \qquad (2\text{-}68)$$

We can then factor the constant f(a) from the integral to get

$$\int_{-\infty}^{\infty} f(t)\delta(t-a)dt = f(a)\int_{a^-}^{a^+} \delta(t-a)dt \qquad (2\text{-}69)$$

Using Eq. (2-65), the integral on the right is 1 and the result is proved. The sampling property finds application in linear system theory.

The integral property. The integral of the unit impulse function is the unit step function. This is true because, from the definition of the impulse function,

$$\int_{-\infty}^{\infty} \delta(x)\,dx = 1 \qquad (2\text{-}70)$$

where we have used x rather than t as the variable of integration. The above integral can be decomposed into two parts,

$$\int_{-\infty}^{\infty} \delta(x)\,dx = \int_{-\infty}^{t} \delta(x)\,dx + \int_{t}^{\infty} \delta(x)\,dx = 1 \qquad (2\text{-}71)$$

The second integral is 0 because for t>0 the interval (t,∞) does not contain δ(t). Hence, we can write

$$\int_{-\infty}^{t} \delta(x)\,dx = 1 \qquad t \geqslant 0 \qquad (2\text{-}72)$$

But this relationship coincides with the definition of the step function, Eq. (2-24), so that

$$u(t) = \int_{-\infty}^{t} \delta(x)\,dx \qquad (2\text{-}73)$$

Differentiating Eq. (2-73) with respect to t reveals that

$$\frac{du(t)}{dt} = \delta(t) \qquad (2\text{-}74)$$

or that the derivative of the unit step is the unit impulse. Equation (2-74) implies that any function which is suddenly "turned on" at t=0 must contain an impulse in its derivative. If, for example, f(t)=u(t)exp(-at), then

$$\dot{f}(t) = -ae^{-at}u(t) + \delta(t)e^{-at} \qquad (2\text{-}75)$$

Equation (2-75) is reasonable because the function exp(-at)u(t) has a jump discontinuity at t=0; hence, the derivative should be undefined at t=0. For values of t>0, the impulse term disappears. This phenomenon is important in applications where we analyze "through" t=0. Fortunately, in most instances we start at t=0 and account for the past history of the system by initial conditions.

Example 2-6. Find (a) df(t)/dt and (b) the integral from 0 to t of the function of Fig. 2-11(a).

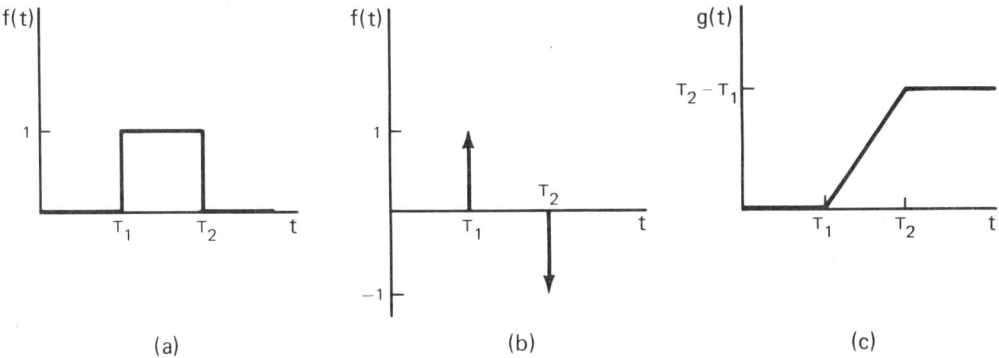

Figure 2-11. Functions for Example 2-6. (a) A rectangular pulse; (b) The derivative of the rectangular pulse is a pair of impulse functions; (c) The integral of the rectangular pulse.

Solution. (a) Write f(t) as

$$f(t) = u(t-T_1) - u(t-T_2) \qquad (2\text{-}76)$$

Differentiating yields

$$\dot{f}(t) = \delta(t-T_1) - \delta(t-T_2) \qquad (2\text{-}77)$$

with the result of Fig. 2-11(b). (b) Let g(t) be the required integral. We must consider three regions. For $t<T_1$

$$g(t) = \int_0^t f(x)dx = \int_0^t (0)dx = 0 \qquad (2\text{-}78)$$

For $T_1 < t < T_2$,

$$g(t) = \int_{T_1}^{t} (1)dx = t - T_1 \qquad (2-79)$$

and for $t > T_2$,

$$g(t) = \int_{T_1}^{T_2} (1)dx + \int_{T_2}^{t} (0)dx = T_2 - T_1 \qquad (2-80)$$

A combined expression for the integral is then

$$g(t) = (t-T_1)\{u(t-T_1) - u(t-T_2)\} + (T_2-T_1)u(t-T_2) \qquad (2-81)$$

or

$$g(t) = (t-T_1)u(t-T_1) - (t-T_2)u(t-T_2) \qquad (2-82)$$

Figure 2-11(c) shows this function.

Combined Signals

The elementary functions described above can be combined to represent a variety of useful waveshapes. Several of these are described in the following paragraphs.

The asymptotic exponential. The asymptotic exponential, Fig. 2-12, is made by combining a step function and an exponential. If $f_1(t) = Au(t)$ and $f_2(t) = A\exp(-t/T_c)$, then

$$f(t) = f_1(t) - f_2(t) \qquad (2-83)$$

and

$$f(t) = Au(t)(1 - e^{-t/T_c}) \qquad (2-84)$$

This function starts at $f(0)=0$, approaches the value $f(\infty)=A$ as $t \to \infty$, and is within one percent of its final value after five time constants. The asymptotic exponential is the solution of the first-order differential equation with the general form

$$T_c \dot{f}(t) + f(t) = Au(t) \qquad (2-85)$$

where $f(0)=0$.

Signals, Functions and Differential Equations

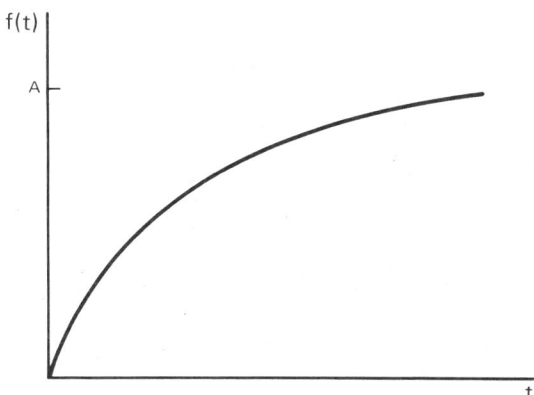

Figure 2-12. The asymptotic exponential $f(t) = Au(t) [1 - \exp(-t/T_c)$ starts at the origin and approaches the final value $f(\infty) = A$.

The damped sinusoid. The product of a sinusoid and an exponential decay gives the damped sinusoid of Fig. 2-13. If $f(t) = f_1(t)f_2(t)$ or

$$f(t) = Au(t)e^{-t/T_c}\{\cos(\omega_o t - \emptyset)\} \qquad (2\text{-}86)$$

This function is a sinusoid whose frequency is constant but whose amplitude decays with time. The function $\exp(-t/T_c)$ is the envelope of the damped sinusoid. After a time span of $5T_c$, the function is within 1% of its final value of 0. The damped sinusoid occurs often in engineering analysis.

Polynomials. Polynomial functions are generated by successively integrating the unit step function. The unit ramp function of Fig. 2-14(a) is defined by

$$r(t) = \int_0^t u(t)dt \qquad (2\text{-}87)$$

The unit ramp rises at 0 and has unit slope. A more general ramp is shown in Fig. 2-14(b). This function, defined by

$$Br(t-T) = (t-T)B \quad \text{for} \quad t \geq T \qquad (2\text{-}88)$$

and $Br(t-T) = 0$ for $t < T$, has slope B and rises at $t = T$.

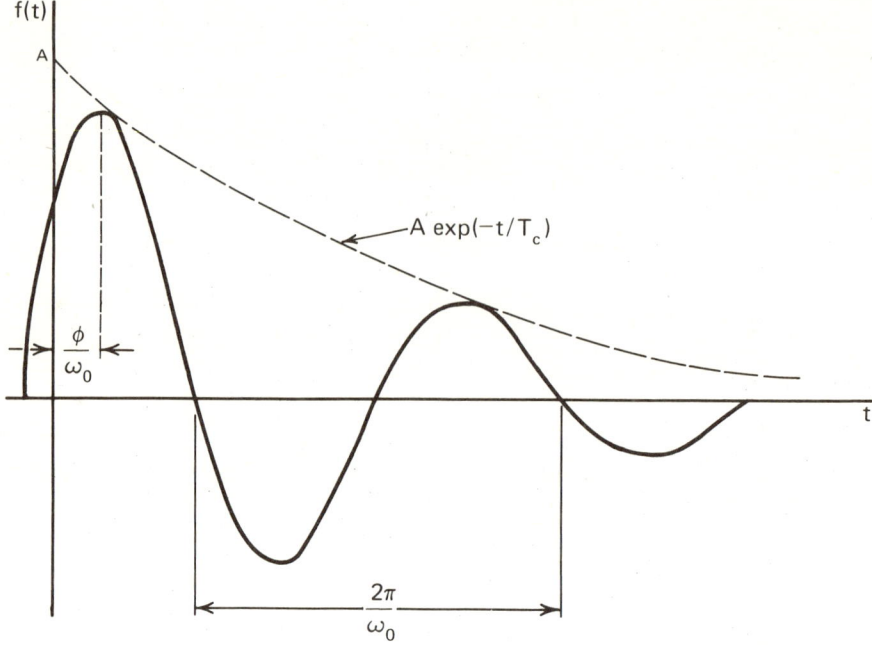

Figure 2-13. The damped sinusoid $f(t) = Au(t)\exp(-t/T_c)[\cos(\omega_0 t - \phi)]$ occurs often in linear systems work.

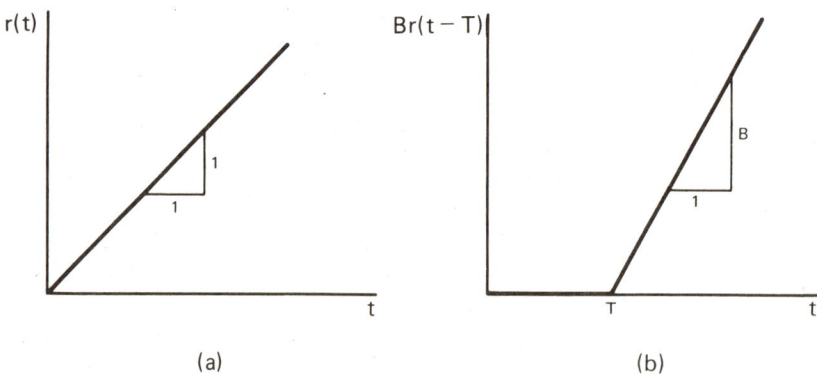

Figure 2-14. Ramp functions such as these can be obtained by integrating step functions. (a) The unit ramp $r(t)$ rises at the origin and has slope = 1; (b) The general ramp $Br(t - T)$ rises at $t = T$ and has slope = B.

Signals, Functions and Differential Equations 41

Example 2-7. Represent the function f(t) of Fig. 2-15(a) using ramp functions.

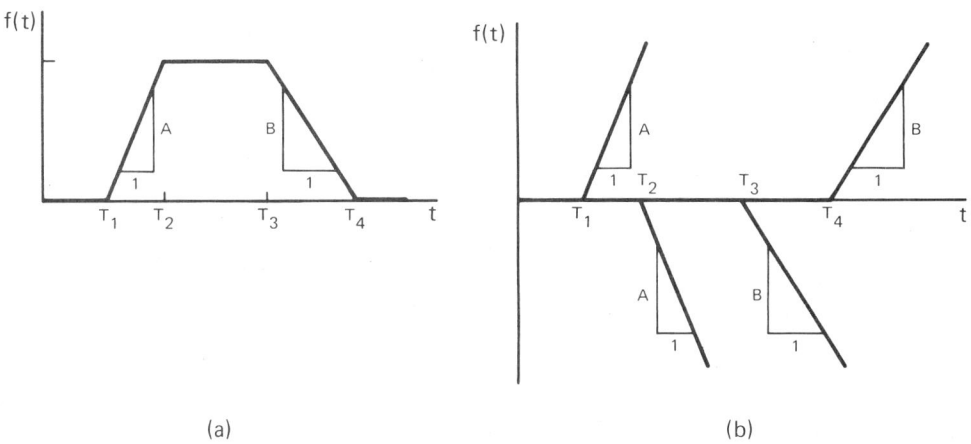

(a) (b)

Figure 2-15. Functions for Example 2-7. (a) A pulse which can be constructed from ramp functions; (b) Ramp functions which can be summed to give (a).

Solution. The function is constructed by combining the four ramps of Fig. 2-15(b). The result is

$$f(t) = Ar(t-T_1) - Ar(t-T_2) - Br(t-T_3) + Br(t-T_4) \qquad (2-89)$$

The ramp which rises at T_1 generates the left-hand side of the waveform. The ramp which rises at T_2 "cancels" this ramp to generate the horizontal top. Similarly, the T_3 ramp generates the required downslope while the T_4 ramp restores the signal to a path along the t-axis.

Fourier series. We can use the Fourier series method to represent a variety of periodic functions as infinite sums of sinusoids.[10] This is one reason why the sinusoidal function is studied so ardently by engineering analysts. Under very general conditions, a periodic function f(t) with period T_o can be written as

$$f(t) = a_o + \sum_{n=1}^{\infty} \{a_n \cos(n\omega_o t) + b \sin(n\omega_o t)\} \qquad (2-90)$$

where $\omega_o = 2\pi/T_o$. The Fourier coefficients a_o, a_n, and b_n are found from the expressions

$$a_o = \frac{1}{T_o} \int_{-T_o/2}^{T_o/2} f(t)dt \qquad (2\text{-}91)$$

$$a_n = \frac{2}{T_o} \int_{-T_o/2}^{T_o/2} f(t)\cos(n\omega_o t)dt \qquad (2\text{-}92)$$

and

$$b_n = \frac{2}{T_o} \int_{-T_o/2}^{T_o/2} f(t)\sin(n\omega_o t)dt \qquad (2\text{-}93)$$

Example 2-8. Express the pulse train of Fig. 2-16 as a Fourier series.

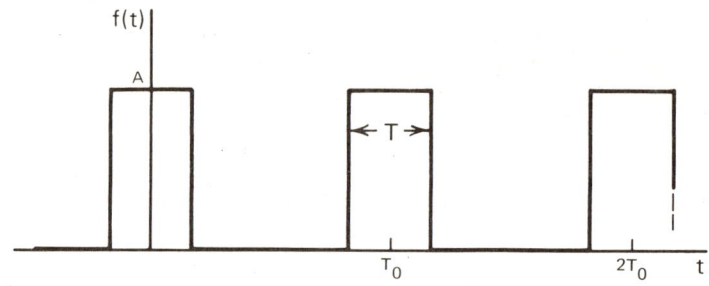

Figure 2-16. Periodic pulse sequence for Example 2-8.

Solution. Use Eqns. (2-91), (2-92), and (2-93) to determine the Fourier coefficients:

$$a_o = \frac{1}{T_o} \int_{-T/2}^{T/2} (A)dt = \frac{AT}{T_o} \qquad (2\text{-}94)$$

$$a_n = \frac{2}{T_o} \int_{-T/2}^{T/2} (A)\cos(n\omega_o t)dt = \frac{2A}{n\omega_o T_o} \Big]_{-T/2}^{T/2} \qquad (2\text{-}95)$$

Signals, Functions and Differential Equations 43

Inserting the limits and using $\omega_o = 2\pi/T_o$ yields

$$a_n = \frac{2A}{n\pi}\sin(n\omega_o T/2) \tag{2-96}$$

Finally,

$$b_n = \frac{2A}{T_o}\int_{-T/2}^{T/2} \sin(n\omega_o t)\,dt = 0 \tag{2-97}$$

The expression for $f(t)$ as a Fourier series is then

$$f(t) = \frac{AT}{T_o} + \frac{2A}{\pi}\sum_{n=0}^{\infty}\frac{1}{n}\sin(n\omega_o T/2)\cos(n\omega_o t) \tag{2-98}$$

2.7 NOTATION, DIMENSIONS, AND UNITS

When possible throughout the remainder of the book, we use the dot notation for derivatives with respect to time; thus $\dot{y}=dy/dt$, $\ddot{y}=d^2y/dt^2$, etc. Lower-case letters near the end of the alphabet are used for time-varying quantities, although the explicit time dependence is often omitted. Lower-case letters near the beginning of the alphabet are usually real constants. Upper-case letters are either real or complex constants, functions of frequency, or transformed quantities.

Dimensions

The physical world can be described in terms of only three dimensions, length (L), force (F), and time (T). All other dimensions, including electric charge (Q), can be expressed in terms of these fundamental quantities. It is convenient, however, to consider two of the derived quantities, mass (M) and charge (Q), as being fundamental and use what is called the FMLTQ system of dimensions.

Our study of continuous systems, being primarily theoretical, makes minimal use of dimensions. Many example problems leave parameters and variables in symbolic rather than numerical form, obviating the need for numerical values, dimensions, or units. Even in numerical examples we often omit units when the problem context makes the situation unambiguous. Dimensions can, however, be of great use even to the theoretical analyst. The arguments of many functions, for example, must be dimensionless. If we encounter the expression exp(-sT) and are given that T is in seconds, we know that the units of s must be (seconds)$^{-1}$. The dimensionless aspect of functional arguments lets us both discover the dimensions of new quantities and check the accuracy of our work.

Similarly, all of the terms in a summation must have the same dimensions. The expression $1+R+R^2$ cannot be dimensionally correct if R has dimensions. Dimensional analysis is thus useful in checking the validity of derivations, but only if parameters are left in symbolic form. If numerical values have been substituted, the term 1+R may be valid since the "1" may have dimensions.

Units

The term "unit" is short for unit of measure. The amount of a quantity is measured by dividing it into subunits which can be counted. A number of systems of units are in wide use. Chemists prefer the cgs (centimeter, gram, second) system, while mechanical engineers in the United States prefer the British engineering system (foot, pound, second). In recent years the trend has been toward the **SI-mks (meter, kilogram, second)** system for all technical work. Appendix B summarizes the dimensions and units of the electrical quantities used in this book.[11] Appendix C lists the power-of-ten prefixes which accompany the units to indicate very small or very large quantities. (The book uses SI units exclusively.)

PROBLEMS FOR CHAPTER 2

2-1. The equation y=mx+b is linear in the variable x because every term on the right-hand side is linear. Prove that

Signals, Functions and Differential Equations 45

the <u>operation</u> of adding a constant to an input function is <u>not</u> a linear operation.

    2-2.   Investigate the linearity of the following operations:

    (a)   Taking the nth derivative of a function.
    (b)   Multiplying a function by a constant.
    (c)   Taking the reciprocal of a function.
    (d)   Rotating a vector counterclockwise by $90°$ assuming conventional vector addition.

    2-3.   The superposition principle can be decomposed into two parts, <u>homogeneity</u> (if y is the response to x then Ay is the response to Ax) and <u>additivity</u> (if y_1 and y_2 are responses to x_1 and x_2, then $y_1 + y_2$ is the response to $x_1 + x_2$). Prove that additivity implies homogeneity if A and B are rational numbers.

    2-4.   Classify the following differential equations as either linear with constant coefficients, linear and time-varying, nonlinear, or nonlinear and time-varying.

(a) $\ddot{y} + 3\dot{y} + 4y = e^{-t^2}$ (2-99)

(b) $\dfrac{\ddot{y}}{y} + 27 + \dfrac{4y}{\dot{y}} = 4$ (2-100)

(c) $(\dot{y})^2 + 5y = u(t)$ (2-101)

(d) $t^2 \ddot{y} + y = \sin(at)$ (2-102)

(e) $\ddot{y} + (ty)^2 = 5$ (2-103)

(f) $\sin(\dot{y}) + y = \sin(t)$ (2-104)

(g) $\dfrac{\ddot{y}}{t} + \dfrac{4\dot{y}}{t} + \dfrac{8y}{t} = 4$ (2-105)

(h) $t^3 \dddot{y} = 0$ (2-106)

(i) $\dfrac{d}{dt}(ty) = \sin(t)$ (2-107)

    2-5.   Show that the function $y = A\exp(-bt)$ is a solution of the differential equation

$$\dot{y} + by = 0 \quad (2\text{-}108)$$

2-6. Show that the functions $y_1(t)=A/b$ and $y_2(t)=A\{1-\exp(-bt)\}/b$ are both solutions of the differential equation

$$\dot{y} + by = A \qquad (2\text{-}109)$$

2-7. Show that the function $y(t)=\cos(t)$ is a solution of the second-order differential equation

$$\ddot{y} + y = 0 \qquad (2\text{-}110)$$

2-8. Show that the change of variables $t=\exp(q)$ transforms the seond-order Cauchy differential equation

$$t^2\ddot{y} + t\dot{y} + y = 0 \qquad (2\text{-}111)$$

into a linear constant-coefficient differential equation with independent variable q.

2-9. The derivative of a function is defined by

$$\frac{dy}{dt} = \lim_{\Delta t \to 0}\left\{\frac{y(t+\Delta t)-y(t)}{\Delta t}\right\} \qquad (2\text{-}112)$$

Let $t=kT$ and $T=\Delta t$; then for small T the derivative is approximately

$$\left.\frac{dy}{dt}\right]_{kT} \approx \frac{y\{(k+1)T\}-y(kT)}{T} \qquad (2\text{-}113)$$

Use the latter expression to transform the continuous differential equation

$$\dot{y} + ay = f(t) \qquad (2\text{-}114)$$

into a difference equation of the form

$$y\{(k+1)T\} + Ay(kT) = Bf(kT) \qquad (2\text{-}115)$$

2-10. Using the method of Problem 2-9, show that the expression

$$\left.\frac{d^2y}{dt^2}\right]_{t=kT} \approx \frac{y\{(k+2)T\}-2y\{(k+1)T\}+y(kT)}{T^2} \qquad (2\text{-}116)$$

is a suitable approximation for the second derivative.

2-11. Use sums and differences of suitably delayed step functions to derive expressions for the following functions.

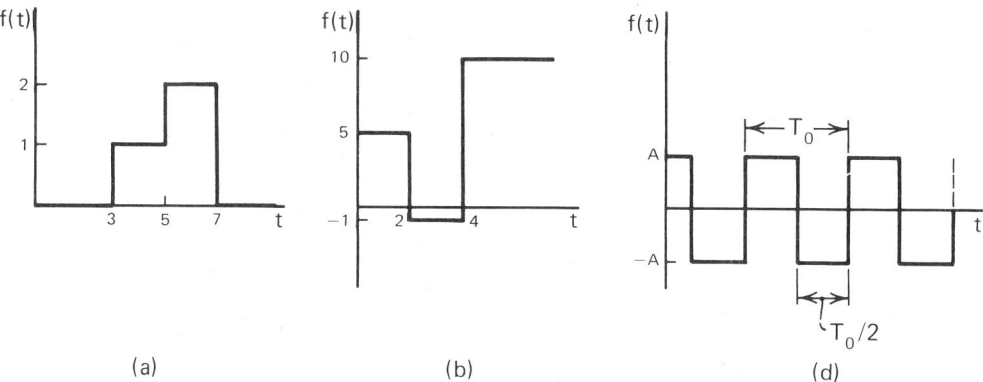

Figure 2-17. Problem 2-11.

(c) $f(t) = 10\cos(20\pi t)$ for $\pi < t < 5\pi$. (2-117)

2-12. A sinusoidal function $f(t)$ with a maximum value of 50 reaches its first positive peak at $t=10$ ms. If the period of the sinusoid is $T_0 = 40$ ms, express the function in the form

$$f(t) = A\cos(\omega_0 t) + B\sin(\omega_0 t) \qquad (2\text{-}118)$$

2-13. Write the function

$$y(t) = -3\cos(50\pi t) + 2\sin(50\pi t) + 2\sin(\pi t/0.02) \qquad (2\text{-}119)$$

as (a) a cosine function with a phase angle, and (b) as a sine function with a phase angle.

2-14. Write the function

$$g(t) = 50\cos(10\pi t - \pi/4) + 10\sin(10\pi t + \pi/8) \qquad (2\text{-}120)$$

as (a) a sum of sines and cosines with zero phase angles, and (b) a cosine function with a phase angle.

2-15. Determine the period, maximum amplitude, and phase angle for each of the following sinusoids, assuming each is expressed as $f(t) = A\cos(2\pi t/T_0 - \varnothing)$.

(a) $f(t) = 5\cos(2\pi t) + 5\sin(2\pi t)$ (2-121)

(b) $f(t) = 30\cos(\pi t) - 20\sin(\pi t)$ (2-122)

(c) $f(t) = 5\cos(2\pi t/50) - 5\sin(2\pi t/50)$ (2-123)

(d) $f(t) = 100\sin\{2\pi(20t)\} + 200\sin\{2\pi(20t)\}$ \hfill (2-124)

2-16. The sinusoid $y(t)=A\cos(\omega_o t-\emptyset)$ is the solution of every differential equation of the form

$$\ddot{y} + By = 0 \qquad (2\text{-}125)$$

with initial conditions $y(0)=C$ and $\dot{y}(0)=D$. Find B, C, and D in terms of A, ω_o, and $\emptyset$.

2-17. In a later chapter we claim that the steady-state solution of a constant-coefficient differential equation has the same form as the forcing function. Find the values of A and B for which $y(t)=A\cos(\omega_o t)+B\sin(\omega_o t)$ is a solution of the differential equation

$$\dot{y} + 25y = 100\cos(\omega_o t) \qquad (2\text{-}126)$$

2-18. Use Eq. (2-44) to show that $de^x/dx = e^x$.

2-19. A certain exponential function decays to half of its initial value in 3 sec. What is the time constant T_c of the exponential?

2-20. For what value of A is the function $y(t)=A\exp(-3t)$ a solution of the differential equation

$$\dot{y} + 10y = 250e^{-3t} \qquad (2\text{-}127)$$

2-21. The exponential $y(t)=A\exp(-pt)$ is a solution of the differential equation $\dot{y}+By=0$ with initial condition $y(0)=C$. Find A and p in terms of B and C.

2-22. If the energy contained in a function is proportional to the integral of the function squared, or

$$W = K \int_0^\infty f(t)^2 dt \qquad (2\text{-}128)$$

where W is the energy and K is a constant of proportionality, find: (a) an expression for the total energy contained in the function $f(t)=\exp(-t/T_c)$ and (b) the number of time constants which correspond to 99% of the total energy.

2-23. Write the real function

$$f(t) = 5e^{-3t}\cos(20\pi t-\pi/4) \qquad (2\text{-}129)$$

as a sum of complex exponentials, or

$$f(t) = K_1 e^{at} + K_2 e^{bt} \tag{2-130}$$

where K_1, K_2, a, and b may be complex.

2-24. Prove the following relationships: (a) $\exp(j\pi/2)=j$, (b) $\exp(j\pi)=-1$, (c) $\exp(j3\pi/2)=-j$, (d) $\exp(j2\pi n)=1$ for $n=0,1,2,\ldots$, (e) $\exp(j\pi/4)=(1+j)/\sqrt{2}$, and (f) $\exp(j1)=0.540+j0.842$.

2-25. Transform the complex exponential $f(t)=5\exp\{j(120\pi t+\pi/2)-1.3t\}$ to an expression of the form $f(t)=\exp(-bt)\{A\cos(\omega t)+B\sin(\omega t)\}$.

2-26. Use series expansions to prove that $\exp(jx)=\cos(x)+j\sin(x)$.

2-27. Using impulse functions as needed, find the derivatives of the functions of Problem 2-11.

2-28. If $f(t)=Au(t)\exp(-t/T_c)$ show that

$$\int_{-\infty}^{t} \dot{f}(t)dt = f(t) \tag{2-131}$$

only if the derivative $f(t)$ includes the impulse at the origin, or

$$\dot{f}(t) = -\frac{A}{T_c}u(t)e^{-t/T_c} + \delta(t)Ae^{-t/T_c} \tag{2-132}$$

2-29. The asymptotic exponential $y(t)=A\{1-\exp(-bt)\}$ is a solution of the differential equation $\dot{y}+5y=20u(t)$ with $y(0)=0$. Find A and b.

2-30. Write expressions for the functions shown in Fig. 2-18.

2-31. Derive expressions for the maximum value y_m and the time t_m at which the maximum occurs for the damped sinusoid

$$y(t) = Au(t)e^{-t/T_c}\{\cos(\omega_0 t-\emptyset)\} \tag{2-133}$$

2-32. How many times does the function $f(t)=\exp(-t)\cos(20\pi t)$ cross the horizontal axis in the first five time constants of the exponential decay?

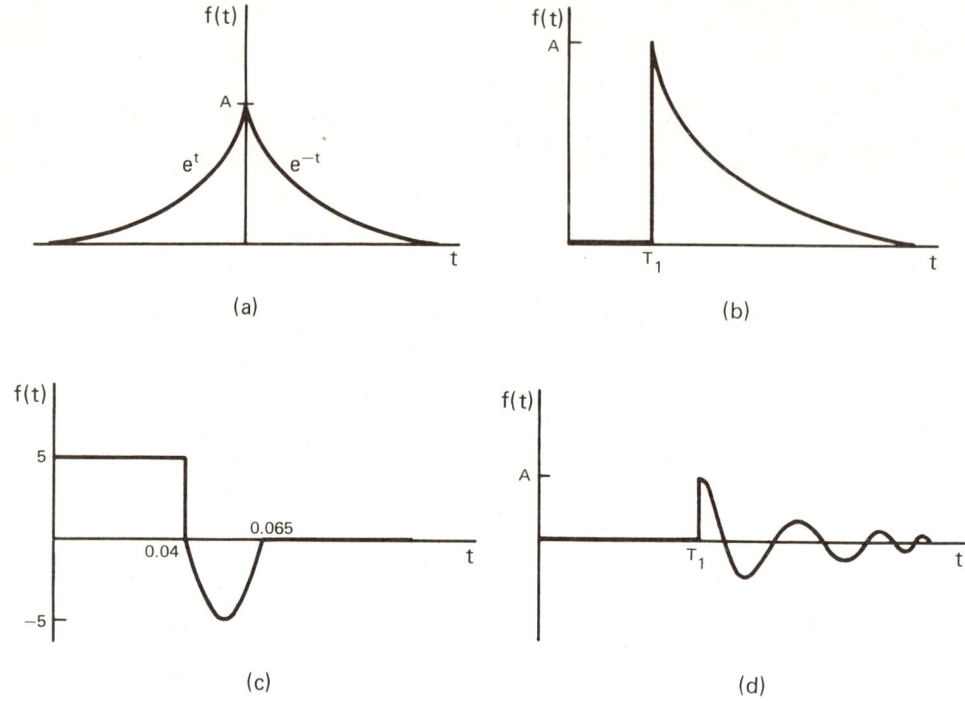

Figure 2-18. Problem 2-30. (b) The function is a delayed exponential with time constant T_c; (c) The "loop" is one arc of a sine function; (d) The function is a delayed damped sinusoid with time constant T_c, period T_0, amplitude A, and phase ϕ.

2-33.   Represent the functions of Fig. 2-19 using ramp and step functions as needed.
  (d) The function $f(t)=3t$ for $4<t<6$ and $f(t)=0$ elsewhere.

2-34.   Sketch the following waveforms:
  (a) $f(t) = 25\{u(t-2)-u(t-6)\}$ (2-134)
  (b) $f(t) = 50u(t)\{1-e^{-4t}\}$ (2-135)
  (c) $f(t) = \{u(t)-u(t-2)\}20\cos(2\pi t)$ (2-136)
  (d) $f(t) = r(t-5)$ (2-137)
  (e) $f(t) = r(t)u(t-5)$ (2-138)
  (f) $f(t) = r(t) - 2r(t-1) + r(t-2)$ (2-139)

2-35.   A certain system has an output $y(t)$ given by $y(t)=\dot{f}(t)+f(t)$ where $f(t)$ is the input function. Find the output if (a) $f(t)=5\cos(2t)$ and (b) $f(t)=u(t)\{1-e^{-5t}\}$.

Signals, Functions and Differential Equations

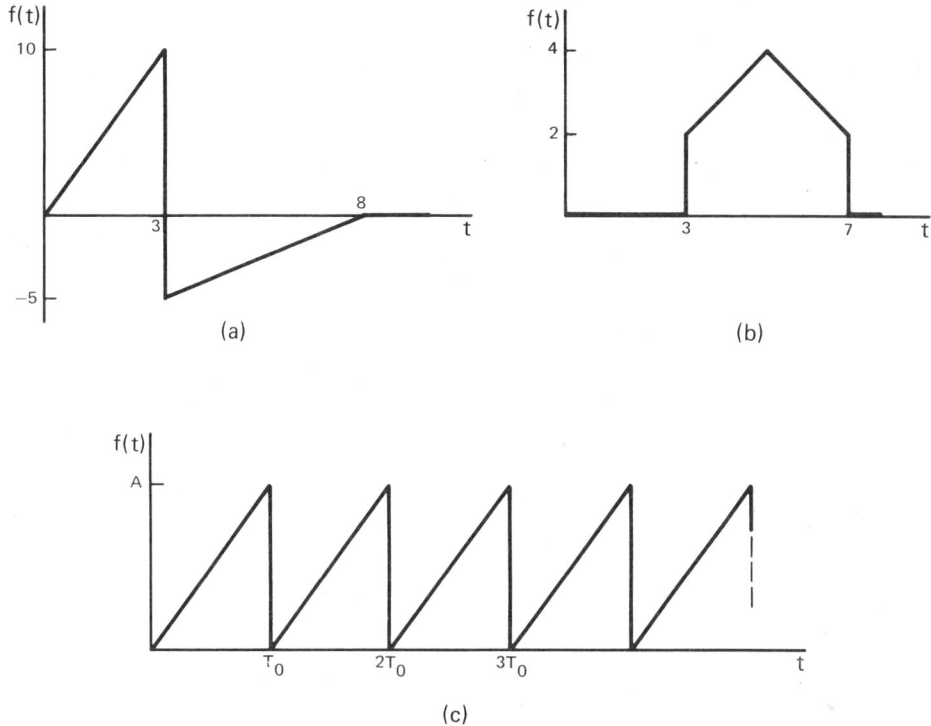

Figure 2-19. Problem 2-33.

2-36. Another form for the Fourier series is the magnitude and phase angle form

$$f(t) = \sum_{n=0}^{\infty} A_n \cos(n\omega_0 t - \phi_n) \qquad (2\text{-}140)$$

Derive expressions which relate A_n and ϕ_n to a_n and b_n of the sine-cosine form.

2-37. Derive the Fourier coefficients a_n and b_n for the functions shown:

2-38. Identify the following units and prefixes:

(a) μF   (f) μH   (k) mV
(b) pF   (g) kΩ   (l) MW
(c) mH   (h) MΩ   (m) μA
(d) mA   (i) kHz   (n) μs
(e) kW   (j) ms   (o) ns

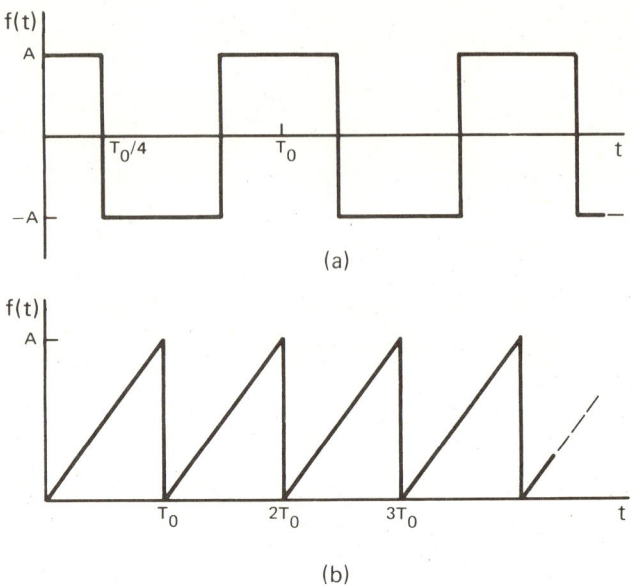

Figure 2-20. Problem 2-37.

REFERENCES FOR CHAPTER 2

(1) Chen, C. T.: <u>Introduction to Linear System Theory</u>, Holt, Rinehart, and Winston, Inc., New York, 1970.

(2) Agnew, R. P.: <u>Differential Equations</u>, McGraw-Hill Book Co., Inc., New York, 1960.

(3) D'Angelo, H.: <u>Linear Time-Varying Systems: Analysis and Synthesis</u>, Allyn and Bacon, Inc., New York, 1970.

(4) Minorsky, N.: <u>Introduction to Nonlinear Oscillations</u>, J. W. Edwards Publishing Co., Inc., Ann Arbor, Michigan, 1947.

(5) Struble, R. A.: <u>Nonlinear Differential Equations</u>, McGraw-Hill Book Co., Inc., New York, 1962.

(6) Dorf, R. C.: <u>Time-Domain Analysis of Control Systems</u> Addison-Wesley Publishing Co., Inc., Reading, Massachusetts, 1965.

(7) Lynch, W. A. and J. G. Truxal: <u>Signals and Systems in Electrical Engineering</u>, McGraw-Hill Book Co., Inc., New York, 1962.

(8) Liu, C. L. and J. W. S. Liu: Linear Systems Analysis, McGraw-Hill Book Co., Inc., New York, 1975.

(9) Brown, B. M.: The Mathematical Theory of Linear Systems, John Wiley and Sons, Inc., New York, 1961.

(10) Kaplan, W.: Operational Methods for Linear Systems, Addison-Wesley Publishing Co., Inc., Reading, Massachusetts, 1962.

(11) Mechtly, E. A.: The International System of Units, U.S. Government Printing Office, Washington, D.C., 1969.

*

CHAPTER 3

Introductory Network Analysis

Electrical engineering is one of the major disciplines in modern technology and is worth studying for that reason alone. Other engineers find that some knowledge of electrical systems is indispensible because many nonelectrical systems are instrumented and controlled by electrical or electronic devices. This chapter presents, in a somewhat condensed form, the fundamental principles of electrical network analysis, with emphasis on writing the differential equations which describe the transient or dynamic behavior of networks. The material is not meant to replace a full-length course in circuit-theory; we have included only those topics needed for an understanding of the system-theoretic aspects of electrical networks. Specialized topics needed by the electronic designer have been omitted. Finally, the chapter is concerned only with writing the differential equations which describe network behavior: solving the equations is the subject of later chapters.

3.1 FUNDAMENTAL LAWS AND QUANTITIES

The first step in acquiring an understanding of the electrical world is to understand the electrical variables and the fundamental physical laws which govern their behavior.[1,2] We do this as briefly as possible without a lengthy excursion into atomic physics.

Electrical Variables

The basic dependent variables associated with electrical networks are <u>voltage</u> and <u>current</u>. Unlike the familiar mechanical quantities, mass, force, displacement, velocity, and acceleration, electrical variables are themselves abstractions. We don't have a "feel" for current or voltage.

<u>Electrical current</u>.[3] An electrical conductor is a material containing free electrons capable of moving from one atom to the next. The unit of electric charge is that of the electron, which has a charge of 1.6021×10^{-19} coulombs (C). As electrons migrate through a conductor, the process of charge transfer is described as a flow of current. Current is defined by the relationship

$$i(t) = \frac{dq(t)}{dt} \qquad (3-1)$$

where $i(t)$ is the current in amperes (A), $q(t)$ is the charge in coulombs, and t is time in seconds. By a universal convention, the direction of current opposes that of electron flow. Figure 3-1 illustrates this concept.
This convention appears to conflict with the reality of the situation, but reality is not always as it appears. Although

current travels at the speed of light, the average velocity of free electrons is only a few millimeters per second. They "drift" rather than flow.

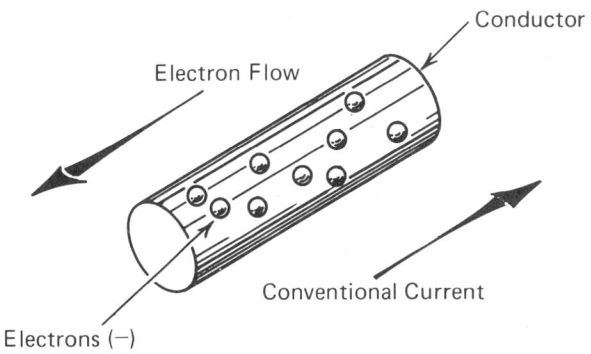

Figure 3-1. Current direction opposes electron flow.

Before an electrical network has been analyzed, we generally do not know the direction of current flow through the various components; we must, however, assign reference directions and indicate these by arrows. If the assumed reference directions are wrong, the subsequent analysis will tell us by algebraic signs. (A minus sign means a current flows opposite to the chosen reference.)

Voltage. Potential difference, or voltage as it is more commonly called, is defined as the work per unit charge expended in moving a quantity of charge from one point to another. The defining equation is

$$v(t) = \frac{dW(t)}{dq} \tag{3-2}$$

where $v(t)$ is the potential difference in volts (V), $W(t)$ is the energy or work in joules (J), and q is charge in coulombs. The joule is the unit of energy in the mks system and is equal to one newton-meter.

A battery or electrical generator creates potential difference by imparting energy to the charge which passes through it. Just as mechanical potential energy has meaning only with respect to a reference potential, electrical potential difference must be assigned reference polarities. If a voltage is identified between two points, then the point at the higher potential is labeled with a plus and that at the lower with a minus (see Fig. 3-3). The reference notation does not mean that the actual voltage has the indicated polarity: if the true voltage has a direction opposite to the reference, it has a negative value. Finally, we usually specify that the voltage and current references must "agree"; by this we mean that the plus of the voltage reference must be at the tail of the current arrow for a passive component and vice versa for a source of voltage or current. (More about references later.)

Power. Power is defined as the time rate of change of energy, or

$$p(t) = \frac{dW(t)}{dt} \tag{3-3}$$

where $p(t)$ is the power in watts (W).[4,5] One watt equals one joule per second. Multiplying Eqns. (3-1) and (3-2) shows that

$$v(t)i(t) = \frac{dq}{dt}\left\{\frac{dW}{dq}\right\} = \frac{dW}{dt} \tag{3-4}$$

and, using Eq. (3-3), we conclude that

$$p(t) = v(t)i(t) \tag{3-5}$$

Equation (3-5) says that the instantaneous power developed in an electrical component is the product of the instantaneous current through the component and the instantaneous voltage across it. In general, voltage, current, and power may be either time-varying or fixed quantities. Equation (3-3) can be integrated between 0 and t_1 to give

$$W(t_1) = \int_0^{t_1} p(t)dt + W(0) \tag{3-6}$$

Introductory Network Analysis

or

$$W(t_1) = \int_0^{t_1} v(t)i(t)dt + W(0) \tag{3-7}$$

as the energy associated with an electrical component at time t_1, provided $v(t)$ is the voltage across the component, $i(t)$ is the current through it, and $W(0)$ is the energy at $t=0$.

Average power. The instantaneous power developed in an electrical network is not a particularly useful quantity. As a result we define the average power over a time interval t_1-t_2 as

$$P_{ave} = \frac{1}{t_2-t_1} \int_{t_1}^{t_2} p(t)dt \tag{3-8}$$

Kirchhoff's Laws [6]

Voltages and currents in an electrical network obey two relationships known as Kirchhoff's voltage law (KVL) and Kirchhoff's current law (KCL). In spite of the sacredness of these relationships to electrical engineers, they are not fundamental laws of nature. The voltage law is a statement of the more basic law of conservation of energy as applied to electrical networks. The current law is a consequence of the principle of conservation of electrical charge--electrical charge cannot accumulate in a dimensionless region in space. Finally, granted that Kirchhoff's laws govern the behavior of voltages and currents in a network, how are they applied? How many times must we write each law to assure ourselves that the resulting equations can be solved for all dependent variables?

Kirchhoff's current law. A node (also called a vertex or junction) in an electrical network is a point where a number of current paths converge. Kirchhoff's current law states that the instantaneous sum of the currents entering a node equals the instantaneous sum of the currents leaving the node. Figure 3-2 illustrates KCL.

Figure 3-2. Kirchhoff's current law (KCL) applied at this node gives Eq. (3-9).

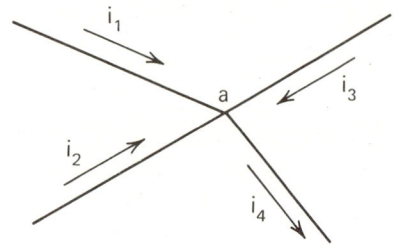

The current directions in Fig. 3-2 are assumed; actual quantities may differ in direction. Thus, i_3 may actually flow out of the node, a fact which will be manifested by its being a negative quantity; $i_3 = -6$ amperes, for example. The current law for the node a of Fig. 3-2 is

$$i_1(t) + i_2(t) + i_3(t) = i_4(t) \tag{3-9}$$

One point about currents needs reemphasizing: it doesn't matter what reference direction you choose for a current; the important consideration is that you <u>must</u> choose one. The reference direction says absolutely nothing about the actual direction of current flow. In fact, if the current is sinusoidal, say $i(t) = I\sin(\omega_0 t)$, then no matter what reference is chosen, the current will agree with the reference half of the time and oppose it the other half.

<u>Kirchhoff's voltage law.</u> Figure 3-3 shows a closed path (also called a <u>loop</u> or <u>mesh</u>) in an electrical network with a reference polarity assigned for each network element. Specific elements are defined in later sections, but their exact nature is unimportant when we are using Kirchhoff's voltage law. According to Kirchhoff's voltage law, the algebriac sum of the voltages around the loop must be zero at every instant. Thus, in Fig. 3-3

$$v_1(t) + v_2(t) + v_4(t) - v_3(t) = 0 \tag{3-10}$$

Introductory Network Analysis   61

where we have circled the loop clockwise and written the voltage with a positive sign if we encounter a "+" sign as we enter a component and with a negative sign if we encounter a "-" sign. Again, the reference polarities are assigned arbitrarily, but they <u>must</u> be chosen; actual voltages may be negative with respect to the assigned polarities.

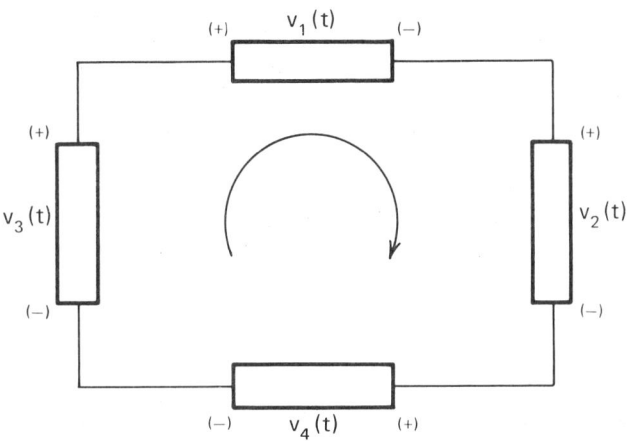

Figure 3-3. Kirchhoff's Voltage Law (KVL) applied to this loop gives Eq. (3-10).

<u>The number of network equations</u>.[7,8] An electrical network (or <u>circuit</u>; the terms are often used interchangeably, although strictly speaking "circuit" is a more practical, functionally oriented, term--an amplifier circuit, for example--while a network is an arbitrary interconnection of electrical components) is a collection of components or elements connected by ideal resistanceless wires. Each element has a voltage reference (a plus at one end and a minus at the other) and a current reference (an arrow).

A network is quantified by the number of nodes and branches where, as noted above, a node is a point at which two or more current paths join. A branch is a path between two nodes, provided the path contains at least one element. The subject of <u>network topology</u> deals with proofs relating the numbers of nodes and branches to the number of equations needed to analyze a network. We will not delve into network topology here as it is a specialized subject beyond our needs. The topological results of interest, which we state without proof, are:

(1) The number of independent KCL equations in a network is N_n-1 where N_n is the number of nodes;

(2) The number of independent KVL equations is $N_b-(N_n-1)$ where N_b is the number of branches.

Example 3-1. Let each numbered component in Fig. 3-4 have an associated voltage and current with the references shown. Component 5 has v_5 and i_5, for example.

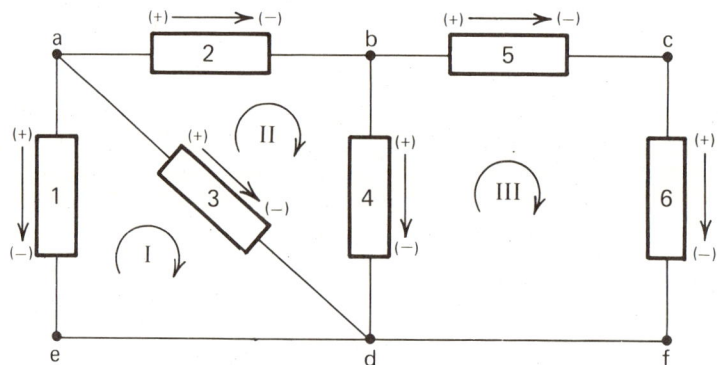

Figure 3-4. Network for Example 3-1, illustrating the application of Kirchhoff's voltage and current laws. Kirchhoff's laws are often called the "connection" equations in network analysis.

Find: (a) N_n, (b) N_b, (c) the number of independent KCL equations, (d) the number of independent KVL equations, (e) the KCL equations at nodes a, b, and c, and (f) the KVL equations around loops I, II, and III.

Solution. (a) There are only four independent nodes, nodes a, b, c, and d. Points d, e, and f are all the same node because they are connected with resistanceless wire;

Introductory Network Analysis

hence, $N_n=4$. (b) As there is one branch for each element, $N_b=6$. The region between nodes d and f is not a branch as it contains no element. (c) $N_n-1=3$ and there are three independent KCL equations--the ones for any three of the four nodes a, b, c, and d. (d) $N_b-(N_n-1)=6-(4-1)=3$. There are three independent loop equations--loops I, II, and III, for example. (The network contains many more loops, the outer loop that contains elements 1, 2, 5, and 6, for example.) (e) Recalling that KCL equates the currents entering a node to those leaving the node, we have

$$\text{Node a:} \qquad 0 = i_1 + i_2 + i_3 \qquad (3\text{-}11)$$

$$\text{Node b:} \qquad i_2 = i_4 + i_5 \qquad (3\text{-}12)$$

$$\text{Node c:} \qquad i_5 = i_6 \qquad (3\text{-}13)$$

(f) Circle each loop clockwise and write the element voltages with a positive sign if a plus is encountered first:

$$\text{Loop I:} \qquad v_3 - v_1 = 0 \qquad (3\text{-}14)$$

$$\text{Loop II:} \qquad v_2 + v_4 - v_3 = 0 \qquad (3\text{-}15)$$

$$\text{Loop III:} \qquad v_5 + v_6 - v_4 = 0 \qquad (3\text{-}16)$$

The KVL and KCL equations cannot be solved for the voltages and currents until we have mathematical models for the components which make up the network. A final observation: while the relationship $N_b-(N_n-1)$ tells us how many independent loop equations there are, we still don't know how to insure that we have an independent set. **This is another job** for network topology, although, as we shall see, there are special cases where the choice of which loops and nodes to use is obvious. The loop and node equations, when written in terms of the branch variables, are often referred to as the <u>connection equations</u> of the network.

Glass–Linear Systems—6

3.2 ELECTRICAL COMPONENTS

Electrical and electronic systems contain a myriad of devices--vacuum tubes, transistors, switches, relays, diodes, motors, generators, resistors, inductors, **capacitors**, transformers, batteries, etc. All of these have mathematical models, some simple, some complex. Cataloging the devices and developing their models would be a formidable task. Fortunately, however, the circuit behavior of any electrical device can be modeled in terms of resistance, inductance, capacitance, ideal sources, and nonlinear transfer characteristics. If we restrict ourselves to networks containing resistance, inductance, capacitance, and sources, the resulting differential equations are linear with constant coefficients.

Passive Electrical Components [9]

The three passive components are the <u>resistor</u>, the <u>inductor</u>, and the <u>capacitor</u>, whose symbols are shown in Fig. 3-5.

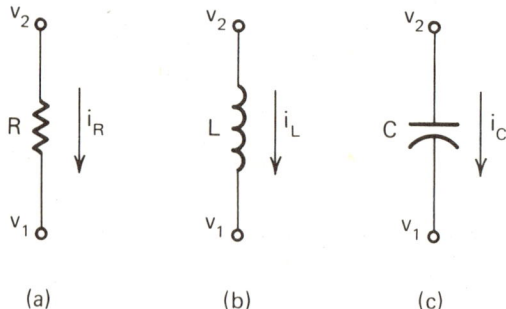

Figure 3-5. Resistors, inductors and capacitors are called passive components because they either consume or store energy. (a) The resistor; (b) The inductor; (c) The capacitor.

Introductory Network Analysis

The <u>resistor</u>. The resistor, Fig. 3-5(a), satisfies the relationship

$$i_R = \frac{v_2 - v_1}{R} \tag{3-17}$$

or

$$v_2 - v_1 = i_R R \tag{3-18}$$

The resistance parameter has the units of ohms (Ω) and can be interpreted as the constant of proportionality between the potential difference across the resistor and the current through it. Note the distinction between the component called a resistor--which is often a cylinder of carbon with wires attached--and the mathematical resistance parameter. There are components other than resistors which have resistance--a long piece of wire, for example. Equation (3-18) is known as <u>Ohm's law</u>. If $v_2 > v_1$ in Fig. 3-5(a) the current has the direction shown. Networks which contain only resistors and sources can be analyzed by algebra; differential equations are not required.

The <u>inductor</u>. The inductor of Fig. 3-5(b) is defined by

$$v_2 - v_1 = L\frac{di_L}{dt} \tag{3-19}$$

The inductance parameter has the units of henries (H) and can be interpreted as the constant of proportionality between the potential difference across the inductor and the rate of change of current through it. If $v_2 > v_1$ in Fig. 3-5(b) the current has the direction shown. Equation (3-19) can be inverted by multiplying by dt and integrating between 0 and t to obtain

$$i_L(t) = \frac{1}{L}\int_0^t v_L(t)dt + i_L(0) \tag{3-20}$$

where $v_L(t) = v_2(t) - v_1(t)$ is the voltage across the inductor.

The current through the inductor at t=0 is $i_L(0)$. Equation (3-20) assumes that $i_L(0)$ has the same direction as the current reference for the inductor. If this is not true $i_L(0)$ is a negative quantity.

Equation (3-20) implies that the inductor may be considered as having memory. If $v_L(t) \equiv 0$ for $0 < t < t_1$, then the current through the inductor at time t_1 is the same as it was at t=0, or

$$i_L(t_1) = i_L(0) \qquad (3\text{-}21)$$

The inductor therefore "remembers" the current it had at t=0.

As with the resistor, we make a distinction between the component called an inductor, which is constructed from a coil of wire, and the inductance parameter. At high enough frequencies the leads of a resistor or of a capacitor have an appreciable inductance, but they are not inductors--at least not intentionally.

<u>The capacitor</u>. The capacitance parameter, defined by

$$i_C = C \frac{d(v_2 - v_1)}{dt} \qquad (3\text{-}22)$$

has the units of farads (F) and can be interpreted as the constant of proportionality between the current through the capacitor and the rate of change of voltage across it. If $v_2 > v_1$ in Fig. 3-5(c), then $i_C(t)$ has the direction shown. Equation (3-22) can be inverted by multiplying by dt and integrating between 0 and t. The result is

$$v_C(t) = \frac{1}{C} \int_0^t i_C(t) dt + v_C(0) \qquad (3\text{-}23)$$

where $v_C(t) = v_2(t) - v_1(t)$ is the voltage across the capacitor at time t and $v_C(0)$ is the initial value. Equation (3-23) assumes that the polarity of $v_C(0)$ agrees with the voltage reference chosen for the capacitor. If this is not true $v_C(0)$ will be a negative quantity. Equation (3-23) indicates

Introductory Network Analysis

that, like the inductor, the capacitor has memory. If no current is drawn from the capacitor, its voltage remains at the initial value, or if $i_C(t)=0$ for $0<t<t_1$ in Eq. (3-23), then

$$v_C(t_1) = v_C(0) \qquad (3-24)$$

and the capacitor remembers its initial voltage. One of the most popular uses of the capacitor is as a memory or storage device.

Again, the capacitor is a physical component which has capacitance. At high enough frequencies two parallel wires have an appreciable capacitance between them, although they are not intentionally designed to be capacitors. Actual capacitors are constructed of parallel plates of conducting material separated by a nonconducting material called a dielectric. Capacitance has the units of farads, but one farad is a very large amount of capacitance. Typical capacitor values lie in the microfarad (μF) and picofarad (pF) range.

The polarity notations for the circuit elements are simplified by the following convention: put a reference arrow to indicate the direction of assumed current flow through the element; then the positive voltage reference is always at the tail of the current arrow and the negative reference is at its head (Fig. 3-6). If this convention is followed, the R, L, and C parameters are always positive.

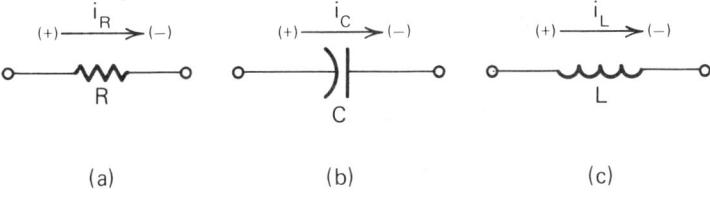

Figure 3-6. The positive voltage reference is at the tail of the current arrow for a passive component. If this convention is violated, we must supply a minus sign in the v—i relationship for the device. (a) Resistor; (b) Capacitor; (c) Inductor.

Sources

In addition to the passive circuit elements, we need two kinds of ideal voltage source and the ideal current source. These are shown in Fig. 3-7.

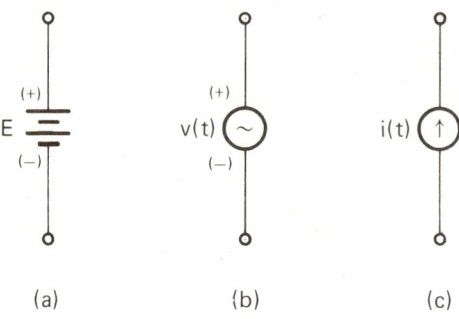

Figure 3-7. Ideal voltage and current sources for electrical networks. These are called active elements since they supply energy to a network. The plus voltage reference is at the head of the current arrow for an active element. (a) A battery or dc source; (b) A variable voltage source; (c) A variable current source.

The variable sources generate time-varying voltages or currents as indicated by v(t) or i(t), respectively. A source of constant voltage is indicated by the battery symbol with the constant V or E beside it. Current is assumed to flow away from the positive terminal of a voltage source, and the current reference arrow is from minus to plus across the source--exactly opposite to the convention for passive elements. Ideal voltage sources have zero resistance to current flow while ideal current sources have infinite resistance. This means that the only current through a current source is that which it generates, and the only voltage across a voltage source is that which it generates.

Actual voltage sources always have a certain amount of internal series resistance (preferably as small as possible). Figure 3-8 shows a practical battery modeled as an ideal battery in series with its internal resistance R_o. The terminal voltage is given by the expression

$$v(t) = E - i(t)R_o \qquad (3-25)$$

Introductory Network Analysis

(Check this by writing KVL around the loop.) As a consequence the terminal voltage v(t) is always less than the ideal value E if a current is drawn from the battery.

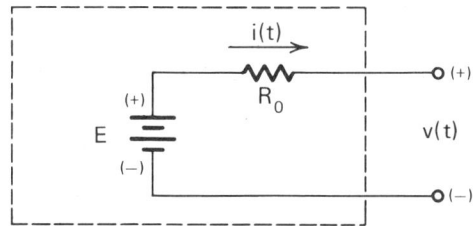

Figure 3-8. A practical battery can be modelled as a constant voltage source in series with a resistor.

Moreover, the terminal voltage is not even constant if the current is time varying. A similar result holds for the current source except that the corrupting resistance is very large and appears in parallel with the ideal current source as shown in Fig. 3-9. The resulting terminal current is

$$i(t) = I - \frac{v(t)}{R_o} \tag{3-26}$$

where I is the ideal source current, R_o is the corrupting resistance, and v(t) is the voltage across the source. This result follows from KCL. While no physical device resembles a current source in the way a battery approximates an ideal voltage source, the current source finds wide use in modeling active devices such as the transistor.

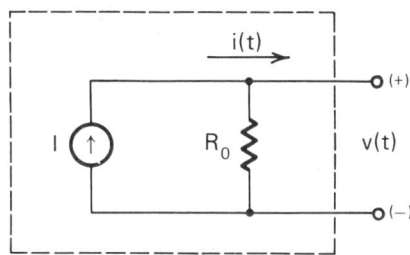

Figure 3-9. A non-ideal current source can be modelled as an ideal current source in parallel with a resistor.

Dependent sources. The value of a generated voltage or current may depend on the voltage or current at some other point in a network. In this case the source is said to be *dependent*. Figure 3-10 shows a dependent voltage source whose output $v_o(t)$ is equal to a constant K multiplied by the voltage $v_i(t)$ across the indicated terminals.

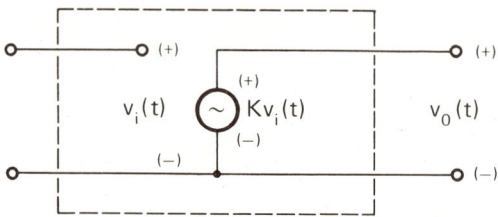

Figure 3-10. This dependent source produces a voltage which depends on the voltage at another point in the network, $v_o(t) = Kv_i(t)$.

Dependent sources are used in modeling transistors, vacuum tubes, operational amplifiers, and other active devices.

Switch Notation

One might wonder why we classify electrical networks as dynamic systems. What moves? The answer is charge q, although we seldom (never in this book) write electrical differential equations in terms of charge, preferring voltage or current instead. We do, however, need a way to set the flow of current in motion. This is done with ideal switches which can be thrown at t=0 or some other time to initiate operation of the network. Figure 3-11 shows a typical network such as will be encountered in later chapters.
The switch S has positions A and B. Each circuit element has a value written by it, either a symbolic value--R_1 or C_2, for example--or a numerical value. Appendices B and C give the unit abbreviations and power-of-ten prefixes for electrical quantities. Appendix B also relates the electrical units to the fundamental dimensions force, mass, length and time.

Introductory Network Analysis

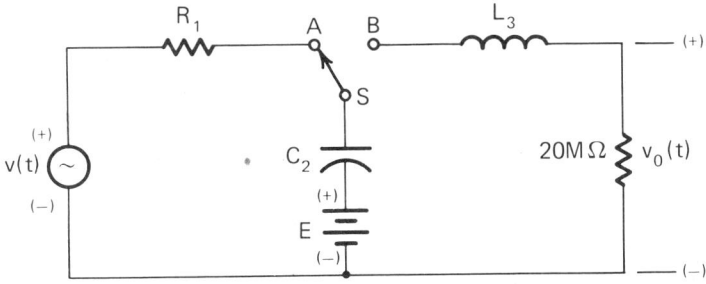

Figure 3-11. This network illustrates the notation and conventions used throughout the remainder of the book.

3.3 LOOP ANALYSIS

Having introduced the electrical variables voltage and current, the fundamental laws KVL and KCL, and the basic components R, L, and C, we are ready to learn how to write the differential equations which describe the dynamic behavior of a network.[10] First, though, we make one further classification. A <u>planar</u> network is one which can be drawn on a two-dimensional surface with no branch crossing another. A <u>nonplanar</u> network, on the other hand, always has at least one wire crossing over another no matter how it is redrawn. The resistor network of Fig. 3-12(a) is planar while that of Fig. 3-12(b) is nonplanar. Both kinds of networks occur in practical applications. For planar networks the number of independent KVL equations equals the number of meshes or <u>windows</u> of the network. Figure 3-12(a) has three such windows labeled I, II, and III. Further, the KVL equation written for the meshes are the required independent equations: they can be solved for all network voltages and currents. For nonplanar networks, if we include each branch in at least one KVL equation, and each equation contains at least one branch that does not appear in the others, we generally have the required independent set of equations. (These are necessary but not sufficient conditions.)

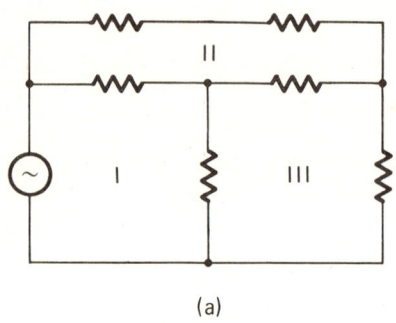
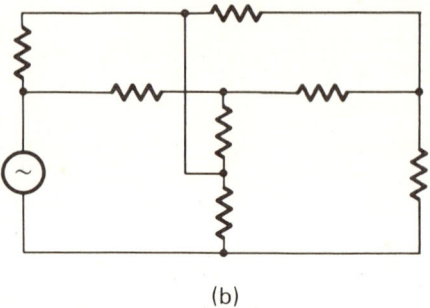

Figure 3-12. A planar network can be drawn without any wires crossing over any other wires. (a) A planar resistor network; (b) A non-planar resistor network.

With these preliminaries disposed of, we present a procedure for analyzing networks by the <u>loop analysis</u> method. The starting point is a circuit diagram with given element and source values but with no references assigned. The procedure is given for planar networks, but it applies in the nonplanar case with minor modifications.

The Loop Analysis Procedure

(1) Number each component consecutively and assign a voltage variable to each--v_1, v_2, v_3, etc. Identify a fictitious loop current in each mesh of the planar network. It is convenient to take all loop currents clockwise.

(2) Put a plus sign where the loop current enters each component and a minus sign where it exits. This agrees with the sign convention proposed in Section 3.2. If a component is shared by two loops it may be possible to satisfy the voltage convention for only one of the loop currents. Take your choice.

(3) Write KVL by proceeding around each loop in the direction of the assumed loop current, starting at a voltage source if there is one. If a plus sign is encountered as the loop current enters an element, write the voltage for that element with a plus sign in the KVL equation for the loop. If a minus is encountered first, write the voltage with a minus sign.

Introductory Network Analysis

(4) Write the v-i relationships for each component, remembering to observe the voltage reference convention. If, for example, resistor R_5 has two loop currents i_1 and i_2 through it and the voltage reference shown in Figure 3-13, then $v_5 = R_5(i_1-i_2)$.

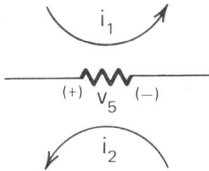

Figure 3-13. According to the voltage reference convention, $v_5 = R_5(i_1 - i_2)$.

The net current is the difference of the two loop currents, and to satisfy the voltage convention we must write the current which agrees with the voltage reference <u>first</u> when writing the v-i relationship. This convention must be followed if the equations are to be correct with R_5 a positive quantity.

(5) Substitute the element v-i relationships into the KVL equations.

(6) If capacitors are present, the loop equations will contain integrals. One must differentiate such equations to eliminate the integrals.

In general, the KVL equations are a set of coupled second-order constant-coefficient differential equations--one for each mesh in the planar network. The loop currents, which are the dependent variables in these differential equations, are fictitious but, once found, can be used to find the actual currents and voltages in the network. If only one loop current passes through a component, that loop current coincides with the actual current. If two loop currents pass through an element, their sum or difference is the true current. The component v-i relationships can be used along with the true currents to find the component voltages.

Example 3-2. Write loop equations for the network of Fig. 3-14.

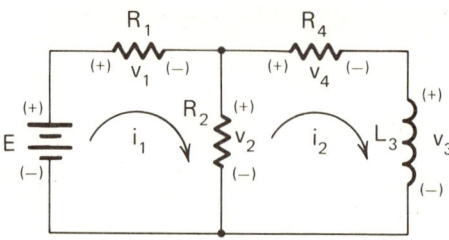

Figure 3-14. This network is used in Example 3-2 to illustrate the procedure for writing loop equations.

Solution. The network is planar because no wires cross. The two loops have assumed currents i_1 and i_2 chosen clockwise as suggested in the procedure. Each element has been assigned a voltage reference which agrees with the "plus at the tail of the current arrow" convention. The voltage reference for R_2 agrees with loop current i_1 but disagrees with current i_2.

Now write KVL for the two loops:

$$\text{Loop 1:} \quad v_1 + v_2 - E = 0 \tag{3-27}$$

$$\text{Loop 2:} \quad v_4 + v_3 - v_2 = 0 \tag{3-28}$$

We write $-v_2$ in the equation for loop 2 because i_2 encounters a minus sign as it enters R_2. Similarly, i_1 encounters a minus sign as it enters the battery, and we write $-E$ in the equation for loop 1.

The v-i relationships for the network elements are:

$$v_1 = R_1 i_1 \tag{3-29}$$

$$v_2 = R_2(i_1 - i_2) \tag{3-30}$$

$$v_3 = \frac{L_3 di_2}{dt} \tag{3-31}$$

$$v_4 = R_4 i_2 \tag{3-32}$$

Introductory Network Analysis

The expression for v_2 requires comment. The net current through R_2 is the difference in the loop currents. As the voltage reference for R_2 agrees with i_1, we write the current difference as i_1-i_2. If we did not adopt this procedure R_2 would have to be negative--and we don't like negative resistance. Writing i_1-i_2 presumes $i_1>i_2$. Of course this may not be true, but then the voltage reference would also fail to coincide with reality. The point is that we have adopted a consistent set of references; the network itself takes care of the rest.

Substituting the v-i relationships into the loop equations yields

$$R_1 i_1 + R_2(i_1-i_2) = E \qquad (3-33)$$

$$\frac{L_3 di_2}{dt} + R_4 i_2 - R_2(i_1-i_2) = 0 \qquad (3-34)$$

These equations specify the behavior of the network. All that remains is to insert numerical values, assign one initial condition, in this case $i_2(0)$, and solve for the loop currents. By inspecting the network diagram we see that the loop currents and the actual element currents are identical except that the true current through R_2 is i_1-i_2.

Example 3-3. Write loop equations for the network of Fig. 3-15.

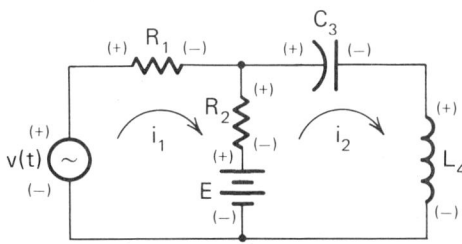

Figure 3-15. Example 3-3 uses this network to illustrate writing loop equations when a loop contains two sources.

Solution. Here we have a two-loop network with two sources. As before, loop currents are indicated and voltage references assigned. The KVL equations for the two loops are

$$\text{Loop 1:} \quad v_1 + v_2 + E - v(t) = 0 \tag{3-35}$$

$$\text{Loop 2:} \quad v_3 + v_4 - E - v_2 = 0 \tag{3-36}$$

where we have assumed that the element voltages are numbered the same as the elements. The v-i relationships are

$$v_1 = R_1 i_1 \tag{3-37}$$

$$v_2 = R_2(i_1 - i_2) \tag{3-38}$$

$$v_3 = \frac{1}{C_3} \int_0^t i_2 \, dt + v_3(0) \tag{3-39}$$

and

$$v_4 = L_4 \frac{di_2}{dt} \tag{3-40}$$

The capacitor voltage is expressed in terms of its current by the integral relationship of Eq. (3-23). Substituting the component v-i relationships into the KVL equations yields

$$R_1 i_1 + R_2(i_1 - i_2) = v(t) - E \tag{3-41}$$

and

$$\frac{1}{C_3} \int_0^t i_2 \, dt + v_3(0) + L_4 \frac{di_2}{dt} - R_2(i_1 - i_2) = E \tag{3-42}$$

Equation (3-42) contains an integral which can be removed by differentiating the entire equation. The result is

$$\frac{i_2(t)}{C_3} + L_4 \frac{d^2 i_2}{dt^2} - R_2 \left\{ \frac{di_1}{dt} - \frac{di_2}{dt} \right\} = 0 \tag{3-43}$$

Introductory Network Analysis

Both E and $v_3(0)$, being constant, have disappeared under the differentiation process. They will reappear as a part of the initial conditions. In this instance we must have $i_1(0)$, $i_2(0)$, and di_2/dt evaluated at t=0. A later section considers the problem of evaluating the initial conditions. Again, the loop currents are the actual currents for all elements except R_2. The true current through R_2 is i_1-i_2.

Example 3-4. As a final and more complex example, consider the network of Fig. 3-16.

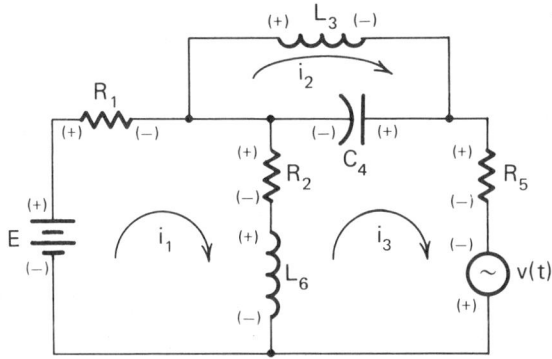

Figure 3-16. Network for Example 3-4.

Solution. The three loop currents are labeled i_1, i_2, and i_3 and are assumed clockwise. With the voltage references assigned as shown, the KVL equations are

$$\text{Loop 1:} \quad v_1 + v_2 + v_6 - E = 0 \qquad (3\text{-}44)$$

$$\text{Loop 2:} \quad v_3 + v_4 = 0 \qquad (3\text{-}45)$$

$$\text{Loop 3:} \quad -v_4 + v_5 - v(t) - v_6 - v_2 = 0 \qquad (3\text{-}46)$$

The component v-i relationships are

$$v_1 = R_1 i_1 \qquad (3\text{-}47)$$

$$v_2 = R_2(i_1 - i_3) \qquad (3\text{-}48)$$

$$v_3 = L_3 \frac{di_2}{dt} \qquad (3-49)$$

$$v_4 = \frac{1}{C_4} \int_0^t (i_2 - i_3) dt + v_4(0) \qquad (3-50)$$

$$v_5 = R_5 i_3 \qquad (3-51)$$

$$v_6 = L_6 \frac{d(i_1 - i_3)}{dt} \qquad (3-52)$$

The current differences in the expressions for v_2, v_4, and v_6 are written so the current which agrees with the voltage references comes first in the difference. Inserting the v-i relationships into the KVL equations results in

$$R_1 i_1 + R_2(i_1 - i_3) + L_6 \frac{d(i_1 - i_3)}{dt} = E \qquad (3-53)$$

$$L_3 \frac{di_2}{dt} + \frac{1}{C_4} \int_0^t (i_2 - i_3) dt + v_4(0) = 0 \qquad (3-54)$$

and

$$-\frac{1}{C_4} \int_0^t (i_2 - i_3) dt - v_4(0) + R_5 i_3 - L_6 \frac{d(i_1 - i_3)}{dt} -$$

$$R_2(i_1 - i_3) = v(t) \qquad (3-55)$$

The latter two equations must be differentiated to remove the integrals. The capacitor initial voltage, which disappears during the differentiation, will reappear as a part of the initial conditions.

We return to loop analysis in Chapter 4 where the procedures for solving differential equations are introduced.

Introductory Network Analysis　　　　　　　　　　　　　　　　　　　　　79

3.4 NODAL ANALYSIS

Just as Kirchhoff's voltage law leads to the loop analysis method of writing network equations, Kirchhoff's current law leads to a method called <u>nodal analysis</u>. The number of loop equations equals the number of meshes or windows in a planar network; the number of node-voltage equations equals the number of independent KCL equations, or one for each independent node in the network. Practical networks are often constructed with a common ground wire connected to a large number of components. This ground wire corresponds to a base or reference node, with the result that fewer nodal equations than loop equations are needed for the analysis. Because of this, nodal analysis is more efficient for a variety of practical networks. Consider the network of Fig. 3-17.

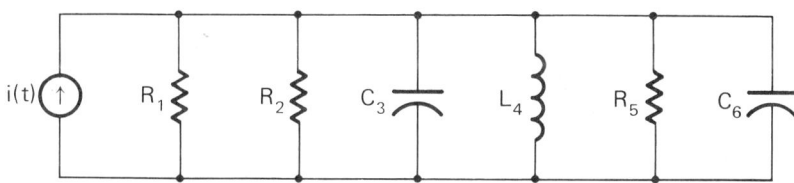

Figure 3-17. Nodal analysis would be preferred for this network which contains six loops but only two nodes.

As mentioned above, a node is a point where two or more branches come together (provided, of course that each branch contains an element). All points of common voltage are connected to the same node. The network of Fig. 3-17 has only two independent nodes because the upper and lower ends of all the network elements are at the same potential. The network is redrawn in Fig. 3-18 to show more clearly that it has only two nodes.

Glass–Linear Systems—7

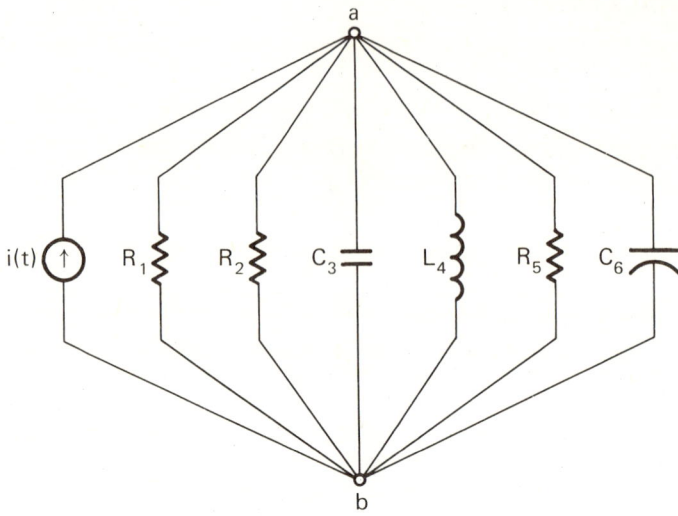

Figure 3-18. Network of Fig. 3-17 redrawn to emphasize that it has only two nodes.

The number of KCL equations needed to analyze a network is one less than the number of independent nodes; hence, the network of Fig. 3-18 requires only one nodal equation. We would need six equations to analyze the same network using loop analysis. The procedure for writing node-voltage equations is presented formally below and then used in a number of examples.

The Nodal Analysis Procedure [11]

(1) Identify the independent nodes and assign each a node voltage. We generally use letters for the nodes and letter subscripts for the node voltages to avoid confusion with component voltages. Unlike the loop currents which are fictitious, node voltages are the actual voltages which exist in the network. Let one node, usually the one connected to the most elements, be the reference or <u>ground</u> node with a voltage of zero. The ground node is often indicated by the ground symbol ⏚.

(2) Assign current and voltage references for the network elements, adhering to the voltage-current reference con-

Introductory Network Analysis 81

vention when possible. By convention, the ends of components connected to the ground node are assigned the minus voltage reference and currents are always directed toward the ground node. If an element is connected between two nonground nodes, then it is impossible to follow the reference convention for both. We will see below how to handle this problem.

(3) Write KCL at each node except for the ground node, taking currents away from a node as positive and towards a node as negative. This is equivalent to equating currents entering and currents leaving a node.

(4) Write the v-i relationships for the network elements. If an element has two node voltages across it, write the voltage which satisfies the current reference convention first when forming the difference. Thus if R_5 is between nodes a and b with the current reference directed from node a toward node b, we write $i_5 = (v_a - v_b)/R_5$.

(5) Insert the v-i relationships into the KCL equations.

(6) If the network contains inductors, integrals appear in the nodal equations. Differentiate as necessary to eliminate these. Here is another reason for preferring nodal over loop analysis. Because of their large size and weight relative to resistors and capacitors, inductors are avoided in network design when possible; hence we are more interested in analyzing networks which contain capacitors. The loop equations for a network which contains capacitors require integrals and thus are more difficult to set up than are nodal equations for the same network.

The node-voltage differential equations are ready for solution after we assign suitable initial conditions. If the network contains only the R, L and C elements and sources, the equations are coupled, linear, constant-coefficient differential equations with the node voltages as dependent variables. The equations are coupled through the elements which share two node voltages.

<u>Example 3-5</u>. Write the node-voltage equation for the network of Fig. 3-19.

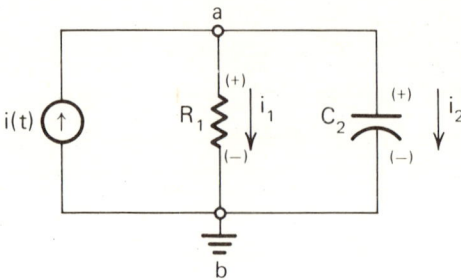

Figure 3-19. This network is used in Example 3-5 to illustrate the procedure for writing node voltage equations.

Solution. The network contains two nodes, nodes a and b. Choose node b as the ground or reference node and assign voltage v_a to node a. Current reference arrows are directed away from node a; voltage references are assigned with the minus at the ground node. Writing KCL for node a yields

$$i(t) = i_1 + i_2 \qquad (3\text{-}56)$$

The component v-i relationships are

$$i_1 = \frac{v_a}{R_1} \qquad (3\text{-}57)$$

$$i_2 = C_2 \dot{v}_a \qquad (3\text{-}58)$$

Substituting these into the KCL equation gives the node-voltage equation

$$\frac{v_a}{R_1} + C\dot{v}_a = i(t) \qquad (3\text{-}59)$$

The initial condition $v_a(0)$ is needed before this equation can be solved for v_a. Once v_a is known, all other voltages and currents can be found.

Example 3-6. Write node-voltage equations for the network of Fig. 3-20.

Solution. Select node d as the reference node and identify node voltages v_a, v_b, and v_c. Choose the voltage

Introductory Network Analysis

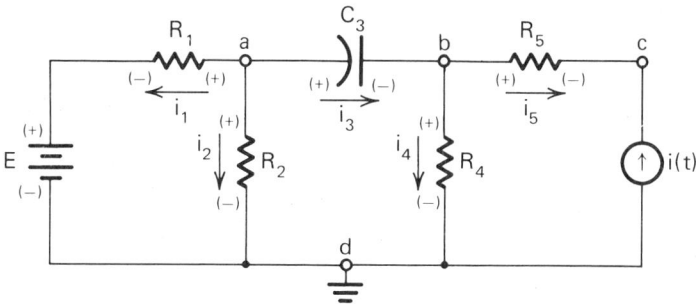

Figure 3-20. This network is used in Example 3-6 to illustrate writing nodal equations when both types of sources are present.

and current references as shown. Writing KCL at the nodes yields

$$\text{Node a:} \quad i_1 + i_2 + i_3 = 0 \tag{3-60}$$

$$\text{Node b:} \quad -i_3 + i_4 + i_5 = 0 \tag{3-61}$$

$$\text{Node c:} \quad -i_5 - i(t) = 0 \tag{3-62}$$

Currents away from the nodes are written with a plus sign; those toward the nodes with a minus sign. The element v-i relationships are

$$i_1 = \frac{v_a - E}{R_1} \tag{3-63}$$

$$i_2 = \frac{v_a}{R_2} \tag{3-64}$$

$$i_3 = C_3(\dot{v}_a - \dot{v}_b) \tag{3-65}$$

$$i_4 = \frac{v_b}{R_4} \tag{3-66}$$

$$i_5 = \frac{v_b - v_c}{R_5} \tag{3-67}$$

The voltage differences are written to agree with the assumed current directions. Notice how we handle the voltage source

E; it was not necessary to put a node between R_1 and E because we know that the voltage across R_1 is $v_a - E$ if the current reference is away from node a. Substituting the element v-i relationships into the node equations, we obtain

$$\frac{v_a - E}{R_1} + \frac{v_a}{R_2} + C_3(\dot{v}_a - \dot{v}_b) = 0 \tag{3-68}$$

$$-C_3(\dot{v}_a - \dot{v}_b) + \frac{v_b}{R_4} + \frac{v_b - v_c}{R_5} = 0 \tag{3-69}$$

$$\frac{-(v_b - v_c)}{R_5} - i(t) = 0 \tag{3-70}$$

These equations contain no integrals and are ready to solve once we know the initial conditions $v_a(0)$ and $v_b(0)$. After v_a, v_b, and v_c are found, we can solve for all other voltages and currents in the network.

Example 3-7. Write the nodal equation for the network of Fig. 3-18.

Solution. Assuming all element currents are directed away from node a, the KCL equation is

$$i_1 + i_2 + i_3 + i_4 + i_5 + i_6 - i(t) = 0 \tag{3-71}$$

The component v-i relationships are

$$i_1 = \frac{v_a}{R_1} \tag{3-72}$$

$$i_2 = \frac{v_a}{R_2} \tag{3-73}$$

$$i_3 = C_3 \dot{v}_a \tag{3-74}$$

$$i_4 = \frac{1}{L_4} \int_0^t v_a \, dt + i_4(0) \tag{3-75}$$

$$i_5 = \frac{v_a}{R_5} \tag{3-76}$$

Introductory Network Analysis

$$i_6 = C_6 \dot{v}_a \tag{3-77}$$

Inserting these into the KCL equation and differentiating to eliminate the integral yields

$$\frac{\dot{v}_a}{R_1} + \frac{\dot{v}_a}{R_2} + C_3 \ddot{v}_a + \frac{v_a}{L_4} + \frac{\dot{v}_a}{R_5} + C_6 \ddot{v}_a = \frac{di(t)}{dt} \tag{3-78}$$

or

$$\ddot{v}_a(C_3+C_6) + \dot{v}_a\left\{\frac{1}{R_1} + \frac{1}{R_2} + \frac{1}{R_5}\right\} + \frac{v_a}{L_4} = \frac{di(t)}{dt} \tag{3-79}$$

Here we need both $v_a(0)$ and $\dot{v}_a(0)$. The initial inductor current, which disappeared when we differentiated to remove the integral, will reappear as a part of the initial conditions.

3.5 OTHER ANALYSIS TECHNIQUES

The loop and nodal analysis methods are formal, structured approaches toward analyzing networks. There are, however, several special techniques which can be used to advantage in reducing the labor of the network analyst.[12] Some of these are discussed in the following paragraphs.

Element Combination Rules

Electrical networks can often be simplifed by combining series and parallel groupings of like components into a single equivalent component. The combination rules are derived by applications of KCL, KVL, and the element v-i relationships.

<u>Resistor combinations</u>. The two series resistors of Fig. 3-21(a) are equivalent to the single equivalent resistor of Fig. 3-21(b) provided

$$R_{eq} = R_1 + R_2 \tag{3-80}$$

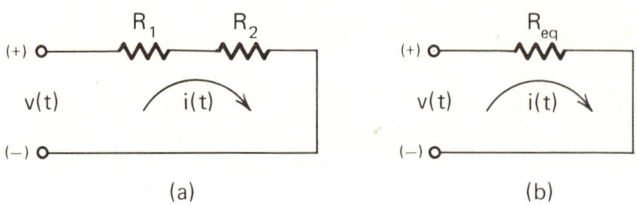

Figure 3-21. Two resistors in series can be replaced by a single resistor, R_{eq} whose value is $R_{eq} = R_1 + R_2$. (a) Resistors R_1 and R_2 are in series; (b) A network equivalent to (a) with a single resistor.

To prove Eq. (3-80), apply KVL to the series combination of resistors and insert the v-i relationship for the resistors to get

$$v(t) = i(t)R_1 + i(t)R_2 = i(t)(R_1+R_2) \qquad (3-81)$$

For the single equivalent resistor

$$v(t) = i(t)R_{eq} \qquad (3-82)$$

Comparing Eqns. (3-81) and (3-82) gives the result indicated in Eq. (3-80). Equation (3-80) can be generalized to any number of series resistors. Their sum is equal to the value of the required equivalent resistor.

For two resistors in parallel we have the situation of Fig. 3-22.

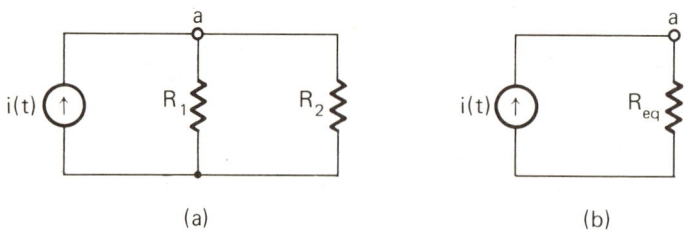

Figure 3-22. Two parallel resistors can be replaced by a single resistor according to Eq. (3-86). (a) Resistors R_1 and R_2 are connected in parallel; (b) A network equivalent to (a) with the proper choice of R_{eq}.

Introductory Network Analysis

Applying KCL to the parallel combination and inserting the v-i relationships yields

$$i(t) = \frac{v_a}{R_1} + \frac{v_a}{R_2} \quad (3\text{-}83)$$

$$= v_a\left\{\frac{1}{R_1} + \frac{1}{R_2}\right\} \quad (3\text{-}84)$$

For the single resistor in Fig. 3-22(b),

$$i(t) = \frac{v_a}{R_{eq}} \quad (3\text{-}85)$$

Equating Eqns. (3-84) and (3-85) shows that

$$\frac{1}{R_{eq}} = \frac{1}{R_1} + \frac{1}{R_2} \quad (3\text{-}86)$$

or

$$R_{eq} = \frac{R_1 R_2}{R_1 + R_2} \quad (3\text{-}87)$$

Equation (3-87) says that the equivalent of two parallel resistors is a single resistor whose value is the product of the two resistances divided by their sum.

Inductor combinations. By a similar procedure we can show that series and parallel inductors follow the same combination rules as resistors: $L_{eq} = L_1 + L_2$ for the series case and $L_{eq} = L_1 L_2/(L_1 + L_2)$ for the parallel case.

Capacitor combinations. Consider the series combination of capacitors shown in Fig. 3-23(a).
Writing KVL for Fig. 3-23(a) and assuming the capacitors are initially uncharged yields

$$v(t) = v_1(t) + v_2(t) \quad (3\text{-}88)$$

$$= \frac{1}{C_1}\int_0^t i(t)dt + \frac{1}{C_2}\int_0^t i(t)dt \quad (3\text{-}89)$$

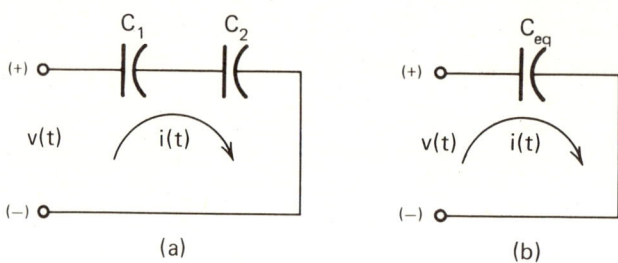

Figure 3-23. Two capacitors in series combine according to Eq. (3-93).

or, after differentiating

$$\dot{v}(t) = \frac{i(t)}{C_1} + \frac{i(t)}{C_2} \qquad (3\text{-}90)$$

For the equivalent capacitor of Fig. 3-23(b)

$$v(t) = \frac{1}{C_{eq}} \int_0^t i(t)dt \qquad (3\text{-}91)$$

or

$$\dot{v}(t) = \frac{i(t)}{C_{eq}} \qquad (3\text{-}92)$$

Equating Eqns. (3-90) and (3-92) shows that

$$\frac{1}{C_{eq}} = \frac{1}{C_1} + \frac{1}{C_2} \qquad (3\text{-}93)$$

or

$$C_{eq} = \frac{C_1 C_2}{C_1 + C_2} \qquad (3\text{-}94)$$

For the parallel capacitors of Fig. 3-24(a), KCL leads to

$$i(t) = C_1 \dot{v}_a + C_2 \dot{v}_a = (C_1 + C_2)\dot{v}_a \qquad (3\text{-}95)$$

and, for the single equivalent capacitor,

$$i(t) = C_{eq} \dot{v}_a \qquad (3\text{-}96)$$

Introductory Network Analysis

Equating these two expressions shows that

$$C_{eq} = C_1 + C_2 \tag{3-97}$$

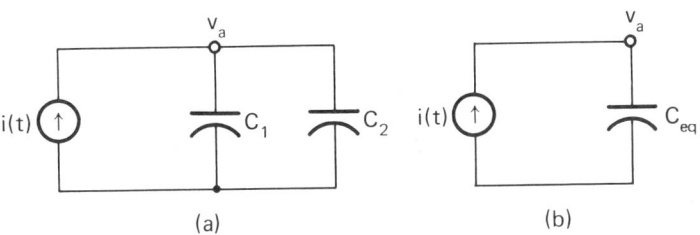

Figure 3-24. Parallel capacitors combine by addition: $C_{eq} = C_1 + C_2$.

The capacitor combination rules are thus seen to be "inverted" with respect to those for resistors and inductors.

Source Combination Rules

Under certain conditions, series and parallel configurations of voltage or current sources can be combined. The various combinations are shown in Fig. 3-25.

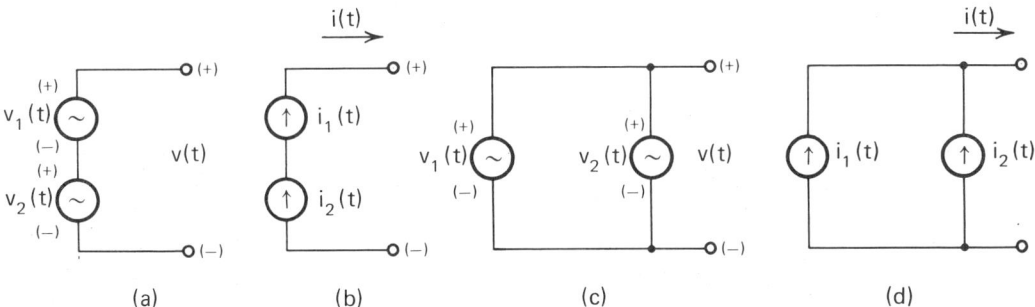

Figure 3-25. Series or parallel combinations of sources can often be replaced by a single equivalent source. (a) Voltage sources in series; (b) Current sources in series; (c) Voltage sources in parallel; (d) Current sources in parallel.

A series combination of voltage sources, Fig. 3-25(a), is equivalent to a single source equal to the sum of the two, or

$$v(t) = v_1(t) + v_2(t) \tag{3-98}$$

Two voltage sources can be connected in parallel only if they are equal; thus in Fig. 3-25(c) $v_1(t)$ must equal $v_2(t)$.

Because only one current can flow in a single wire, the series combination of current sources, Fig. 3-25(b), is possible only if $i_1(t)=i_2(t)$. Finally, two current sources in parallel, Fig. 3-25(d), can be replaced by a single current source whose value is their sum or

$$i(t) = i_1(t) + i_2(t) \tag{3-99}$$

Extraneous Elements

Consider the source configuration of Fig. 3-26(a). The voltage across R is $v(t)$. The current through R is $i(t) = v(t)/R$ regardless of the network connected to the right of terminals a-a'. For this reason the resistor is extraneous and can be removed without affecting any analysis involving the network to the right of a-a'.

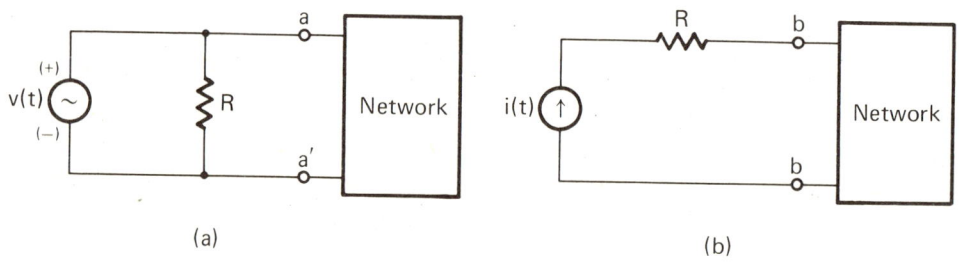

Figure 3-26. A resistor in parallel with a voltage source or in series with a current source is extraneous and can often be removed.

A similar argument holds for a resistor in series with a current source. As the current through R in Fig. 3-26(b) is always $i(t)$ no matter what network is connected to

Introductory Network Analysis

terminals b-b', the resistor can be removed without affecting any analysis problem involving the current source. Because extraneous elements have presumably been omitted in practical network analysis problems, voltage sources always appear in series with resistors and current sources appear in parallel with resistors.

Source Conversions

Loop equations are easier to write if all of the sources in the network are voltage sources; and nodal equations are easier to write if all sources are current sources. For these reasons it is advantageous to know how to convert one kind of source to the other without altering the network variables. Figure 3-27 shows standard configurations for voltage and current sources. The resistances are present because, as noted above, a practical voltage source is modeled by an ideal voltage source in series with an internal resistance, and a practical current source is modeled by an ideal current source in parallel with a large internal resistance.

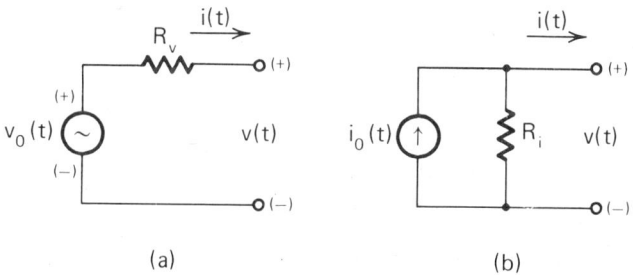

Figure 3-27. A voltage source can be replaced by an equivalent current source and vice versa. (a) A voltage source in series with a resistor; (b) This configuration is equivalent to (a) if $R_i = R_v$ and $i_0(t) = v_0(t)/R_v$.

To convert the voltage source of Fig. 3-27(a) to an equivalent current source, we must have the same voltage and current at the output terminals. Writing the KVL equation for Fig. 3-27(a) yields

$$v_o(t) = i(t)R_v + v(t) \qquad (3\text{-}100)$$

Solving for $i(t)$ yields

$$i(t) = \frac{v_o(t)}{R_v} - \frac{v(t)}{R_v} \qquad (3\text{-}101)$$

The node-voltage equation for Fig. 3-27(b) is

$$i_o(t) = i(t) + \frac{v(t)}{R_i} \qquad (3\text{-}102)$$

or

$$i(t) = i_o(t) - \frac{v(t)}{R_i} \qquad (3\text{-}103)$$

Comparing Eqns. (3-101) and (3-103) reveals that for equivalence we must have

$$i_o(t) = \frac{v_o(t)}{R_v} \qquad (3\text{-}104)$$

and

$$R_i = R_v \qquad (3\text{-}105)$$

Equations (3-104) and (3-105) can be used in reverse to transform a current source to an equivalent voltage source.

We leave it as an **exercise for the reader to verify the** source conversions involving inductors and capacitors. These are shown in Figs. 3-28 and 3-29.

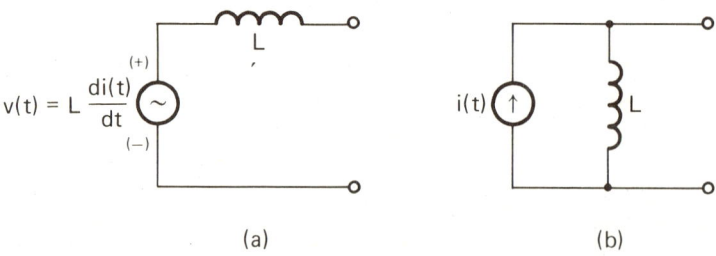

Figure 3-28. Source conversion with an inductor.

Introductory Network Analysis

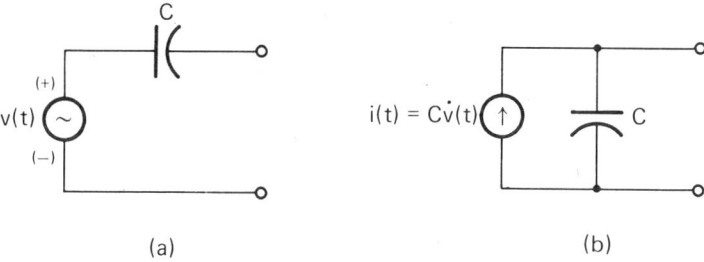

Figure 3-29. Source conversion with a capacitor.

The Voltage-Divider Principle

A simple but powerful network analysis technique, the voltage-divider principle, can be proved by elementary loop analysis. Consider the network of Fig. 3-30.

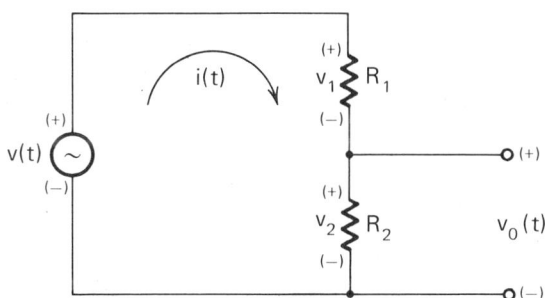

Figure 3-30. Voltage divides in proportion to resistance. The voltage-divider principle is a useful analysis tool.

The problem is to determine the ratio $v_o(t)/v(t)$, or the fraction of the input voltage which appears at the output. The loop equation for the network is

$$v_1(t) + v_2(t) - v(t) = 0 \qquad (3\text{-}106)$$

The component relationships,

$$v_1(t) = R_1 i(t) \qquad (3\text{-}107)$$

$$v_2(t) = R_2 i(t) \qquad (3\text{-}108)$$

lead to

$$R_1 i(t) + R_2 i(t) = v(t) \qquad (3\text{-}109)$$

Equation (3-109) can be solved for

$$i(t) = \frac{v(t)}{R_1 + R_2} \qquad (3\text{-}110)$$

As $v_o(t) = v_2(t) = R_2 i(t)$ we have

$$v_o(t) = \frac{R_2 v(t)}{R_1 + R_2} \qquad (3\text{-}111)$$

or

$$\frac{v_o(t)}{v(t)} = \frac{R_2}{R_1 + R_2} \qquad (3\text{-}112)$$

Equation (3-112) indicates that voltage divides in proportion to resistance. The <u>potentiometer</u> is a variable resistor based on the voltage-divider principle. A potentiometer is a resistor with a slider or wiper arm which picks off a fraction of the voltage applied to the ends of the resistor.

Current Division

Analogous to the voltage divider is the <u>current divider</u> of Fig. 3-31.

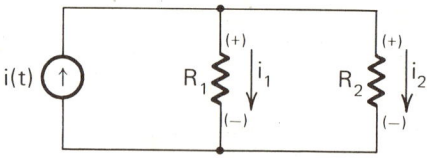

Figure 3-31. The currents in parallel branches are in inverse proportion to the resistances.

Because the resistors are in parallel, $v_1 = v_2$ and

$$R_1 i_1 = R_2 i_2 \qquad (3\text{-}113)$$

Introductory Network Analysis

or

$$\frac{i_1}{R_2} = \frac{i_2}{R_1} \tag{3-114}$$

Equation (3-114) shows that the currents through parallel resistors are in inverse proportion to the resistances.

Thevenin's Theorem for Resistive Networks

Thevenin's theorem states that any combination of resistors and sources can be replaced by a single voltage source in series with a single resistor.[12,13] The source-resistor combination, called the <u>Thevenin equivalent circuit</u>, produces the same results as the original network when connected to another device or network. We are not concerned here with proving Thevenin's theorem and will consider only the problem of finding the Thevenin equivalent circuit.

To find the Thevenin equivalent to the left of a pair of terminals such as a-a' in Fig. 3-32(a): (1) Replace all voltage sources by short circuits and all current sources by open circuits; then find the equivalent resistance between the terminals using the combination rules for series and parallel resistors. This resistance R_T is the Thevenin resistance; (2) Remove the network to the right of terminals a-a', leaving the output or a-a' terminals open circuited. Use loop or nodal analysis to find the open-circuit voltage V_T across the output terminals. This is the Thevenin voltage. The procedure is illustrated by the following example.

<u>Example 3-8</u>. Find the Thevenin equivalent circuit for the network of Fig. 3-32(a).

<u>Solution</u>. Replace the battery by a short circuit as in Fig. 3-32(b). Then the 50 Ω, 200 Ω, and 300 Ω resistors are all in parallel. Combining the 200 Ω and the 300 Ω resistors using Eq. (3-87) results in $(300)(200)/(300+200)=120$ Ω. Combining this equivalent resistance with the 50 Ω resistor yields $(120)(50)/(120+50)=35.29$ Ω. As this equivalent resis-

tance is in series with the 25 Ω resistor, the Thevenin resistance R_T is

$$R_T = 25 + 35.29 = 60.29 \text{ Ω} \qquad (3\text{-}115)$$

Because terminals a-a' are open circuited and no current is drawn, the voltage across terminals a-a' is the same as that across the 120 Ω equivalent of the 200 Ω and the 300 Ω resistors (Fig. 3-32(c)). This voltage, which is the Thevenin voltage V_T, can be found by applying the voltage-divider principle. Thus,

$$V_T = \frac{150(120)}{120+50} = 105.9 \text{ V} \qquad (3\text{-}116)$$

Figure 3-32(d) is the final Thevenin equivalent circuit with $V_T=105.9$ V and $R_T=60.29$ Ω. Any network connected to the right of terminals a-a' will be unable to distinguish whether it is connected to the original network of Fig. 3-32(a) or the Thevenin equivalent of Fig. 3-32(d).

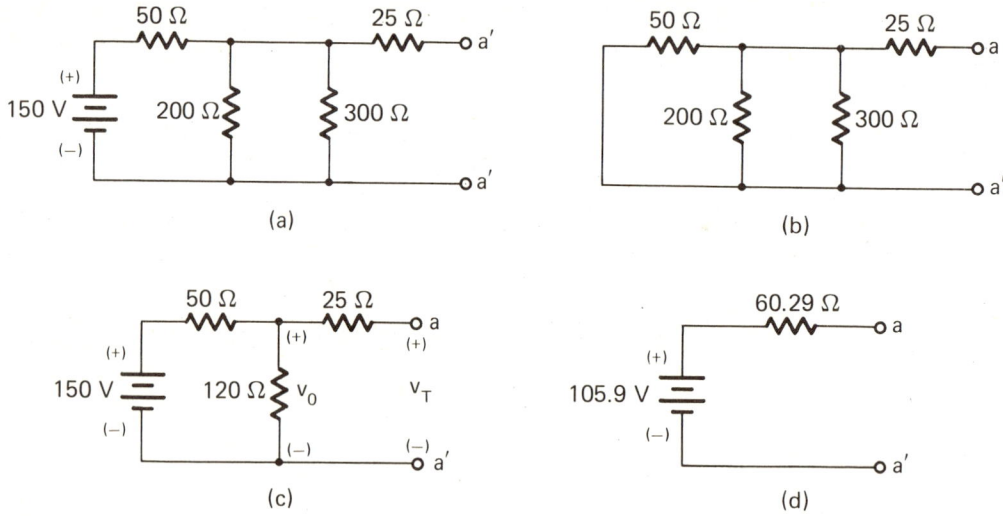

Figure 3-32. Networks for Example 3-8, illustrating the Thevenin equivalent circuit. (a) The original network; (b) Voltage sources are replaced by short circuits when R_T is evaluated; (c) The 200 Ω and 300 Ω resistors in parallel be replaced by a single 120 Ω resistor; (d) A Thevenin equivalent of the original network.

Introductory Network Analysis

In a later chapter we show how to apply Thevenin's theorem to networks containing inductance and capacitance as well as resistance. The theorem also applies if the sources produce variable voltages and currents.

Norton's Theorem for Resistive Networks

Norton's theorem states that a network containing resistors and sources can be replaced by a single current source in parallel with a suitable resistor.[9] The Norton current I_N is the current which flows through the output terminals a-a' when they are short circuited (connected together). The Norton resistance R_N is the equivalent resistance to the left of terminals a-a' after all voltage sources have been replaced with short circuits and all current sources by open circuits. Rather than apply this procedure, it is often easier to find the Thevenin equivalent circuit, then apply the source conversion of Fig. 3-27 to get the Norton equivalent. Thus the Thevenin equivalent of Fig. 3-32(d) becomes the Norton equivalent of Fig. 3-33.

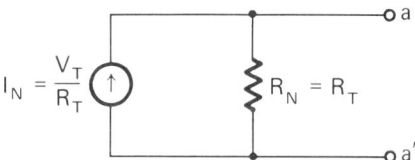

Figure 3-33. The Norton equivalent circuit can be found from the Thevenin equivalent by a source conversion.

For complex networks it may not be possible to find the Thevenin resistance by repeated series and parallel combinations. In such cases we may use the relationship

$$R_N = R_T = \frac{V_T}{I_N} \qquad (3\text{-}117)$$

where R_T is the Thevenin resistance, V_T is the Thevenin voltage (the open-circuit voltage across terminals a-a'), and I_N

is the Norton current (the short-circuit current through terminals a-a'). The quantities V_T and I_N can be found by loop and/or nodal analysis.

Example 3-9. Verify Eq. (3-117) for the network of Example 3-8.

Solution. In Example 3-8 we found the open-circuit Thevenin voltage V_T=105.9 V. To find the Norton current, connect terminals a-a' together to get the network of Fig. 3-34(a).

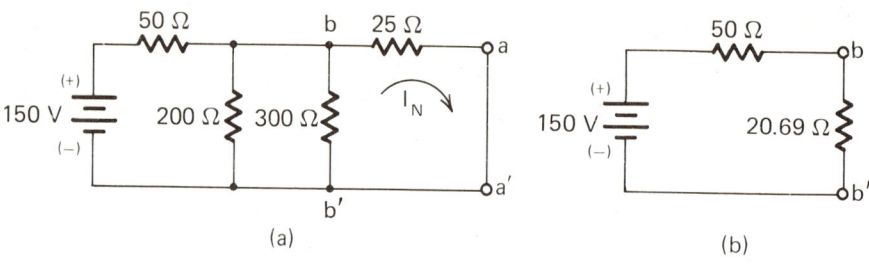

Figure 3-34. The short-circuit current I_N is needed when one wants to replace a network by its Norton equivalent. (a) The short-circuit current I_N is the current which flows when terminals a-a' are connected; (b) The 200 Ω, 300 Ω, and 25 Ω resistors can be replaced by a single 20.69 Ω resistor.

The 200 Ω, 300 Ω, and 25 Ω resistors are now in parallel. Combining them gives the network of Fig. 3-34(b). The voltage-divider principle yields the voltage across terminals b-b':

$$V_{bb'} = \frac{150(20.69)}{50+20.69} = 43.9 \text{ V} \tag{3-118}$$

The short-circuit current I_N is therefore

$$I_N = \frac{V_{bb'}}{25} = \frac{43.9}{25} = 1.756 \text{ A} \tag{3-119}$$

As V_T/I_N=105.9/1.756=60.32 Ω, which is approximately the same value for R_T found in Example 3-8, Eq. (3-117) is verified.

Introductory Network Analysis

3.6 TRANSIENT MODELS AND INITIAL CONDITIONS

Our treatment of network analysis to this point has been concerned primarily with writing the differential equations which describe network behavior. The differential equations can be solved only if we have a correct set of initial conditions. These are the values of the dependent variables and their rates of change which prevail at the beginning of the solution interval. With no loss of generality we can assume that dynamic network problems start with the closing of one or more switches which may have been in a previous position for a very long time. The initial conditions needed by the network differential equations are then the values of certain voltages and currents and their derivatives just after switch closure.[14,15]

Models for Initial Conditions

One of the best ways to approach the problem of finding initial conditions is to derive network models for times just before and just after t=0. We will see that inductors and capacitors have a simplified behavior when the voltages or currents associated with them attempt to change abruptly due to a switch-induced change in the network configuration.

Initial-value models. In the following material the expressions $t=0^+$ and $t=0^-$ designate quantities just after and just before t=0. Thus, $v_1(0^+)$ is the voltage across element 1 an infinitesimal time after t=0. The reason for introducing this notation is that some network variables can change instantaneously--the current through a resistor, for example--while others cannot; hence the values of some variables differ at $t=0^-$ and $t=0^+$. In fact, the only variables which cannot change instantaneously are capacitor voltages and inductor currents. We can see this by examining the inverted definitions of these elements,

$$v_C(t) = \frac{1}{C}\int_{0^-}^{t} i_C(t)dt + v_C(0^-) \qquad (3-120)$$

and

$$i_L(t) = \frac{1}{L}\int_{0^-}^{t} v_L(t)dt + i_L(0^-) \qquad (3-121)$$

The quantities $v_C(t)$ and $i_L(t)$ are determined by integrals. As a result, even though the integrands may undergo abrupt changes, the quantities themselves cannot change instantaneously; they can change only after a finite time has passed so the integrals can accumulate some "area". Thus

$$v_C(0^+) = \frac{1}{C}\int_{0^-}^{0^+} i_C(t)dt + v_C(0^-) \qquad (3-122)$$

and, because the integral is zero if 0^+ and 0^- differ by an infinitesimal amount,

$$v_C(0^+) = v_C(0^-) \qquad (3-123)$$

In evaluating initial conditions then we can have Eq. (3-123) and, for inductor currents,

$$i_L(0^+) = i_L(0^-) \qquad (3-124)$$

The initial conditions for differential equations are always the $t=0^+$ values of variables and their derivatives.

Consider next the defining equation for the capacitor

$$i_C(t) = C\dot{v}_C(t) \qquad (3-125)$$

If the voltage is some function which is zero at t=0, but which has a nonzero slope, then i(t) jumps to a finite value at $t=0^+$ even though the voltage remains nearly zero. If, for example,

$$v_C(t) = Ktu(t) \tag{3-126}$$

then

$$i_C(t) = C\frac{d}{dt}\{Ktu(t)\} = CK\{u(t) + t\delta(t)\} \tag{3-127}$$

and at $t=0^+$

$$i_C(0^+) = CK \tag{3-128}$$

even though $v_C(0^+)=0$. The only circuit model which has a nonzero current with zero voltage drop is the <u>short circuit</u> shown in Fig. 3-35(a). A short circuit has zero resistance between the terminals, $v(t)=0$, and $i(t)=$anything. We can, therefore, model the capacitor as a short circuit at time $t=0^+$ when attempting to evaluate initial conditions.

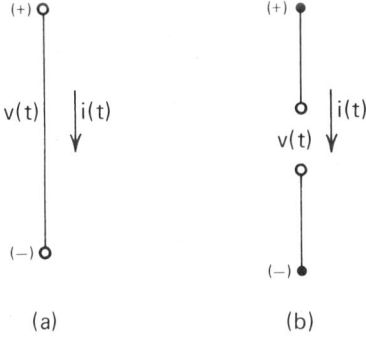

Figure 3-35. The voltage across a short circuit is zero but the current can have any value. The current through an open circuit is zero but the voltage can have any value. (a) A short circuit; (b) An open circuit.

Following a similar line of reasoning, we can use the inductance definition

$$v_L = \frac{L di_L}{dt} \tag{3-129}$$

to convince ourselves that an inductor voltage can undergo a jump change at $t=0^+$ even though current has not begun to flow. If for example

$$i_L(t) = Ktu(t) \qquad (3\text{-}130)$$

then

$$v_L(t) = LK\{u(t) + t\delta(t)\} \qquad (3\text{-}131)$$

so that $v_L(0^+)=LK$ even though $i_L(0^+)=0$. This type of behavior is adequately described by the <u>open circuit</u> of Fig. 3-35(b). The open circuit has infinite resistance between the terminals, $i(t)=0$, with $v(t)$ having an arbitrary value. This model allows the terminal voltages to change at will without regard to current flow. As a result, we can replace inductors by open circuits when evaluating initial conditions.

<u>Example 3-10.</u> Find the current $i_2(0^+)$ which flows in resistor R_2 at time $t=0^+$, just after switch S closes, in the network of Fig. 3-36(a).

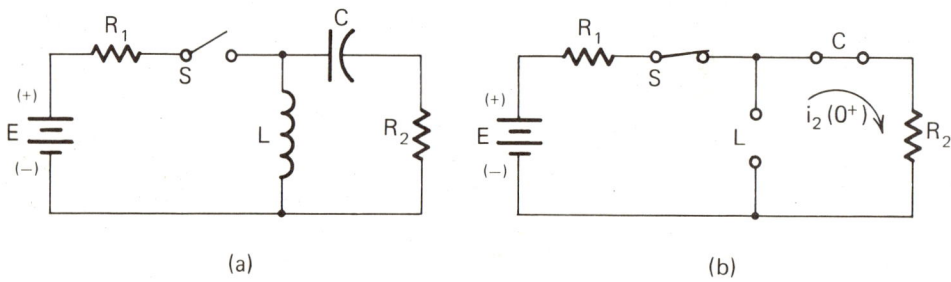

(a) (b)

Figure 3-36. These networks are used in Example 3-10 to illustrate the procedure for finding initial voltages and currents. (a) Network at $t = 0^-$; (b) The network of (a) at $t = 0^+$ with switch S closed.

<u>Solution.</u> As argued above, at time $t=0^+$ the inductor resembles an open circuit and the capacitor a short circuit. Redrawing the network for $t=0^+$ using these models gives the

Introductory Network Analysis

initial-value network of Fig. 3-36(b). Using Ohm's law, we have by inspection

$$i_2(0^+) = \frac{E}{R_1+R_2} \quad (3\text{-}132)$$

The network with capacitors replaced by short circuits and inductors by open circuits applies only at $t=0^+$. For finite times after $t=0^+$ the inductors and capacitors must, of course, be replaced by their true network models.

Final-value models. If a network has been in a given condition for a long time, and if there are final **equilibrium** values, it is possible to find those values from the differential equations by setting all derivatives to zero and solving the remaining algebraic equations for the steady-state values. It is also possible to develop network models for the $t \to \infty$ case. A network has constant steady-state voltages and currents if all the voltage and current sources are constant and if the network is <u>stable</u>. Stability is discussed in Chapter 6.

If a network is in **equilibrium, all changes of current** and voltage must have ceased: the network variables have "settled down" to their final values. Recall the definitions of inductance and capacitance,

$$i_C(t) = \frac{C dv_C}{dt} \quad (3\text{-}133)$$

and

$$v_L(t) = \frac{L di_L}{dt} \quad (3\text{-}134)$$

It follows from these expressions that if all change has ceased, which means that the derivatives are zero as $t \to \infty$, then $i_C(\infty)=0$ and $v_L(\infty)=0$. Because zero current implies an open circuit and zero voltage a short circuit, capacitors and inductors can be modeled by open and short circuits respectively in the steady state.

Example 3-11. Assume the switch in the network of Fig. 3-36(a) has been closed long enough for steady-state conditions to prevail. Find the currents $i_1(\infty)$ and $i_2(\infty)$ through R_1 and R_2 in the steady state.

Solution. Replace the inductor by a short circuit and the capacitor by an open circuit. The resulting <u>final-value</u> network is shown in Fig. 3-37.

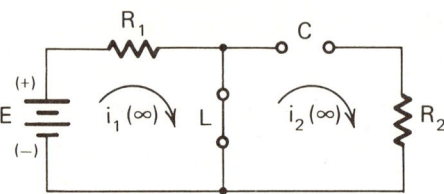

Figure 3-37. Final-value network for Example 3-11.

By inspection, the steady-state current through R_2 is $i_2(\infty)=0$; through R_1 we have $i_1(\infty)=E/R_1$.

The $t=0^+$ and $t \to \infty$ models for inductors and capacitors are summarized in Table 3-1.

Table 3-1.

Inductor and Capacitor Models For Transient Problems

Time	Inductor	Capacitor
$t=0^+$	Open Circuit	Short Circuit
$t \to \infty$	Short Circuit	Open Circuit

The physical interpretation of Table 3-1 is as follows. At $t=0^+$, if a capacitor is uncharged, it will accept an inrush of charge and appear as a short circuit. After it is fully charged (in the steady state), the capacitor accepts no more charge and resembles an open circuit. At $t=0^+$, an inductor develops a large opposing magnetic field which resists current flow, causing the device to resemble an open circuit.

Introductory Network Analysis

When the current no longer changes (the steady state), the magnetic field is collapsed, and the ideal inductor is a resistanceless coil of wire--a short circuit.

<u>Source models for initial conditions</u>. If Eq. (3-120) is interpreted as a loop equation written with Kirchhoff's voltage law, we see that a capacitor with an initial voltage can be represented by an uncharged capacitor in series with a battery whose voltage is $v_C(0^+)$. See Fig. 3-38(a). The positive terminal of the initial-condition battery is at the tail of the reference arrow for the capacitor current.

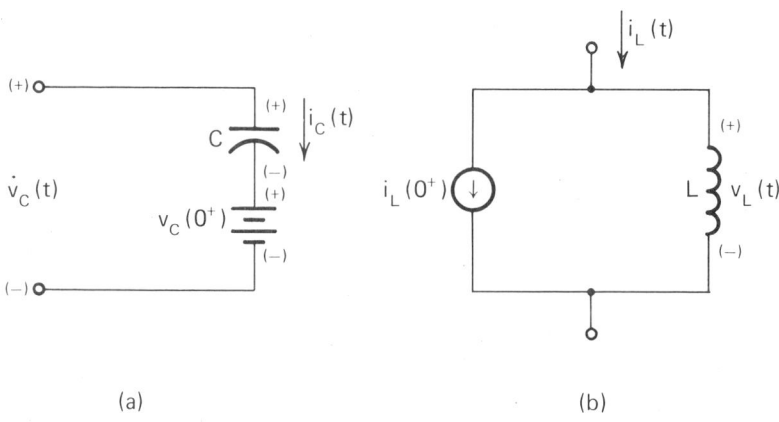

Figure 3-38. Initial capacitor voltages or inductor currents can be treated as sources. (a) A charged capacitor can be replaced by an uncharged capacitor in series with a suitable voltage source; (b) An inductor with an initial current can be replaced by an inert inductor in parallel with a suitable current source.

Similarly, Eq. (3-121), interpreted as a statement of Kirchhoff's current law, shows that an inductor with an initial current can be modeled as an inductor with no initial current in parallel with a current source whose value is $i_L(0^+)$ as in Fig. 3-38(b). The direction of the initial condition source current is the same as the reference current for the inductor; that is, the inductor tries to sustain its current flow.

These source models are useful in deriving initial conditions for transient problems. In contrast with the short-

and open-circuit models for the inductor and capacitor, which prevail only at $t=0^+$ and $t\to\infty$, the source models for the initial conditions remain in the network for all time.

Example 3-12. The capacitor of Fig. 3-39(a) is initially charged to 10 V with the polarity shown. The inductor has an initial current of 2 A downward. Find (1) the initial values of the currents through the 25 Ω and 75 Ω resistors after switch S is closed at t=0, and (2) the final values of these quantities after switch S has been closed long enough for steady-state conditions to result.

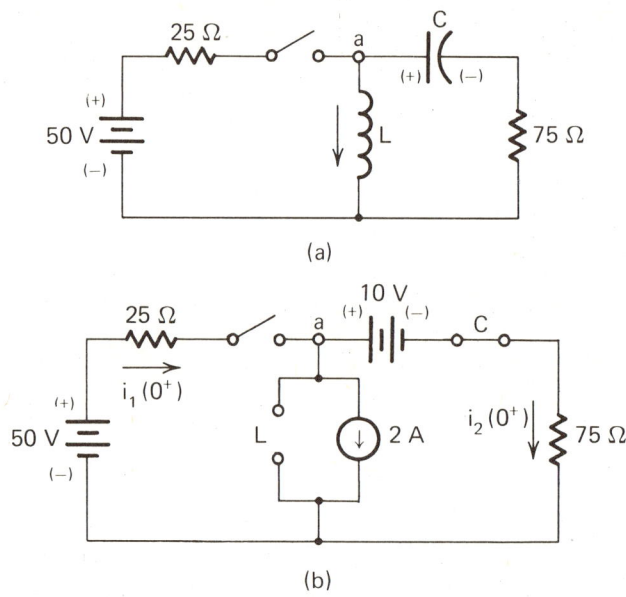

Figure 3-39. Networks for Example 3-12. (a) Original network; (b) Initial-value network model.

Solution. (1) The network model for $t=0^+$ is shown in Fig. 3-39(b). The inductor is replaced by an open circuit in parallel with a 2 A current source which accounts for the current which flowed before the switch was closed. This is valid because the current through an inductor cannot change instantaneously. The capacitor is modeled by a short circuit in series with a battery whose value is equal to the initial capacitor voltage. As the voltage across a capacitor cannot change instantaneously, $v_C(0^+)=v_C(0^-)$. The KCL equation at node a is

$$i_1(0^+) = i_2(0^+) + 2 \tag{3-135}$$

The element v-i relationships are

$$i_1(0^+) = \frac{50-v_a}{25} \tag{3-136}$$

and

$$i_2(0^+) = \frac{v_a-10}{75} \tag{3-137}$$

Substituting these expressions into Eq. (3-135) gives

$$\frac{50-v_a}{25} = \frac{v_a-10}{75} + 2 \tag{3-138}$$

Equation (3-138) has the solution $v_a=2.5$ V. Using this value in the expressions for i_1 and i_2 yields

$$i_1(0^+) = \frac{50-2.5}{25} = 1.9 \text{ A} \tag{3-139}$$

and

$$i_2(0^+) = \frac{2.5-10}{75} = -0.1 \text{ A} \tag{3-140}$$

The minus sign with $i_2(0^+)$ indicates that the initial current flows opposite to the assumed reference.

(2) Figure 3-40 shows the final-value network in which the inductor has been replaced by a short circuit and the capacitor by an open circuit.

We see by inspecting Fig. 3-40 that $i_2(\infty)=0$ and $i_1(\infty)= 50/25=2.0$ A. In the final-value model, the sources which account for the initial inductor current and the initial capacitor voltage are effectively removed--the inductor current source is short circuited and the capacitor battery open circuited. This source removal accounts for the dissipation of the energy initially stored in those devices.

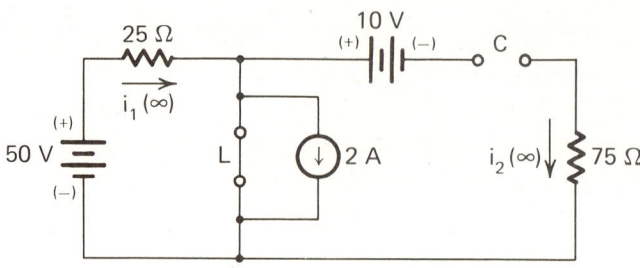

Figure 3-40. Final-value network for Example 3-12.

Examples of Evaluating Initial Conditions

Equipped with the techniques of the preceeding paragraphs, we are now in a position to evaluate initial conditions for more complicated transient problems. The procedure is to develop a network model for $t=0^-$, solve for the appropriate variables, then use these in a similar model for $t=0^+$. (The $t=0^-$ values are the $t\to\infty$ values before the switch is moved to the $t=0^+$ position.) The model for $t=0^+$ is then solved for the initial values to be used in the main problem. While this procedure sounds complex, it should be remembered that the $t=0^+$ and $t=0^-$ models are algebraic: it is not necessary to solve differential equations.

Example 3-13. In the network of Fig. 3-41 the switch has been in position A long enough that steady-state conditions prevail. At t=0 the switch is moved to position B. (a) Write differential equations for the loop currents $i_2(t)$ and $i_3(t)$ for t>0. (b) Determine the initial condition for the differential equations of part (a).

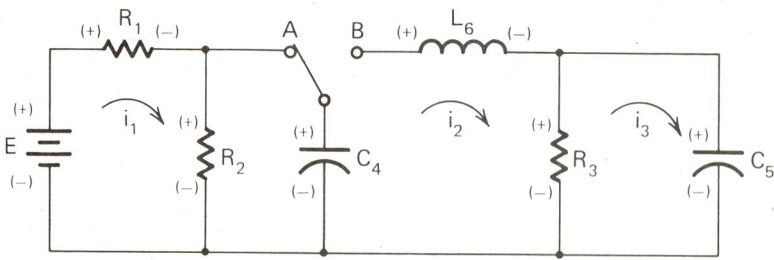

Figure 3-41. Network for Example 3-13.

Solution. Assume each numbered element has an associated voltage with the polarities of Fig. 3-41. Thus, v_6 appears across L_6, etc. Using these reference polarities the KVL equations for the i_2 and i_3 loops with the switch at B are

$$\text{Loop 2:} \quad v_6 + v_3 - v_4 = 0 \qquad (3\text{-}141)$$

$$\text{Loop 3:} \quad v_5 - v_3 = 0 \qquad (3\text{-}142)$$

The element v-i relationships are

$$v_3 = R_3(i_2 - i_3) \qquad (3\text{-}143)$$

$$v_4 = -\frac{1}{C_4} \int_0^t i_2 \, dt - v_4(0^+) \qquad (3\text{-}144)$$

$$v_5 = \frac{1}{C_5} \int_0^t i_3 \, dt + v_5(0) \qquad (3\text{-}145)$$

and

$$v_6 = L_6 \frac{di_2}{dt} \qquad (3\text{-}146)$$

Substituting these expressions into the loop equations yields

$$L_6 \frac{di_2}{dt} + R_3(i_2 - i_3) + \frac{1}{C_4} \int_0^t i_2 \, dt + v_4(0^+) = 0 \qquad (3\text{-}147)$$

$$\frac{1}{C_5} \int_0^t i_3 \, dt + v_5(0^+) - R_3(i_2 - i_3) = 0 \qquad (3\text{-}148)$$

Differentiating Eqns. (3-147) and (3-148) to remove the integrals and rearranging slightly results in

$$\frac{d^2 i_2}{dt^2} + \frac{R_3}{L_6} \frac{di_2}{dt} + \frac{i_2}{L_6 C_4} - \frac{R_3}{L_6} \frac{di_3}{dt} = 0 \qquad (3\text{-}149)$$

$$\frac{di_3}{dt} + \frac{i_3}{C_5 R_3} - \frac{di_2}{dt} = 0 \qquad (3\text{-}150)$$

(b) To solve Eqns. (3-149) and (3-150) for $i_2(t)$ and $i_3(t)$, we need the initial conditions $i_2(0^+)$, $i_3(0^+)$, and $di_2(0^+)/dt$. The first step is to find the $t=0^-$ values, those which prevailed before the switch was moved to position B. Figure 3-42(a) is the steady-state network model for the network to the left of the switch. Capacitor C_4 resembles an open circuit under steady-state conditions. The KVL equation for loop 1 is

$$E = v_1(0^-) + v_2(0^-) \qquad (3\text{-}151)$$

with

$$v_1(0^-) = i_1(0^-)R_1 \quad \text{and} \quad v_2(0^-) = i_1(0^-)R_2 \qquad (3\text{-}152)$$

from which

$$E = i_1(0^-)(R_1 + R_2) \qquad (3\text{-}153)$$

and

$$i_1(0^-) = \frac{E}{R_1 + R_2} \qquad (3\text{-}154)$$

Thus,

$$v_2(0^-) = i_2(0^-)R_2 = \frac{ER_2}{R_1 + R_2} \qquad (3\text{-}155)$$

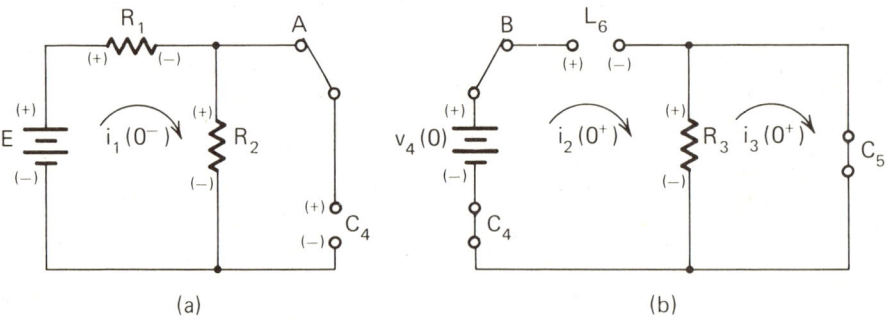

Figure 3-42. Initial and final value networks for Example 3-13. (a) Network for $t = 0^-$; (b) Network for $t = 0^+$.

Writing a loop equation around the loop containing R_2 and C_4 yields

$$v_4(0^-) = v_2(0^-) \tag{3-156}$$

hence

$$v_4(0^-) = \frac{ER_2}{R_1+R_2} \tag{3-157}$$

Because the voltage across a capacitor cannot change instantaneously, $v_4(0^+)=v_4(0^-)$. When the switch is moved to position A, we can therefore replace C_4 by a short circuit in series with an initial condition battery whose value is

$$v_4(0^+) = \frac{ER_2}{R_1+R_2} \tag{3-158}$$

In the $t=0^+$ network model of Fig. 3-42(b), C_4 and C_5 have been replaced by short circuits and L_6 by an open circuit. The currents $i_2(0^+)$ and $i_3(0^+)$ are clearly zero. To find $di_2(0^+)/dt$, write KVL around loop 2 to get

$$v_6(0^+) + v_3(0^+) - v_4(0^+) = 0 \tag{3-159}$$

But

$$v_3(0^+) = i_3(0^+)R_3 = 0 \tag{3-160}$$

hence,

$$v_6(0^+) = \frac{ER_2}{R_1+R_2} \tag{3-161}$$

Finally, since

$$v_6(t) = L_6 \frac{di_2(t)}{dt} \tag{3-162}$$

we have

$$\frac{di_2(t)}{dt} = \frac{v_6(t)}{L_6} \tag{3-163}$$

or, evaluating at $t=0^+$ and substituting Eq. (3-161),

$$\frac{di_2(0^+)}{dt} = \frac{ER_2}{L_6(R_1+R_2)} \qquad (3-164)$$

Example 3-14. The network of Fig. 3-43 has reached steady state with the switch in position A. Find expressions for the voltages $v_a(t)$ and $v_b(t)$ immediately after the switch is moved to position B.

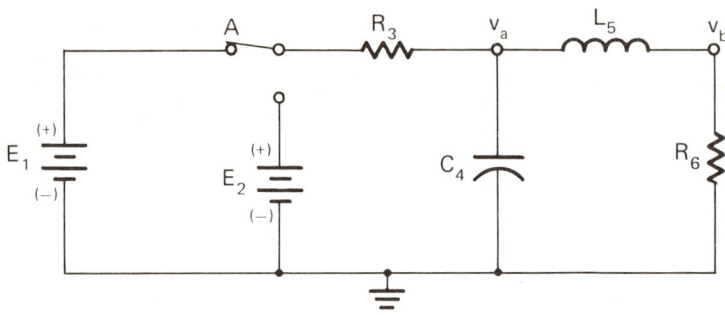

Figure 3-43. Network for Example 3-14.

Solution. Figure 3-44(a) is the final-value network with the switch in position A. Analyzing this network gives the $t=0^-$ values.
Using voltage division we have

$$v_a(0^-) = v_b(0^-) = \frac{E_1 R_6}{R_6+R_3} \qquad (3-165)$$

Figure 3-44(b) is a $t=0^+$ model for the network after the switch has been moved to position B. The capacitor is modeled by a short circuit in series with a battery whose voltage is $v_4(0^+)$. Because $v_a = v_4$ at $t=0^+$, and v_a, being the voltage across a capacitor, cannot change instantaneously, we have

$$v_4(0^+) = v_a(0^+) = v_a(0^-) = \frac{E_1 R_6}{R_6+R_3} \qquad (3-166)$$

Introductory Network Analysis

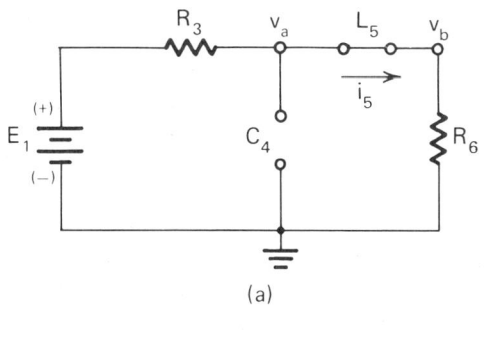

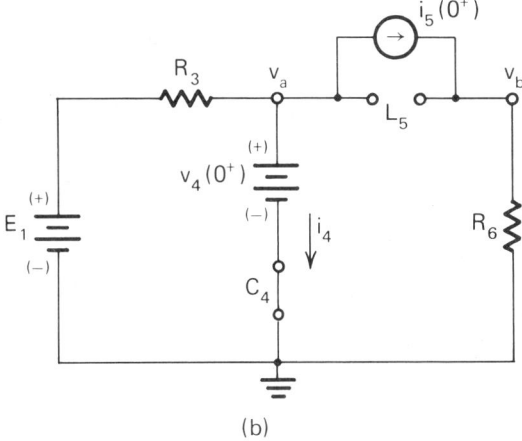

Figure 3-44. Initial-value networks for Example 3-14. (a) Network at $t = 0^-$; (b) Network at $t = 0^+$.

The inductor is represented by an open circuit in parallel with a current source whose value $i_5(0^+)$ is, from Fig. 3-44(a),

$$i_5(0^+) = i_5(0^-) = \frac{E_1}{R_6 + R_3} \qquad (3\text{-}167)$$

The voltage $v_b(0^+)$ is given by

$$v_b(0^+) = i_5(0^+) R_6 \qquad (3\text{-}168)$$

$$= \frac{E_1 R_6}{R_6 + R_3} \qquad (3\text{-}169)$$

This completes our study of how to write the differential equations which describe the transient behavior of electrical networks. The reader will, however, find electrical networks treated repeatedly in the remaining chapters as we consider such topics as solving differential equations, state variables, impedance, Laplace transforms, and the sinusoidal steady state.

3.7 OPERATIONAL AMPLIFIERS

As the name implies, an amplifier is a device which amplifies or makes something bigger. Amplifiers, which can be mechanical (a lever is a force amplifier), hydraulic, magnetic pneumatic, or electronic, are useful building blocks in designing dynamic systems. Because they come in neat little packages (and have neat little mathematical models) we will restrict our discussion to one particular kind of electronic amplifier--the operational amplifier.

The Operational Amplifier

Integrated-circuit operational amplifiers have become so widely used that they now constitute fundamental electronic building blocks along with resistors, inductors, capacitors, and sources.[16] Ideal operational amplifiers (or OP AMP's as they are affectionately called by electrical engineers) are useful as mathematical models, while practical OP AMPS are used in the design of instrumentation systems, data acquisition devices, transducers, and analog computers. The operational amplifier performs a number of useful jobs in electrical and electronics systems; among these are summation, integration, amplification, and signal isolation. This section shows how the OP AMP does some of these jobs. We will assume the existence of an ideal device, leaving to books on electronic components the study of deviations of actual operational amplifiers from the ideal.

__Symbols and relationships__.[17,18] Figure 3-45(a) is the circuit model for an ideal operational amplifier; Fig. 3-45(b)

Introductory Network Analysis

shows the most popular symbol for the circuit model. Notice the absence of the ground wire in Fig. 3-45(b); this connection is often omitted as it conveys no information.

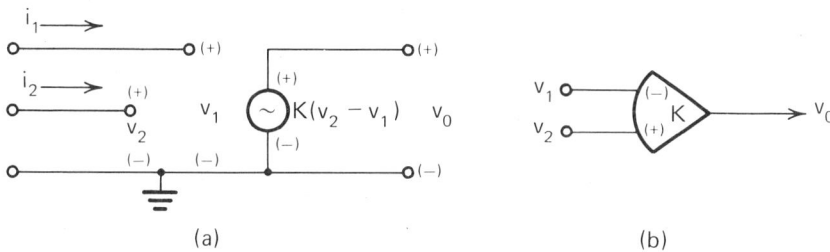

Figure 3-45. A circuit model and a common symbol for the operational amplifier. (a) Circuit Model; (b) OP AMP symbol.

The voltage source in the circuit model of Fig. 3-45(a) is a dependent source: its output depends on the constant K and the difference between the input voltages. The inputs v_1 and v_2 are called the <u>inverting</u> and <u>noninverting</u> inputs, respectively. The mathematical relationships which describe the ideal operational amplifier are

$$i_1 = i_2 = 0 \qquad (3\text{-}170)$$

$$v_o = K(v_2 - v_1) \qquad (3\text{-}171)$$

$$R_{in} \to \infty \qquad (3\text{-}172)$$

and

$$R_{out} = 0 \qquad (3\text{-}173)$$

The voltages and currents may of course be time varying. Equations (3-170) through (3-173) are derived from Fig. 3-45(a). The input resistance, R_{in}, is the resistance to current flow seen by looking into the input terminals; it is clearly infinite because both inputs are modeled as open circuits. The output resistance, R_{out}, is the resistance seen looking back into the device from its output terminals. As the internal resistance of an ideal voltage source is zero, it follows that $R_{out}=0$ for the operational amplifier. A final property of the ideal operational amplifier is that its gain K is infinite. If v_o is to remain finite with infinite gain K, then clearly

$$v_2 - v_1 = 0 \qquad (3\text{-}174)$$

at all times. Because the user is free to choose v_1 and v_2, one may well ask what happens if unequal voltages are applied to the input terminals. The answer is that the mathematical model of Eqns. (3-170) through (3-173) breaks down and is no longer valid. When this happens, the actual OP AMP saturates and goes into nonlinear operation (Fig. 3-46(a)).

The ideal operational amplifier is indeed a queer animal. It has infinite gain, draws no current, has infinite input resistance, zero output resistance, and requires that equal voltages be connected to the inputs. Can such a strange device exist and if so, what good is it?

Practical operational amplifiers only approximate the ideal properties but the degree of correspondence is surprisingly close. One can design with high-quality operational amplifiers as if Eqns. (3-170) through (3-174) were completely valid. In practice the gain might fall in the range $10^3 < K < 10^8$ and the output v_o might be limited to $v_o < 10$ volts for linear operation. As a consequence, v_2-v_1 would fall between 10^{-2} volts and 10^{-7} volts, probably close enough to zero relative to other voltages used in the system.

<u>Range of operation</u>. In order for Eqns. (3-170) through (3-174) to remain approximately true, the amplifier must stay within its prescribed operating range. By this we mean that both the magnitude and frequency of the input voltage difference should be restricted to the so-called linear range of operation. Figure 3-46 shows a typical input-output characteristic and a frequency response for an operational amplifier. For the relationship $v_o=K(v_2-v_1)$ to remain valid, the input voltage difference must be less than the value shown as v_{max}

in Fig. 3-46(a). If $|v_2-v_1|>v_{max}$ the device saturates and v_o remains at the value $v_o=v_{sat}$.

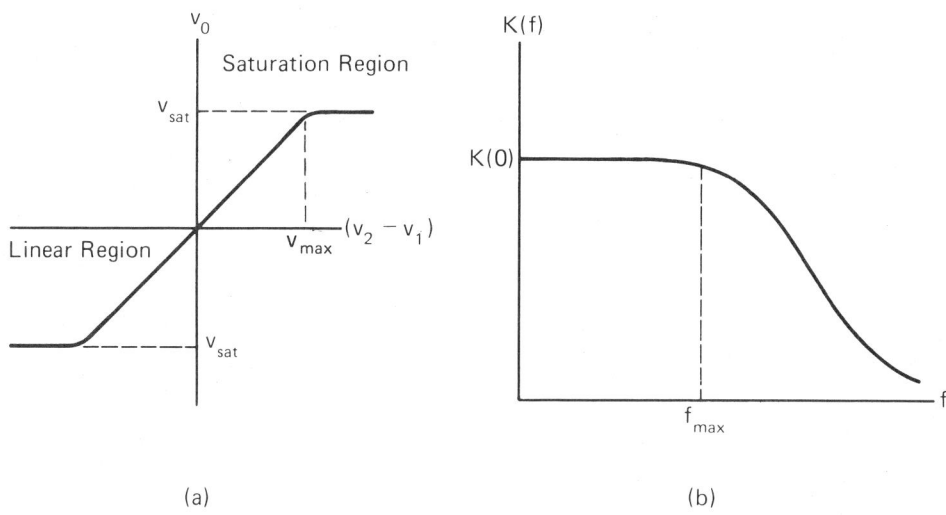

Figure 3-46. OP AMP transfer characteristic and frequency response. Frequency response is discussed at length in Chapter 9. (a) Transfer characteristic; (b) Frequency response.

Similarly, if the frequency of the input voltage difference greatly exceeds the value f_{max} in Fig. 3-46(b), the gain K (which is a decreasing function of frequency) falls and Eq. (3-174) becomes less valid. The frequency response of linear systems is explored at length in Chapter 9. Here we merely observe that the behavior of many devices depends on the frequency content of the input functions. In the following discussions we assume linear, low-frequency operation of all operational amplifiers.

Operational Amplifier Applications

The operational amplifier gets its name from its utility as a building block for electrical networks which perform useful mathematical operations.[18] The analog computer, for example, uses operational amplifiers to solve differential equations.[19,20] Some of the operations which can be done by OP AMP's are discussed in the following paragraphs.

Multiplication by a constant. Figure 3-47 shows an operational amplifier with the v_2 input connected to ground and the v_1 input connected to an input resistor R_i. A feedback resistance R_f is connected from the output terminal to the point J. Because v_2 is connected to ground, and hence is always zero, the input-output relationship is, from Eq. (3-171)

$$v_o = -Kv_1 \qquad (3\text{-}175)$$

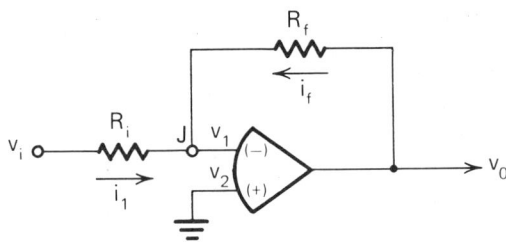

Figure 3-47. Operational amplifier circuit for multiplication by a constant. Here the input-output relationship is $v_o = -(R_f/R_i)v_i$.

Applying KCL at node J gives

$$i_1 + i_f = 0 \qquad (3\text{-}176)$$

Substituting the element v-i relationships gives

$$\frac{v_i - v_1}{R_i} + \frac{v_o - v_1}{R_f} = 0 \qquad (3\text{-}177)$$

If v_o is restricted to a few volts and K is nearly infinite, say $K \approx 10^8$, then v_1 is negligibly small and we may safely take it as zero. Thus, Eq. (3-177) becomes

$$\frac{v_i}{R_i} + \frac{v_o}{R_f} = 0 \qquad (3\text{-}178)$$

Introductory Network Analysis

Solving for the ratio v_o/v_i yields

$$\frac{v_o}{v_i} = -\frac{R_f}{R_i} \qquad (3-179)$$

The overall gain of the device, $-R_f/R_i$, can be set accurately by using precision resistors for R_f and R_i. The exact value of K is immaterial as long as it is large (say greater than 10^4). If $R_f=R_i$ the device becomes a unity-gain <u>inverter</u>. The unity-gain inverter is widely used in analog computers.

The summing amplifier. Figure 3-48 shows an operational amplifier configured as a <u>summing amplifier</u>.

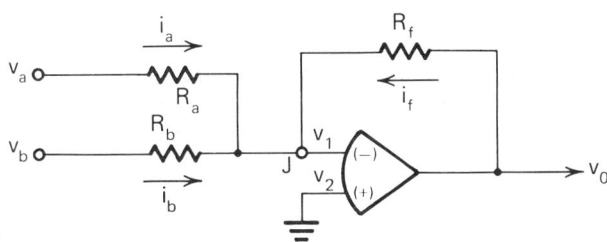

Figure 3-48. An operational amplifier configured as a summing amplifier. Equation (3-182) is the input-output relationship.

Sum currents at node J to get

$$i_a + i_b + i_f = 0 \qquad (3-180)$$

and, after inserting the v-i relationships with $v_1=0$, we have

$$\frac{v_a}{R_a} + \frac{v_b}{R_b} + \frac{v_o}{R_f} = 0 \qquad (3-181)$$

Solving for v_o yields

$$v_o = -\frac{R_f}{R_a}v_a - \frac{R_f}{R_b}v_b \qquad (3-182)$$

The output v_o is the negative sum of the inputs, each multiplied by a resistance ratio (which may, of course, be different for each input). In theory we may sum as many inputs as we like with a single amplifier just by adding more input paths. Summing amplifiers are used in analog computers and in a variety of special purpose signal processors.

The integrator. Probably the most useful operation which can be performed by the OP AMP is integration. Figure 3-49 shows an operational amplifier connected as an integrator.

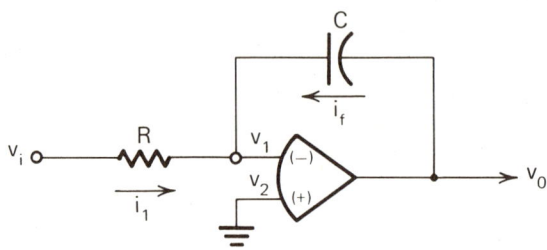

Figure 3-49. An operational amplifier configured as an integrator. Equation (3-186) is the input-output relationship.

Sum currents at node J to get

$$i_1 + i_f = 0 \tag{3-183}$$

and, after inserting the v-i relationships with $v_1=0$,

$$\frac{v_i}{R} + C\frac{dv_o}{dt} = 0 \tag{3-184}$$

Rearranging Eq. (3-184) and integrating between 0 and t yields

$$\int_0^t dv_o(t) = -\frac{1}{RC}\int_0^t v_i(t)dt \tag{3-185}$$

or

$$v_o(t) = -\frac{1}{RC}\int_0^t v_i(t)dt + v_o(0) \tag{3-186}$$

The term $v_o(0)$ corresponds to an initial voltage on the feedback capacitor C. The term 1/RC is the gain of the integrator;

it lets us simultaneously integrate and multiply by a constant. Integrators are used in analog computers, as building blocks for electrical filters, and in other signal processors.

Networks Containing Operational Amplifiers

The following examples illustrate the use of operational amplifiers in signal isolation and to perform complex operations.

The voltage follower. Recall our earlier model of a practical battery as an ideal battery in series with a corrupting voltage source, Fig. 3-8. The terminal voltage of the battery was a function of the current drawn from the battery. For the case where the load resistance R_L equals the battery internal resistance, the terminal voltage v(t) is only half of the source voltage! This is the situation shown in Fig. 3-50(a); if $R_L = R_O$ then, by the voltage-divider principle, v(t)=E/2.

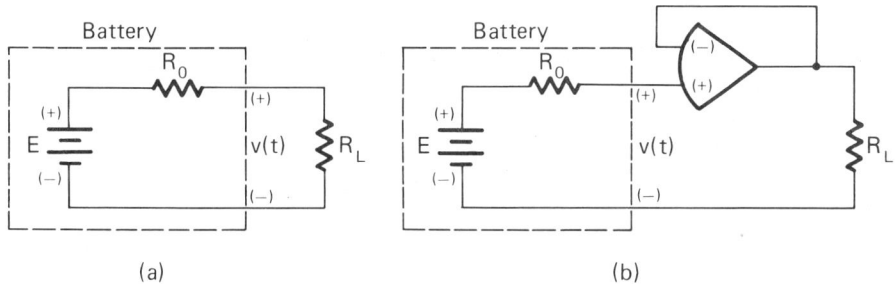

(a) (b)

Figure 3-50. The operational amplifier used in load isolation. (a) A battery with load resistor R_L; (b) An operational amplifier configured as a voltage follower has been inserted between source and load.

If an operational amplifier is correctly inserted between the battery and the load, then the terminal voltage v(t) is unaffected by the load. Figure 3-50(b) is the required network. From the operational amplifier model, Eq. (3-174), the voltages connected to the "+" and "-" terminals must be equal. As the battery terminal voltage v(t) is applied to the "+"

terminal of the OP AMP, it also appears across R_L, regardless of the value of R_L. Moreover, because ideal operational amplifiers draw no current, v(t)=E. This in turn means that the voltage across R_L is always E.

Signal isolation. Figure 3-51(a) shows a two-stage RC network.

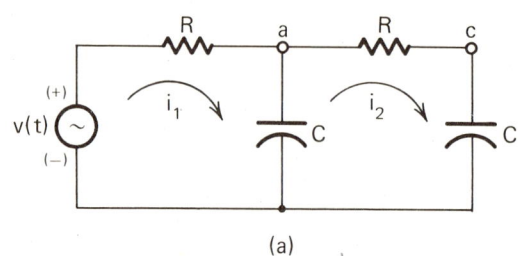

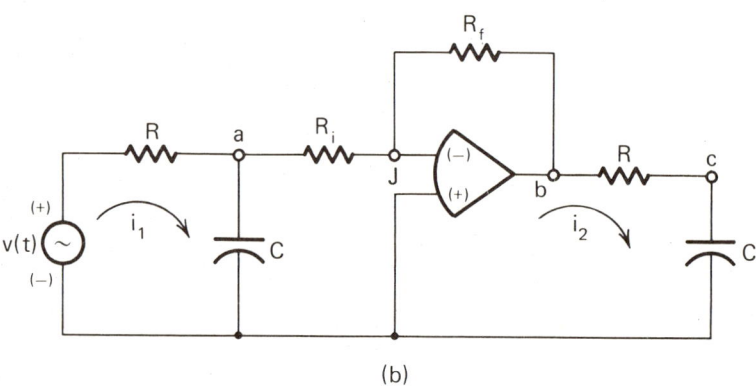

Figure 3-51. A unity-gain OP AMP can be used to de-couple parts of a network, in this case, to isolate the transient behavior of the two loops. (a) A two-loop RC network with coupling through a capacitor; (b) A unity-gain amplifier has been inserted to isolate the two loops.

The KCL equations at nodes a and c in Figure 3-51(a) are

$$\frac{v_a - v(t)}{R} + C\dot{v}_a + \frac{v_a - v_c}{R} = 0 \tag{3-187}$$

and

$$\frac{v_c - v_a}{R} + C\dot{v}_c = 0 \tag{3-188}$$

Introductory Network Analysis

Because v_a appears in the v_c equation and v_c in the v_a equation, the equations are __coupled__ and must be solved as simultaneous differential equations. We will now show how introducing a unity-gain operational amplifier uncouples the differential equations.

If $R_f = R_i$ the gain of the operational amplifier in Fig. 3-51(b) is unity and the network at first glance appears similar to that of Fig. 3-51(a). Writing nodal equations for nodes a and c in Fig. 3-51(b), and using the OP AMP constraint that $v_J = 0$, yields

$$\text{Node a:} \quad \frac{v_a - v(t)}{R} + C\dot{v}_a + \frac{v_a}{R_i} = 0 \qquad (3\text{-}189)$$

$$\text{Node c:} \quad \frac{v_c - v_b}{R} + C\dot{v}_c = 0 \qquad (3\text{-}190)$$

But if $R_f = R_i$, $v_b = -v_a$; hence, Eq. (3-190) becomes

$$\frac{v_c + v_a}{R} + C\dot{v}_c = 0 \qquad (3\text{-}191)$$

or

$$\dot{v}_c + \frac{v_c}{RC} = -\frac{v_a}{RC} \qquad (3\text{-}192)$$

Equations (3-189) and (3-192) are __uncoupled__: Equation (3-189) can be solved for v_a directly as it does not contain v_c or any term which depends on v_c. After v_a is found, it is used as an input function to Eq. (3-192). The situation is entirely different in Eqns. (3-187) and (3-188) where each node voltage depends on the other. When operational amplifiers are thus used to uncouple systems, they are called __isolation amplifiers__.

__The resonator__. Figure 3-52 is the circuit diagram for a so-called __resonator__.

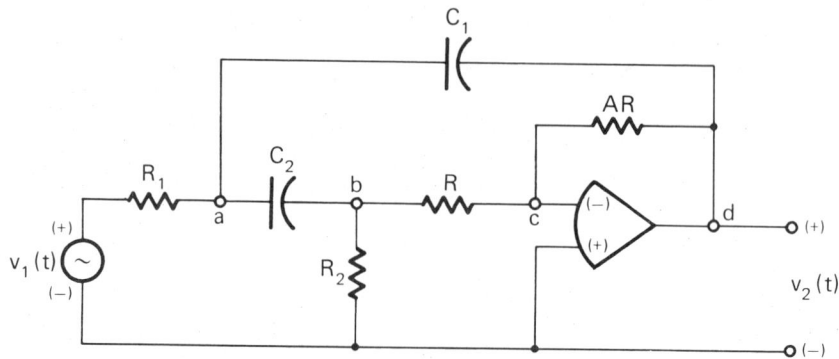

Figure 3-52. This OP AMP network is called a "resonator." It can be used to produce a variety of output waveshapes depending on the input and the choice of resistors and capacitors.

The network is best analyzed using nodal equations. Identify the nodes a, b, and c and associate with each its node voltage v_a, v_b, and v_c. The nodal equations are then

$$\text{Node a:} \quad \frac{v_a - v_1}{R_1} + C_1(\dot{v}_a - \dot{v}_d) + C_2(\dot{v}_a - \dot{v}_b) = 0 \quad (3\text{-}193)$$

$$\text{Node b:} \quad C_2(\dot{v}_b - \dot{v}_a) + \frac{v_b}{R_2} + \frac{v_b - v_c}{R} = 0 \quad (3\text{-}194)$$

and

$$\text{Node c:} \quad \frac{v_c - v_b}{R} + \frac{v_c - v_d}{AR} = 0 \quad (3\text{-}195)$$

Clearly, $v_d = v_2$, and from the properties of the ideal operational amplifier, $v_c = 0$. Moreover, as the gain of the amplifier is $-R_f/R_i = -AR/R = -A$, we have $v_2 = -Av_b$. Substituting these relationships into Eqns. (3-193) and (3-194) yields

$$\frac{v_a - v_1}{R_1} + C_1(\dot{v}_a - \dot{v}_2) + C_2\left\{\dot{v}_a + \frac{\dot{v}_2}{A}\right\} = 0 \quad (3\text{-}196)$$

and

Introductory Network Analysis

$$C_2 \left\{ \frac{\dot{v}_2}{A} + \dot{v}_a \right\} + \frac{v_2}{A} \left\{ \frac{1}{R_2} + \frac{1}{R} \right\} = 0 \qquad (3\text{-}197)$$

or, rearranging in standard form,

$$\dot{v}_a + \frac{v_a}{R_1(C_1+C_2)} + \dot{v}_2 \frac{(C_2-AC_1)}{A(C_1+C_2)} = \frac{v_1}{R_1(C_1+C_2)} \qquad (3\text{-}198)$$

and

$$\dot{v}_2 + v_2 \left\{ \frac{1}{R_2 C_2} + \frac{1}{RC_2} \right\} + A\dot{v}_a = 0 \qquad (3\text{-}199)$$

These simultaneous differential equations can be solved for the response $v_2(t)$ in terms of the driving voltage $v_1(t)$. In a later chapter we will see how to derive an overall transfer function which would represent the network equations in a more useful form.

PROBLEMS FOR CHAPTER 3

3-1. A certain electric heater uses constant voltages and currents. If the current is I=20 A and the voltage is E=200 V, find (a) the power used by the device in kilowatts (kW), and (b) the energy in joules (J) consumed during an operation of one hour.

3-2. The kilowatt-hour is a unit of energy familiar to people who pay utility bills. Derive the conversion factor between joules and kilowatt-hours.

3-3. The voltage and current waveforms associated with a certain device are shown in Fig. 3-53. Find (a) the average power developed in the device over one period, and (b) the energy consumed or stored in the first 2.5 seconds. The units in the figure are volts, amperes, and seconds.

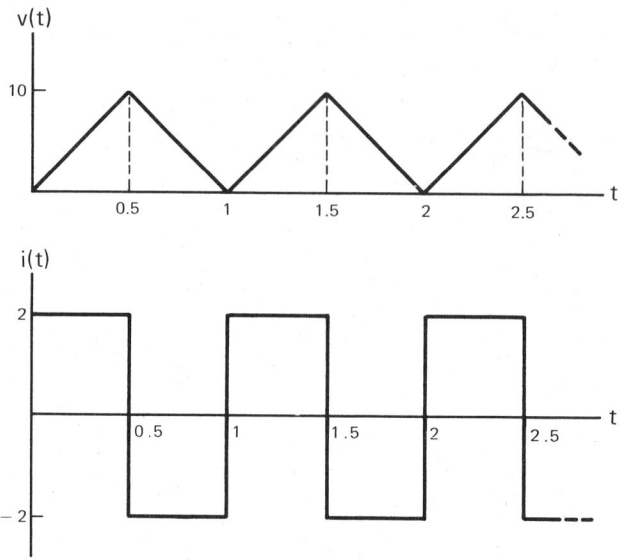

Figure 3-53. Problem 3-3.

3-4. Use KCL to find the unknown current i_x at each of the nodes shown:

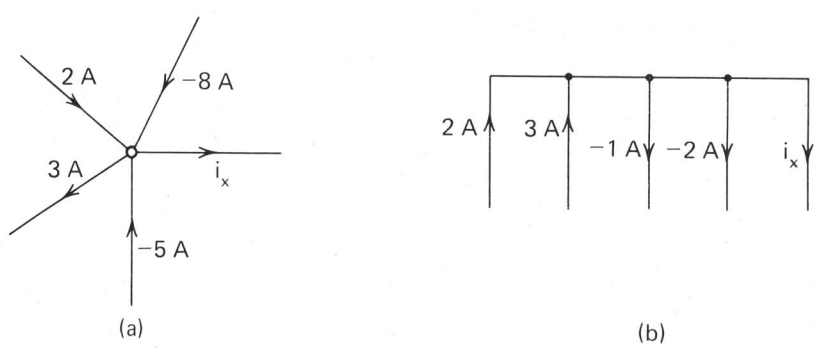

Figure 3-54. Problem 3-4.

3-5. Use KVL to find the unknown voltage v_x in each of the loops shown.

Introductory Network Analysis

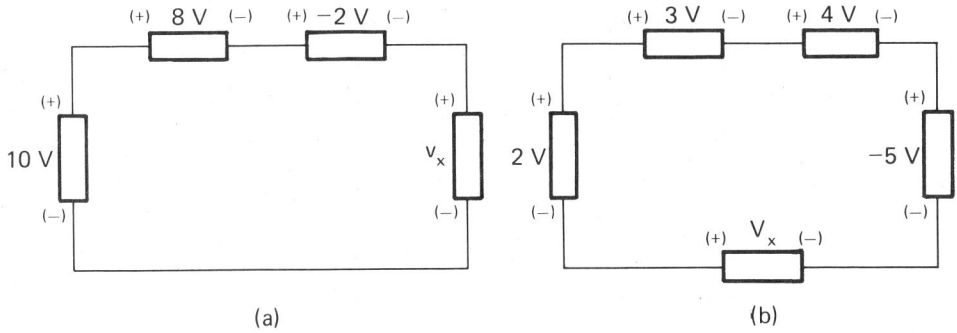

(a)   (b)

Figure 3-55. Problem 3-5.

3-6. Find the number of independent KCL and KVL equations for each of the networks shown:

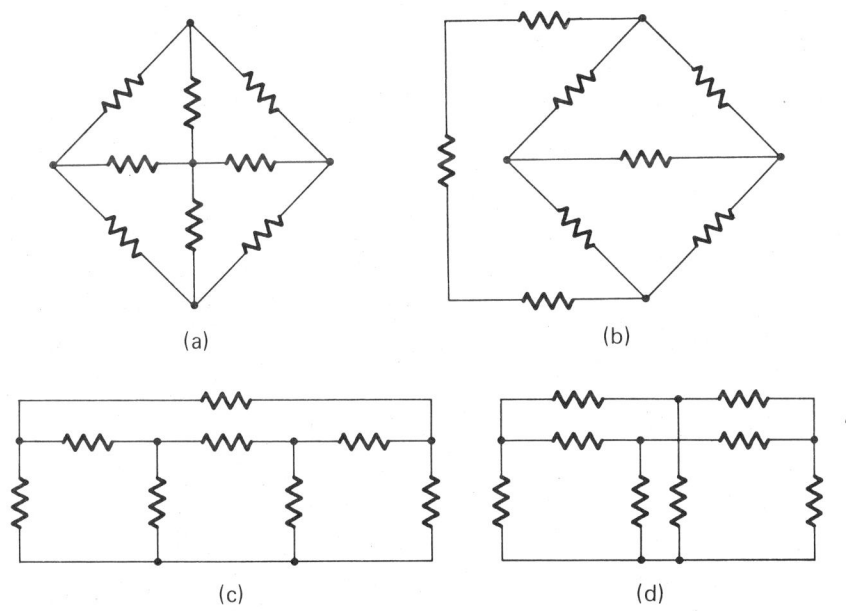

Figure 3-56. Problem 3-6.

3-7. Using the given current reference directions, write KCL for the networks shown:

Glass–Linear Systems—10

128   *Introductory Network Analysis*

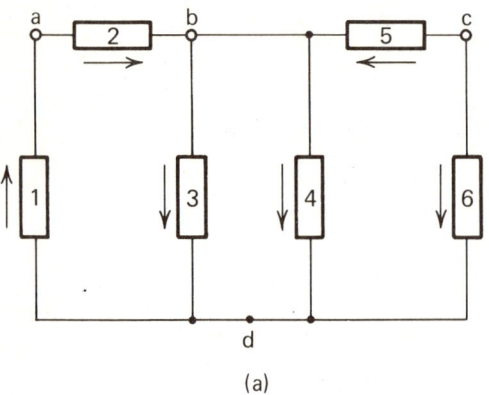
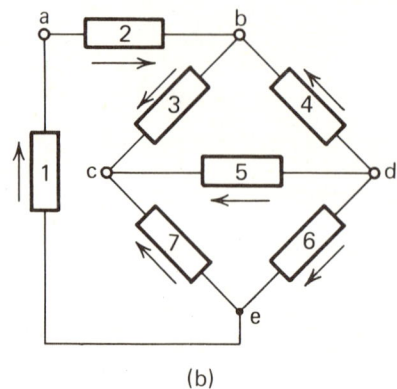

Figure 3-57. Problem 3-7.

3-8. Assume a positive voltage reference at the tail of each current arrow in the networks of Problem 3-7; then write KVL for each of the meshes.

3-9. Resistance can be interpreted as the slope of a voltage-versus-current curve. Find the resistance of the devices with the following v-i characteristics. The units are amperes and volts.

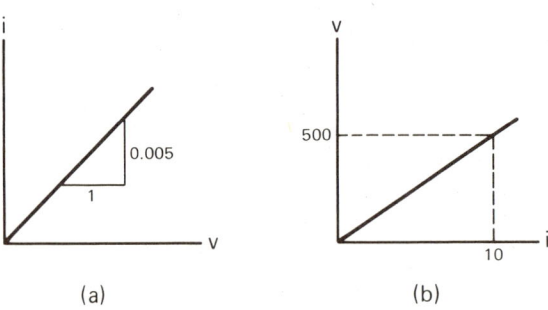

Figure 3-58. Problem 3-9.

3-10. A certain nonlinear electrical component has a current determined by the function

$$i = 0.01 e^{v/100} \tag{3-200}$$

where i is the current through the device in amperes and v is the voltage across it in volts. Use the slope interpretation of resistance to find the resistance of this device when v=50 V.

3-11. Find the equivalent resistance between the terminals a-a' for each of the networks shown:

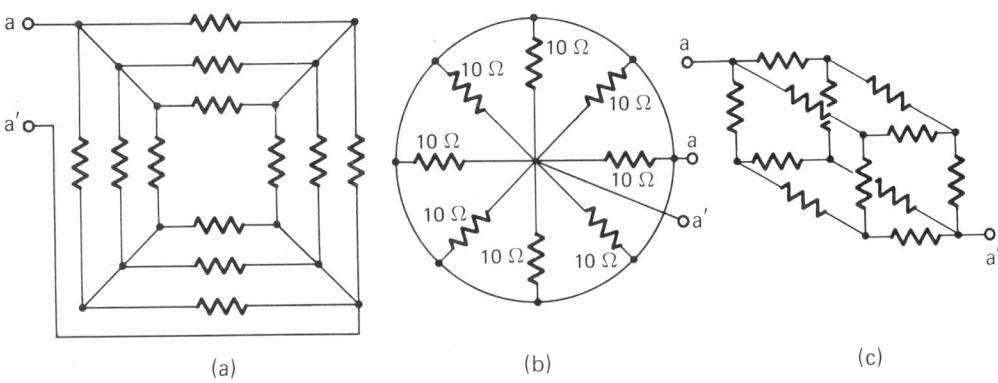

(a)   (b)   (c)

Figure 3-59. Problem 3-11. (a) All resistors are 1 Ω; (c) All resistors are 1 Ω.

3-12. The current $i(t)=0.025\cos(500t)$ A passes through an inductor while the voltage $v(t)=-500\sin(500t)$ V appears across its terminals. Find the inductance in henries.

3-13. Find the voltage across the inductor after the switch S is closed.

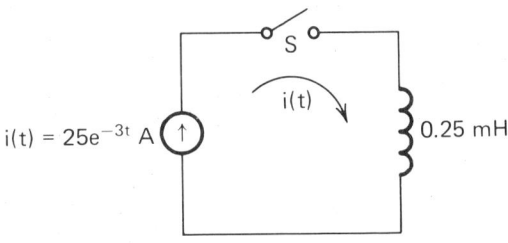

Figure 3-60. Problem 3-13.

3-14. Find the indicated quantities for the capacitor circuit shown:

(a) Find $v_C(t_1)$ if $f(t)=i_C(t)$. Assume $v_C(0)=0$.

(b) Find $v_C(t_1)$ if $g(t)=i_C(t)$. Assume $v_C(0)=0$.

(c) Find $v_C(t_1)$ if $h(t)=i_C(t)$. Assume $v_C(0)=0$.

(d) Find $i_C(t_1)$ if $g(t)=v_C(t)$.

(e) Find $i_C(t_1)$ if $h(t)=v_C(t)$.

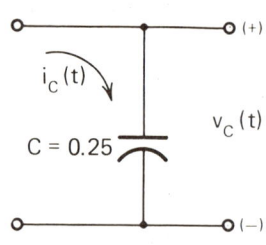
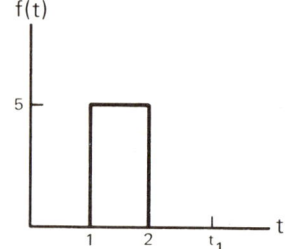
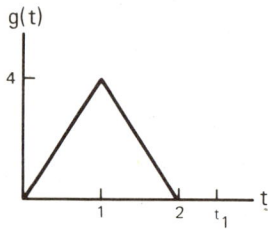
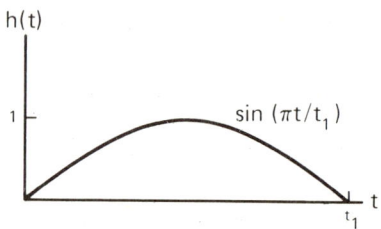

Figure 3-61. Problem 3-14.

3-15. The functions $f(t)$, $g(t)$, and $h(t)$ refer to Problem 3-14. Find the indicated quantities for the inductor circuit shown:

(a) Find $v_L(t_1)$ if $i_L(t)=g(t)$.

(b) Find $v_L(t_1)$ if $i_L(t)=h(t)$.

(c) Find $i_L(t_1)$ if $v_L(t)=f(t)$. Assume $i_L(0)=0$.

(d) Find $i_L(t_1)$ if $v_L(t)=g(t)$. Assume $i_L(0)=0$.

(e) Find $i_L(t_1)$ if $v_L(t)=h(t)$. Assume $i_L(0)=0$.

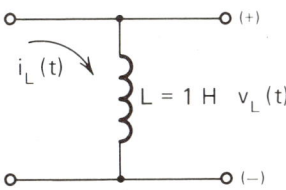

Figure 3-62. Problem 3-15.

3-16. Show that the current through an inductor cannot change instantaneously provided the voltage across the inductor is finite.

3-17. Show that the change in the energy stored in an inductor during the time interval t_1-t_2 is given by

$$W_L = \frac{L}{2}(i_2^2 - i_1^2) \tag{3-201}$$

where i_2 and i_1 are the currents at times t_2 and t_1 respectively.

3-18. Prove the combination rules for series and parallel inductors.

3-19. Show that one henry is equivalent to one volt-second per ampere.

3-20. Show that the change in stored energy in a capacitor during the time interval t_1-t_2 is given by

$$W_C = \frac{C}{2}(v_2^2 - v_1^2) \tag{3-202}$$

where v_2 and v_1 are the voltages at times t_2 and t_1 respectively.

3-21. One of the traditional uses for a capacitor is to block the flow of direct current. Show that if the voltage across a capacitor is constant, then the current through the device must be zero.

3-22. Show that the voltage across a capacitor cannot change instantaneously provided the current supplied to the capacitor is finite.

3-23. Show that one farad is equivalent to one ampere-second per volt.

3-24. Verify the dimensions for the following quantities. Use Appendix B.

Quantity	Dimensions
RC	Time
L/R	Time
$\sqrt{LC}$	Time
$R^2 C$	Inductance
$\sqrt{L/C}$	Resistance
L/R^2	Capacitance

3-25. The device shown below is an idealization of a so-called track-store unit. When the switch is closed, the capacitor voltage equals the source voltage (verify this with KVL). When the switch is open, the capacitor holds or maintains the voltage it had at the instant the switch opened. If the switch is closed during the first and every odd second, sketch the response of the track-store unit to the given source voltage $v(t)$.

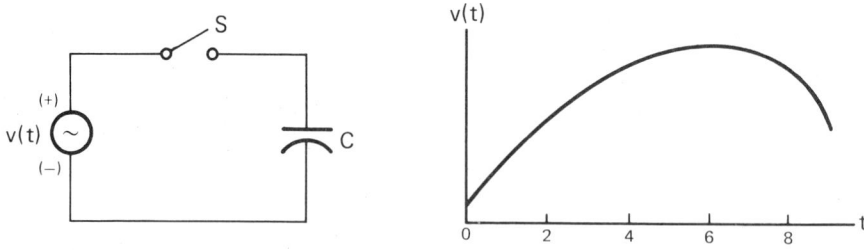

Figure 3-63. Problem 3-25.

Introductory Network Analysis

3-26. Use loop analysis to solve for the loop currents and the output voltage v_o in the following resistor networks:

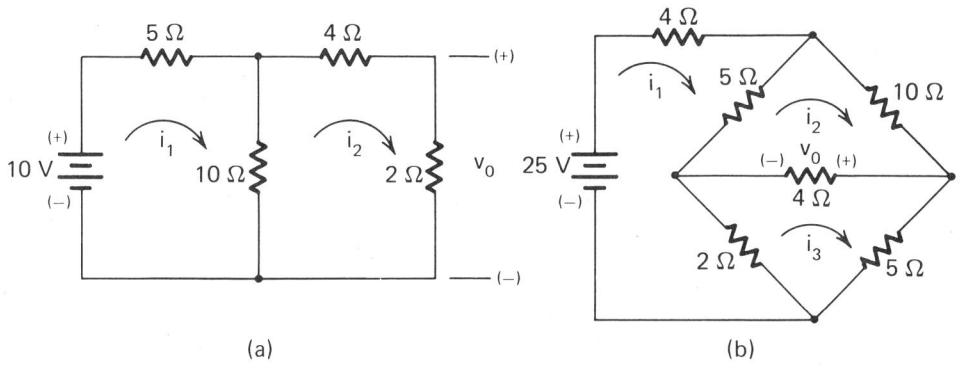

Figure 3-64. Problem 3-26.

3-27. Write loop equations for the following networks. Do not attempt to solve the resulting differential equations. Differentiate to remove any integrals.

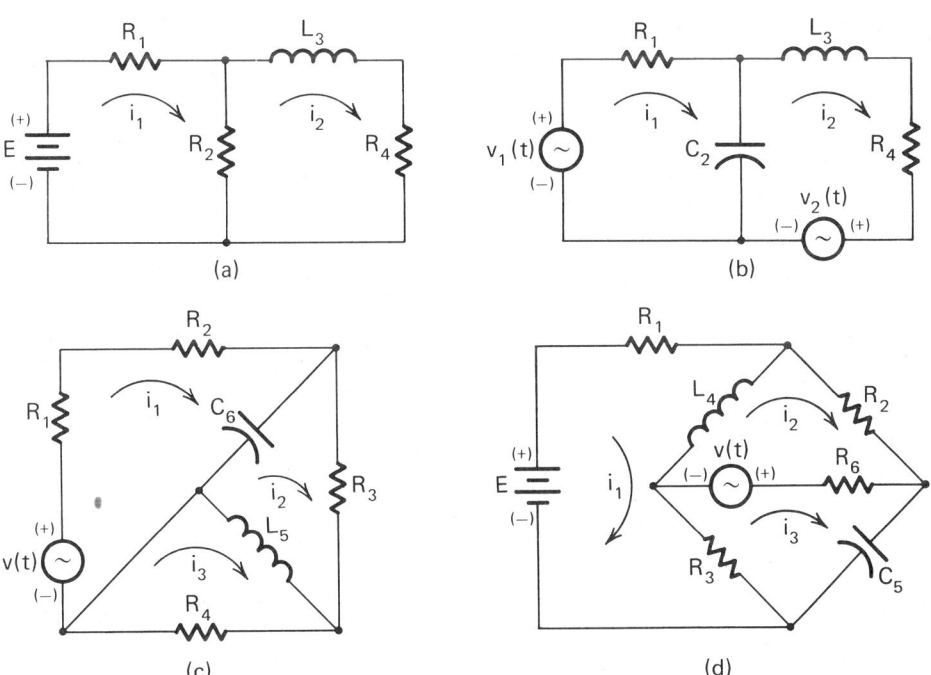

Figure 3-65. Problem 3-27.

3-28. Use nodal analysis to solve for the lettered node voltages in the following resistor networks. Use ground as the reference node.

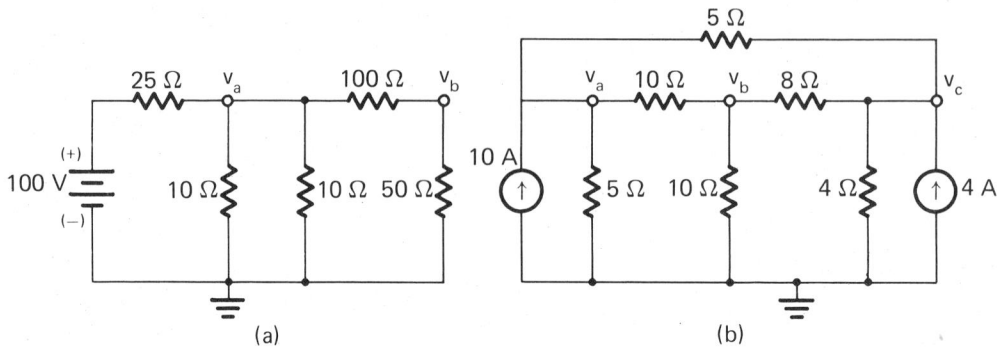

Figure 3-66. Problem 3-28.

3-29. Write node voltage equations for the following networks. Differentiate to remove integrals but do not attempt to solve the equations.

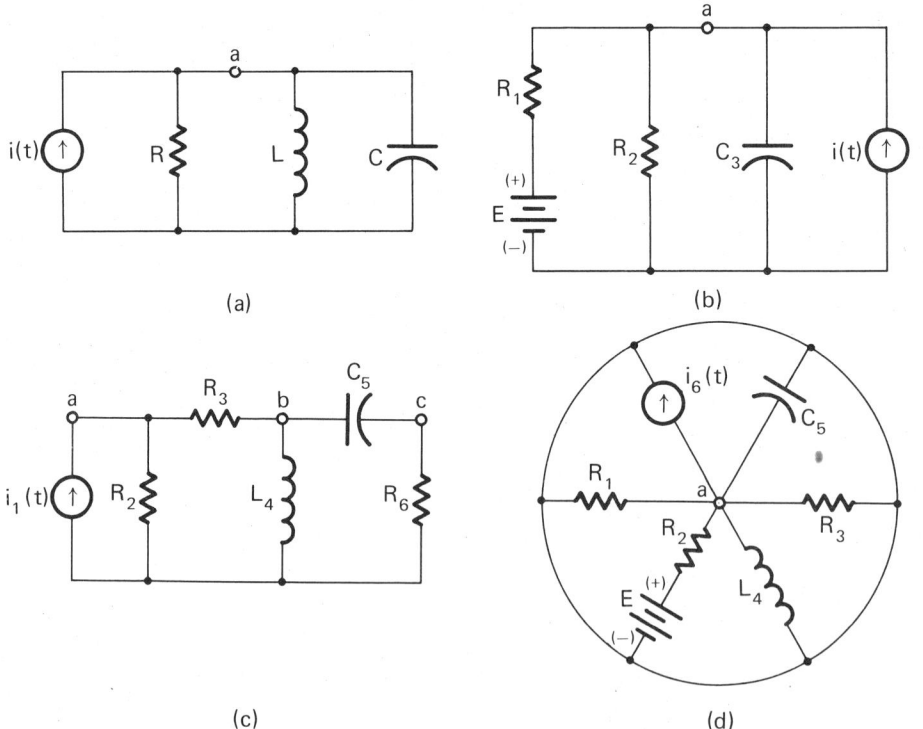

Figure 3-67. Problem 3-29.

3-30. Write loop equations for the following network after converting the current sources to voltage sources.

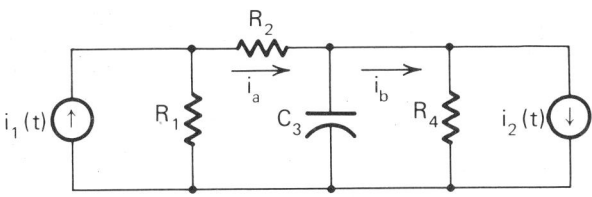

Figure 3-68. Problem 3-30.

3-31. Write a node voltage equation for the following network after converting the voltage sources to equivalent current sources.

Figure 3-69. Problem 3-31.

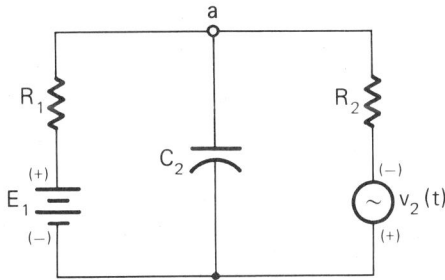

3-32. Find the output voltage in the following network by applying the voltage-divider principle.

Figure 3-70. Problem 3-32.

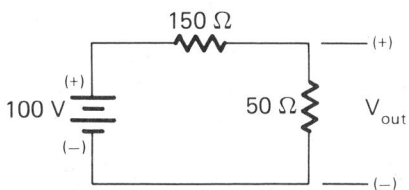

3-33. Show that resistor R_1 is extraneous by finding the output voltage v_o with and without R_1 in the network. Use $E=100$ V, $R_2=5$ Ω, $R_3=10$ Ω, $R_4=8$ Ω and $R_5=25$ Ω.

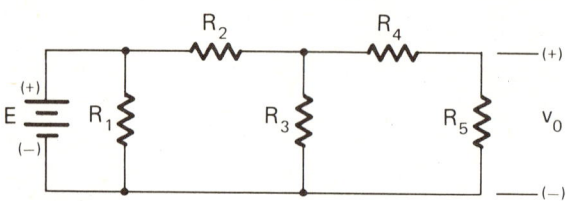

Figure 3-71. Problem 3-33.

3-34. Find Thevenin equivalent circuits to the left of terminals a-a' for the following networks:

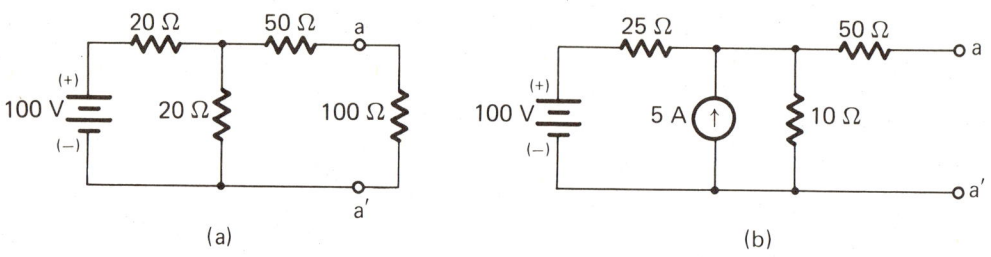

Figure 3-72. Problem 3-34.

3-35. Find Norton equivalent circuits for the networks of Problem 3-34.

3-36. Find a Thevenin equivalent circuit for the network of Problem 3-26(a) assuming the 2 Ω resistor is the load resistor (the one which is removed to find the open-circuit voltage).

3-37. Repeat Problem 3-36 if the 2 Ω resistor is a part of the network to be reduced to a Thevenin equivalent, i.e., the network is already open circuited. Compare the results with those of Problem 3-36.

3-38. Find a Thevenin equivalent circuit for the bridge network shown. (This is a fairly difficult problem because the Thevenin resistance cannot be found by combining resistors. Instead, we must find the open-circuit voltage and short-circuit current and take their ratio to determine R_T.)

Figure 3-73. Problem 3-38.

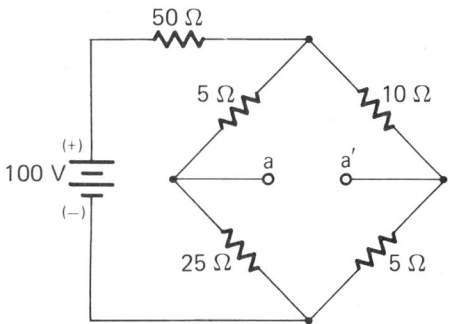

3-39. The switch has been in position A for a long time and is moved to position B at t=0. Find $v_o(0^+)$, $i_o(0^+)$, and $i_1(0^+)$ if $R_3 = 50\ \Omega$.

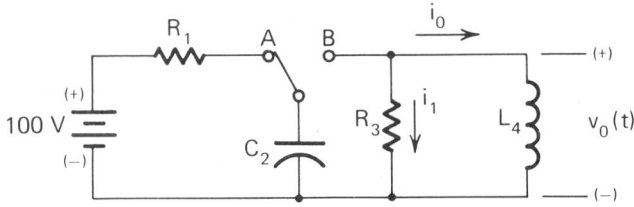

Figure 3-74. Problem 3-39.

3-40. The network shown has reached steady state before switch S is opened at t=0. Find the values at $t=0^+$ for the currents $i_C(t)$ and $i_L(t)$ and their derivatives.

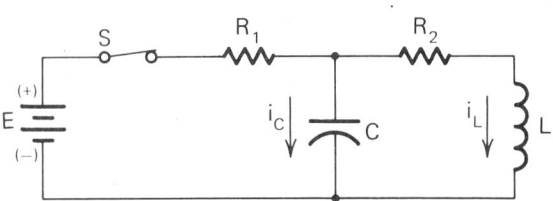

Figure 3-75. Problem 3-40.

3-41. The network shown has reached steady state before the switch is moved from A to B. Find the initial conditions for $v_C(t)$, $i_L(t)$ and their derivatives. Find also the new steady-state values for $v_C(t)$ and $i_L(t)$. Use $E_1=100$ V, $E_2=50$ V, $L=2$ H, $C=0.5$ F, and $R=5$ Ω.

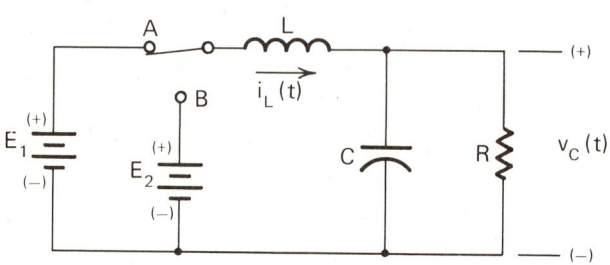

Figure 3-76. Problem 3-41.

3-42. The network has reached steady state before the switch is moved from A to B. Find initial values for $v_1(t)$, $v_2(t)$ and the derivative of the capacitor voltage. Also find new steady-state values for $v_1(t)$ and $v_2(t)$.

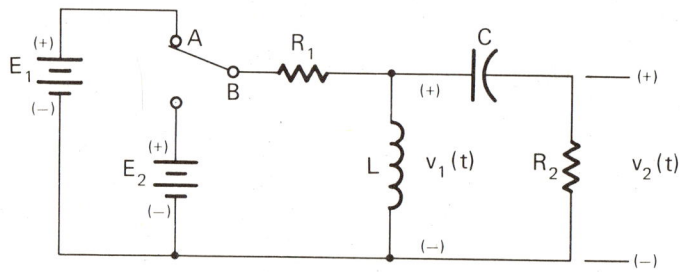

Figure 3-77. Problem 3-42.

3-43. The switch has been in position A for a long time and is moved to position B at t=0. Calculate $i_L(0^-)$, $i_L(0^+)$, and $v_0(0^+)$.

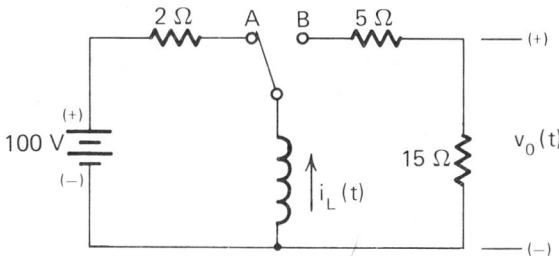

Figure 3-78. Problem 3-43.

3-44. Write, but do not solve, node-voltage equations which describe the operation of the following operational-amplifier networks. Assume ideal operational amplifiers.

Figure 3-79. Problem 3-44.

(a)

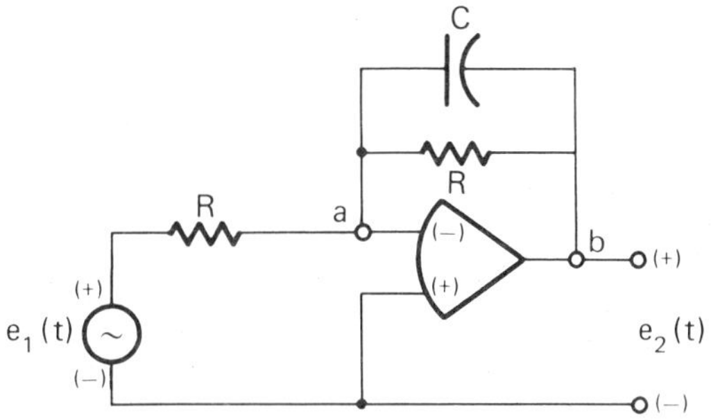

(b)

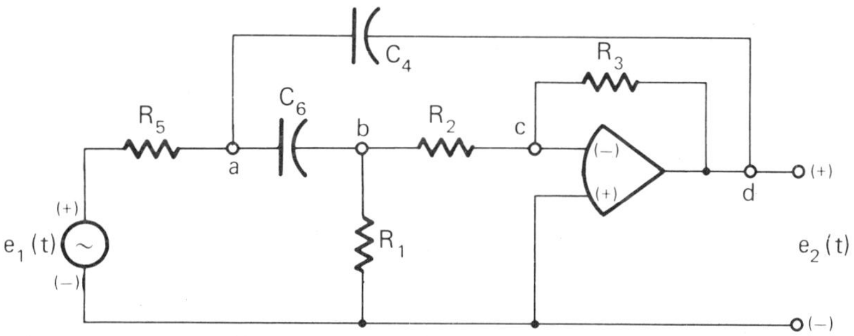

(c)

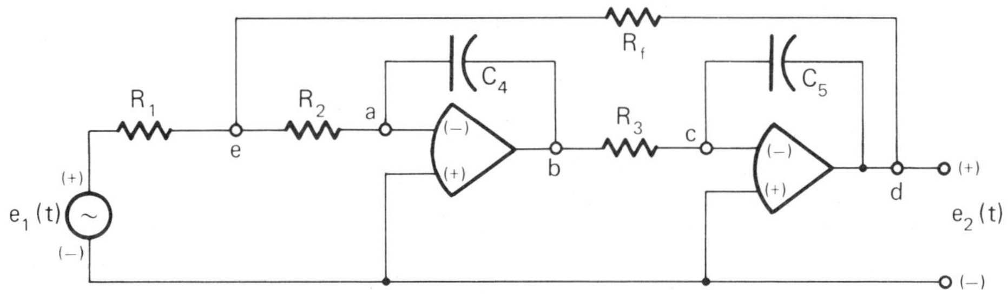

3-45. It is a well-known phenomena in heavy-current electrical engineering that dangerous "arcs" can result if an inductive circuit is suddenly interrupted. Assume the switch is suddenly opened. If the resistance of the air gap is 1000 MΩ, what initial voltage appears across the air gap?

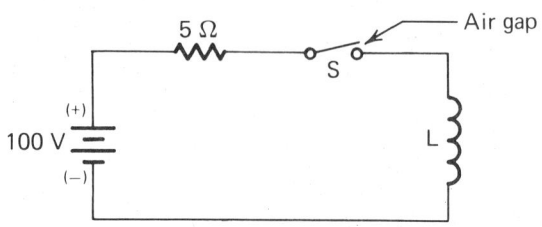

Figure 3-80. Problem 3-45.

3-46. Show that the operational-amplifier network of Fig. 3-81 is a summing integrator.

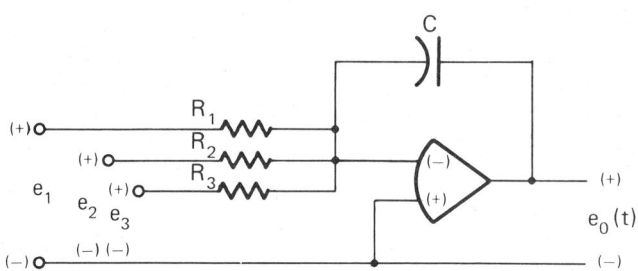

Figure 3-81. Problem 3-46.

3-47. Derive expressions for the operations performed by the following operational-amplifier networks, that is, find v_o in terms of v_a and v_b.

REFERENCES FOR CHAPTER 3

(1) Constant, F. W.: <u>Theoretical Physics: Electromagnetics</u>, Addison-Wesley Publishing Co., Inc., Reading, Massachusetts, 1954.

Introductory Network Analysis 141

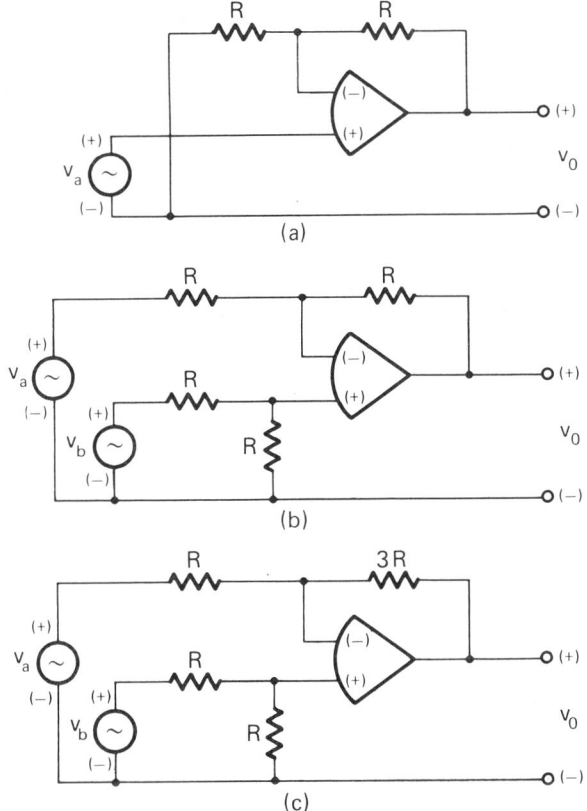

Figure 3-82. Problem 3-47.

(2) Nussbaum, A.: <u>Electronic and Magnetic Behavior of Materials</u>, Prentice-Hall Inc., Englewood Cliffs, New Jersey, 1967.

(3) Sears, F. W. and M. W. Semanski: <u>University Physics</u>, Addison-Wesley Publishing Co., Inc., Reading, Massachusetts, 1964.

(4) Kraybill, E. K.: <u>Electric Circuits for Engineers</u>, The MacMillian Company, Inc., New York, 1951.

(5) Gehmlich, D. K. and S. B. Hammond: <u>Electromechanical Systems</u>, McGraw-Hill Book Co., Inc., New York, 1967.

(6) Van Valkenburg, M.E.: <u>Network Analysis</u>, Prentice-Hall Inc., Englewood Cliffs, New Jersey, 1964.

(7) Cassell, W. L.: <u>Linear Electric Circuits</u>, John Wiley and Sons, Inc., New York, 1965.

(8) Calahan, D. A. and Macnee, A. B., and E. L. McMahon, <u>Introduction to Modern Circuit Analysis</u>, Holt, Rinehart and Winston, Inc., New York, 1974.

(9) Balabanian, N.: <u>Fundamentals of Circuit Theory</u>, Allyn and Bacon, Inc., Boston, Massachusetts, 1961.

(10) Gullemin, E. A.: <u>Introductory Circuit Theory</u>, John Wiley and Sons, Inc., New York, 1953.

(11) Seshu, S. and N. Balabanian: <u>Linear Network Analysis</u>, John Wiley and Sons, Inc., New York, 1959.

(12) Hammond, S. B.: <u>Electrical Engineering</u>, McGraw-Hill Book Co., Inc., New York, 1961.

(13) Fitzgerald, A. E. and D. E. Higginbotham, <u>Basic Electrical Engineering</u>, McGraw-Hill Book Co., Inc., New York, 1957.

(14) Kuo, F. F.: <u>Network Analysis and Synthesis</u>, John Wiley and Sons, Inc., New York, 1962.

(15) Gardner, M. F. and J. L. Barnes: <u>Transients in Linear Systems</u>, John Wiley and Sons, Inc., New York, 1942.

(16) Huelsman, L. P.: <u>Theory and Design of Active RC Circuits</u>, McGraw-Hill Book Co., Inc., New York, 1968.

(17) Budak, A.: <u>Passive and Active Network Analysis and Synthesis</u>, Houghton and Mifflin Co., Inc., Boston, Massachusetts, 1974.

(18) Tobey, G. E., Graeme, J. G., and L. P. Huelsman, eds.: <u>Operational Amplifiers--Design and Applications</u>, McGraw-Hill Book Co., Inc., New York, 1971.

(19) Blum, J. J.: <u>Introduction to Analog Computation</u>, Harcourt, Brace and World, Inc., New York, 1968.

(20) Rogers, A. E. and T. W. Connolly: <u>Analog Computation in Engineering Design</u>, McGraw-Hill Book Co., Inc., New York, 1960.

CHAPTER **4**

Transient Response of Dynamic Systems

Chapter 4 presents techniques for determining the transient response of a dynamic system starting with its mathematical model, in other words, how to solve the differential equations of motion. We assume that one intends to use pencil-and-paper methods, but many of the solution techniques are algorithmic and hence can be implemented on a digital computer. In any event, a firm background in the theory which underlies the transient response of linear systems is essential before we can proceed further.

Transient response studies fall into two categories: the response of a system to stored energy, called the <u>natural</u> or <u>force-free response</u>, and the response to an external stimulus, called the <u>forced response</u>.[1,2] The energy stored in a system manifests itself as initial conditions on the differential equations. In all of our work we assume, with no loss of generality, that the solution interval starts at t=0, and that the initial conditions are the conditions which existed an infinitesimal time after t=0. If a system is linear, we can apply the principle of superposition and add the solutions due to initial conditions and a forcing function to obtain the response due to a combination of inputs. For this reason, considering forcing functions and initial conditions separately does not limit the generality of our derivations.[3,4]

4.1 THE NATURAL RESPONSE

The general linear, constant-coefficient differential equation takes the form

$$\frac{d^n y(t)}{dt^n} + a_{n-1}\frac{d^{n-1} y(t)}{dt^{n-1}} + \ldots a_1 \frac{dy(t)}{dt} + a_0 y(t) = f(t)$$
(4-1)

where the a_i are real constants. Accompanying Eq. (4-1) is a set of n initial conditions, $y(0)$, $\dot{y}(0)$, ... $y(0)^{(n-1)}$. We need initial conditions on the dependent variable y(t) and all of its derivatives through the (n-1)-th derivative. The initial value of the nth derivative is not needed as it can be found by substituting the other initial conditions into Eq. (4-1).

If the forcing function f(t) in Eq. (4-1) is zero, the system has only initial conditions as inputs. In this situation the system is said to "relax" from its initial conditions. The term relax is appropriate because the initial conditions represent stored energy which is dissipated as time passes. The solution y(t) in response to the initial

Transient Response of Dynamic Systems

conditions with no forcing function is called the natural response or the force-free response. The natural response is therefore the solution to the homogeneous equation,(5) which is Eq. (4-1) with $f(t) \equiv 0$. This section treats the problem of finding the natural response of an unforced differential equation.

The Characteristic Equation

If a function $y(t)$ is a solution to the homogeneous differential equation

$$\frac{d^n y(t)}{dt^n} + a_{n-1} \frac{d^{n-1} y(t)}{dt^{n-1}} + \ldots a_1 \frac{dy(t)}{dt} + a_0 y(t) = 0 \quad (4-2)$$

then it must be a function which, when repeatedly differentiated and added to itself, gives zero. The most general function of this type is the exponential; hence, we hopefully assume a solution of the form

$$y(t) = Ae^{pt} \quad (4-3)$$

where A and p are constants which may be complex. Substituting Eq. (4-3) into Eq. (4-2) gives

$$\frac{d^n}{dt^n}(Ae^{pt}) + a_{n-1} \frac{d^{n-1}}{dt^{n-1}}(Ae^{pt}) + \ldots a_1 \frac{d}{dt}(Ae^{pt}) + a_0 Ae^{pt} = 0 \quad (4-4)$$

Performing the indicated differentiations and factoring $A\exp(pt)$ from the resulting expression yields

$$Ae^{pt}\{p^n + a_{n-1}p^{n-1} + \ldots a_1 p + a_0\} = 0 \quad (4-5)$$

If $y(t) = A\exp(pt)$ is a valid solution, then Eq. (4-5) must be true for all t. This is possible under three conditions: (1) $A=0$; (2) $p \to -\infty$; or (3) p is a root of the equation

$$p^n + a_{n-1}p^{n-1} + \ldots a_1 p + a_0 = 0 \quad (4-6)$$

We disregard the first two conditions as they imply that the solution is y(t)≡0 (which is a valid solution to the homogeneous equation only if all of the initial conditions are zero). Equation (4-6) is called the <u>characteristic equation</u> of the homogeneous differential equation.[6] The characteristic equation is a polynomial in the parameter p; its roots determine the natural response of the differential equation. A different solution corresponds to each of the n roots of the characteristic equation. Superposition lets us sum these to obtain the most general solution. With this brief introduction we turn now to several special cases of the natural response.

Natural Response of the First-Order System [7, 8]

The homogeneous first-order equation has the form

$$\dot{y}(t) + ay(t) = 0 \qquad (4\text{-}7)$$

with the initial condition y(0). Consulting Eq. (4-6), we see that the characteristic equation is p+a=0. The only root of this characteristic equation is p=-a, so the solution is

$$y(t) = Ae^{-at} \qquad (4\text{-}8)$$

To determine the constant A, we evaluate Eq. (4-8) at t=0 to obtain y(0)=A. The natural response is then

$$y(t) = y(0)e^{-at} \qquad (4\text{-}9)$$

Figure 4-1 shows the natural reponse for a>0 and a<0. If a<0, the solution grows without bound as t→∞; in such cases the system is said to be <u>unstable</u>.

<u>Example 4-1</u>. The rotor of an electrical motor has moment of inertia J and is turning with angular velocity Ω rad/sec when the power is abruptly removed. It can be shown that the subsequent motion is described by the differential equation

$$J\dot{\omega}(t) + b\omega(t) = 0 \qquad (4\text{-}10)$$

Transient Response of Dynamic Systems

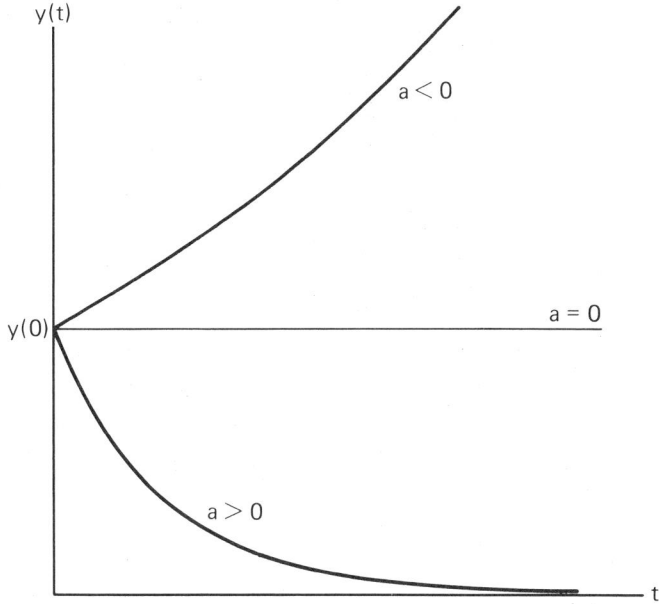

Figure 4-1. First-order natural response, y(t) = y(0)exp (−at).

where $\omega(t)$ is the instantaneous velocity of the rotor and b is a coefficient of damping to account for air friction. The initial condition is $\omega(0)=\Omega$. Derive an expression for $\omega(t)$.

Solution. The characteristic equation is $p+b/J=0$ with the root $p=-b/J$. Using Eq. (4-9) the natural response is

$$\omega(t) = \Omega e^{-bt/J} \qquad (4-11)$$

and the rotor slows exponentially. One might argue that the rotor never stops turning because exponentials never decay completely to zero. What actually happens is that when the velocity becomes sufficiently small, the linear model no longer holds, nonlinear effects take over, and the rotor halts.

Example 4-2. The capacitor of Fig. 4-2 has an initial charge of V volts as modelled by the initial-condition battery. At t=0 the switch is closed and the capacitor discharges through resistor R. Find an expression for the current i(t).

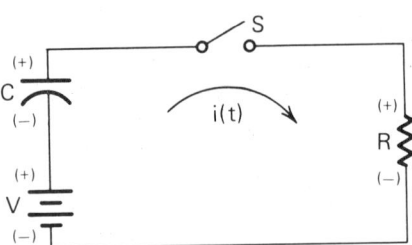

Figure 4-2. This RC circuit for Example 4-2 is excited by a charged capacitor. The current i(t) is an example of first-order natural response.

<u>Solution</u>. Writing KVL around the loop results in

$$0 = i(t)R + \frac{1}{C}\int_0^t i(t)dt - V \qquad (4-12)$$

Differentiating to eliminate the integral yields

$$R\frac{di(t)}{dt} + \frac{i(t)}{C} = 0 \qquad (4-13)$$

with the initial condition i(0)=V/R (Why?). The characteristic equation is p+1/RC=0 with the root p=-1/RC. The natural response is then

$$i(t) = \frac{V}{R}e^{-t/RC} \qquad (4-14)$$

The current starts at i(0)=V/R and decays exponentially as the capacitor discharges. The energy stored in the capacitor is dissipated in the resistor as heat.

<u>Example 4-3</u>. Figure 4-3 is a feedback amplifier excited by a charged capacitor. The initial voltage on the capacitor is V volts as indicated by the initial-condition battery. Find an expression for $v_a(t)$ after switch S closed at t=0.

<u>Solution</u>. With the switch closed, the node-voltage equation at node a is

$$\frac{v_a - e_2}{R_1} + C\left\{\frac{d(v_a - V)}{dt}\right\} + \frac{v_a - v_b}{R} = 0 \qquad (4-15)$$

Transient Response of Dynamic Systems

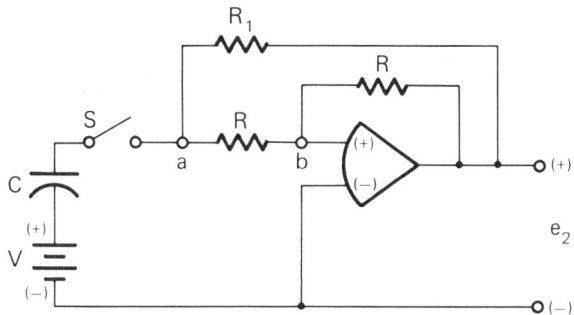

Figure 4-3. Example 4-3 shows that this feedback amplifier, when excited by a charged capacitor, illustrates the idea of first-order natural response.

For an ideal operational amplifier $v_b=0$. Further, because the feedback and input resistances are equal, the overall amplifier gain is -1 and $e_2=-v_a$. Using this information, the nodal equation becomes

$$C\dot{v}_a + \left\{\frac{2}{R_1} + \frac{1}{R}\right\}v_a = 0 \qquad (4\text{-}16)$$

with $v_a(0)=V$. The characteristic equation is $p+b=0$ where

$$b = \frac{1}{C}\left\{\frac{2}{R_1} + \frac{1}{R}\right\} \qquad (4\text{-}17)$$

Using the root $p=-b$ and the initial condition, the natural response is $v_a(t)=V\exp(-bt)$.

Natural Response of the Second-Order System

Although first-order systems are important in dynamic analysis, their natural response can be determined by inspection. We seldom need computer assistance. Second-order systems are more complex--and more important.[9,10] Newton's second law, $\Sigma f=m\ddot{y}$, gives rise naturally to second-order equations, as do Kirchhoff's voltage and current laws when applied to networks with both inductance and capacitance. For these reasons we examine second-order differential equations in some detail.

The second-order homogeneous equation is

$$\ddot{y} + a_1\dot{y} + a_0 y = 0 \qquad (4\text{-}18)$$

with initial conditions $y(0)$ and $\dot{y}(0)$. The characteristic equation is $p^2 + a_1 p + a_0 = 0$. The quadratic formula gives the roots

$$p_1, p_2 = -\frac{a_1}{2} \pm \sqrt{\left\{\frac{a_1}{2}\right\}^2 - a_0} \qquad (4\text{-}19)$$

where the "+" sign goes with the root p_1. The three principle cases are (1) real, unequal roots, (2) identical real roots, and (3) complex-conjugate roots (with the subcase of imaginary roots). These cases are treated in the following paragraphs.

<u>Real unequal roots</u>. Designating the roots by the real constants p_1 and p_2, we have two solutions to the homogeneous equation; these are

$$y_1(t) = A_1 e^{p_1 t} \qquad \text{and} \qquad y_2(t) = A_2 e^{p_2 t} \qquad (4\text{-}20)$$

Because the differential equation is linear and we seek the most general form of the response, we apply superposition and use as the solution $y(t) = y_1(t) + y_2(t)$ or

$$y(t) = A_1 e^{p_1 t} + A_2 e^{p_2 t} \qquad (4\text{-}21)$$

To determine the constants A_1 and A_2, evaluate Eq. (4-21) at $t=0$ to get $y(0) = A_1 + A_2$. As this expression alone is insufficient to find both A_1 and A_2, differentiate Eq. (4-21) to obtain

$$\dot{y}(t) = p_1 A_1 e^{p_1 t} + p_2 A_2 e^{p_2 t} \qquad (4\text{-}22)$$

and evaluate at $t=0$ to find $\dot{y}(0) = p_1 A_1 + p_2 A_2$. The expressions for $y(0)$ and $\dot{y}(0)$ are solved simultaneously for A_1 and A_2 in

Transient Response of Dynamic Systems

terms of the known quantities $y(0)$, $\dot{y}(0)$, p_1, and p_2 with the results

$$A_1 = \frac{\dot{y}(0)-p_2 y(0)}{p_1 - p_2} \quad \text{and} \quad A_2 = \frac{p_1 y(0)-\dot{y}(0)}{p_1 - p_2} \quad (4\text{-}23)$$

Substituting Eqns. (4-23) into (4-21) gives the natural response of the second-order system with real roots. Graphically, the natural response for this case is the sum of two real exponentials. If p_1 and p_2 are both negative, the response approaches zero as $t\to\infty$ and the system is <u>stable</u>. If at least one of the roots is positive, the response tends to infinity as $t\to\infty$ and the system is <u>unstable</u>.

<u>Example 4-4</u>. Figure 4-4 is a model for an automobile tire. The spring and the damper represent the air in the tire and the elasticity of its walls. The center of the tire is depressed 0.05 meters and released. Solve for the displacement $y(t)$ of the subsequent motion if k=4000, b=500, and m=10, all in consistent units, and the differential equation of motion (see Appendix H) is

$$m\ddot{y} + b\dot{y} + ky = 0 \quad (4\text{-}24)$$

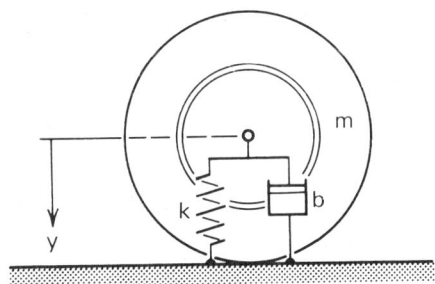

Figure 4-4. Mechanical model of an automobile tire for Example 4-4. Appendix H shows how to write the differential equations of motion for mechanical systems.

<u>Solution</u>. Insert numerical values to obtain

$$\ddot{y} + 50\dot{y} + 400y = 0 \quad (4\text{-}25)$$

with initial conditions $y(0)=0.05$ and $\dot{y}(0)=0$. The characteristic equation, $p^2+50p+400=0$, has roots $p_1=-40$ and $p_2=-10$. Consequently, the natural response takes the form

$$y(t) = A_1 e^{-40t} + A_2 e^{-10t} \qquad (4-26)$$

At $t=0$, $0.05=A_1+A_2$. Differentiating $y(t)$ yields

$$\dot{y}(t) = -40A_1 e^{-40t} - 10A_2 e^{-10t} \qquad (4-27)$$

Evaluating at $t=0$ gives $0=-40A_1-10A_2$. After solving for $A_1=-0.0167$ and $A_2=0.0668$, we have the natural response

$$y(t) = -0.0167 e^{-40t} + 0.0668 e^{-10t} \qquad (4-28)$$

Real equal roots. If $p_1=p_2$, the method just outlined fails because the denominators of Eq. (4-23) become zero. To analyze this case we assume that p_1 and p_2 differ by a small quantity ε and explore the resulting solution. If $p_2-p_1=\varepsilon$, then Eq. (4-21) takes the form

$$y(t) = e^{p_1 t}\left\{A_1 + A_2 e^{\varepsilon t}\right\} \qquad (4-29)$$

where, from Eq. (4-23),

$$A_1 = \frac{\dot{y}(0)-(\varepsilon+p_1)y(0)}{-\varepsilon} \quad \text{and} \quad A_2 = \frac{p_1 y(0)-\dot{y}(0)}{-\varepsilon} \qquad (4-30)$$

The power series expansion for $\exp(\varepsilon t)$ is

$$e^{\varepsilon t} = 1 + \varepsilon t + \frac{(\varepsilon t)^2}{2!} + \frac{(\varepsilon t)^3}{3!} + \ldots \qquad (4-31)$$

Using this expansion in Eq. (4-29) gives

$$y(t) = e^{p_1 t}\left\{A_1 + A_2\left(1 + \varepsilon t + \frac{\varepsilon^2 t^2}{2!} + \ldots\right)\right\} \qquad (4-32)$$

After substituting Eqns. (4-30) for A_1 and A_2 and letting $\varepsilon \to 0$, Eq. (4-32) reduces to

$$y(t) = e^{p_1 t}\left\{y(0) + \{\dot{y}(0) - p_1 y(0)\}t\right\} \qquad (4-33)$$

Transient Response of Dynamic Systems

Repeated real roots are the borderline case between systems which have oscillations (complex-roots) and those which are overdamped (real roots). Practical systems seldom have roots which are precisely equal.

Example 4-5. Figure 4-5 shows an RLC "tank" circuit. Determine the value of R as a function of L and C in order for the roots of the characteristic equation to be real and equal.

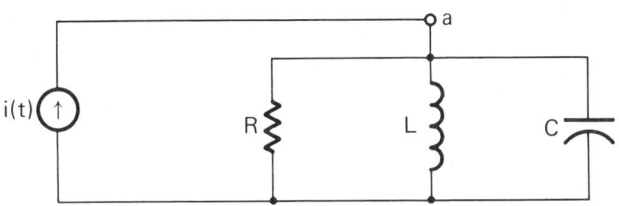

Figure 4-5. An RLC "tank" circuit for Example 4-5. The proper choice of R results in real, equal roots for the characteristic equation.

Solution. Writing a node-voltage equation at node a gives

$$i(t) = \frac{v_a(t)}{R} + C\dot{v}_a(t) + \frac{1}{L}\int_0^t v_a(t)dt + i_L(0) \qquad (4-34)$$

and, after differentiating to remove the integral,

$$\frac{di(t)}{dt} = \frac{\dot{v}_a(t)}{R} + C\ddot{v}_a(t) + \frac{v_a(t)}{L} \qquad (4-35)$$

The homogeneous differential equation is obtained by setting the forcing function, in this case $di(t)/dt$, to zero. The characteristic equation is then $p^2 + p/RC + 1/LC = 0$ with roots

$$p_1, p_2 = -\frac{1}{2RC} \pm \frac{1}{2}\sqrt{\left\{\frac{1}{RC}\right\}^2 - \frac{4}{LC}} \qquad (4-36)$$

If the roots are to be equal, the quantity under the radical must be zero, a condition which leads to $R = \sqrt{L/4C}$.

Complex-conjugate roots. Analysis of this case is simplified by introducing for the second-order differential equation the new notation

$$\ddot{y} + 2\zeta\omega_n\dot{y} + \omega_n^2 y = 0 \qquad (4\text{-}37)$$

We will see the significance of the parameters ζ and ω_n below. The characteristic equation for Eq. (4-37) is

$$p^2 + 2\zeta\omega_n p + \omega_n^2 = 0 \qquad (4\text{-}38)$$

with roots

$$p_1, p_2 = -\zeta\omega_n \pm \omega_n\sqrt{\zeta^2 - 1} \qquad (4\text{-}39)$$

For the roots to be complex, ζ must be less than 1. Under this condition the roots become

$$p_1, p_2 = -\zeta\omega_n \pm j\omega_n\sqrt{1 - \zeta^2} \qquad (4\text{-}40)$$

where $j=\sqrt{-1}$. Define the quantities σ and ω_d such that $p_1, p_2 = -\sigma \pm j\omega_d$ where $\sigma = \zeta\omega_n$ and

$$\omega_d = \omega_n\sqrt{1 - \zeta^2} \qquad (4\text{-}41)$$

Using σ, ω_d and Eq. (4-21) the natural response takes the form

$$y(t) = e^{-\sigma t}\left\{A_1 e^{j\omega_d t} + A_2 e^{-j\omega_d t}\right\} \qquad (4\text{-}42)$$

Though correct, Eq. (4-42) is not too useful as it contains complex exponentials. We can remove these by using Euler's identities, $\exp(\pm jx) = \cos(x) \pm j\sin(x)$. The result is

$$y(t) = e^{-\sigma t}\left\{A_1\{\cos(\omega_d t) + j\sin(\omega_d t)\} + A_2\{\cos(\omega_d t) - j\sin(\omega_d t)\}\right\}$$

$$(4\text{-}43)$$

Transient Response of Dynamic Systems

or

$$y(t) = e^{-\sigma t}\left\{(A_1+A_2)\cos(\omega_d t) + j(A_1-A_2)\sin(\omega_d t)\right\} \tag{4-44}$$

Next determine the constants A_1 and A_2 in terms of the initial conditions $y(0)$ and $\dot{y}(0)$ and the system parameters ζ and ω_n (or, equivalently, σ and ω_d). Evaluate Eq. (4-44) for t=0 to obtain $y(0)=A_1+A_2$. Differentiate Eq. (4-44) to get

$$\dot{y}(t) = -\sigma y(t) + \omega_d e^{-\sigma t}\left\{-(A_1+A_2)\sin(\omega_d t)+j(A_1-A_2)\cos(\omega_d t)\right\} \tag{4-45}$$

Evaluate Eq. (4-45) at t=0 to arrive at

$$\dot{y}(0) = -\sigma y(0) + j\omega_d(A_1 - A_2) \tag{4-46}$$

Equation (4-46) can be solved for A_1-A_2 with the result

$$A_1 - A_2 = \frac{\dot{y}(0) + \sigma y(0)}{j\omega_d} \tag{4-47}$$

Substituting $A_1+A_2=y(0)$ and Eq. (4-47) into Eq. (4-44) gives $y(t)$ in terms of known quantities and real functions:

$$y(t) = e^{-\sigma t}\left\{y(0)\cos(\omega_d t) + \frac{\dot{y}(0) + \sigma y(0)}{\omega_d}\sin(\omega_d t)\right\} \tag{4-48}$$

Equation (4-48) can be expressed as a cosine with a phase angle using the trigonometric substitutions of Fig. 4-6.

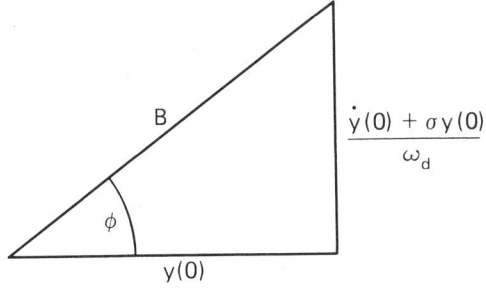

Figure 4-6. Triangle for the trigonometric substitution which transforms Eq. (4-48) into Eq. (4-52).

From the triangle

$$B = \sqrt{y(0)^2 + \left\{\frac{\dot{y}(0) + \sigma y(0)}{\omega_d}\right\}^2} \qquad (4\text{-}49)$$

Write Eq. (4-48) as

$$y(t) = Be^{-\sigma t}\left[\left\{\frac{y(0)}{B}\right\}\cos(\omega_d t) + \left\{\frac{\dot{y}(0) + \sigma y(0)}{B\omega_d}\right\}\sin(\omega_d t)\right] \qquad (4\text{-}50)$$

The angle $\emptyset$ in Fig. 4-6 has the property that $\cos(\emptyset)=y(0)/B$ and $\sin(\emptyset)=\{\dot{y}(0)+\sigma y(0)\}/B\omega_d$. Using these relationships, Eq. (4-50) becomes

$$y(t) = Be^{-\sigma t}\{\cos(\emptyset)\cos(\omega_d t) + \sin(\emptyset)\sin(\omega_d t)\} \qquad (4\text{-}51)$$

Applying the trigonometric formula for the cosine of the difference of two angles yields the final expression

$$y(t) = Be^{-\sigma t}\cos(\omega_d t - \emptyset) \qquad (4\text{-}52)$$

where, again from Fig. 4-6, $\emptyset = \cos^{-1}\{y(0)/B\}$. Equation (4-52) can be rewritten in terms of ζ and ω_n if we recall that $\sigma = \zeta\omega_n$ and $\omega_d = \omega_n\sqrt{1-\zeta^2}$. Figure 4-7 is a plot of Eq. (4-52) for several values of ζ. In the derivations we have assumed that ζ is positive and less than 1. This gives a stable system with complex-conjugate roots. Section 4.4 describes the response for other values of ζ. The figure assumes that $\dot{y}(0)=0$ and $y(0)=1$.

Example 4-6. Find the ζ and ω_n for the active filter of Fig. 4-8 if RC=50.

Solution. Write node-voltage equations at nodes a and b:

$$\text{Node a:} \quad \frac{v_a-v_1}{R} + \frac{v_a-v_2}{R} + \frac{v_a-v_b}{R} + C\dot{v}_a = 0 \qquad (4\text{-}53)$$

$$\text{Node b:} \quad \frac{v_b-v_a}{R} + \frac{v_b}{R} + C\dot{v}_b = 0 \qquad (4\text{-}54)$$

Transient Response of Dynamic Systems

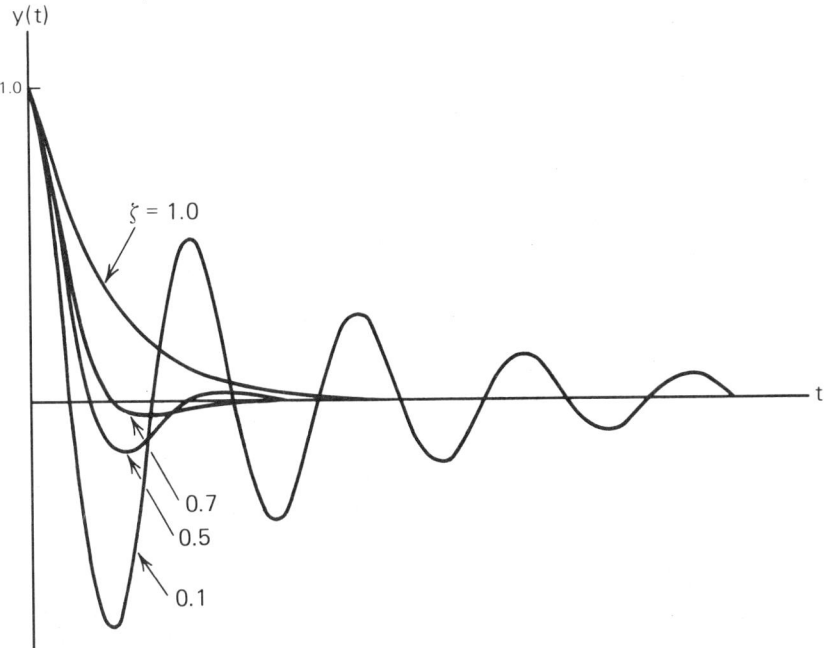

Figure 4-7. Natural response of a second-order system. The roots of the characteristic equation determining how oscillatory the response is.

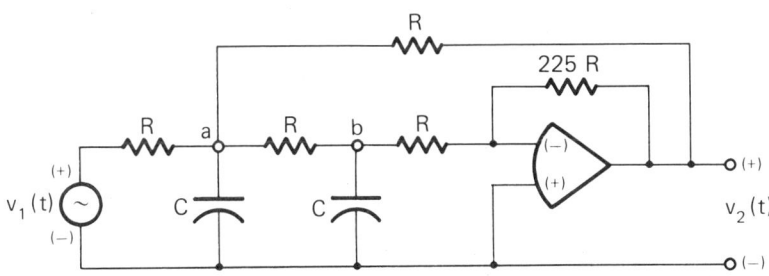

Figure 4-8. Example 4-6 shows that the characteristic equation of this OP AMP network is of the second order.

Recalling from Chapter 3 that the gain of an operational amplifier configured in this fashion is $-R_f/R_i$, where R_f and R_i are the feedback and input resistances respectively, we see that $v_2 = -225v_b$. This result allows us to rewrite the node-voltage equations as

$$50\dot{v}_a + 3v_a + 224v_b = v_1 \qquad (4\text{-}55)$$

and

$$50\dot{v}_b + 2v_b - v_a = 0 \qquad (4\text{-}56)$$

where we have also used RC=50. Before we can identify ζ and ω_n, we must reduce the simultaneous first-order equations to a single second-order equation; to do this, differentiate Eq. (4-56) to obtain

$$50\ddot{v}_b + 2\dot{v}_b - \dot{v}_a = 0 \qquad (4\text{-}57)$$

Solve for $\dot{v}_a$ and substitute the result into Eq. (4-55) with the result

$$2500\ddot{v}_b + 100\dot{v}_b + 224v_b + 3v_a = v_1 \qquad (4\text{-}58)$$

Use Eq. (4-56) to eliminate v_a and collect terms:

$$2500\ddot{v}_b + 250\dot{v}_b + 230v_b = v_1 \qquad (4\text{-}59)$$

The characteristic equation associated with Eq. (4-59) is

$$p^2 + \frac{p}{10} + \frac{23}{250} = 0 \qquad (4\text{-}60)$$

Comparing this expression with the standard form, $p^2 + 2\zeta\omega_n p + \omega_n^2 = 0$, leads to $\omega_n = 0.303$ and $\zeta = 0.165$. If we had reduced the nodal equations to a single second-order equation in v_a, the charactcristic equation would have been the same. In a later chapter we will see that Laplace transforms give us a better way to find characteristic equations than the somewhat roundabout method used here.

Example 4-7. The capacitor C in Fig. 4-9 is initially charged to 10 V with the polarity shown. The switch is closed at t=0 causing a transient current i(t). Solve for i(t) if R=2 kΩ, L=4.0 H, and C=1.0 μF.

Transient Response of Dynamic Systems

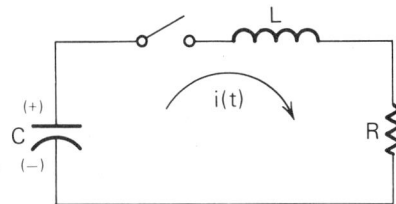

Figure 4-9. Network for Example 4-7. After the switch is closed, the current response is the damped sinusoid of Eq. (4-63).

Solution. Writing KVL around the loop and differentiating to remove the integral yields

$$L\frac{di}{dt} + Ri + \frac{i}{C} = 0 \qquad (4\text{-}61)$$

with the initial conditions (see Chapter 3) $i(0)=0$ and $di(0)/dt=E/L=2.5$ A/s. Dividing Eq. (4-61) by L and inserting numerical values yields the characteristic equation $p^2+500p+250,000=0$. Comparison with the standard form, Eq. (4-38), reveals that $\omega_n=500$ rad/s and $\zeta=0.5$. Using these values $\sigma=\zeta\omega_n=250$ and

$$\omega_d = \omega_n\sqrt{1-\zeta^2} = 500\sqrt{1-0.25} = 433.0 \text{ rad/s} \qquad (4\text{-}62)$$

The solution resembles Eq. (4-52). From $\cos(\emptyset)=i(0)/B$ we have $\emptyset=0$, and from Eq. (4-50), B=5.774 mA. The expression for $i(t)$ is then

$$i(t) = 5.774e^{-2.5t}\cos(433t-\pi/2) \text{ mA} \qquad (4\text{-}63)$$

A Special Case: Pure Oscillations

If $\zeta=0$ the homogeneous second-order differential equation takes the form $\ddot{y}+\omega_n^2 y=0$. The characteristic equation is then $p^2+\omega_n^2=0$ with roots $p_1=j\omega_n$ and $p_2=-j\omega_n$. The associated solution is

$$y(t) = A_1 e^{j\omega_n t} + A_2 e^{-j\omega_n t} \qquad (4\text{-}64)$$

Using Euler's identities,

$$y(t) = A_1\{\cos(\omega_n t)+j\sin(\omega_n t)\} + A_2\{\cos(\omega_n t)-j\sin(\omega_n t)\} \quad (4\text{-}65)$$

$$= (A_1+A_2)\cos(\omega_n t) + j(A_1-A_2)\sin(\omega_n t) \quad (4\text{-}66)$$

To satisfy the initial conditions, $A_1+A_2=y(0)$ and $A_1-A_2=-j\dot{y}(0)/\omega_n$. Using these values, the solution becomes

$$y(t) = y(0)\cos(\omega_n t) + \frac{\dot{y}(0)}{\omega_n}\sin(\omega_n t) \quad (4\text{-}67)$$

This result can be expressed as a cosine with a phase angle using the results for the general case. The final expression is

$$y(t) = B\cos(\omega_n t - \emptyset) \quad (4\text{-}68)$$

where $B^2 = y(0)^2 + \{\dot{y}(0)/\omega_n\}^2$ and $\emptyset = \tan^{-1}\{\dot{y}(0)/\omega_n y(0)\}$. Equation (4-68) is an undamped sinusoid whose magnitude and phase depend on the initial conditions. Pure oscillations with no forcing function represent perpetual motion, and hence cannot exist in nature. Practical oscillators always have power sources which make up the losses which would cause the oscillations to decay. It is often advantageous to model lightly damped systems ($\zeta \approx 0$) as undamped ones to simplify the analysis. Consider the following example.

Example 4-8. A wooden cylinder of radius R and height H is partially immersed in water as shown in Fig. 4-10. The cylinder is depressed D units from the equilibrium point and released. Derive an expression for the subsequent motion. Use m for the mass of the cylinder and d for the density of the water.

Solution. When the cylinder is depressed, the restoring force is proportional to the weight of the displaced water, which is $\pi R^2 y d$. Newton's second law yields $m\ddot{y}=-Ky$ where $K=\pi R^2 d$. The equation of motion is then

$$\ddot{y}(t) + y(t)K/m = 0 \quad (4\text{-}69)$$

Transient Response of Dynamic Systems 161

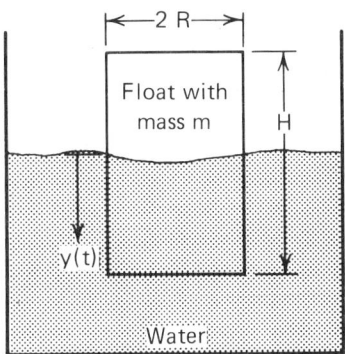

Figure 4-10. As shown in Example 4-8, if the float is depressed and then released, the displacement y(t) will exhibit sinusoidal oscillations, provided we neglect the viscosity of the water.

with initial conditions $y(0)=D$ and $\dot{y}(0)=0$. The motion consists of pure oscillations because the $\dot{y}(t)$ term is missing, indicating that $\zeta=0$. Using Eq. (4-68), the solution is

$$y(t) = D\cos\left\{\sqrt{K/m}\ t\right\} \qquad (4-70)$$

indicating that the cylinder will bob up and down forever. Clearly, the actual motion would not persist indefinitely. As we have neglected the drag force due to the water in setting up the equation of motion, the mathematical model is only an approximation (as are all models). The frequency of the oscillations would be somewhat different if damping were present because ω_d depends on ζ.

Example 4-9. Figure 4-11(a) shows a series RLC circuit connected to a battery through a switch. How must R, L, and C be related in order for the circuit to exhibit pure oscillations?
Solution. Writing a loop equation for i(t) gives

$$Eu(t) = L\frac{di(t)}{dt} + \frac{1}{C}\int_0^t i(t)dt + v_C(0) + i(t)R \qquad (4-71)$$

where the battery-switch combination produces Eu(t). Differentiate to eliminate the integral:

$$0 = L\frac{d^2 i(t)}{dt^2} + \frac{i(t)}{C} + R\frac{di(t)}{dt} \qquad (4\text{-}72)$$

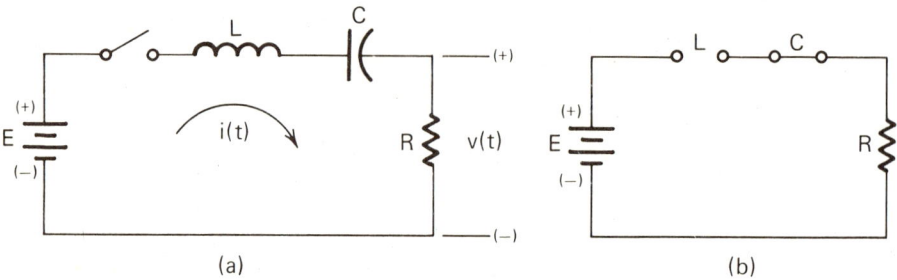

Figure 4-11. Example 4-9 shows that the RLC network will exhibit pure sinusoidal oscillations only if r = 0. (a) An RLC network for Example 4-9; (b) An initial-value network model to determine initial conditions.

The derivative of the step input does not result in an impulse because we have assumed operation for t⩾0. To obtain initial conditions, consider the $t=0^+$ equivalent shown in Fig. 4-11(b), where we have replaced the inductor by an open circuit and the capacitor by a short circuit. Clearly, $i(0^+)=0$ as a consequence of the open-circuit inductor model. In addition, we need an initial condition on di/dt. To get this, note that the voltage $v_L(t)$ across the inductor is $v_L(t)=Ldi/dt$ so that

$$\left.\frac{di(t)}{dt}\right]_{t=0^+} = \left.\frac{v_L(t)}{L}\right]_{t=0^+} \qquad (4\text{-}73)$$

Because the inductor equivalent at $t=0^+$ is an open circuit, all of the voltage E appears across the inductor and $v_L(0^+)=E$. Hence, $di(0^+)/dt=E/L$. The differential equation in standard form is

$$\frac{d^2 i}{dt^2} + \frac{R}{L}\frac{di}{dt} + \frac{i}{LC} = 0 \qquad (4\text{-}74)$$

Transient Response of Dynamic Systems

with initial conditions $i(0)=0$ and $di(0)/dt=E/L$. The characteristic equation is $p^2+Rp/L+1/LC=0$. Comparing this with the standard form, $p^2+2\zeta\omega_n p+\omega_n^2=0$, reveals that $\omega_n=1/\sqrt{LC}$ and $\zeta=R\sqrt{C/4L}$.

Pure oscillations at the frequency $\omega_n=1/\sqrt{LC}$ occur only if $\zeta=0$, which in turn implies that $R=0$. Even if the resistor is removed, the wires connecting the other components will have a small amount of resistance and damping is not zero. For this reason it is impossible to build a passive RLC oscillator. Practical sinusoidal oscillators have active elements such as operational amplifiers to "revive" the output as it begins to decay.

Higher-Order Systems

For systems of order higher than the second, the same techniques can be used to determine the natural response, although the algebra becomes more difficult. For one thing, finding the roots of the characteristic equation presents problems. Exact formulas are available for factoring polynomials up to and including the fourth order. For characteristic equations of order five or greater, a computer can be used to find approximate values for the roots. Moreover, inserting the initial conditions becomes tedious, requiring solution of n simultaneous algebraic equations with complex coefficients for an nth-order system. This is by no means a trivial task for three or more equations. For such reasons, an analog or a digital computer becomes a valuable aid for evaluating the natural response of a higher-order system. Most digital computer centers have in their program libraries numerical analysis packages which contain algorithms for determining point-by-point solutions of differential equations. See Chapters 11, 12, and 13.

4.2 FORCED RESPONSE

This section shows how to determine the solution of a constant-coefficient differential equation with a forcing function, f(t), but with all initial conditions zero. We concentrate on first- and second-order equations with step or sinusoidal functions as inputs. The response to an impulse input is treated by Laplace-transform methods in Chapter 5.

To determine the complete response, we need the following three results:

Rule 1. In the steady state (as t→∞), the forced response has the same <u>functional form</u> as the forcing function. This means that if the forcing function is a sinusoid, then the forced response is a sinusoid with the same frequency, but with a different amplitude and phase.

Rule 2. The complete response is the sum of (a) the forced response and (b) the natural response <u>with the arbitrary constants left unevaluated</u> (in symbolic form).

Rule 3. The initial conditions are inserted to evaluate the arbitrary constants in the natural response <u>after</u> it has been added to the forced response.

The interested reader is referred to a more complete treatise on differential equations for justification of these three observations.[1,12,13] The following sections show how to use the rules to find the forced response, and although we have not proved the rules, the fact that they do indeed work should be reassuring.

First-Order Systems

<u>The step-function input</u>. Consider the first-order equation

$$\dot{y} + ay = bu(t) \qquad (4-75)$$

with the **initial condition** y(0)=0. According to Rule 1 above, the forced response must be a step function, or, for t≥0, a constant; hence, we assume a forced response $y_f(t)$ of the form

$$y_f(t) = Bu(t) \qquad (4-76)$$

Transient Response of Dynamic Systems

This method of solution, wherein we assume a forced response of the form of the input function but with arbitrary coefficients, is called the <u>method of undetermined coefficients</u>.[14] Substituting Eq. (4-76) into Eq. (4-75) for t>0 gives

$$\cancelto{0}{\frac{dB}{dt}} + aB = b \qquad (4\text{-}77)$$

and B=b/a. From Eq. (4-8), the natural response $y_n(t)$ with the arbitrary constant left unevaluated is $y_n(t)=A\exp(-at)$, where we have introduced the "n" subscript to denote natural response. The complete response is the sum of $y_n(t)$ and $y_f(t)$, or $y(t)=y_f(t)+y_n(t)$. Using B=b/a and the natural response yields

$$y(t) = \frac{b}{a} + Ae^{-at} \qquad (4\text{-}78)$$

Finally, the initial condition y(0)=0 is used to determine A. At t=0, Eq. (4-78) becomes 0=b/a+Aexp(-0) and A=-b/a. The complete solution is then

$$y(t) = \frac{b}{a}(1-e^{-at}) \qquad (4\text{-}79)$$

Observe that the natural response affects the complete response even though y(0)=0. Equation (4-79) satisfies both the initial condition and the requirement (Rule 1) that in the steady state the response is a constant. Figure 4-12 is a plot of Eq. (4-79), assuming the parameters a and b are both positive.

Example 4-10. Figure 4-13 shows an operational amplifier configured as a pseudo-integrator. The switch is closed at t=0, resulting in a step-function input. The capacitor is initially uncharged. Calculate the time necessary for the output $e_o(t)$ to rise to 90% of its final value. Use R=1.0 MΩ and C=1.0 μF.

166   *Transient Response of Dynamic Systems*

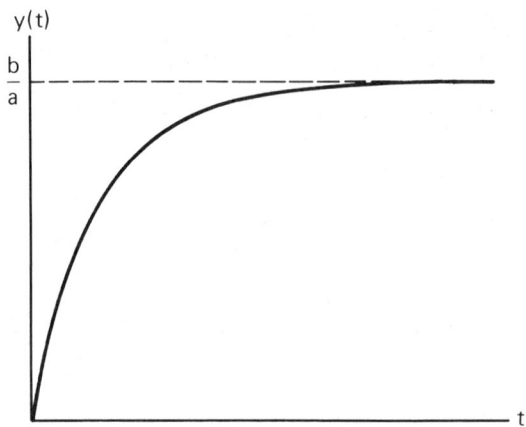

Figure 4-12. First-order step response, $y(t) = (b/a)[1 - \exp(-at)]$.

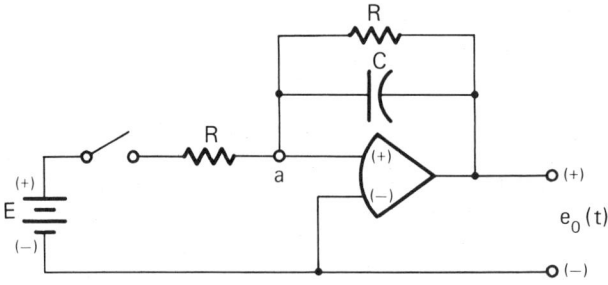

Figure 4-13. As shown in Example 4-10, the forced response of this OP AMP network is the asymptotic exponential $e_o(t) = -E[1 - \exp(-t/RC)]$.

Solution. The node-voltage equation at node a is

$$\frac{v_a - E}{R} + \frac{v_a - e_o}{R} + C\frac{d}{dt}(v_a - e_o) = 0 \qquad (4\text{-}80)$$

Assuming an ideal operational amplifier, $v_a = 0$, and Eq. (4-80) becomes

$$-\frac{E}{R} - \frac{e_o}{R} - C\dot{e}_o = 0 \qquad (4\text{-}81)$$

Transient Response of Dynamic Systems 167

or $\dot{e}_o + e_o/RC = -E/RC$ with the initial condition $e_o(0)=0$ because C is initially uncharged. Using Eq. (4-79), the complete solution is

$$e_o(t) = -E(1 - e^{-t/RC}) \qquad (4-82)$$

As $t \to \infty$, $e_o(t) \to -E$, which is the steady-state value. Setting $e_o(t_1) = -.9E$ at $t = t_1$ gives

$$-.9E = -E(1 - e^{t_1/RC}) \qquad (4-83)$$

Solving Eq. (4-83) for t_1 yields $t_1 = 2.3$ s.

The sinusoidal input. The first-order differential equation with a sinusoidal input is

$$\dot{y} + ay = b\cos(\omega t) \qquad (4-84)$$

Again assume $y(0)=0$. The forced response has the same form as the input, a sinusoid with frequency ω but with a different magnitude and phase than the original input. We assume, therefore, a forced resonse $y_f(t) = C\cos(\omega t - \emptyset)$ or, using the trigonometric identity for the cosine of the sum of two angles,

$$y_f(t) = A_1\cos(\omega t) + A_2\sin(\omega t) \qquad (4-85)$$

where $A_1 = C\cos(\emptyset)$ and $A_2 = C\sin(\emptyset)$. To evaluate the constants A_1 and A_2, substitute Eq. (4-85) into Eq. (4-84) to obtain

$$\frac{d}{dt}\{A_1\cos(\omega t) + A_2\sin(\omega t)\} + a\{A_1\cos(\omega t) + A_2\sin(\omega t)\} = b\cos(\omega t) \qquad (4-86)$$

Performing the indicated differentiation and collecting the coefficients of the sine and cosine terms gives

$$(aA_2 - A_1\omega)\sin(\omega t) + (A_2\omega + aA_1)\cos(\omega t) \equiv b\cos(\omega t) \qquad (4-87)$$

To satisfy the identity we match coefficients to obtain $aA_2-A_1\omega=0$ and $A_2\omega+A_1 a=b$. Solving these expressions for A_1 and A_2 yields

$$A_1 = \frac{ab}{a^2+\omega^2} \quad \text{and} \quad A_2 = \frac{b\omega}{a^2+\omega^2} \qquad (4\text{-}88)$$

Combine the steady-state and natural responses to get the complete solution

$$y(t) = Ae^{-at} + A_1\cos(\omega t) + A_2\sin(\omega t) \qquad (4\text{-}89)$$

Evaluating at $t=0$ with $y(0)=0$ results in $0=A+A_1$ from which $A=-A_1$. The complete solution is then

$$y(t) = A_1\{\cos(\omega t)-e^{-at}\} + A_2\sin(\omega t) \qquad (4\text{-}90)$$

Combining the sine and cosine terms into a cosine with a phase angle gives the alternate form

$$y(t) = \sqrt{A_1^2 + A_2^2}\,\cos(\omega t-\emptyset) - A_1 e^{-at} \qquad (4\text{-}91)$$

where $\emptyset=\tan^{-1}(\omega/a)$. For sinusoidal inputs the procedure has begun to get complicated, even for a first-order system.

Other input functions. An endless variety of other forms are theoretically possible for the input function, but if we restrict our investigation to those commonly encountered in linear systems work, the list is pretty-well exhausted by step functions, sinusoids, exponentials, polynomials, and sums and products thereof. Table 4-1 lists a number of typical input functions and the form of the forced response associated with each.

The table indicates the approach which must be used when constructing functional forms for the forced response. If one term of a polynomial is present in the input function, then all lower-order terms must also appear in the assumed forced response; and whenever a sine or a cosine is present, then both sine and cosine terms must be in the assumed forced response.

Transient Response of Dynamic Systems 169

Table 4-1.

Functional Forms For The Method Of Undetermined Coefficients

Input Function	Form For Undetermined-Coefficient Solution
$u(t)$	$Au(t)$
$\sin(\omega t)$	$A_1 \sin(\omega t) + A_2 \cos(\omega t)$
$\cos(\omega t)$	$A_1 \sin(\omega t) + A_2 \cos(\omega t)$
e^{-at}	Ae^{-at}
t	$A_1 t + A_2$
t^2	$A_1 t^2 + A_2 t + A_3$
t^n	$A_1 t^n + A_2 t^{n-1} + \ldots A_{n-1} t + A_n$
te^{-at}	$(A_1 t + A_2) e^{-at}$

Example 4-11. A network consists of a resistor and a capacitor connected in series to a voltage source through a switch. When the switch is closed the source produces $v(t) = 5\exp(-2t)$ V. Solve for the loop current $i(t)$ if R=10 kΩ and C=100 μF. The capacitor is initially uncharged.

Solution. With the switch closed the KVL equation for the loop is

$$R\frac{di(t)}{dt} + \frac{i(t)}{C} = \frac{dv(t)}{dt} \tag{4-92}$$

where we have differentiated to remove the capacitor integral. The initial current is $i(0) = v(0)/R = 0.5$ mA since a capacitor looks like a short circuit at switch closure. The natural response $i_n(t)$ takes the form $i_n(t) = B\exp(-t/RC) = B\exp(-t)$. The forced response is the exponential $i_f(t) = K\exp(-2t)$.

Insert the numerical values for R and C into Eq. (4-92) to obtain

$$\frac{di(t)}{dt} + i(t) = -0.001 e^{-2t} \tag{4-93}$$

Substitute $i_f(t) = K\exp(-2t)$ into Eq. (4-93) and solve for K=0.001 A. Combining the forced and natural responses gives

$$i(t) = i_n(t) + i_f(t) = Be^{-t} + 0.001 e^{-2t} \tag{4-94}$$

Evaluating at t=0 yields

$$i(0) = B + 0.001 \qquad (4\text{-}95)$$

from which

$$B = i(0) - 0.001 \qquad (4\text{-}96)$$

or, with $i(0)=0.5$ mA, $B=-0.5$ mA. The complete solution is then

$$i(t) = -0.5e^{-t} + e^{-2t} \text{ mA} \qquad (4\text{-}97)$$

Example 4-12. Find the complete response of the differential equation $\dot{y}+y=t$ with $y(0)=2$.

Solution. The natural response has the form $y_n(t)=A\exp(-t)$, and the assumed forced response for a first-order polynomial is $y_f(t)=A_1t+A_2$. Substituting $y_f(t)$ into the differential equation gives

$$A_1 + A_1 t + A_2 = t \qquad (4\text{-}98)$$

and, by matching coefficients of like powers of t, $A_1+A_2=0$, $A_1=1$, and $A_2=-1$. The complete response is then $y(t)=A\exp(-t)+t-1$. Evaluating at t=0 and inserting the initial condition gives $y(0)=2=A-1$ from which $A=3$. The complete solution is

$$y(t) = 3e^{-t} + t - 1 \qquad (4\text{-}99)$$

Second-Order Systems

The procedure for finding the forced response of a second-order system parallels that for the first-order case. Unfortunately, however, the algebra gets out of hand for all but the simplest problems. For this reason we consider only the most tractable (and generally the most important) case-- the step-function input.

The step-function input. The second-order equation with a step input and zero initial conditions is

$$\ddot{y} + 2\zeta\omega_n\dot{y} + \omega_n^2 y = bu(t) \qquad (4\text{-}100)$$

Transient Response of Dynamic Systems

with $y(0)=0$ and $\dot{y}(0)=0$. The forced response is found by assuming $y_f(t)=Au(t)$ which, when substituted into Eq. (4-100), leads to $A=b/\omega_n^2$ and

$$y_f(t) = \frac{b}{\omega_n^2} u(t) \qquad (4\text{-}101)$$

We consider the natural response for two cases, <u>over-damped response</u>, characterized by $\zeta>1$ and real roots for the characteristic equation, and <u>underdamped response</u>, for which $0<\zeta<1$ and the roots are complex conjugates.

Real unequal roots. The characteristic equation associated with Eq. (4-100) is $p^2+2\zeta\omega_n p+\omega_n^2=0$ with roots p_1 and p_2 given by

$$p_1, p_2 = \left\{-\zeta\omega_n \pm \sqrt{\zeta^2-1}\right\}\omega_n \qquad (4\text{-}102)$$

If $\zeta>1$, the roots are real and unequal, and by the method of Section 4.1, the natural response is the sum of two real exponentials:

$$y_n(t) = A_1 e^{p_1 t} + A_2 e^{p_2 t} \qquad (4\text{-}103)$$

The complete response is the sum of $y_n(t)$ and $y_f(t)$, Eqns. (4-101) and (4-103):

$$y(t) = \frac{b}{\omega_n^2} + A_1 e^{p_1 t} + A_2 e^{p_2 t} \qquad (4\text{-}104)$$

When $t=0$, $y(0)=0$, and Eq. (4-104) becomes $0=b/\omega_n^2+A_1+A_2$. Differentiating Eq. (4-104) gives

$$\dot{y}(t) = p_1 A_1 e^{p_1 t} + p_2 A_2 e^{p_2 t} \qquad (4\text{-}105)$$

When $t=0$, Eq. (4-105) becomes $0=p_1A_1+p_2A_2$. Solving for A_1 and A_2 yields

$$A_1 = \frac{-p_2 b/\omega_n^2}{p_2 - p_1} \qquad (4\text{-}106)$$

and

$$A_2 = \frac{p_1 b/\omega_n^2}{p_2 - p_1} \qquad (4\text{-}107)$$

The complete solution is then

$$y(t) = \frac{b}{\omega_n^2}\left\{1 + \frac{p_1 e^{p_2 t} - p_2 e^{p_1 t}}{p_2 - p_1}\right\} \qquad (4\text{-}108)$$

Equation (4-108) satisfies both the initial condition, $y(0)=0$, and the requirement that in the steady state the form of the solution must be a constant to match the input. This latter result holds only if the exponential terms decay to zero as t approaches infinity, which in turn implies that p_1 and p_2 are negative. An equivalent statement is that the system must be <u>stable</u> (see Chapter 6). Observe too that even though the initial conditions are both zero, the natural response terms are still present in the complete solution. Qualitatively, the natural response is how the system "wants" to behave, and it is generally impossible to suppress the natural response by setting the initial conditions to zero. If the system is stable, however, the natural response terms must eventually decay to zero.

<u>Example 4-13</u>. Figure 4-14 is the circuit diagram for a two-stage active filter using operational amplifiers. Derive the differential equation for $v_3(t)$ and show that the roots of the characteristic equation are always real, no matter what we choose for R_1 and R_2.

<u>Solution</u>. Writing node-voltage equations at nodes a and b and assuming $v_a = v_b = 0$ (ideal operational amplifiers) results in

$$\frac{v_1}{R} + \frac{v_2}{R_2} + C\dot{v}_2 = 0 \qquad (4\text{-}109)$$

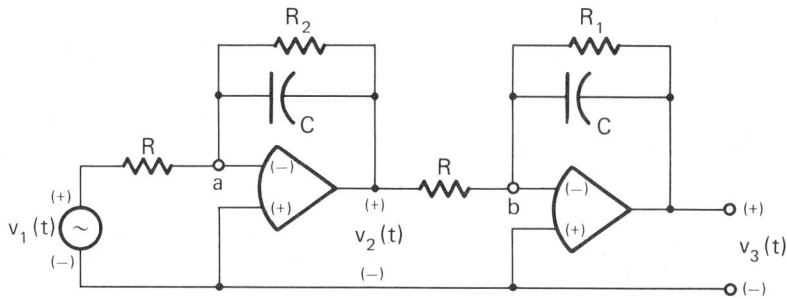

Figure 4-14. A two-stage active filter using OP AMP's. The characteristic equation, Eq. (4-115), is derived in Example 4-13.

and

$$\frac{v_2}{R} + \frac{v_3}{R_1} + C\dot{v}_3 = 0 \qquad (4\text{-}110)$$

Solve Eq. (4-109) for $\dot{v}_2$:

$$\dot{v}_2 = -\frac{v_1}{CR} - \frac{v_2}{CR_2} \qquad (4\text{-}111)$$

Differentiate Eq. (4-110) to obtain

$$\frac{\dot{v}_2}{R} + \frac{\dot{v}_3}{R_1} + C\ddot{v}_3 = 0 \qquad (4\text{-}112)$$

Substitute Eq. (4-111) into Eq. (4-112) with the result

$$-\frac{1}{CR}\left\{\frac{v_1}{R} + \frac{v_2}{R_2}\right\} + \frac{\dot{v}_3}{R_1} + C\ddot{v}_3 = 0 \qquad (4\text{-}113)$$

Finally, solve Eq. (4-110) for $v_2 = -Rv_3/R_1 - RC\dot{v}_3$ and use the result to eliminate v_2 from Eq. (4-113); the result is

$$C\ddot{v}_3 + \left\{\frac{1}{R_1} + \frac{1}{R_2}\right\}\dot{v}_3 + \frac{1}{R_1 R_2 C} v_3 = \frac{1}{R^2 C} v_1 \qquad (4\text{-}114)$$

The characteristic equation

$$p^2 + \frac{p}{C}\left\{\frac{1}{R_1} + \frac{1}{R_2}\right\} + \frac{1}{R_1 R_2 C^2} = 0 \qquad (4\text{-}115)$$

can be factored as $(p+1/R_1C)(p+1/R_2C)=0$ so that the roots are always real, regardless of the values of R_1 and R_2. From a design standpoint this result is interesting: we can control the roots of the characteristic equation independently. This is an example of the isolation property of operational amplifiers.

Complex-conjugate roots. If the roots of the characteristic equation are complex, a situation which arises when $0<\zeta<1$, then, from Eq. (4-52), the second-order natural response takes the form

$$y_n(t) = Be^{-\sigma t}\cos(\omega_d t - \emptyset) \qquad (4\text{-}116)$$

where $\sigma = \zeta\omega_n$ and $\omega_d = \omega_n\sqrt{1-\zeta^2}$. The constants B and $\emptyset$ again depend on the initial conditions, but now they cannot be evaluated until after we combine the natural and forced responses.

Combining Eqns. (4-101) and (4-116) gives the complete response

$$y(t) = \frac{b}{\omega_n^2} + Be^{-\sigma t}\cos(\omega_d t - \emptyset) \qquad (4\text{-}117)$$

Evaluating Eq. (4-117) at t=0 with y(0)=0 results in $0 = b/\omega_n^2 + B\cos(\emptyset)$. Differentiating Eq. (4-117) gives

$$\dot{y}(t) = -\sigma Be^{-\sigma t}\cos(\omega_d t - \emptyset) - \omega_d Be^{-\sigma t}\sin(\omega_d t - \emptyset) \qquad (4\text{-}118)$$

and, at t=0, $0 = -\sigma B\cos(\emptyset) + B\omega_d\sin(\emptyset)$. From the latter expression $\emptyset = \tan^{-1}(\sigma/\omega_d)$ or $\cos(\emptyset) = \omega_d/\omega_n$. Using this value, the y(0) expression $B\cos(\emptyset) + b/\omega_n^2 = 0$ yields

$$B = \frac{-b}{\omega_n \omega_d} \qquad (4\text{-}119)$$

The complete response is then

$$y(t) = \frac{b}{\omega_n^2}\left\{1 - \frac{\omega_n}{\omega_d}e^{-\sigma t}\cos(\omega_d t - \emptyset)\right\} \qquad (4\text{-}120)$$

As t→∞, the exponential decays to zero and the steady-state solution is a constant as required. The natural response term is present even though the initial conditions are both zero. If $\zeta=0$, the roots of the characteristic equation are imaginary and Eq. (4-120) reduces to

$$y(t) = \frac{b}{\omega_n^2}\{1-\cos(\omega_n t)\} \qquad (4-121)$$

Sinusoidal inputs. The case where a second or higher order system is forced by a sinusoid is the subject of an entire chapter, Chapter 8, first, because the algebra becomes complicated for systems of order greater than the first, and second, because the steady-state response to a sinusoidal input is of great importance in electrical engineering. Chapter 8 shows how to find the steady-state response to a sinusoidal input without solving differential equations. We will defer the subject until then.

Combined Inputs

If the input function is the sum of several elementary input functions, we can use superposition to find the forced response. If, for example, we have the equation

$$\dot{y}(t) + 2y(t) = 5u(t) + 6e^{-3t} \qquad (4-122)$$

with $y(0)=12$, the forced response assumes the form $y_f(t) = A+B\exp(-3t)$. Substituting this assumed solution into the differential equation yields

$$-3Be^{-3t} + 2A + 2Be^{-3t} = 5u(t) + 6e^{-3t} \qquad (4-123)$$

and, after equating coefficients, we find that A=2.5 and B=-6. The complete response is then

$$y(t) = Ce^{-2t} + 2.5 - 6e^{-3t} \qquad (4-124)$$

Inserting the initial condition results in 12=C+2.5-6 from which C=15.5. The complete response is

$$y(t) = 15.5e^{-2t} - 6e^{-3t} + 2.5 \qquad (4-125)$$

The values for A and B are the same as would have resulted if the forced response for each input function had been found separately. In summary, when the input is a sum of functions, find the response to each separately and then sum the results to get the overall forced response. Just remember to insert the initial conditions <u>after</u> the partial responses have been summed.

Combined Response

If a system has both initial conditions and a forcing function, the procedure for determining the response is identical to that given above except that the algebraic equations to be solved for the arbitrary constants will contain the initial conditions.

4.3 SECOND-ORDER TRANSIENTS

Second-order systems arise so frequently that a standard set of quantitative descriptors has evolved for second-order transients.[11,15,16] The second-order differential equation with a unit step input and zero initial conditions is

$$\ddot{y} + 2\zeta\omega_n \dot{y} + \omega_n^2 y = \omega_n^2 u(t) \tag{4-126}$$

with $y(0)=0$ and $\dot{y}(0)=0$. With $b=\omega_n^2$ Eq. (4-120) becomes

$$y(t) = 1 - \frac{1}{\sqrt{1-\zeta^2}} e^{-\sigma t} \cos(\omega_d t - \phi) \tag{4-127}$$

where $\sigma=\zeta\omega_n$, $\omega_d=\omega_n\sqrt{1-\zeta^2}$, and $\phi=\sin^{-1}(\zeta)$. Figure 4-15 is a plot of Eq. (4-127) for $\zeta \approx 0.5$.

The quantitative descriptors of second-order transients, based on Eq. (4-127) and Fig. 4-15, are as follows:

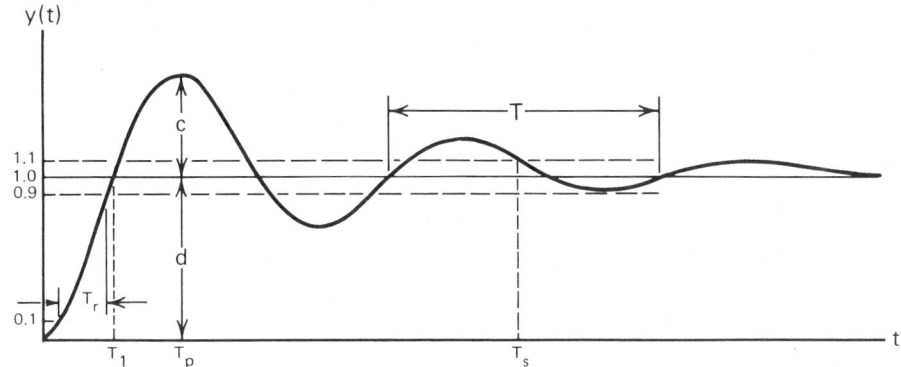

Figure 4-15. Second-order step response used to define transient performance factors.

Damping Ratio

The damping ratio ζ is a measure of how oscillatory the system is. The value of ζ determines the type of response as follows:

$\zeta<0$. The system is <u>unstable</u>. If it is excited by either a forcing function or by initial conditions, the response grows without bound until the system destroys itself, or, more realistically, until the linear mathematical model no longer represents the system. Two subcases for an unstable response are unstable and overdamped for $\zeta<-1$ and unstable and underdamped for $-1<\zeta<0$.

$\zeta=0$. The system is oscillatory. A step or initial condition input gives rise to a sustained sinusoidal response. The roots of the characteristic equation are imaginary.

$0<\zeta<1$. The system is underdamped. The response, which is the sum of two real exponentials, decays to zero as $t\to\infty$.

$\zeta=1$. The system is <u>critically damped</u>. This is the borderline case between a damped sinusoidal response and one which is the sum of two real exponentials. The roots of the characteristic equation are real and equal.

178 Transient Response of Dynamic Systems

ζ>1. The system is **overdamped**. The response is the sum of two real exponentials. The roots of the characteristic equation are real and unequal.

Figure 4-16 shows a typical response for each of these cases. The responses in the figure assume y(0)=1, ẏ(0)=0 and f(t)=0, but the general form is the same for other combinations of initial conditions and a step forcing function.

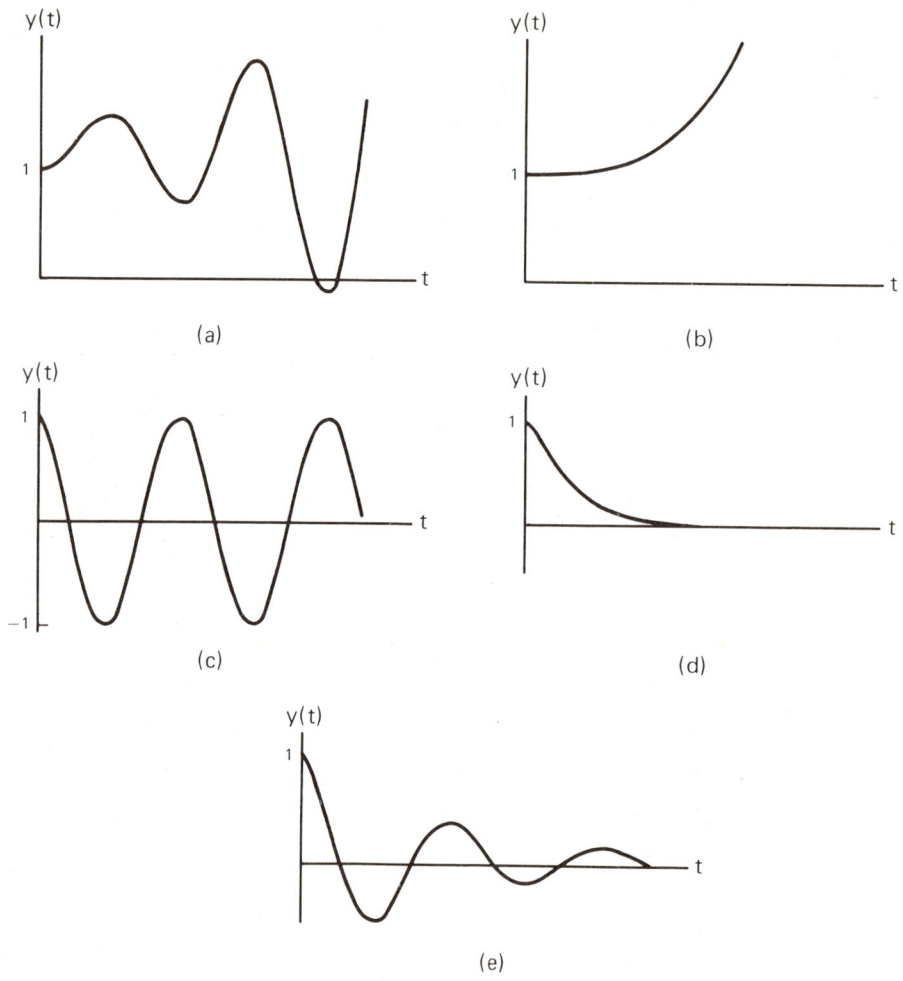

Figure 4-16. Typical responses to illustrate the role of ζ for a second-order system. (a) Unstable and underdamped, $-1 < ζ < 0$; (b) Unstable and overdamped, $ζ < -1$; (c) Undamped, $ζ = 0$; (d) Stable and overdamped, $ζ > 1$, or critically damped, $ζ = 1$; (e) Stable and underdamped, $0 < ζ < 1$.

Undamped Natural Frequency

The undamped natural frequency ω_n is the frequency of the sinusoidal oscillations that occur if $\zeta=0$. If $\zeta \neq 0$ the oscillations are at the <u>damped natural frequency</u>, ω_d, defined by $\omega_d = \omega_n \sqrt{1-\zeta^2}$. The <u>period</u> T of the oscillations is constant and given by $T = 2\pi/\omega_d$. T/2 is the time between successive zero crossings.

Time Constant

The quantity $T_c = 1/\zeta\omega_n = 1/\sigma$ has units of time (ζ is dimensionless and ω_n is in **rad/s**). T_c is the <u>time constant</u> of the second-order transient. It determines the rate at which the envelope $\exp(-\sigma t)$ approaches zero. After $5T_c$, the response has effectively died out. Both ζ and ω_n affect the duration of the transient response.

Peak Overshoot

The peak overshoot C_p, defined only for the unit step function response with zero-initial conditions, is the percentage by which the peak value of the transient response exceeds the steady-state value. It is found from Fig. 4-15 by expressing the ratio c/d in percent. To compute the peak overshoot analytically from Eq. (4-127), compute $\dot{y}(t)$, set $\dot{y}(T_p) = 0$, then solve for the time T_p. Use T_p in Eq. (4-127) to find the value of the overshoot. Following this procedure yields

$$T_p = \frac{\pi}{\omega_d} \tag{4-128}$$

and

$$C_p = e^{-\zeta\omega_n \pi/\omega_d} \tag{4-129}$$

where, as usual, $\sigma = \zeta\omega_n$ and $\omega_d = \omega_n\sqrt{1-\zeta^2}$.

Time to First Steady State

The time T_1 between t=0 and the first instant when the steady-state value is reached is shown in Fig. 4-15. This time is a measure of how quickly the response rises.

Rise Time

The rise time T_r of the system is the length of time it takes the response to rise from 10% to 90% of the steady-state value. Rise time is also shown in Fig. 4-15.

Settling Time

The settling time T_s is the time required for the transient response to settle to within a certain percentage of its final value and remain within that percentage or less. We normally use either the 5% or 10% settling times. Figure 4-15 shows the 10% settling time.

Although defined for second-order systems, the descriptors listed above are often used with damped oscillations from higher-order systems. Exact analytical expressions could be found for T_1, T_r, and T_s but they are not too useful. The quantities are more frequently taken empirically from graphical plots of the response y(t).

4.4 COUPLED SYSTEMS

Many systems give rise to <u>coupled differential equations</u>. Equations (4-130) and (4-131) are coupled because y_2 appears in the y_1 equation and vise versa.

$$\ddot{y}_1 + a_1 \dot{y}_1 + a_0 y_1 - k_1 y_2 = f_1(t) \qquad (4\text{-}130)$$

$$\ddot{y}_2 + b_1 \dot{y}_2 + b_0 y_2 - k_2 y_1 = f_2(t) \qquad (4\text{-}131)$$

While coupled equations can be solved by classical methods in the time domain, the process is somewhat easier if Laplace transforms are used. We will, therefore, postpone our discussion of coupled differential equations until Chapter 5. Further, coupled equations can also be treated in the more general context of <u>state-variable</u> methods which are addressed in Chapter 10.

4.5 NONLINEAR AND TIME-VARYING SYSTEMS

Nonlinear and/or time-varying models for dynamic systems are often unavoidable.[17,18] As a consequence, the analyst must be familiar with the classes of nonlinear and time-varying differential equations which can be solved. (By a solution we mean a closed-form expression in terms of elementary functions--polynomials, sinusoids, exponentials, etc.)

First-Order Nonlinear and Time-Varying Equations

The first-order, nonlinear and time-varying homogeneous differential equation is

$$\dot{y} + f(t,y) = 0 \qquad (4\text{-}132)$$

with the initial condition $y(0)$. Such equations can often be solved by separating the variables and integrating from 0 to t (provided, of course, that we can do the indicated integration). By this technique, if Eq. (4-132) can be written as $\dot{y}(t)+g(y)h(t)=0$, then we can separate the variables and obtain

$$\int_{y(0)}^{y(t)} \frac{dy}{g(y)} = - \int_0^t h(t)dt \qquad (4\text{-}133)$$

Assuming the integration can be done--and if pencil-and-paper methods fail we might use numerical analysis and a digital computer--we can develop an expression for the solution $y(t)$, or, if the computer is used, a list of solution points.

Example 4-14. Solve the nonlinear differential equation

$$\dot{y} + \cos(t)\sqrt{1 - y^2} = 0 \qquad (4\text{-}134)$$

with initial condition $y(0)$.

Solution. The variables can be separated, giving

$$\int_{y(0)}^{y(t)} \frac{dy}{\sqrt{1 - y^2}} = -\int_0^t \cos(t)dt \qquad (4\text{-}135)$$

Performing the indicated integration yields

$$\sin^{-1}\{y(t)\} = \sin^{-1}\{y(0)\} - \sin(t) \qquad (4\text{-}136)$$

or

$$y(t) = \sin\left\{\sin^{-1}\{y(0)\} - \sin(t)\right\} \qquad (4\text{-}137)$$

The separation-of-variables technique fails if the differential equation has a forcing function, as in $\dot{y}+f(t,y)=g(t)$ or if the variables cannot be separated as in $\dot{y}+\cos(ty)=0$. Moreover, separation of variables fails for all nontrivial equations of order greater than the first.

Classical Nonlinear and Time-Varying Differential Equations

A number of nonlinear or time-varying differential equations have been investigated by mathematicians, either because they have properties which make them solvable, or because they arise naturally in the study of physical systems. Some of the equations have infinite-series solutions, and because the equations occur frequently, the series solutions have been abbreviated as functions--Bessel functions are the solution to Bessel's equation, for example. Table 4-2 lists a few of these famous equations.[19] The initial conditions are omitted for brevity. Some of the equations are nonlinear and others are time-varying; none are both nonlinear and time-varying. Fortunately, such monstrosities seldom occur in practical systems work.

Transient Response of Dynamic Systems

Table 4-2.

Classical Nonlinear and Time-Varying Differential Equations

NAME	EQUATION	APPLICATION
Van der Pol's Eq.	$\ddot{y} - \varepsilon(1-y^2)\dot{y} + y = 0$	Electronic Oscillators
Duffing's Eq.	$\ddot{y} + \dot{y} + g(y) = f(t)$	Nonlinear Springs
Bessel's Eq.	$t^2\ddot{y} + t\dot{y} + (\lambda^2 t^2 - v^2)y = 0$	Heat Transfer
Legendre's Eq.	$(1-t^2)\ddot{y} - 2t\dot{y} + n(n+1)y = 0$	
Hermite's Eq.	$\ddot{y} - 2t\dot{y} + 2ny = 0$	
Laguerre's Eq.	$t\ddot{y} + (1-t)\dot{y} + ny = 0$	
Euler's Eq.	$t^2\ddot{y} + t\dot{y} + y = 0$	
Riccati's Eq.	$\dot{y} + ay^2 = bt^n$	Optimal Control
Generalized Riccati Eq.	$\dot{y} + a + by + cy^2 = 0$	
Chebyshev's Eq.	$(1-t^2)\ddot{y} - t\dot{y} + n^2 y = 0$	
Mathieu's Eq.	$\ddot{y} + a^2\{1+\varepsilon\cos(bt)\}y = 0$	
	$L\ddot{\theta} + g\sin(\theta) = 0$	Pendulum Motion
	$\ddot{y} + a\dot{y}/|\dot{y}| + by = 0$	Dry Friction

The classical equations are discussed in various mathematics texts.[14,17,20] We will not investigate them further except for the following example.

Example 4-15. Show that the substitution $z = \ln(t)$ will reduce the time-varying equation

$$t^2\ddot{y} + a_1 t\dot{y} + a_0 y = 0 \qquad (4\text{-}138)$$

to a constant-coefficient differential equation which can be solved by the methods of this chapter.

Solution. If $z=\ln(t)$ then $dz/dt=1/t$. By the chain rule for derivatives,

$$\frac{dy}{dt} = \frac{dy}{dz}\frac{dz}{dt} = \frac{dy}{dz}\frac{1}{t} \tag{4-139}$$

and

$$\frac{d^2y}{dt^2} = \frac{d}{dt}\left\{\frac{dy}{dz}\frac{1}{t}\right\} = -\frac{1}{t^2}\frac{dy}{dz} + \frac{1}{t^2}\frac{d^2y}{dz^2} \tag{4-140}$$

Substituting these expressions into Eq. (4-138) yields

$$t^2\left\{-\frac{1}{t^2}\frac{dy}{dz} + \frac{1}{t^2}\frac{d^2y}{dz^2}\right\} + a_1 t\left\{\frac{1}{t}\frac{dy}{dz}\right\} + a_0 y = 0 \tag{4-141}$$

and

$$\frac{d^2y}{dz^2} + \frac{dy}{dz}(a_1 - 1) + a_0 y = 0 \tag{4-142}$$

Equation (4-142) is a constant-coefficient differential equation with characteristic equation $p^2+(a_1-1)p+a_0=0$. It can be solved by the methods of this chapter, after which the substitution $z=\ln(t)$ must be reversed to get a time-domain solution; terms of the form $A\exp(-bz)$ become At^{-b} and so on. Unfortunately, substitutions which convert nonlinear or time-varying differential equations to constant coefficient ones are rare.

PROBLEMS FOR CHAPTER 4

4-1. Show that $y_1=\exp(-3t)$ and $y_2=\exp(-2t)$ are solutions of the differential equation $\ddot{y}+5\dot{y}+6y=0$.

4-2. Write the characteristic equation for each of the following differential equations: (a) $\ddot{\phi}=0$, (b) $\dddot{y}+y=0$, (c) $\dot{q}/q+16=0$, (d) $\ddot{y}+4\dot{y}+8y=\cos(2t)$, and (e) $\dddot{\theta}+3\ddot{\theta}+2\dot{\theta}=\exp(10t)$.

4-3. Find the natural response for each of the following differential equations:

Transient Response of Dynamic Systems 185

(a) $\dot{y} + 10y = 0$ $y(0) = 4$ (4-143)

(b) $\dot{v}(t) = -3v(t)$ $v(0) = -5$ (4-144)

(c) $\dfrac{\dot{q}}{q} + 4 = 2$ $q(0) = 5$ (4-145)

(d) $\dot{y} = 4$ $y(0) = 5$ (4-146)

4-4. Find ω_n and ζ for the systems described by the following differential equations.

(a) $\ddot{y} + 5\dot{y} + 50y = 0$ (4-147)

(b) $2\ddot{\theta} + 8\dot{\theta} + 128\theta = 0$ (4-148)

(c) $\ddot{q} + 10\dot{q} + 25q = 0$ (4-149)

(d) $\ddot{v}_c + 100v_c = 0$ (4-150)

(e) $\ddot{y} + 40\dot{y} = 0$ (4-151)

4-5. Find $e_0(t)$ for the network shown. The switch has been at position A long enough for steady-state to be achieved and is moved to B at t=0.

Figure 4-17. Problem 4-5.

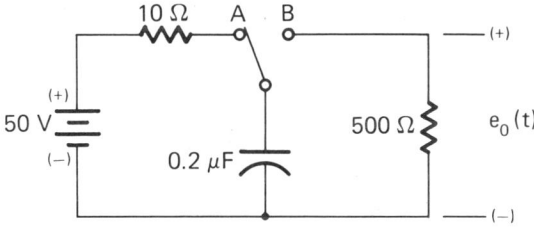

4-6. Assuming the voltage response v(t) is underdamped, derive expressions for ζ and ω_n in terms of the parameters R_1, R_2, L, and C for the network of Fig. 4-18. (Hint: use nodal analysis.)

Figure 4-18. Problem 4-6.

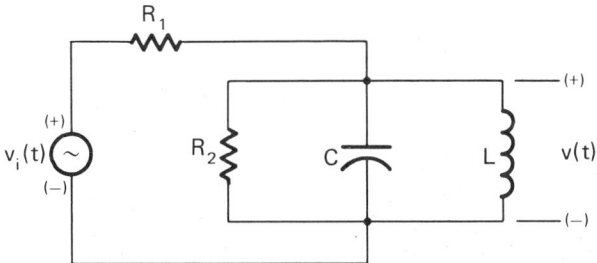

4-7. Derive an expression for $e_o(t)$ in terms of E, R_1, R_2, and L. Assume the switch has been in position A for a long time and is moved to B at t=0.

Figure 4-19. Problem 4-7.

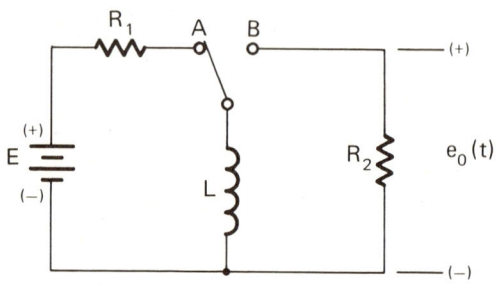

4-8. Steady-state conditions prevailed in the network of Fig. 4-20 before the switch was moved from A to B. Solve for i(t).

Figure 4-20. Problem 4-8.

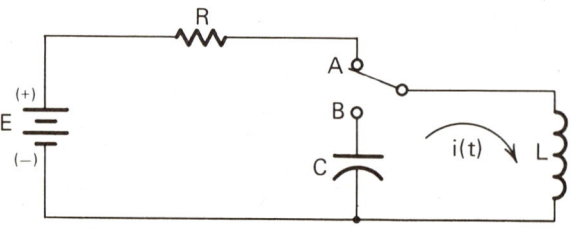

4-9. The network of Fig. 4-21 was in steady state with the switch S closed. Solve for i(t) after the switch is opened at t=0.

Figure 4-21. Problem 4-9.

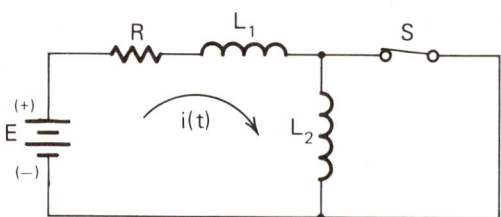

4-10. Refer to Example 4-9. Choose values of L and C that result in $\omega_n = 2000$ rad/s and $\zeta = 0.01$ if R=1000 Ω.

4-11. In Fig. 4-22 the switch has been in position A long enough for steady state to result. It is moved to position B at t=0. If C=2.0 µF and R_1=0.5 Ω, choose values of L and R_2 to make ω_n=5000 rad/s and ζ=0.01.

Figure 4-22. Problem 4-11.

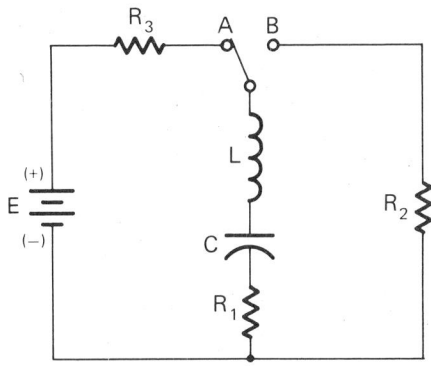

4-12. Find the correct *form* for the forced response of the following differential equations:

(a) $\ddot{y} + a\dot{y} + by = cu(t)$ (4-152)

(b) $\dot{v} + 5v = 10e^{-4t}$ (4-153)

(c) $\ddot{\theta} + 2\zeta\omega_n\dot{\theta} + \omega_n^2\theta = \cos(4t)$ (4-154)

(d) $\dot{y} + 4y = t^2$ (4-155)

(e) $\ddot{q} + q = te^{-3t}$ (4-156)

(f) $\ddot{r} + 2\dot{r} = t^2\sin(t)$ (4-157)

4-13. Find the complete solution for each of the following differential equations:

(a) $\dot{y} + 3y = 2u(t)$ $y(0) = 0$ (4-158)

(b) $\dot{\theta} + \theta = e^{-2t}$ $\theta(0) = 2$ (4-159)

(c) $\dot{q} + 2q = tu(t)$ $q(0) = -5$ (4-160)

4-14. A mechanical system is described by the equation

$$\ddot{\theta} + 5\dot{\theta} + 125\theta = 500u(t) \tag{4-161}$$

with zero initial conditions. Find (a) the forced response, and (b) the complete response.

4-15. Derive the expressions for T_p and C_p given by Eqns. (4-128) and (4-129).

4-16. Solve the following nonlinear differential equations:

(a) $\dot{y} + t^2 y = 0$ $y(0) = y_o$ (4-162)

(b) $\dot{\theta} + e^{\theta} e^t = 0$ $\theta(0) = \theta_o$ (4-163)

(c) $\ln(\dot{y}) + \ln(ty) = -2$ $y(1) = 3$ (4-164)

(d) $\ddot{q} + (\dot{q})^2 = 0$ $\dot{q}(0) = 1$ (4-165)
                                $q(0) = 2$

4-17. Find a second-order differential equation which has $y(t)=2+e^{-3t}$ as its solution.

4-18. The operational amplifier circuit below solves a certain differential equation. Find the differential equation (with initial conditions) and its solution.

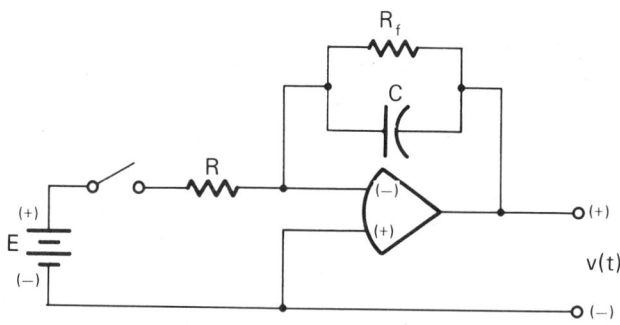

Figure 4-23. Problem 4-18.

4-19. Derive an expression for the complete response of the equation $\ddot{y}+2\zeta\omega_n\dot{y}+\omega_n^2 y=bu(t)$ with $y(0)=\dot{y}(0)=0$ when the roots of the characteristic equation are real and equal.

REFERENCES FOR CHAPTER 4

(1) Rainville, E. D.: Elementary Differential Equations, The Macmillan Company, New York, 1949.

(2) Ross, S. L.: Differential Equations, Blaisdell Publishing Co., New York, 1964.

(3) Schwarz, R. J. and B. Friedland: Linear Systems, McGraw-Hill Book Co., New York, 1965.

(4) Cheng, D. K.: Analysis of Linear Systems, Addison-Wesley Publishing Co., Inc., Reading, Massachusetts, 1959.

(5) Kaplan, W.: Ordinary Differential Equations, Addison-Wesley Publishing Co., Inc., Reading, Massachusetts, 1958.

(6) Pennisi, L.L.: Elements of Ordinary Differential Equations, Holt, Rinehart, and Winston, Inc., New York, 1972.

(7) Spiegel, M. R.: Applied Differential Equations, Prentice-Hall, Inc., Englewood Cliffs, New Jersey, 1958.

(8) Wylie, C. R., Jr.: Advanced Engineering Mathematics, McGraw-Hill Book Co., Inc., New York, 1960.

(9) Kuo, B. C.: Automatic Control Systems, Prentice-Hall, Inc., Englewood Cliffs, New Jersey, 1962.

(10) Hsu, J. C. and A. U. Meyers: Modern Control Principles and Applications, McGraw-Hill Book Co., Inc., New York, 1968.

(11) D'Azzo, J. J. and C. H. Houpis: Feedback Control System Analysis and Synthesis, McGraw-Hill Book Co., Inc., New York, 1960.

(12) Tenenbaum, M. and H. Pollard: Ordinary Differential Equations, Harper and Row Publishers, Inc., New York, 1963.

(13) Murphy, G. M.: Ordinary Differential Equations and Their Solutions, D. Van Nostrand Co., Inc., Princeton, New Jersey, 1960.

(14) Rabenstein, A. L.: Introduction to Ordinary Differential Equations, Academic Press, Inc., New York, 1972.

(15) Chestnut, H. and R. W. Mayer: Servomechanisms and Regulating Systems Design (vol. I), John Wiley and Sons, Inc., New York, 1951.

(16) Thaler, G. J.: Design of Feedback Systems, Dowden, Hutchinson, and Ross, Inc., Stroudsburg, Pennsylvania, 1973.

(17) Minorsky, N.: Nonlinear Oscillations, D. Van Nostrand Co., Inc., Princeton, New Jersey, 1962.

(18) Gibson, J. E.: Nonlinear Automatic Control, McGraw-Hill Book Co., Inc., New York, 1963.

(19) Scarborough, J. B.: Differential Equations and Applications, Waverly Press, Inc., Baltimore, Maryland, 1965.

(20) Cunningham, W. J.: Introduction to Nonlinear Analysis, McGraw-Hill Book Co., Inc., New York, 1958.

(21) Thaler, G. J. and M. P. Pastel: Analysis and Design of Nonlinear Feedback Control Systems, McGraw-Hill Book Co., Inc., New York, 1962.

CHAPTER 5

The Laplace Transform

Chapter 4 discussed classical time-domain methods for solving differential equations. The process consisted of four steps:

(1) Factoring the characteristic equation to find the natural response.

(2) Determining the forced response using the method of undetermined coefficients.

(3) Adding the natural response (with the arbitrary constants left in symbolic form) to the forced response to obtain the complete response.

(4) Using the initial conditions to evaluate the arbitrary constants.

This process often involves some tedious algebra because factoring the characteristic equation is a formidable task if the differential equation is of an order higher than the second. Chapter 5 introduces the Laplace transformation, a method whereby we can transform linear, constant-coefficient differential equations into algebraic equations. More specifically, a differential equation whose independent variable is time becomes an algebraic equation in terms of a complex frequency variable s.

The four-step process for solving a constant-coefficient differential equation with dependent variable y(t) and forcing function f(t) can be replaced by the following Laplace transform procedure.[1,2]

(1) Take the Laplace transformation of the given differential equation. The result is an algebraic equation which relates the transformed dependent variable Y(s) to the transformed input function F(s).

(2) Solve for Y(s) in terms of F(s).

(3) Take the inverse Laplace transform of Y(s) to obtain the complete solution y(t).

The Laplace procedure automatically incorporates the initial conditions, and the form of the forced response is inherent in the method--it need not be guessed at as in the method of undetermined coefficients. Unfortunately, the Laplace method still requires factoring the characteristic equation, and in that sense the law of conservation of difficulty applies: the amount of "dog work" involved in solving differential equations is about the same either way. The chief advantage in using the Laplace method to solve a differential equation is that both the natural and forced responses are found simultaneously, and the initial conditions are inserted, all in a single unified process.

Laplace methods are not, however, introduced primarily as a means of solving differential equations. Their main use is to provide a framework within which we can discuss and represent linear systems.[3,4] Laplace transforms enable us to model complicated linear systems as interconnections of functional blocks which can be manipulated by an algebra of block diagrams. Each block in such a representation consists

The Laplace Transform   193

of an s-domain <u>transfer function</u> which relates the input and output variables associated with that block through a function of the Laplace variable s. The transformed output of such a block is the transformed input variable multiplied by the transfer function: it is not necessary to solve any differential equations. Chapter 7 is an introductory treatment of systems analysis in terms of transfer functions.

A final reason for introducing the Laplace transform is that s-domain functions contain a great deal of information about the performance of a system, information which can be gleaned without solving for the actual output variables of the system. We can, for example, often tell whether a system will have a sluggish, overdamped response or an oscillatory response by inspecting its s-domain transfer function.[5]

A historical note. Laplace transforms are named after the French mathematician Pierre Simon Laplace (1749-1827), but they were used for solving differential equations by Leonhard Euler (1707-1783) long before Laplace was born.[6]

5.1 THE LAPLACE TRANSFORM DEFINED

The direct Laplace transform of a time function f(t) is given by

$$L\{f(t)\} = F(s) = \int_0^\infty f(t)e^{-st}dt \qquad (5-1)$$

where upper-case letters represent the Laplace transforms of time functions; thus Y(s) is the transform of y(t). The alternate notation, L{ }, for the Laplace transform of the bracketed quantity is also used in the following chapters. Because the lower limit of integration in Eq. (5-1) is 0, values of the function for t<0 are ignored. In some textbooks this information is emphasized by multiplying f(t) by a unit step function u(t). We avoid the need for this cumbersome notation by restricting our discussions to problems which "start" at t=0. The two-sided Laplace transform takes negative time into account, but it is seldom needed in dynamic systems analysis because the operation of such systems can always be assumed to start at a reference time which can be designated as t=0.

In Eq. (5-1) the complex frequency variable s has the form

$$s = \sigma + j\omega \qquad (5-2)$$

where σ and ω are real constants.

Equation (5-1) contains an improper integral (the upper limit is infinite), and for the equation to give a useful result the integral must converge to a definite functional value. The condition for convergence is that f(t) be piecewise continuous and of exponential order.[7,8] A function is piecewise continuous if it is defined everywhere and has both right- and left-hand limits at points of discontinuity. Figure 5-1 shows a piecewise-continuous function.

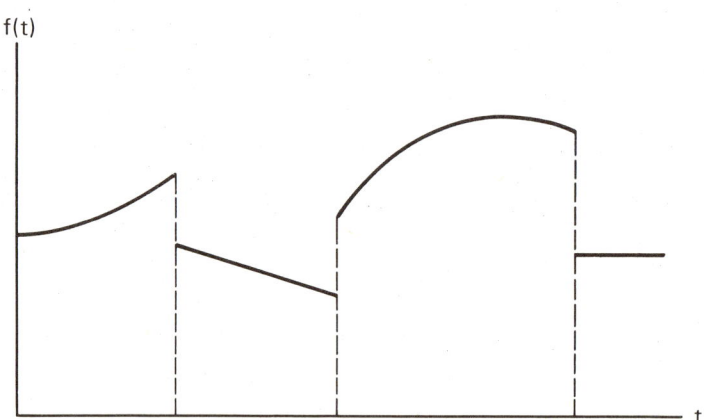

Figure 5.1. Piecewise-continuous functions such as this one are Laplace transformable.

A function f(t) is of exponential order if it goes to infinity "slower" than exp(-σt), where σ is a positive constant called the radius of convergence. All of the time functions needed in dynamic analysis are piecewise continuous and of exponential order and hence are Laplace transformable.

The Laplace Transform

Finally, it should be evident that Eq. (5-1) yields a function of s because time is "integrated out".

Example 5-1. Show that the function $f(t)=t^2$ is of exponential order.

Solution. It can be shown that a function is of exponential order if a constant σ exists such that the limit L equals zero where

$$L = \lim_{t \to \infty} \{e^{-\sigma t} |f(t)|\} \tag{5-3}$$

In this example we need only show that

$$L = \lim_{t \to \infty} \{t^2 e^{-\sigma t}\} = 0 \tag{5-4}$$

Rearranging this expression yields

$$L = \lim_{t \to \infty} \frac{t^2}{e^{\sigma t}} \tag{5-5}$$

Equation (5-5) can be evaluated by L'Hopital's rule from elementary calculus as

$$L = \lim_{t \to \infty} \frac{t^2}{e^{\sigma t}} = \lim_{t \to \infty} \frac{2t}{\sigma e^{\sigma t}} = \lim_{t \to \infty} \frac{2}{\sigma^2 e^{\sigma t}} = 0 \tag{5-6}$$

and the result is proved.

The inverse Laplace transform, which can be used to transform s-domain functions back to the time domain, is given by

$$f(t) = L^{-1}\{F(s)\} = \int_{\sigma_1-j\infty}^{\sigma_1+j\infty} F(s) e^{st} ds \tag{5-7}$$

where the notation $L^{-1}\{\ \}$ identifies the inverse Laplace transform of the bracketed quantity, and σ_1 is a real constant which satisfies the condition given above for the exponential order of f(t). Equation (5-7) involves <u>contour integration</u>

in the complex plane, a topic beyond the scope of this book. Happily, there is another method for finding inverse transforms which obviates the need for contour integration. One can use Eq. (5-1) to develop a table of Laplace-transform pairs for frequently encountered functions, and then use partial-fraction expansions to reduce complicated s-domain functions to sums of terms which can be found in the table.

5.2 LAPLACE TRANSFORMS OF USEFUL FUNCTIONS

The most commonly encountered functions in systems analysis are step functions, exponentials, sinusoids, polynomials, and impulses. We now proceed to calculate the transforms of these functions by using the defining integral, Eq. (5-1).

The Step Function

The Laplace transform of the step function $Au(t)$ is given by

$$L\{Au(t)\} = \int_0^\infty (1)Ae^{-st}dt \qquad (5\text{-}8)$$

Evaluating Eq. (5-8) gives

$$L\{Au(t)\} = -\left.\frac{Ae^{-st}}{s}\right]_0^\infty = \frac{A}{s} \qquad (5\text{-}9)$$

provided $\sigma>0$ so that $\exp(-st)\to 0$ as $t\to\infty$. (Recall that $s=\sigma+j\omega$). In future derivations we will omit the condition on σ needed for convergence; this information is useful only if one plans to use the complex inversion integral, Eq. (5-7).

Example 5-2. Find the Laplace transform of the delayed step function $Au(t-T)$.

The Laplace Transform

Solution. Using the defining integral,

$$L\{Au(t-T)\} = A \int_0^\infty u(t-T)e^{-st} dt \qquad (5\text{-}10)$$

Because the step function is 0 until t=T, we can change the lower limit of integration from 0 to T, obtaining

$$L\{Au(t-T)\} = A \int_T^\infty (1)e^{-st} dt = \left. \frac{-A}{s}e^{-st} \right|_T^\infty \qquad (5\text{-}11)$$

or, finally,

$$L\{Au(t-T)\} = \frac{Ae^{-sT}}{s} \qquad (5\text{-}12)$$

The Decaying Exponential

The function f(t)=exp(-at) has the Laplace transform

$$L\{e^{-at}\} = \int_0^\infty e^{-at} e^{-st} dt \qquad (5\text{-}13)$$

$$= \int_0^\infty e^{-t(s+a)} dt = \left. \frac{-1}{s+a} e^{-t(s+a)} \right|_0^\infty \qquad (5\text{-}14)$$

or

$$L\{e^{-at}\} = \frac{1}{s+a} \qquad (5\text{-}15)$$

Example 5-3. Find the Laplace transform of the composite function $f(t) = 5 + e^{-3t} + 3e^{2t}$.

Solution. Because we take Laplace transforms for t⩾0, we assume that f(t) is multiplied by a unit step function or that the constant term 5 is actually 5u(t). Thus

$$L\{f(t)\} = L\{5u(t)\} + L\{e^{-3t}\} + L\{3e^{2t}\} \qquad (5\text{-}16)$$

Equations (5-9) and (5-15) apply and the result is found by inspection:

$$L\{f(t)\} = F(s) = \frac{5}{s} + \frac{1}{s+3} + \frac{3}{s-2} \qquad (5\text{-}17)$$

Sinusoidal Functions

The Laplace transform of $f(t)=\cos(\omega t)$ is given by

$$L\{\cos(\omega t)\} = \int_0^\infty \cos(\omega t) e^{-st} dt \qquad (5\text{-}18)$$

Writing the cosine in terms of exponentials gives

$$\cos(\omega t) = \frac{e^{j\omega t} + e^{-j\omega t}}{2} \qquad (5\text{-}19)$$

Substituting Eq. (5-19) into Eq. (5-18) yields

$$L\{\cos(\omega t)\} = \frac{1}{2} \int_0^\infty (e^{j\omega t} + e^{-j\omega t}) e^{-st} dt \qquad (5\text{-}20)$$

$$= \frac{1}{2} \int_0^\infty e^{(j\omega-s)t} dt + \frac{1}{2} \int_0^\infty e^{-(j\omega+s)t} dt$$

$$(5\text{-}21)$$

- from which

$$L\{\cos(\omega t)\} = \frac{1}{2} \left\{ \frac{-1}{j\omega-s} + \frac{1}{j\omega+s} \right\} \qquad (5\text{-}22)$$

and, after some rearrangement,

$$L\{\cos(\omega t)\} = \frac{s}{s^2+\omega^2} \qquad (5\text{-}23)$$

A similar derivation shows that the Laplace transform of $\sin(\omega t)$ is

$$L\{\sin(\omega t)\} = \frac{\omega}{s^2+\omega^2} \qquad (5\text{-}24)$$

The Laplace Transform

Example 5-4. Find the Laplace transform of a general sinusoid with a phase angle, $f(t)=A\sin(\omega t-\emptyset)$.

Solution. Use the trigonometric identity

$$\sin(x-y)=\sin(x)\cos(y)-\cos(x)\sin(y) \qquad (5-25)$$

to obtain

$$f(t) = A\{\sin(\omega t)\cos(\emptyset) - \cos(\omega t)\sin(\emptyset)\} \qquad (5-26)$$

The expression for the Laplace transform is then

$$F(s) = A\cos(\emptyset)L\{\sin(\omega t)\} - A\sin(\emptyset)L\{\cos(\omega t)\} \qquad (5-27)$$

Applying Eqns. (5-23) and (5-24) gives the result

$$F(s) = \frac{\omega A\cos(\emptyset)}{s^2+\omega^2} - \frac{sA\sin(\emptyset)}{s^2+\omega^2} \qquad (5-28)$$

Polynominals

The Laplace transform of the first-order polynomial $f(t)=t$ is given by

$$L\{t\} = \int_0^\infty te^{-st}dt \qquad (5-29)$$

This integral can be evaluated using the formula for integration by parts

$$\int_a^b udv = uv\Big]_a^b - \int_a^b vdu \qquad (5-30)$$

Let $u=t$ and $dv=\exp(-st)dt$; then $du=dt$ and $v=-\exp(-st)/s$. Hence we have

$$\int_0^\infty te^{-st}dt = \frac{-te^{-st}}{s}\Big]_0^\infty + \frac{1}{s}\int_0^\infty e^{-st}dt \qquad (5-31)$$

and, using Eq. (5-8),

$$L\{t\} = \frac{1}{s^2} \qquad (5-32)$$

By a similar procedure it can be shown that

$$L\{t^n\} = \frac{n}{s}L\{t^{n-1}\} \tag{5-33}$$

where n is an integer. Equation (5-33) can be used iteratively with Eq. (5-32) to transform polynomials of any order.

Example 5-5. Find G(s), the Laplace transform of the polynomial

$$g(t) = 5t^3 + 3t^2 - 8t + 4 \tag{5-34}$$

Solution. Using Eq. (5-33), the transform of t^2 is

$$L\{t^2\} = \frac{2L\{t\}}{s} = \frac{2}{s^3} \tag{5-35}$$

Similarly, the transform of t^3 is

$$L\{t^3\} = \frac{3L\{t^2\}}{s} = \frac{6}{s^4} \tag{5-36}$$

Using these results gives

$$G(s) = \frac{30}{s^4} + \frac{6}{s^3} - \frac{8}{s^2} + \frac{4}{s^1} \tag{5-37}$$

The Impulse Function

The unit impulse function $\delta(t)$ has the Laplace transform

$$L\{\delta(t)\} = \int_0^\infty \delta(t) e^{-st} dt \tag{5-38}$$

Using the sampling property of the impulse function, Eq. (2-66), the integral in Eq. (5-38) is nonzero only at t=0 and

$$L\{\delta(t)\} = 1 \tag{5-39}$$

The unit impulse function thus assumes the role of unity in the transform domain; it is the time function whose Laplace transform is 1.

The Laplace Transform

Example 5-6. Find the Laplace transform of the "gust" function, $f(t)=t\exp(-at)$.

Solution. Using Eq. (5-1), we have

$$L\{te^{-at}\} = \int_0^\infty te^{-at}e^{-st}dt \qquad (5\text{-}40)$$

Integrate by parts letting $u=t$ and $dv=\exp\{-(s+a)t\}dt$. This results in

$$L\{te^{-at}\} = \left\{\frac{-t}{s+a}\right\}e^{-(s+a)t}\Big|_0^\infty + \frac{1}{s+a}\int_0^\infty e^{-(s+a)t}dt \qquad (5\text{-}41)$$

$$= -\frac{1}{(s+a)^2}e^{-(s+a)t}\Big|_0^\infty \qquad (5\text{-}42)$$

and, finally,

$$L\{te^{-at}\} = \frac{1}{(s+a)^2} \qquad (5\text{-}43)$$

The transforms of more complicated functions and of differential equations can be found readily after we have studied some further properties of the Laplace transform. Appendix D contains many of the transform pairs needed for dynamic systems work.

5.3 PROPERTIES OF THE LAPLACE TRANSFORM

The following properties of the Laplace transform extend its utility as a tool for analyzing dynamic systems. Where proofs are trivial or extremely complex they are omitted. The interested reader can consult one of the references listed at the end of the chapter.[4,7,9] Appendix E summarizes the properties for the reader who prefers to skip this section.

Linearity

Because integration is a linear operation and the Laplace transform is defined by an integral, we have

$$L\{Af(t)+Bg(t)\} = AL\{f(t)\} + BL\{g(t)\} \qquad (5\text{-}44)$$

and

$$L\{Af(t)+Bg(t)\} = AF(s) + BG(s) \qquad (5\text{-}45)$$

where A and B are constants. This property applies for any number of terms.

Translation in Time

If $f(t)$ transforms to $F(s)$ and if T is a positive constant, then

$$L\{f(t-T)u(t-T)\} = e^{-sT}F(s) \qquad (5\text{-}46)$$

Equation (5-46) says that delay in the time domain by T units corresponds to multiplication by $\exp(-sT)$ in the frequency domain. The unit step function $u(t-T)$ in Eq. (5-46) is needed to insure that the function is 0 for $t<T$. To prove Eq. (5-46) we write

$$L\{f(t-T)u(t-T)\} = \int_0^\infty f(t-T)u(t-T)e^{-st}dt \qquad (5\text{-}47)$$

$$= \int_T^\infty f(t-T)e^{-st}dt \qquad (5\text{-}48)$$

Let $p=t-T$ so that Eq. (5-48) becomes

$$L\{f(t-T)u(t-T)\} = \int_{p=0}^\infty f(p)e^{-s(p+T)}dt \qquad (5\text{-}49)$$

$$= e^{-sT}\int_{p=0}^\infty f(p)e^{-sp}dp \qquad (5\text{-}50)$$

The Laplace Transform

But the integral in Eq. (5-50) is the defining integral for the Laplace transform with t replaced by p. For this reason, the integral gives F(s) and Eq. (5-46) is proved.

Translation in Frequency

The frequency translation theorem states that

$$L\{e^{at}f(t)\} = F(s-a) \qquad (5\text{-}51)$$

where $L\{f(t)\}=F(s)$. Multiplication by exp(at) in the time domain thus corresponds to a shift of -a units in the s-domain. The quantity a may be real or complex. Equation (5-51) is proved by writing

$$L\{e^{at}f(t)\} = \int_0^\infty e^{at}f(t)e^{-st}dt \qquad (5\text{-}52)$$

$$= \int_0^\infty f(t)e^{-(s-a)t}dt \qquad (5\text{-}53)$$

But Eq. (5-53) is identical to Eq. (5-1) with s replaced by s-a; therefore, it results in F(s-a) and Eq. (5-51) is proved.

Example 5-7. Use frequency translation to derive an expression for the Laplace transform of the damped sinusoid

$$f(t) = Ae^{-\sigma t}\cos(\omega_d t) \qquad (5\text{-}54)$$

<u>Solution</u>. The transform of the pure cosine is

$$L\{A\cos(\omega_d t)\} = \frac{As}{s^2+\omega_d^2} \qquad (5\text{-}55)$$

To take the exponential multiplier into account we use Eq. (5-51), obtaining

$$L\{Ae^{-\sigma t}\cos(\omega_d t)\} = \frac{A(s+\sigma)}{(s+\sigma)^2+\omega_d^2} \qquad (5\text{-}56)$$

Real Differentiation

The key to using Laplace transforms to solve differential equations is to develop expressions for the derivatives of unknown functions in terms of the transforms of the undifferentiated functions. Consider, for example, a function y(t) with derivative $\dot{y}(t)$. According to Eq. (5-1) the transform of $\dot{y}(t)$ is given by

$$L\{\dot{y}(t)\} = \int_0^\infty \dot{y}(t)e^{-st}dt \qquad (5\text{-}57)$$

This expression can be evaluated using integration by parts. Let u=exp(-st) and dv=$\dot{y}$(t)dt; then v=y(t), du=-sexp(-st)dt, and Eq. (5-57) becomes

$$L\{\dot{y}(t)\} = y(t)e^{-st}\Big]_0^\infty + s\int_0^\infty y(t)e^{-st}dt \qquad (5\text{-}58)$$

$$= -y(0^+) + sL\{y(t)\} \qquad (5\text{-}59)$$

or

$$L\{\dot{y}(t)\} = sY(s) - y(0^+) \qquad (5\text{-}60)$$

In words, Eq. (5-60) states that the Laplace transform of the derivative of a function is s times the transform of the undifferentiated function minus the function evaluated at t=0$^+$ (The 0$^+$ notation tells us to take the limit from the right, or slightly after 0 if y(t) is discontinuous at t=0.) By a similar analysis it can be shown that the Laplace transform of the second derivative is

$$L\{\ddot{y}(t)\} = s^2 Y(s) - sy(0^+) - \dot{y}(0^+) \qquad (5\text{-}61)$$

These results are extended to the nth-order derivative by the relationship

$$L\{y(t)^{(n)}\} = s^n Y(s) - s^{n-1}y(0^+) - s^{n-2}\dot{y}(0^+) - \ldots - y(0^+)^{(n-1)}$$

$$(5\text{-}62)$$

The Laplace Transform

The transform of a derivative includes the initial conditions. This is why Laplace transform solutions of differential equations automatically take into account the initial conditions. Equation (5-62) is the property which makes Laplace transforms useful in analyzing dynamic systems; it enables us to transform time-domain differential equations into s-domain algebraic equations. A consequence of Eq. (5-60) is that multiplication by s in the frequency domain corresponds to differentiation in the time domain. Similarly, division by s corresponds to integration in the time domain.

Complex Differentiation

Multiplication by t in the time domain implies differentiation with respect to s in the frequency domain, or

$$L\{tf(t)\} = \frac{-dF(s)}{ds} \qquad (5-63)$$

This relationship, which is proved by differentiating Eq. (5-1), is useful in evaluating the transforms of complicated functions.

Example 5-8. Use the complex differentiation relationship to find the transform of $f(t) = t\sin(\omega t)$.

Solution. Starting with $L\{\sin(\omega t)\} = \omega/(s^2 + \omega^2)$, we apply Eq. (5-63) to obtain

$$L\{t\sin(\omega t)\} = -\frac{d}{ds}\left\{\frac{\omega}{s^2 + \omega^2}\right\} \qquad (5-64)$$

$$= \frac{2\omega s}{(s^2 + \omega^2)^2} \qquad (5-65)$$

This result would have been difficult to obtain by a direct application of Eq. (5-1). (Try it!)

Real Integration

If $g(t)$ is defined by the integral

$$g(t) = \int_0^t f(t)dt + g(0^+) \tag{5-66}$$

then $L\{g(t)\}=G(s)$ is given by

$$G(s) = \frac{F(s)}{s} + \frac{g(0^+)}{s} \tag{5-67}$$

To prove Eq. (5-67) we write

$$G(s) = \int_0^\infty \left\{\int_0^t f(t)dt\right\} e^{-st}dt + \int_0^\infty g(0^+)e^{-st}dt \tag{5-68}$$

The first integral on the right can be evaluated by parts if we define

$$u = \int_0^t f(t)dt \qquad dv = e^{-st}dt \tag{5-69}$$

from which

$$du = f(t) \qquad v = \frac{-e^{-st}}{s} \tag{5-70}$$

Equation (5-68) now becomes

$$G(s) = \frac{-e^{-st}}{s}\int_0^t f(t)dt \Big]_0^\infty + \frac{1}{s}\int_0^\infty f(t)e^{-st} + \frac{g(0^+)}{s} \tag{5-71}$$

$$= \frac{1}{s}F(s) + \frac{g(0^+)}{s} \tag{5-72}$$

where we have used the result that $g(0^+)$ is a constant which can be interpreted as a unit step function. The real integration property, which can be extended to multiple integrals, is useful for transforming integral equations in the time domain to algebraic equations in the s-domain.

The Laplace Transform

The Final-Value Theorem

If a function $f(t)$ and its first derivative $\dot{f}(t)$ are Laplace transformable, and if the limit of $f(t)$ as $t\to\infty$ exists, then

$$\lim_{s\to 0}\{sF(s)\} = \lim_{t\to\infty}\{f(t)\} = f(\infty) \tag{5-73}$$

where $L\{f(t)\}=F(s)$. This result is useful for finding the steady-state values of variables in a dynamic system without solving the differential equations which describe the system.

The Initial-Value Theorem

If $f(t)$ and $\dot{f}(t)$ are Laplace transformable, and if the limit of $sF(s)$ as $s\to\infty$ exists, then

$$\lim_{s\to\infty}\{sF(s)\} = \lim_{t\to 0}\{f(t)\} = f(0) \tag{5-74}$$

Example 5-9. It can be shown that for a piecewise continuous function $f(t)$ which is of exponential order with radius of convergence σ_o the relationship

$$|L\{f(t)\}| < \frac{M}{s-\sigma_o} \tag{5-75}$$

holds for all values of $s > \sigma_o$. M is a constant. And as a consequence, $L\{f(t)\} \to 0$ as $s\to\infty$. We see the evidence of this result by inspecting the Laplace transforms of the functions in Appendix D and observing that they all approach 0 as $s\to\infty$ -- except for the impulse function which is not piecewise continuous and of exponential order. To prove the initial-value theorem,

$$\lim_{t\to 0}\{f(t)\} = \lim_{s\to\infty}\{sF(s)\} \tag{5-76}$$

start with the Laplace transform of the derivative

$$L\{\dot{f}(t)\} = sF(s) - f(0^+) \tag{5-77}$$

and take the limit as $s \to \infty$:

$$\lim_{s \to \infty} L\{\dot{f}(t)\} = \lim_{s \to \infty}\{sF(s)\} - f(0^+) \tag{5-78}$$

The left-hand side of this expression is the limit of the Laplace transform of a function as $s \to \infty$ and, according to Eq. (5-75), must be 0. Rewriting $f(0^+)$ as

$$f(0^+) = \lim_{t \to 0}\{f(t)\} \tag{5-79}$$

completes the proof.

Example 5-10. Find the initial value $f(0^+)$ of the function

$$F(s) = \frac{s^2 + 3s + 4}{5s^3 + 2s^2 + 8s} \tag{5-80}$$

Solution. Use the initial-value theorem, Eq. (5-76), to obtain

$$f(0^+) = \lim_{t \to 0}\{f(t)\} = \lim_{s \to \infty}\{sF(s)\} \tag{5-81}$$

$$= \lim_{s \to \infty} \frac{s(s^2+3s+4)}{5s^3+2s^2+8s} \tag{5-82}$$

and $f(0^+) = 1/5$.

Example 5-11. Find the final value $f(\infty)$ for the function of Example 5-10.

Solution. Use the final-value theorem, Eq. (5-73):

$$\lim_{t \to \infty}\{f(t)\} = \lim_{s \to 0}\{sF(s)\} = \left.\frac{s(s^2+3s+4)}{5s^3+2s^2+8s}\right|_{s=0} \tag{5-83}$$

$$= \left.\frac{s^2+3s+4}{5s^2+2s+8}\right|_{s=0} \tag{5-84}$$

The Laplace Transform

and

$$\lim_{t \to \infty} \{f(t)\} = 1/2 \tag{5-85}$$

Laplace Transforms of Linear Differential Equations [10,11]

If we apply the definition of the Laplace transform to the second-order constant-coefficient differential equation

$$\ddot{y} + a_1 \dot{y} + a_0 y = f(t) \tag{5-86}$$

we obtain

$$\int_0^\infty \{\ddot{y} + a_1 \dot{y} + a_0 y\} e^{-st} dt = \int_0^\infty f(t) e^{-st} dt \tag{5-87}$$

Taking transforms on the left using Eqns. (5-60) and (5-61) we obtain

$$s^2 Y(s) - s y(0^+) - \dot{y}(0^+) + a_1 \{s Y(s) - y(0^+)\} + a_0 Y(s) = F(s) \tag{5-88}$$

Solving Eq. (5-88) for $Y(s)$ yields

$$Y(s) = \frac{F(s) + y(0^+)(s + a_1) + \dot{y}(0^+)}{s^2 + a_1 s + a_0} \tag{5-89}$$

Equation (5-89) is the s-domain solution of the differential equation. The time-domain solution is found by taking the inverse Laplace transform of Eq. (5-89). This procedure can be extended to constant-coefficient differential equations of any order. (Nonlinear and/or time-varying differential equations cannot be solved by Laplace methods.)

Example 5-12. Solve the differential equation

$$\dot{y} + ay = 0 \tag{5-90}$$

with $y(0) = b$ using Laplace methods.

Solution. Transforming the given differential equation results in

$$sY(s) - y(0^+) + aY(s) = 0 \tag{5-91}$$

Using $y(0^+)=b$ and solving for $Y(s)$ yields

$$Y(s) = \frac{b}{s+a} \qquad (5\text{-}92)$$

Equation (5-92) recalls Eq. (5-15), the transform of an exponential function, and the solution $y(t)$, which is the inverse transform of Eq. (5-92), is

$$y(t) = be^{-at} \qquad (5\text{-}93)$$

This solution is good only for $t \geqslant 0$.

Example 5-13. Find a differential equation for which the function

$$g(t) = 5e^{-3t} - 4e^{-2t} \qquad (5\text{-}94)$$

is a possible solution.

Solution. Transforming $g(t)$ yields

$$G(s) = \frac{5}{s+3} - \frac{4}{s+2} \qquad (5\text{-}95)$$

$$= \frac{5s+10-4s-12}{(s+3)(s+2)} \qquad (5\text{-}96)$$

and

$$G(s) = \frac{s-2}{s^2+5s+6} \qquad (5\text{-}97)$$

Comparing Eqns. (5-97) and (5-89) reveals that $a_1=5$, $a_0=6$, $g(0^+)=1$, and $\dot{g}(0^+)=-7$. The original differential equation is then

$$\ddot{g} + 5\dot{g} + 6g = 0 \qquad (5\text{-}98)$$

with $g(0)=1$ and $\dot{g}(0)=-7$. (Henceforth we will omit the "+" superscript on initial conditions if there is no danger of ambiguity.)

Example 5-14. Derive an expression for the s-domain response of a general second-order system to a step input;

The Laplace Transform

that is, solve for $Y(s)$ if $y(t)$ satisfies the differential equation

$$\ddot{y} + 2\zeta\omega_n \dot{y} + \omega_n^2 y = Au(t) \tag{5-99}$$

Assume zero initial conditions.

Solution. Laplace transform the given equation term by term to obtain

$$s^2 Y(s) + 2\zeta\omega_n s Y(s) + \omega_n^2 Y(s) = \frac{A}{s} \tag{5-100}$$

The solution for $Y(s)$ is

$$Y(s) = \frac{A}{s(s^2 + 2\zeta\omega_n s + \omega_n^2)} \tag{5-101}$$

5.4 THE INVERSE TRANSFORM: PARTIAL FRACTIONS

The ease with which we can use Laplace transforms depends on how efficiently we can find direct and inverse transforms. While a table of Laplace transform pairs is useful, a table could not possibly list all of the inverse transforms one needs for systems work. This section shows how to find inverse Laplace transforms by the method of partial fractions.[12,13]

Rational Functions

Fortunately, when we transform a linear differential equation into an algebraic equation and solve for the dependent variable, the result is a rational function in s. (A rational function is the ratio of two polynomials.) The general form, then, of the s-domain functions for which we seek inverse transforms is

$$F(s) = \frac{a_0 + a_1 s + a_2 s^2 + \ldots + a_m s^m}{b_0 + b_1 s + b_2 s^2 + \ldots + b_n s^n} \tag{5-102}$$

where the a's and b's are real constants. In come cases m=n, but the most frequently encountered situation has n>m. When n⩽m Eq. (5-102) is called an improper rational function. We can always convert an improper rational function into a proper rational function plus a polynomial in s by long division. For example, the improper rational function

$$F(s) = \frac{s^2+4s+6}{s^2+2s+4} \tag{5-103}$$

becomes

$$F(s) = 1 + \frac{2s+2}{s^2+2s+4} \tag{5-104}$$

If we take the inverse Laplace transform of Eq. (5-104) the s-domain term 1 becomes the impulse function $\delta(t)$ in the time domain. It can be shown that the inverse transform of $G(s)=s$ is a mathematical oddity called a "doublet," which is a pair of impulses, both occuring at t=0, but with one positive and the other negative. Higher powers of s transform into "triplets" and other higher-order impulses. As these bizarre functions rarely occur in real-world dynamic systems, we seldom need to treat functions like Eq. (5-102) with m⩾n. For this reason we will consider only the problem of finding inverse Laplace transforms of proper rational functions.

Partial Fractions

Because a polynomial can always be expressed as the product of its roots, any rational function can be written as

$$F(s) = \frac{N(s)}{(s-p_0)(s-p_1)\ldots(s-p_n)} \tag{5-105}$$

where $N(s)$ is a numerator polynomial and the p_i are the roots of the denominator polynomial. Values of s which make the denominator of Eq. (5-105) go to zero are called the poles of $F(s)$; values which make the numerator become zero are called the zeros of $F(s)$. We shall have more to say of poles and zeros later.

The Laplace Transform

The partial-fraction theorem from algebra tells us that Eq. (5-105) can always be expressed as

$$F(s) = \frac{A_0}{s-p_0} + \frac{A_1}{s-p_1} + \cdots \frac{A_n}{s-p_n} \quad (5\text{-}106)$$

where the A_i are constants called <u>residues</u> which may be complex. If any of the poles p_i are repeated, the form of Eq. (5-106) is slightly different. The technique for transforming Eq. (5-105) into Eq. (5-106) is called a <u>partial-fraction expansion</u>. Once we have achieved Eq. (5-106), the remainder of the process is easy. We just take inverse Laplace transforms term by term on the right-hand side using Eq. (5-15). The result is

$$f(t) = A_0 e^{p_0 t} + A_1 e^{p_1 t} + \cdots + A_n e^{p_n t} \quad (5\text{-}107)$$

Equation (5-107) is awkward if any of the A_i are complex. In this case we can combine some of the exponentials into real sines and cosines, eliminating the complex constants.

<u>The general method</u>. The general approach to expanding a function by partial fractions is to write the expanded form with the residues left in symbolic form. Then find a common denominator and equate the coefficients of like powers of s in the numerator to find the unknown residues. The procedure is illustrated below by several examples which show how to handle (a) non-repeated or simple real poles, (b) repeated real poles, (c) imaginary poles, and (d) complex-conjugate poles.

<u>Example 5-15</u>: <u>simple real poles</u>. Expand the function

$$F(s) = \frac{2}{s(s+5)} \quad (5\text{-}108)$$

by partial fractions and find its inverse Laplace transform $f(t)$.

<u>Solution</u>. According to Eq. (5-106), Eq. (5-108) can be written

$$F(s) = \frac{A_1}{s} + \frac{A_2}{(s+5)} \quad (5\text{-}109)$$

Finding a common demoninator on the right and replacing $F(s)$ by its functional value yields

$$\frac{2}{s(s+5)} = \frac{A_1(s+5)+A_2 s}{s(s+5)} \tag{5-110}$$

or

$$\frac{2}{s(s+5)} = \frac{s(A_1+A_2)+5A_1}{s(s+5)} \tag{5-111}$$

Because the right- and left-hand sides of Eq. (5-111) must be the same, the coefficients of like powers of s in the numerators of both sides must be equal. Thus

$$(0)s + 2 = (A_1+A_2)s + 5A_1 \tag{5-112}$$

Equation (5-112) leads to the simultaneous equations $A_1+A_2=0$ and $5A_1=2$ from which $A_1=2/5$ and $A_2=-2/5$. (The partial-fraction method always results in a set of simultaneous algebraic equations which must be solved for the residues.) Substituting the residues into Eq. (5-109) yields

$$F(s) = \frac{2}{5}\frac{1}{s} - \frac{2}{5(s+5)} \tag{5-113}$$

and $f(t)$ is

$$f(t) = \frac{2}{5}u(t) - \frac{2}{5}e^{-5t} \tag{5-114}$$

Example 5-16: repeated real poles. Find the inverse Laplace transform of the function

$$F(s) = \frac{20(s+2)}{s(s+5)^2} \tag{5-115}$$

Solution. If a real root, say p_i, has multiplicity r, so that the factor $(s-p_1)^r$ appears in the denominator of $F(s)$, then the partial-fraction expansion must contain a sequence of terms of the form

$$\frac{A_{i1}}{(s-p_i)^1} + \frac{A_{i2}}{(s-p_i)^2} + \frac{A_{i3}}{(s-p_i)^3} + \cdots \frac{A_{ir}}{(s-p_i)^r} \tag{5-116}$$

The Laplace Transform

In this example we write

$$F(s) = \frac{20(s+2)}{s(s+5)^2} = \frac{A_1}{s} + \frac{A_2}{s+5} + \frac{A_3}{(s+5)^2} \qquad (5\text{-}117)$$

Proceed as in the previous example by finding a common denominator on the right. The result is

$$\frac{20(s+2)}{s(s+5)^2} = \frac{A_1(s+5)^2 + A_2 s(s+5) + A_3 s}{s(s+5)^2} \qquad (5\text{-}118)$$

Equating numerators leads to

$$(0)s^2 + 20s + 40 = A_1(s^2+10s+25) + A_2(s^2+5s) + A_3 s \qquad (5\text{-}119)$$

$$= s^2(A_1+A_2) + s(10A_1+5A_2+A_3) + 25A_1 \qquad (5\text{-}120)$$

Equating the coefficients of like powers of s leads to the simultaneous equations

$$0 = A_1 + A_2 \qquad (5\text{-}121)$$

$$20 = 10A_1 + 5A_2 + A_3 \qquad (5\text{-}122)$$

and

$$40 = 25A_1 \qquad (5\text{-}123)$$

These are solved for $A_1=8/5$, $A_2=-8/5$, and $A_3=12$. The partial-fraction expansion is then

$$F(s) = \frac{8/5}{s} - \frac{8/5}{s+5} + \frac{12}{(s+5)^2} \qquad (5\text{-}124)$$

Using Entries 2, 5, and 6 of Appendix D, the inverse transform is

$$f(t) = \tfrac{8}{5}u(t) - \tfrac{8}{5}e^{-5t} + 12te^{-5t} \qquad (5\text{-}125)$$

Example 5-17: imaginary poles. Find $f(t)$ if

$$F(s) = \frac{5(s+2)}{(s+1)(s^2+4)} \qquad (5\text{-}126)$$

Solution. If $F(s)$ contains a pair of imaginary poles, or a denominator term such as $(s^2+\omega^2)$, we must insert a partial-fraction term with the form

$$\frac{A_1 s + A_2}{s^2 + \omega^2} \qquad (5\text{-}127)$$

after which the procedure follows that of the other examples. In this instance we write

$$F(s) = \frac{5(s+2)}{(s+1)(s^2+4)} = \frac{A_1}{s+1} + \frac{A_2 s + A_3}{s^2+4} \qquad (5\text{-}128)$$

$$= \frac{A_1(s^2+4) + (A_2 s + A_3)(s+1)}{(s+1)(s^2+4)} \qquad (5\text{-}129)$$

Expanding both numerators and equating coefficients of like powers of s yields

$$(0)s^2 + 5s + 10 = s^2(A_1+A_2) + s(A_3+A_2) + 4A_1 + A_3 \qquad (5\text{-}130)$$

from which the simultaneous equations are

$$0 = A_1 + A_2 \qquad (5\text{-}131)$$

$$5 = A_2 + A_3 \qquad (5\text{-}132)$$

and

$$10 = 4A_1 + A_3 \qquad (5\text{-}133)$$

These are solved to yield $A_1=1$, $A_2=-1$, and $A_3=6$. The partial-fraction expansion is therefore

$$F(s) = \frac{1}{s+1} + \frac{-s+6}{s^2+4} \qquad (5\text{-}134)$$

The Laplace Transform

or

$$F(s) = \frac{1}{s+1} - \frac{s}{s^2+4} + 3\frac{2}{s^2+4} \quad (5\text{-}135)$$

Using Entries 5, 8 and 9 of Appendix D, we have

$$f(t) = e^{-t} - \cos(2t) + 3\sin(2t) \quad (5\text{-}136)$$

Repeated imaginary poles are treated as are repeated real poles (see Problem 5-44).

Example 5-18: complex-conjugate poles. Find $f(t)$ if

$$F(s) = \frac{2}{s(s^2+s+6)} \quad (5\text{-}137)$$

Solution. The partial-fraction expansion for $F(s)$ is

$$F(s) = \frac{A_1}{s} + \frac{A_2}{s-p_1} + \frac{A_3}{s-p_2} \quad (5\text{-}138)$$

where p_1 and p_2, the factors of s^2+s+6, are found by applying the quadratic formula: $p_1, p_2 = -0.5 \pm j2.4$. Equation (5-138) can be written

$$F(s) = \frac{A_1(s-p_1)(s-p_2) + A_2 s(s-p_2) + A_3 s(s-p_1)}{s(s^2+s+6)} \quad (5\text{-}139)$$

Finding a common denominator and equating numerators yields

$$(0)s^2 + (0)s + 2 = s^2(A_1+A_2+A_3) + s(A_1-A_2 p_2 - A_3 p_1) + 6A_1 \quad (5\text{-}140)$$

The simultaneous equations are

$$0 = A_1 + A_2 + A_3 \quad (5\text{-}141)$$

$$0 = A_1 - A_2 p_2 - A_3 p_1 \quad (5\text{-}142)$$

and

$$2 = 6A_1 \quad (5\text{-}143)$$

These are solved to yield $A_1=0.33$, $A_2=0.17\exp(2.94)$, and $A_3=0.17\exp(-j2.94)$. The expression for $F(s)$ is then

$$F(s) = \frac{0.33}{s} + 0.17\left\{\frac{e^{j2.94}}{s+0.5-j2.4} + \frac{e^{-j2.94}}{s+0.5+j2.4}\right\} \qquad (5\text{-}144)$$

with the inverse transform

$$f(t) = 0.33u(t) + 0.17e^{-0.5t}\left\{e^{j(2.94+2.4t)} + e^{-j(2.94+2.4t)}\right\} \qquad (5\text{-}145)$$

This result is somewhat unwieldy as it contains complex coefficients and complex exponentials. The next paragraph treats the problem of reducing such terms to sums of real sinusoids and exponentials.

<u>Simplifying complex exponentials</u>. A partial-fraction expansion which contains complex-conjugate poles will contain a pair of terms in the inverse transform with the form

$$y_1(t) = A_1 e^{p_1 t} + A_1^* e^{p_1^* t} \qquad (5\text{-}146)$$

where the * notation denotes the complex conjugate. To put Eq. (5-146) in a more useful form, let $A_1 = C\exp(j\emptyset)$ and $A_1^* = C\exp(-j\emptyset)$. Then with $p_1 = -\sigma + j\omega$

$$y_1(t) = Ce^{j\emptyset}e^{-(\sigma-j\omega)t} + Ce^{-j\emptyset}e^{-(\sigma+j\omega)t} \qquad (5\text{-}147)$$

$$= Ce^{-\sigma t}\{e^{j(\omega t+\emptyset)} + e^{-j(\omega t+\emptyset)}\} \qquad (5\text{-}148)$$

or, using Euler's identity for the cosine

$$y_1(t) = 2Ce^{-\sigma t}\cos\{\omega t + \emptyset\} \qquad (5\text{-}149)$$

In this expression C is the magnitude of the residue of either of the poles and $\emptyset$ is the angle of the residue of the pole with the positive imaginary part, the one of the form $p_1=-\sigma+j\omega$ where ω is a positive constant. The function $y_1(t)$ is added to the time function obtained in the remainder of the partial-fraction expansion to get the complete solution.

The Laplace Transform

Using Eq. (5-149), the result for Example 5-18 is

$$f(t) = \frac{1}{3}u(t) + 0.34e^{-t/2}\cos\{2.4t + 168.5°\} \qquad (5\text{-}150)$$

where $168.5°$ is the degree equivalent of 2.94 radians.

Multiple complex poles are handled in the same manner as multiple real poles, but the algebra becomes tedious. Fortunately, multiple complex poles are rarely encountered in the analysis of practical systems.

A short cut. The following short-cut method is useful for finding the residues of non-repeated poles. Consider

$$F(s) = \frac{A_1}{s-p_1} + \frac{A_2}{s-p_2} + \ldots \frac{A_n}{s-p_n} \qquad (5\text{-}151)$$

Multiply both sides of Eq. (5-151) by $s-p_1$ to obtain

$$(s-p_1)F(s) = A_1 + A_2\frac{(s-p_1)}{(s-p_2)} + \ldots A_n\frac{(s-p_1)}{(s-p_n)} \qquad (5\text{-}152)$$

If we now let $s=p_1$ in Eq. (5-152), all terms on the right except A_1 become zero and we have

$$A_1 = (s-p_1)F(s)\Big]_{s=p_1} \qquad (5\text{-}153)$$

The residues of the other poles can be found in the same way. When applicable, this procedure saves a great deal of labor as it avoids the need to solve simultaneous equations for the residues.

Example 5-19. Use the short-cut method to find the inverse transform of

$$F(s) = \frac{5(s+1)}{s(s+2)(s+3)} = \frac{A_1}{s} + \frac{A_2}{s+2} + \frac{A_3}{s+3} \qquad (5\text{-}154)$$

Using Eq. (5-153),

$$A_1 = sF(s)\Big|_{s=0} = \frac{5(s+1)}{(s+2)(s+3)}\Big|_{s=0} = \frac{5}{6} \qquad (5\text{-}155)$$

$$A_2 = (s+2)F(s)\Big|_{s=-2} = \frac{5(s+1)}{s(s+3)}\Big|_{s=-2} = \frac{5}{2} \qquad (5\text{-}156)$$

$$A_3 = (s+3)F(s)\Big|_{s=-3} = \frac{5(s+1)}{s(s+2)}\Big|_{s=-3} = \frac{-10}{3} \qquad (5\text{-}157)$$

The inverse transform is then

$$f(t) = \frac{5}{6}u(t) + \frac{5}{2}e^{-2t} - \frac{10}{3}e^{-3t} \qquad (5\text{-}158)$$

The short-cut method can be used to find the residues of individual poles even if all of the poles in F(s) are not simple. It can also be used to find the complex residues for complex-conjugate poles.

Example 5-20. The inverse transform of the function

$$F(s) = \frac{s^2+8}{(s+2)(s^3+8s^2+14s+6)} \qquad (5\text{-}159)$$

contains a term of the form Aexp(-2t). Calculate A.

Solution. We can apply Eq. (5-153) without finding the other roots of the numerator or denominator. The result is

$$A = (s+2)F(s)\Big|_{s=-2} = \frac{s^2+8}{(s^3+8s^2+14s+6)}\Big|_{s=-2} \qquad (5\text{-}160)$$

and A=6.

When finding the inverse transforms of complicated functions, we can compute the residues of non-repeated poles by the short-cut method and use these known values to simplify solving the simultaneous equations for the other residues.

Example 5-21 illustrates this idea.

Example 5-21. Find the inverse Laplace transform of

$$F(s) = \frac{25(s+2)}{s(s+1)(s^2+16)} = \frac{A_1}{s} + \frac{A_2}{s+1} + \frac{A_3 s + A_4}{s^2+16} \tag{5-161}$$

Solution. Find a common denominator on the right and equate numerators to obtain

$$25s+50 = A_1(s+1)(s^2+16) + A_2 s(s^2+16) + s(A_3 s + A_4)(s+1) \tag{5-162}$$

or

$$25s+50 = s^3(A_1+A_2+A_3) + s^2(A_1+A_3+A_4) + s(16A_1+16A_2+A_4) + 16A_1 \tag{5-163}$$

From this expression we get the simultaneous equations

$$0 = A_1 + A_2 + A_3 \tag{5-164}$$

$$0 = A_1 + A_3 + A_4 \tag{5-165}$$

$$25 = 16A_1 + 16A_2 + A_4 \tag{5-166}$$

and

$$50 = 16A_1 \tag{5-167}$$

To simplify solving these equations we can solve for A_1 and A_2 by the short-cut method:

$$A_1 = sF(s)\Big]_{s=0} = \frac{25(2)}{(1)(16)} = \frac{25}{8} \tag{5-168}$$

$$A_2 = (s+1)F(s)\Big]_{s=-1} = \frac{25(1)}{(-1)(17)} = \frac{-25}{17} \tag{5-169}$$

Using these two values and Eq. (5-164), we have $A_3 = -225/136$, and from Eq. (5-165), $A_4 = -25/17$. Equation (5-166) was not used; hence it can be used as an independent check on the values. The inverse transform is

$$f(t) = \frac{25}{8}u(t) - \frac{25}{17}e^{-t} - \frac{225}{136}\cos(4t) - \frac{25}{68}\sin(4t) \quad (5\text{-}170)$$

An Alternate Method for Quadratic Factors

The following alternate method for handling quadratic factors in a partial fraction expansion eliminates the need for manipulating complex quantities at the expense of more algebra. Recall from Example 5-7, Eq. (5-56), that

$$\mathcal{L}\left\{e^{-\sigma t}\cos(\omega_d t)\right\} = \frac{s + \sigma}{(s+\sigma)^2 + \omega_d^2} \quad (5\text{-}171)$$

In a similar fashion we can show that

$$\mathcal{L}\left\{e^{-\sigma t}\sin(\omega_d t)\right\} = \frac{\omega_d}{(s+\sigma)^2 + \omega_d^2} \quad (5\text{-}172)$$

To exploit Eqns. (5-171) and (5-172) we observe that the function

$$F(s) = \frac{As + B}{(s+\sigma)^2 + \omega_d^2} \quad (5\text{-}173)$$

which might appear as a term in a partial-fraction expansion, can be rearranged as

$$F(s) = A\left\{\frac{s + \sigma}{(s+\sigma)^2 + \omega_d^2}\right\} + \frac{B - A\sigma}{\omega_d}\left\{\frac{\omega_d}{(s+\sigma)^2 + \omega_d^2}\right\} \quad (5\text{-}174)$$

with the inverse transform

$$f(t) = e^{-\sigma t}\left\{A\cos(\omega_d t) + \frac{B - A\sigma}{\omega_d}\sin(\omega_d t)\right\} \quad (5\text{-}175)$$

Finally, if desired, Eq. (5-175) can be converted to the amplitude-phase form

$$f(t) = De^{-\sigma t}\cos(\omega_d t - \phi) \quad (5\text{-}176)$$

where

$$D = \sqrt{A^2 + \left\{\frac{B-A\sigma}{\omega_d}\right\}^2} \qquad (5\text{-}177)$$

and

$$\phi = \tan^{-1}\left\{\frac{B-A\sigma}{A\omega_d}\right\} \qquad (5\text{-}178)$$

As an example of this approach, consider Eq. (5-137) from Example 5-18. We can write F(s) as the partial-fraction expansion

$$F(s) = \frac{2}{s(s^2+s+6)} = \frac{A}{s} + \frac{Bs+C}{s^2+s+6} \qquad (5\text{-}179)$$

Finding a common denominator and equating like powers of s in the numerators leads to the simultaneous equations A+B=0, A+C=0, and 6A=2 with solutions A=1/3, B=-1/3, and C=-1/3.

The next step is to identify σ and ω_d. We do this by completing the square in the denominator. (To complete the square in a quadratic we add and subtract the square of half the coefficient of the middle term.) Thus,

$$s^2 + s + 6 = s^2 + s + 6 + \tfrac{1}{4} - \tfrac{1}{4}$$

$$= (s+0.5)^2 + 5.75 \qquad (5\text{-}180)$$

from which $\sigma=0.5$ and $\omega_d=2.4$. Using these values along with Eqns. (5-174) and (5-179), the partial fraction expansion becomes

$$F(s) = \frac{0.33}{s} - 0.33\left\{\frac{s+0.5}{(s+0.5)^2 + (2.4)^2}\right\} -$$

$$0.069\left\{\frac{2.4}{(s+0.5)^2 + (2.4)^2}\right\} \qquad (5\text{-}181)$$

with the inverse transform

$$f(t) = 0.33u(t) - e^{-0.5t}\left\{0.33\cos(2.4t) + 0.069\sin(2.4t)\right\} \qquad (5\text{-}182)$$

The interested reader can use Eqns. (5-177) and (5-178) to verify that Eq. (5-182) transforms to Eq. (5-150), the amplitude-phase form.

We conclude this section with two examples which show how to apply Laplace transform methods to simple electrical circuits.

Example 5-22. Figure 5-2 shows an LC "tank" circuit. The transient response is initiated by a switch which connects the charged capacitor to the inductor at t=0. Derive an expression for the inductor voltage v(t) if the capacitor is initially charged to V_o volts with the polarity shown.

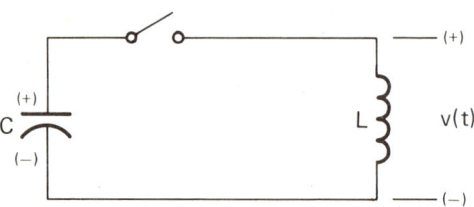

Figure 5-2. LC network for Example 5-22. When the switch is closed, the energy stored in the capacitor is endlessly transferred back and forth between the capacitor and the inductor.

Solution. Assuming a node between C and L, the KCL equation is

$$C\dot{v}(t) + \frac{1}{L}\int_0^t v(x)dx + v(0) = 0 \qquad (5\text{-}183)$$

where $v(0)=V_o$. Differentiating to remove the integral yields $C\ddot{v}(t)+v(t)/L=0$. To obtain $\dot{v}(0)$ we use $\dot{v}(t)=i(t)/C$ where $i(t)$ is the capacitor current. As the inductor resembles an open circuit at t=0, i(0)=0 and $\dot{v}(0)=0$. Laplace transforming $\ddot{v}+v/LC=0$ yields

$$s^2 V(s) - sV_o + \frac{V(s)}{LC} = 0 \qquad (5\text{-}184)$$

and

$$V(s) = \frac{V_o s}{s^2 + 1/LC}$$

Taking the inverse Laplace transform of this expression results in

$$v(t) = V_o \cos(t\sqrt{LC}) \tag{5-185}$$

The voltage oscillates eternally although the circuit is supplied with the finite energy of a stored capacitor. This phenomena is, of course, impossible and would not happen due to the nonzero resistance of the wires connecting the network elements.

Example 5-23. Figure 5-3 is a more realistic model for an RLC tank circuit. Show that the larger the value of R the more sustained are the oscillations. Also derive expressions for ω_d and σ in terms of R, L, and C.

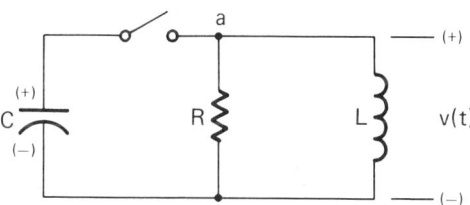

Figure 5-3. Network for Example 5-23. When the switch is closed, damped oscillations result, with the energy stored in the capacitor eventually dissipated in the resistor as heat.

Solution. Writing a KCL equation for node a and differentiating to remove the integral yields

$$C\ddot{v}(t) + \frac{\dot{v}(t)}{R} + \frac{v(t)}{L} = 0 \tag{5-186}$$

with $v(0)=V_o$ and $\dot{v}(0)=-V_o/RC$. Taking Laplace transforms and solving for $V(s)$ yields

$$C\left\{s^2 V(s) - sV_o + \frac{V_o}{RC}\right\} + \frac{1}{R}\left\{sV(s) - V_o\right\} + \frac{V(s)}{L} = 0 \tag{5-187}$$

with

$$V(s) = V_o \frac{s}{s^2 + s/RC + 1/LC} = V_o \frac{s}{(s+1/2RC)^2 + \{1/LC - 1/4R^2C^2\}} \tag{5-188}$$

Examining the denominator reveals that

$$\omega_d = \sqrt{\frac{1}{LC} - \frac{1}{4R^2C^2}} \quad \text{and} \quad \sigma = \frac{1}{2RC} \tag{5-189}$$

As R increases σ decreases and the exponential envelope decays more slowly, allowing the sinusoidal oscillations to persist longer. Also, as R increases, $\omega_d \to \omega_n = 1/\sqrt{LC}$.

5.5 s-DOMAIN MODELS OF ELECTRICAL COMPONENTS

Electrical networks lend themselves naturally to interpretation in the s-domain, especially since the initial conditions can be interpreted as conventional voltage or current sources.[16,17]

The resistor. The resistor definition,

$$v_R(t) = i_R(t)R \tag{5-190}$$

becomes

$$V_R(s) = I_R(s)R \tag{5-191}$$

when represented in the transform domain. The time-domain and frequency-domain models are identical, both R (Fig. 5-4).

The Laplace Transform

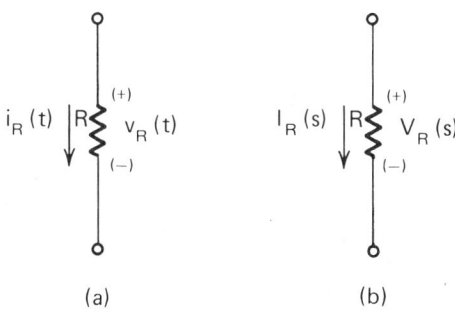

Figure 5-4. The resistor is modelled by the parameter R in either the time domain or the s-domain.

The inductor. The inductor is defined in the time domain by

$$v_L(t) = L\frac{di_L(t)}{dt} \qquad (5\text{-}192)$$

When Laplace transformed Eq. (5-192) becomes

$$V_L(s) = LsI_L(s) - Li_L(0) \qquad (5\text{-}193)$$

To handle the initial current more easily, solve Eq. (5-193) for $I_L(s)$, or

$$I_L(s) = \frac{V_L(s)}{Ls} + \frac{i_L(0)}{s} \qquad (5\text{-}194)$$

The term $i_L(0)/s$ corresponds to the transform of a step function of magnitude $i_L(0)$. Interpreting Eq. (5-194) as Kirchhoff's current law at a node yields the s-domain inductor of Fig. 5-5(b).
The current source is omitted if the initial current is known to be zero.

The capacitor. The capacitor is defined by the expression

$$i_C(t) = C\dot{v}_C(t) \qquad (5\text{-}195)$$

whose Laplace transform is

$$I_C(s) = CsV_C(s) - Cv_C(0) \qquad (5\text{-}196)$$

or, solving for $V_C(s)$,

$$V_C(s) = \frac{I_C(s)}{sC} + \frac{v_C(0)}{s} \qquad (5-197)$$

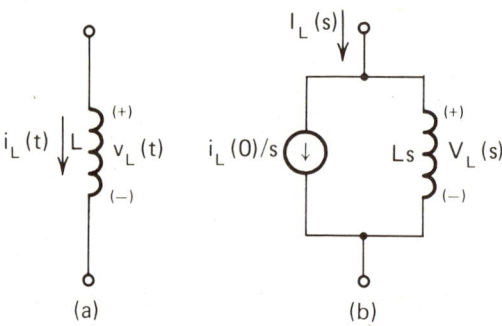

Figure 5-5. In the s-domain the inductor is modelled by the function Ls. The parallel current source accounts for the initial current through the inductor. (a) A time-domain inductor; (b) An s-domain inductor.

Interpreting Eq. (5-197) as a statement of Kirchhoff's voltage law around a loop yields the frequency-domain capacitor model of Fig. 5-6(b).

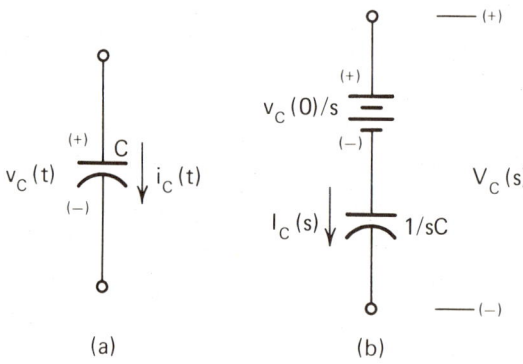

Figure 5-6. In the s-domain a capacitor is modelled by the function 1/sC in series with a voltage source which accounts for the initial capacitor voltage. (a) A time-domain capacitor; (b) An s-domain capacitor.

<u>Sources</u>. The ideal voltage and current sources which generate the time functions v(t) and i(t) are replaced by

The Laplace Transform

their Laplace transformed equivalents, V(s) and I(s). A battery with voltage E becomes E/s because the transient solution is always assumed to start at t=0; this means that the battery voltage is a step function in the time domain. Similarly, a voltage source which produces the sinusoidal voltage v(t)=Asin(ωt) in the time domain is replaced by an s-domain source which produces

$$V(s) = \frac{A\omega}{s^2+\omega^2} \qquad (5\text{-}198)$$

Finally, the sources introduced in the L and C models to account for initial conditions remain in the transformed circuit models for the entire analysis, just as if they were true sources. Table 5-1 summarizes the s-domain models for electrical components, assuming zero initial conditions.

Table 5-1.

s-Domain Models For Systems Components

Name	Symbol	s-Domain Relationship
Resistor	R	$V_R(s)=I_R(s)R$
Capacitor	1/sC	$V_C(s)=I_C(s)/sC$
Inductor	Ls	$V_L(s)=LsI_L(s)$

s-Domain Network Models

The first step in analyzing an electrical network using Laplace methods is to relabel the network elements with their s-domain equivalents and insert appropriate sources to account for initial capacitor voltages and inductor currents. One can then use loop or nodal analysis to write the equations which describe the behavior of the network. Kirchhoff's voltage and current laws still apply, even though the voltages and currents have been Laplace transformed. An advantage of s-domain analysis is that the network behavior is described by algebraic rather than differential equations. The overall procedure is best illustrated by examples.

Example 5-24. The switch in the network of Fig. 5-7(a) is closed at t=0. Derive an expression for $v_o(t)$ using Laplace methods. The capacitor is uncharged before the switch is closed.

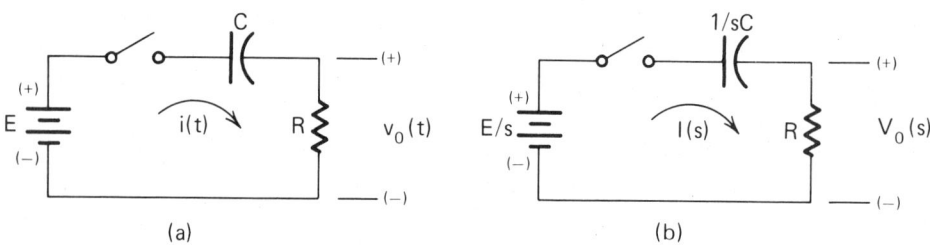

Figure 5-7. Networks for Example 5-24. The transient output voltage $v_o(t)$ can be found by Laplace transform methods without solving any differential equations. (a) Time-domain network; (b) s-domain network.

Solution. Figure 5-7(b) shows the transformed network in which the network elements and the source have been replaced by their s-domain models. As the voltage across a capacitor cannot charge instantaneously, $v_C(0^+)=0$, and we do not need an initial condition source for the capacitor. Write KVL around the loop to get

$$\frac{E}{s} = \frac{I(s)}{sC} + I(s)R \tag{5-199}$$

and

$$I(s) = \frac{EC}{RCs + 1} \tag{5-200}$$

But $V_o(s) = RI(s)$ so that

$$V_o(s) = \frac{ERC}{RCs + 1} = \frac{E}{s + 1/RC} \tag{5-201}$$

and, by inspection,

$$v_o(t) = E e^{-t/RC} \tag{5-202}$$

The Laplace Transform

The output voltage initially equals the battery voltage, but as the capacitor charges $v_o(t)$ decays exponentially to zero.

Example 5-25. The switch S in the network of Fig. 5-8(a) is moved to position B at t=0 after having been in position A for a long time. Draw the Laplace transformed network. Assume zero initial conditions for C_2 and L.

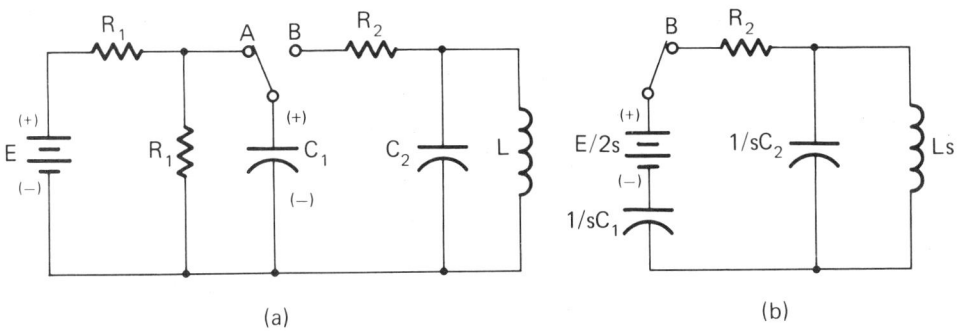

(a) (b)

Figure 5-8. Networks for Example 5-25. (a) The switch has been at A long enough for steady state conditions to prevail. Assume the loop containing L and C_2 is not oscillating; (b) Network with the switch at B showing the initial C_2 voltage as an initial condition source.

Solution. If the switch has been at A long enough for steady state conditions to prevail, then by the voltage-divider principle, capacitor C_1 is charged to E/2 volts with the polarity shown. The inductor has no current at $t=0^-$ because it was not in the circuit. When the switch is moved to position B, the conditions at $t=0^+$ are

$$v_{C_1}(0^+) = v_{C_1}(0^-) = \frac{E}{2} \qquad (5-203)$$

$$i_L(0^+) = i_L(0^-) = 0 \qquad (5-204)$$

(Capacitor voltages and inductor currents cannot change instantaneously.) The initial voltage on C_1 is modeled by a battery with a step function voltage in the time domain or as E/2s in the frequency domain. Figure 5-8(b) shows the Laplace-transformed network for $t=0^+$. This network can be analyzed to find the transient voltages and currents without solving any differential equations.

5.6 s-DOMAIN NETWORK ANALYSIS

In the last section we Laplace transformed the defining relationships of the network elements to arrive at the s-domain elements R, Ls, and 1/sC for the resistor, inductor, and capacitor. By virtue of having eliminated the need for differential equations in describing network behavior, we can develop a new family of analysis tools which are based on algebraic manipulation of s-domain functions. These analysis techniques enable us in many instances to investigate network performance without recourse to loop or nodal analysis--providing even more savings in labor over the classical methods of Chapter 4.[18,19] Similar techniques are available for mechanical systems, but they are seldom needed because of the relative simplicity of linear mechanical systems. Complex mechanical systems tend to be nonlinear, in which case Laplace methods fail.

Impedance And Admittance

The concept of _impedance_ is a valuable aid in analyzing both electrical and mechanical systems. We can generalize the idea that a resistor "resists" the flow of electrical current and define the property of impedance by which an element "impedes" the flow of current (or, in the case of mechanical elements, by which the element impedes motion). Whereas Ohm's law for the resistor is $v(t)=i(t)R$, for an impedance we have

$$V(s) = I(s)Z(s) \qquad (5-205)$$

where $V(s)$ and $I(s)$ are the transformed voltage and current associated with the impedance and $Z(s)$ is the impedance function itself. While resistance R is a constant, impedance $Z(s)$ is a function (always a rational function) in the s-domain. And, as Laplace-transformed voltages and currents

The Laplace Transform

have the units of volt-seconds and ampere-seconds, Eq. (5-205) shows us that impedance must have the units of ohms.

Single resistors, capacitors, and inductors have impedance values which can be found by comparing Eq. (5-205) with the first three entries of Table 5-1, or

$$Z_R(s) = R \tag{5-206}$$

$$Z_L(s) = Ls \tag{5-207}$$

$$Z_C(s) = \frac{1}{sC} \tag{5-208}$$

A Laplace-transformed network can be thought of as an interconnection of impedances. Moreover, just as we can combine series and parallel combinations of resistors into a single equivalent resistor, we can combine entire networks into a single equivalent impedance. In fact, we can combine a complex RLC network into a "black box" containing the impedance function and a pair of input terminals (see Fig. 5-9).

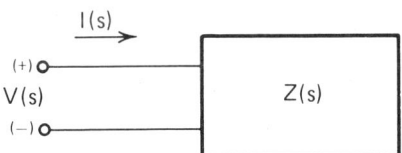

Figure 5-9. An impedance Z(s) represented as a "black box".

If we are interested only in the terminal voltage and current, we can ignore the exact configuration of the network as long as we know its impedance.

It is also useful to define the <u>admittances</u> Y(s) of the network elements as

$$Y_R = \frac{1}{R} \tag{5-209}$$

$$Y_L = \frac{1}{Ls} \tag{5-210}$$

$$Y_C = sC \tag{5-211}$$

and, noting that $Y(s)=1/Z(s)$, write the equation

$$I(s) = Y(s)V(s) \qquad (5-212)$$

Impedance has the units of <u>ohms</u>; admittance has the units of reciprocal ohms or <u>mhos</u>.

Example 5-26. Find the input impedance, $Z_i(s)=V(s)/I(s)$, for the network shown in Fig. 5-10.

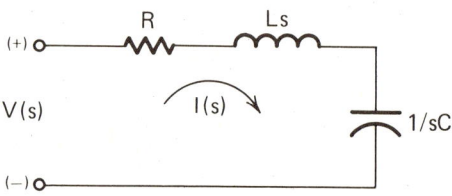

Figure 5-10. Example 5-26. The input impedance, defined by $Z_i(s) = V(s)/I(s)$ is given by Eq. (5-214).

Solution. Write KVL for the loop to obtain

$$V(s) = RI(s) + LsI(s) + \frac{I(s)}{sC} \qquad (5-213)$$

and solve for $Z_i(s)$ as

$$Z_i(s) = R + Ls + \frac{1}{sC} \qquad (5-214)$$

Network Simplification

The concepts of impedance and admittance are useful because they let us analyze networks without writing loop or nodal equations. By combining impedances and admittances we can often reduce a network to a single loop or node whose equation can be solved by inspection.

<u>Combining impedances</u>. Figure 5-11(a) shows a series connection of impedances. Because of the analogy between resistance and impedance, they are combined by similar rules.

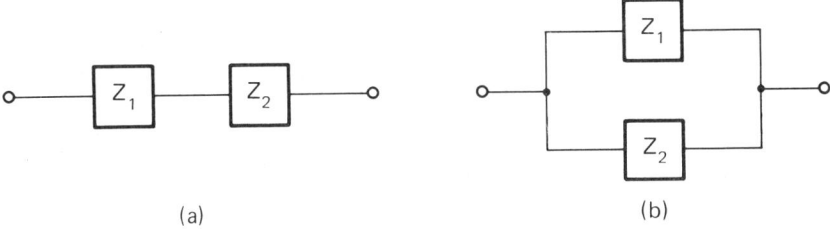

(a) (b)

Figure 5-11. Impedance connections which can be reduced to a single equivalent impedance Z_{eq}. (a) The series connection satisfies $Z_{eq} = Z_1 + Z_2$. (b) The parallel connection satisfies $Z_{eq} = Z_1 Z_2/(Z_1 + Z_2)$.

Two impedances in series can be replaced by an equivalent impedance Z_{eq} whose value is their sum, or

$$Z_{eq} = Z_1 + Z_2 \tag{5-215}$$

This addition rule holds for any number of impedances in series. The resultant impedance of the parallel connection of Fig. 5-11(b) is

$$\frac{1}{Z_{eq}} = \frac{1}{Z_1} + \frac{1}{Z_2} \tag{5-216}$$

or

$$Z_{eq} = \frac{Z_1 Z_2}{Z_1 + Z_2} \tag{5-217}$$

Example 5-27. Find the input impedance $Z(s)$ of the network of Fig. 5-12.

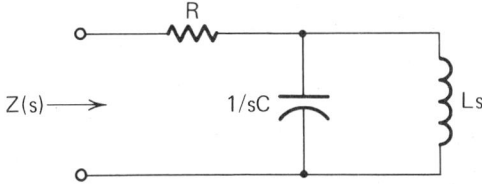

Figure 5-12. Example 5-27. The equivalent impedance of this series-parallel network is given by Eq. (5-219).

Solution. Using Eqns. (5-215) and (5-217),

$$Z(s) = R + \frac{(Ls)(1/sC)}{Ls + 1/sC} \tag{5-218}$$

or

$$Z(s) = R + \frac{Ls}{LCs^2 + 1} \tag{5-219}$$

Combining admittances. Two admittances in series are combined by

$$\frac{1}{Y_{eq}} = \frac{1}{Y_1} + \frac{1}{Y_2} = \frac{Y_1 Y_2}{Y_1 + Y_2} \tag{5-220}$$

while two parallel admittances are combined by

$$Y_{eq} = Y_1 + Y_2 \tag{5-221}$$

Example 5-28. Reduce the network of Fig. 5-13(a) to a single s-domain impedance connected to the source. The impedances are $Z_1 = 1$, $Z_2 = 1/(s+a)$, $Z_3 = 1/s$, and $Z_4 = 1/(s+b)$.

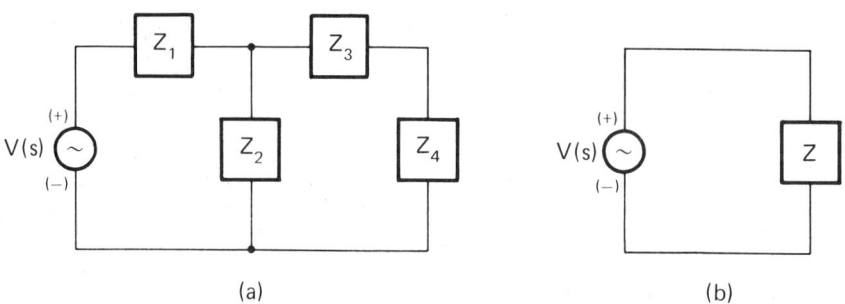

Figure 5-13. Example 5-28 shows how the rules for combining series and parallel impedances can be used to reduce a complicated network to a source connected to single equivalent impedance. (a) Original network; (b) Reduced network.

Solution. Impedances Z_3 and Z_4 are in series; they can be added to give

$$Z_5 = Z_3 + Z_4 \tag{5-222}$$

$$= \frac{1}{s} + \frac{1}{s+b} = \frac{2s + b}{s(s+b)} \tag{5-223}$$

The resultant Z_5 is in parallel with Z_2; these two can be combined by the product-over-sum rule for parallel impedances to give Z_6:

$$Z_6 = \frac{Z_5 Z_2}{Z_5 + Z_2} = \frac{\frac{2s+b}{s(s+b)} \cdot \frac{1}{s+a}}{\frac{2s+b}{s(s+b)} + \frac{1}{s+a}} \tag{5-224}$$

or

$$Z_6 = \frac{2s+b}{3s^2 + 2(a+b)s + ab} \tag{5-225}$$

Impedances Z_6 and Z_1 are in series and can be added to give the overall impedance Z as

$$Z = Z_1 + Z_6 = 1 + \frac{2s+b}{3s^2 + 2(a+b)s + ab} \tag{5-226}$$

or

$$Z = \frac{3s^2 + 2(a+b+1)s + b(a+1)}{3s^2 + 2(a+b)s + ab} \tag{5-227}$$

Figure 5-13(b) shows the reduced network.

<u>Thevenin's theorem for impedances</u>. In Section 3.5 we presented Thevenin's theorem for resistive networks. According to this theorem, any electrical network consisting of resistors and sources can be replaced, at a given pair of terminals, by a single resistor and a single voltage source. Thevenin's theorem can be extended to networks containing inductors and capacitors as well as resistors if we use the impedances of the elements. Figure 5-14 introduces this concept.

The value of the equivalent impedance $Z_T(s)$ is the value one arrives at by looking back into the network from the output terminals a-a' after removing Z_L and applying the combining rules for impedances. When doing this simplification, replace all voltage sources by short circuits and all current sources by open circuits. The Thevenin voltage source, $V_T(s)$ in Fig. 5-14, is the voltage one obtains at the output terminals with those terminals open-circuited.

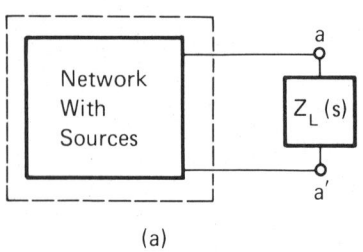 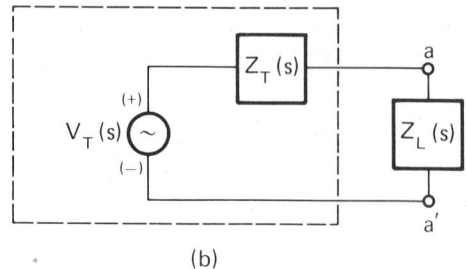

(a)　　　　　　　　　　　　　　　(b)

Figure 5-14. Thevenin's theorem lets us replace a general network by a voltage source $V_T(s)$ in series with an impedance $Z_T(s)$. (a) A network with load impedance $Z_L(s)$; (b) Thevenin's equivalent network.

A similar **theorem** due to Norton lets one replace a network by a single admittance in parallel with a current source. The Norton admittance $Y_N(s)$ is the reciprocal of the Thevenin impedance, or

$$Y_N(s) = \frac{1}{Z_T(s)} \qquad (5\text{-}228)$$

The Norton current is the current that would flow through the output terminals if they were connected together (the short-circuit current). The Norton source can also be found by converting the Thevenin source by the relationship

$$I_N(s) = \frac{V_T(s)}{Z_T(s)} \qquad (5\text{-}229)$$

<u>Example 5-29</u>. Find the Thevenin equivalent network to the left of terminals a-a' for the network of Fig. 5-15(a).

<u>Solution</u>. If network operation is assumed to start when the switch is closed, we can model the battery by E/s in the s-domain. Figure 5-15(b) shows the Laplace-transformed network with the output terminals open-circuited. The Thevenin impedance $Z_T(s)$ is found by replacing the battery by a short circuit and combining impedances. The result is

$$Z_T(s) = Ls + \frac{R(1/sC)}{R + 1/sC} = Ls + \frac{R}{RCs + 1} \qquad (5\text{-}230)$$

The Laplace Transform

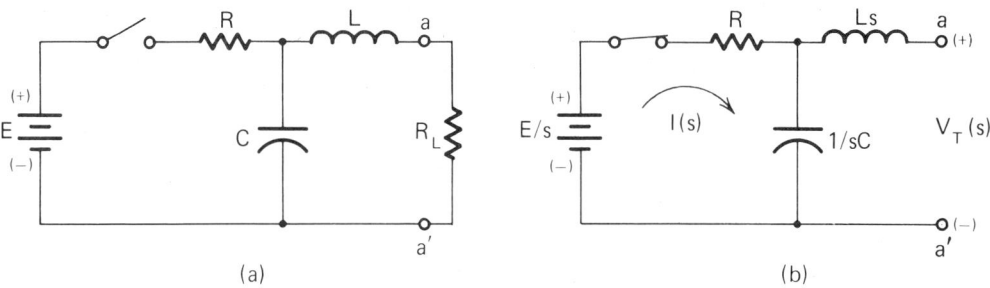

Figure 5-15. Example 5-29. Figure (a) is the given network. Figure (b) is the s-domain network with the load removed to find the Thevenin voltage $V_T(s)$.

and

$$Z_T(s) = \frac{RLCs^2 + Ls + R}{RCs + 1} \tag{5-231}$$

To determine the Thevenin voltage $V_T(s)$ we solve for $I(s)$ in Fig. 5-15(b):

$$I(s) = \frac{E/s}{R + 1/sC} \tag{5-232}$$

Observing that under open-circuit conditions the voltage across the inductor is zero, we have

$$V_T(s) = \frac{I(s)}{sC} = \frac{E}{s(RCs + 1)} \tag{5-233}$$

Figure 5-16 shows the Thevenin equivalent network.

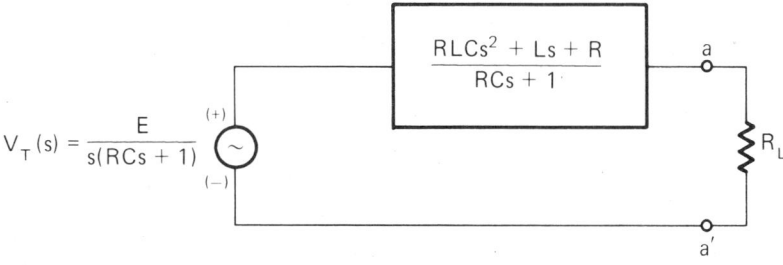

Figure 5-16. Thevenin equivalent network for Example 5-29.

Glass–Linear Systems—17

<u>Voltage division</u>. The voltage-divider principle also applies to impedances. Thus in Fig. 5-17 we have

$$V_o(s) = \frac{V(s)Z_2(s)}{Z_1(s) + Z_2(s)} \qquad (5-234)$$

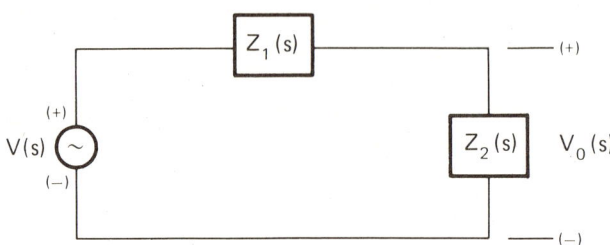

Figure 5-17. The voltage-divider principle also applies to impedances. In this instance $V_o(s) = V(s)Z_2/(Z_1 + Z_2)$.

Voltages divide in proportion to the impedances across which they appear. The current-division principle also applies to impedances and admittances. In Fig. 5-18,

$$I_o(s) = \frac{I(s)Y_2(s)}{Y_1(s) + Y_2(s)} \qquad (5-235)$$

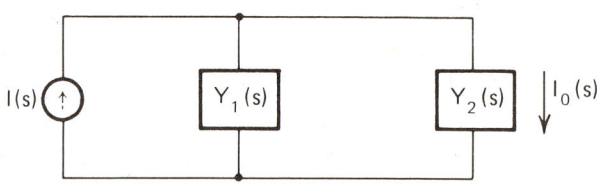

Figure 5-18. Current division applies readily to parallel admittances. In this instance $I_o(s) = I(s)Y_2/(Y_1 + Y_2)$.

Equation (5-235) is verified by observing that the same voltage appears across both admittances, or

$$V(s) = \frac{I(s)}{Y_1 + Y_2} = \frac{I_o(s)}{Y_2} \qquad (5-236)$$

The Laplace Transform

where we have used Ohm's law for admittances, Eq. (5-212), and the result that parallel admittances combine by direct addition, Eq. (5-221).

Example 5-30. Find a Thevenin equivalent circuit for the network of Fig. 5-19(a) where $Z_1=4/(s+2)$, $Z_2=3s$, and $Z_3=5/s$.

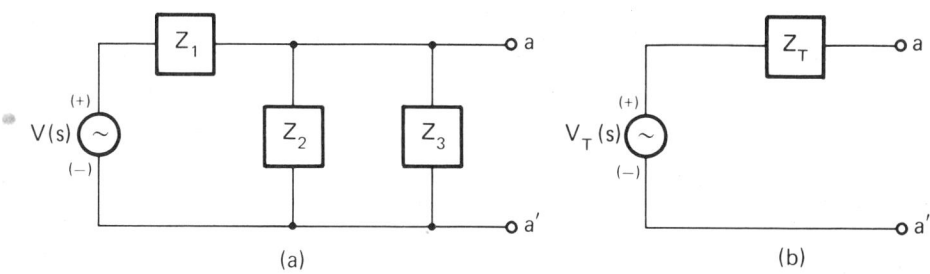

Figure 5-19. Network for Example 5-30. In the Thevenin equivalent network $V_T(s)$ and Z_T are given by Eqns. (5-240) and (5-238), respectively. (a) Original network; (b) Thevenin equivalent network.

Solution. Replacing $V(s)$ by a short circuit reveals that all three impedances are in parallel. First combining Z_2 and Z_3 gives

$$\frac{Z_2 Z_3}{Z_2 + Z_3} = \frac{15}{3s + 5/s} = \frac{15s}{3s^2 + 5} \qquad (5\text{-}237)$$

Combining this impedance in parallel with Z_1 yields the Thevenin impedance

$$Z_T(s) = \frac{60s}{27s^2 + 30s + 20} \qquad (5\text{-}238)$$

The open-circuit output voltage is obtained by observing that the source voltage divides across Z_1 and the parallel combination of Z_2 and Z_3. The voltage-divider principle yields the Thevenin voltage $V_T(s)$:

$$V_T(s) = V(s) \frac{15s/(3s^2 + 5)}{15s/(3s^2 + 5) + 4/(s + 2)} \qquad (5\text{-}239)$$

or

$$V_T(s) = \frac{15s(s+2)V(s)}{27s^2 + 30s + 20} \qquad (5\text{-}240)$$

Equations (5-238) and (5-240) along with Fig. 5-19(b) define the Thevenin equivalent circuit.

5.7 COUPLED DIFFERENTIAL EQUATIONS

To this point we have used Laplace transforms to transform single differential equations, and we have seen that the result is often a modest savings in effort. Systems of coupled differential equations can also be handled using Laplace methods, but here we have a considerable advantage over classical time-domain methods.[20] When Laplace transformed, sets of simultaneous linear differential equations become sets of algebraic equations which can be solved by the methods of ordinary algebra. We can use matrices, determinants, and Cramer's rule, or algebraic substitution. Chapter 10 shows how to treat coupled systems using state-variable methods in the s-domain.

Coupled second-order equations arise often in both mechanical and electrical analysis. Each mass in a mechanical system generates a second-order differential equation by way of Newton's second law, and there is often coupling to other masses through springs, dampers, or other types of linkages. Each loop in an electrical network can be described by a differential equation, and if the loop contains resistance, inductance and capacitance, the equation is of the second order. Coupling to the differential equations which describe other loops is through shared components.

Consider the fourth-order system comprised of the two second-order differential equations

$$\ddot{y}_1 + a_1 \dot{y}_1 + a_0 y_1 - k_2 y_2 = f(t) \qquad (5\text{-}241)$$

The Laplace Transform

and

$$\ddot{y}_2 + b_1\dot{y}_2 + b_0 y_2 - k_1 y_1 = 0 \qquad (5\text{-}242)$$

with zero initial conditions. Taking Laplace transforms gives

$$(s^2 + a_1 s + a_0)Y_1(s) - k_2 Y_2(s) = F(s) \qquad (5\text{-}243)$$

$$(s^2 + b_1 s + b_0)Y_2(s) - k_1 Y_1(s) = 0 \qquad (5\text{-}244)$$

Solving Eq. (5-244) for $Y_2(s)$ and substituting the result into Eq. (5-243) yields

$$\left\{ s^2 + a_1 s + a_0 - \frac{k_1 k_2}{s^2 + b_1 s + b_0} \right\} Y_1(s) = F(s) \qquad (5\text{-}245)$$

from which

$$Y_1(s) = \frac{F(s)(s^2 + b_1 s + b_0)}{(s^2 + a_1 s + a_0)(s^2 + b_1 s + b_0) - k_1 k_2} \qquad (5\text{-}246)$$

If $F(s)$ were known, we could apply the inverse Laplace transform to the latter expression to obtain $y_1(t)$. The other dependent variable $y_2(t)$ is found by a similar approach.

Example 5-31. The coupled equations

$$\dot{q}(t) = r(t) \quad \text{and} \quad \dot{r}(t) = -q(t) \qquad (5\text{-}247)$$

with $r(0)=1$ and $q(0)=0$ are often implemented on an analog computer to generate sinusoidal functions. Use Laplace transform methods to show that $q(t)=\sin(t)$ and $r(t)=\cos(t)$.

Solution. Laplace transform the given equations to obtain

$$sQ(s) - R(s) = 0 \qquad (5\text{-}248)$$

$$sR(s) - 1 + Q(s) = 0 \qquad (5\text{-}249)$$

Solving for $Q(s)$ yields

$$s^2 Q(s) + Q(s) = 1 \qquad (5\text{-}250)$$

from which

$$Q(s) = \frac{1}{s^2 + 1} \tag{5-251}$$

and $q(t)=\sin(t)$. Similarly, we can write

$$s\{1-sR(s)\} - R(s) = 0 \tag{5-252}$$

which yields

$$R(s) = \frac{s}{s^2 + 1} \tag{5-253}$$

and $r(t)=\cos(t)$.

Example 5-32. Two RLC networks are coupled through a resistor and are energized by the current source I_o as shown in Fig. 5-20. Derive expressions for the Laplace transformed node voltages $V_a(s)$ and $V_b(s)$ after the switch is closed at t=0. Assume RC=1.0 sec, $R_cC=0.5$ sec, LC=0.25 sec^2, and $I_o/C=3.0$ A/F. Also assume zero initial conditions.

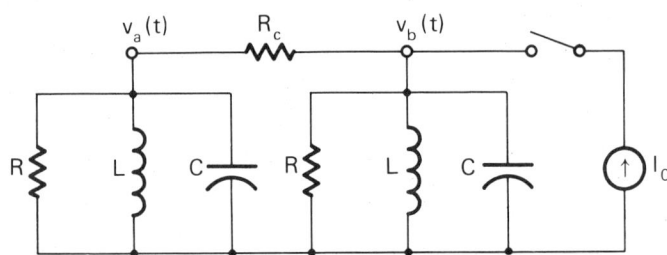

Figure 5-20. Example 5-32. This network gives rise to coupled differential equations which can be solved by Laplace transforms.

Solution. Using Laplace transformed network elements we can write the node voltage equations by inspection:

$$\left\{\frac{1}{R} + \frac{1}{Ls} + sC\right\}V_a(s) + \frac{V_a(s)-V_b(s)}{R_c} = 0 \tag{5-254}$$

$$\left\{\frac{1}{R} + \frac{1}{Ls} + sC\right\}V_b(s) + \frac{V_b(s)-V_a(s)}{R_c} = \frac{I_o}{s} \tag{5-255}$$

The Laplace Transform 245

These can be rearranged as

$$\left\{s^2+s\left\{\frac{1}{RC}+\frac{1}{R_cC}\right\}+\frac{1}{LC}\right\}V_a(s) - \left\{\frac{s}{R_cC}\right\}V_b(s) = 0 \qquad (5\text{-}256)$$

$$\left\{-\frac{s}{R_cC}\right\}V_a(s) + \left\{s^2+s\left\{\frac{1}{RC}+\frac{1}{R_cC}\right\}+\frac{1}{LC}\right\}V_b(s) = \frac{I_o}{C} \qquad (5\text{-}257)$$

Inserting numerical values yields

$$(s^2+3s+4)V_a(s) - 2sV_b(s) = 0 \qquad (5\text{-}258)$$

$$-2sV_a(s) + (s^2+3s+4)V_b(s) = 3 \qquad (5\text{-}259)$$

Solving Eq. (5-258) for $V_b(s)$ and substituting the result into Eq. (5-259) gives

$$V_a(s) = \frac{6s}{(s^2+3s+4)^2 - 4s^2} \qquad (5\text{-}260)$$

Similarly,

$$V_b(s) = \frac{3(s^2+3s+4)}{(s^2+3s+4)^2 - 4s^2} \qquad (5\text{-}261)$$

To find the time-domain node voltages $v_a(t)$ and $v_b(t)$ we would need to factor the fourth-order denominators of Eqns. (5-260) and (5-261) and then do partial-fraction expansions.

5.8 THE IMPULSE RESPONSE

Finding the response of a linear system to an impulse function input can be tricky using time-domain methods, but the problem is almost trivial in the s-domain.[21] Consider the following second-order differential equation with an impulse input and zero initial conditions:

$$\ddot{y} + a_1\dot{y} + a_0y = A\delta(t) \qquad (5\text{-}262)$$

Taking Laplace transforms, and recalling that $L\{\delta(t)\}=1$, yields

$$(s^2 + a_1 s + a_0)Y(s) = A \qquad (5\text{-}263)$$

and

$$Y(s) = \frac{A}{s^2 + a_1 s + a_0} \qquad (5\text{-}264)$$

We see from Eq. (5-264) that the impulse response is the inverse transform of the reciprocal of the characteristic equation.

Suppose Eq. (5-262) had a zero input but a nonzero initial condition on the first derivative. The transformed equation would be

$$\{s^2 + a_1 s + a_0\}Y(s) - y(0) = 0 \qquad (5\text{-}265)$$

and

$$Y(s) = \frac{y(0)}{s^2 + a_1 s + a_0} \qquad (5\text{-}266)$$

Comparing Eqns. (5-264) and (5-266) reveals that <u>an impulse input is equivalent to an initial condition on the next to the highest-order derivative</u>. This result, which applies to differential equations of any order, is useful when we attempt to find the impulse response by either analog or digital computer methods. Impulse functions are difficult to simulate: initial conditions are not.

Example 5-33. Find the impulse response of the differential equation

$$\dot{y} + 6y = 4\delta(t) \qquad (5\text{-}267)$$

with $y(0)=16$.

Solution. Taking Laplace transforms gives

$$sY(s) - y(0) + 6Y(s) = 4 \qquad (5\text{-}268)$$

and

$$Y(s) = \frac{20}{s+6} \tag{5-269}$$

from which $y(t)=20\exp(-6t)$.

Example 5-34. Solve the differential equation

$$\ddot{q} + 2\dot{q} + 8q = 4\delta(t) \tag{5-270}$$

with $q(0)=0$ and $\dot{q}(0)=-4$.

Solution. Transforming Eq. (5-270) results in

$$s^2 Q(s) - \dot{q}(0) + 2sQ(s) + 8Q(s) = 4 \tag{5-271}$$

and, using $\dot{q}(0)=-4$, we have

$$Q(s) = \frac{0}{s^2 + 2s + 8} \tag{5-272}$$

The solution is therefore $q(t)=0$. The impulse input cancelled the initial condition and $q(t)=0$ for all time!

5.9 CONVOLUTION

The concept of convolution is fundamentally important in the theory of linear systems.[22,23] We will approach convolution by seeking to find the time-domain equivalent of s-domain multiplication. Suppose

$$H(s) = F(s)G(s) \tag{5-273}$$

From Eq. (5-1) we can write

$$F(s) = \int_0^\infty f(t)e^{-st}dt \tag{5-274}$$

and

$$G(s) = \int_0^\infty g(y)e^{-sy}dy \tag{5-275}$$

where we have used y rather than t in the second integral to avoid confusion. (Because the integrals are definite integrals, the symbol used for the dummy variable of integration is unimportant.) Substitute Eq. (5-275) into Eq. (5-273) to obtain

$$H(s) = \int_0^\infty F(s)g(y)e^{-sy}dt \qquad (5\text{-}276)$$

where we have brought F(s) inside the integral because it is not a function of y. The inverse transform of F(s)exp(-sy) can be found using the shifting theorem, Eq. (5-46):

$$L^{-1}\{F(s)e^{-sy}\} = f(t-y)u(t-y) \qquad (5\text{-}277)$$

Applying Eq. (5-1) to Eq. (5-277) gives

$$L\{L^{-1}F(s)e^{-sy}\} = \int_0^\infty f(t-y)u(t-y)e^{-st}dt \qquad (5\text{-}278)$$

Substitute the right-hand side of Eq. (5-278) for F(s)exp(-sy) in Eq. (5-276) to obtain

$$H(s) = \int_0^\infty g(y) \left\{ \int_0^\infty e^{-st} f(t-y)u(t-y)dt \right\} dy \qquad (5\text{-}279)$$

As u(t-y)=0 for 0<t<y, and u(t-y)=1 for t⩾y, the lower limit of integration on the second integral can be changed from 0 to y so that

$$H(s) = \int_0^\infty g(y) \left\{ \int_y^\infty (1)f(t-y)e^{-st}dt \right\} dy \qquad (5\text{-}280)$$

It can be shown using the techniques of advanced calculus that if f(t) and g(t) are functions of exponential order, then it is permissible to change the order of integration (we can first bring g(y) under the second integral sign because g(y) is not a function of t). We must, however, pay attention to

The Laplace Transform

the limits of integration when we reverse the order of integration. If we temporarily define a function h(t,y) by

$$h(t,y) = g(y)f(t-y)e^{-st} \qquad (5\text{-}281)$$

then Eq. (5-280) becomes

$$H(s) = \int_{y=0}^{\infty}\int_{t=y}^{\infty} h(t,y)\,dt\,dy \qquad (5\text{-}282)$$

Equation (5-282) can be interpreted as the volume under the function h(t,y). This volume lies above the region in the ty-plane (see Fig. 5-21) defined by the limits of integration; thus t goes from y to ∞ and y goes from 0 to ∞.

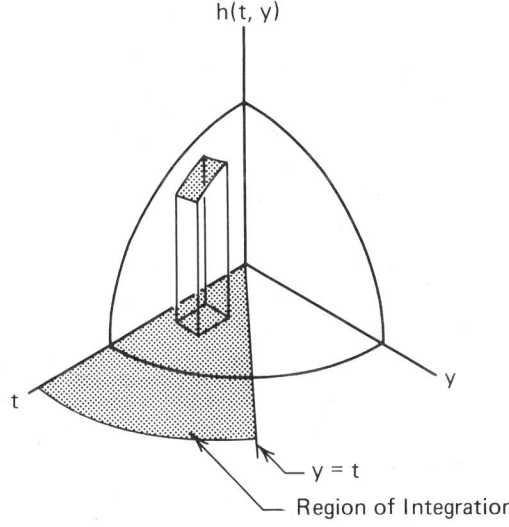

Figure 5-21. The integral equation Eq. (5-282) can be interpreted as the volume under the surface defined by h(t, y).

In order to cover the same volume with the order of integration reversed, y must go from 0 to t, and t must go from 0 to ∞. Hence, we can rewrite Eq. (5-282) as

$$H(s) = \int_{t=0}^{\infty}\int_{y=0}^{t} g(y)f(t-y)e^{-st}\,dy\,dt \qquad (5\text{-}283)$$

or

$$H(s) = \int_{t=0}^{\infty} e^{-st} \left\{ \int_{y=0}^{t} g(y)f(t-y)dy \right\} dt \qquad (5\text{-}284)$$

Because Eq. (5-284) is, by Eq. (5-1), the Laplace transform of the bracketed quantity, we have

$$L^{-1}\{H(s)\} = L^{-1}\{F(s)G(s)\} = \int_{0}^{t} g(y)f(t-y)dy \qquad (5\text{-}285)$$

The integral in Eq. (5-285) is called the <u>convolution integral</u>; when it is evaluated, the result is the convolution of the two functions $f(t)$ and $g(t)$. To convolve two functions we can either find their Laplace transforms, multiply them together, and then take the inverse transform; or we can apply the convolution integral directly in the time domain. The Laplace method is almost always easier.

<u>Example 5-35</u>. Evaluate the convolution of $f(t)=\exp(-2t)$ and $g(t)=tu(t)$ by both methods and show that the results are identical.

<u>Solution</u>. Taking Laplace transforms of the given functions gives

$$F(s) = \frac{1}{s+2} \quad \text{and} \quad G(s) = \frac{1}{s^2} \qquad (5\text{-}286)$$

Then

$$H(s) = F(s)G(s) = \frac{1}{s^2(s+2)} \qquad (5\text{-}287)$$

and, as can be found using partial fractions,

$$h(t) = L^{-1}\{H(s)\} = \tfrac{1}{4}(2t - 1 + e^{-2t}) \qquad (5\text{-}288)$$

Alternatively, Eq. (5-285) gives

$$h(t) = \int_{0}^{t} y e^{-2(t-y)} dy = e^{-2t} \int_{0}^{t} y e^{2y} dy \qquad (5\text{-}289)$$

and

$$h(t) = e^{-2t}\left\{\frac{e^{2y}}{4}(2y-1)\right\}\Big]_0^t$$

which, when evaluated, gives the same result as Eq. (5-288).

PROBLEMS FOR CHAPTER 5

5-1. Determine whether or not the following functions are Laplace transformable by showing that they either are or are not of exponential order: (a) $f(t)=t^3 u(t)$, (b) $f(t)=\exp(t^2)$, (c) $f(t)=\exp(-j\omega t)$, (d) $f(t)=t^2 \sin(kt)$, and (e) $f(t)=\sinh(kt)$.

5-2. Use Eq. (5-1) to find the Laplace transforms of the following functions: (a) $f(t)=3e^{-5t}$, (b) $f(t)=u(t)-u(t-a)$, (c) $f(t)=tu(t-a)$, and (d) $f(t)=e^{-at}\sin(bt)$.

5-3. Use Eq. (5-1) and integration by parts to show that $L\{t^n\}=L\{t^{n-1}\}n/s$.

5-4. Use the results of Problem 5-3 to show that $L\{t^n\}=n!/s^{n+1}$.

5-5. Express $F(s)=L\{\exp(-at)-\exp(-bt)\}$ as a ratio of polynomials in s.

5-6. Find the Laplace transform of $f(t)=\cos^2(\omega t)$ and express the result as a rational function. (Hint: Use $\cos^2(x)=\{1+\cos(2x)\}/2$.)

5-7. Laplace transform the function $f(t)=t$ for $0<t<5$ and $f(t)=3$ for $t\geqslant 5$.

5-8. Laplace transform function $y(t)=2$ for $0<t<1$ and $y(t)=5$ for $t\geqslant 1$.

5-9. The analytical expression for a rectangle of height 1/a and duration a is

$$f(t) = \frac{u(t) - u(t-a)}{a} \qquad (5-290)$$

Show that the Laplace transform of the impulse function is 1 by first taking the Laplace transform of f(t) and then letting

$a \to 0$. (Hint: Use a series expansion for $\exp(-as)$.)

5-10. Write a mathematical proof that the process of taking the Laplace transform of a function is linear.

5-11. Derive Eq. (5-61) using Eqns. (5-1) and (5-60).

5-12. Prove Eq. (5-63).

5-13. Prove the final-value theorem, Eq. (5-73).

5-14. Use the **real differentiation theorem**, Eq. (5-60), to show that the derivative of a unit step function is a unit impulse function.

5-15. Find $f(\infty)$ for the time-domain functions which have the following Laplace transforms:

(a) $F(s) = \dfrac{s + 2}{s(s^2 + 3s + 4)}$ \hfill (5-291)

(b) $F(s) = \dfrac{s + 5}{s + 6}$ \hfill (5-292)

(c) $F(s) = \dfrac{5}{s^2(s + 2)}$ \hfill (5-293)

(d) $F(s) = \dfrac{s^2 + 5s + 25}{s(s + 2)(s + 5)}$ \hfill (5-294)

5-16. Find the initial values $f(0^+)$ for the functions which have the following Laplace transforms:

(a) $F(s) = \dfrac{s + 2}{s^2 + 5s + 4}$ \hfill (5-295)

(b) $F(s) = \dfrac{s(s + 8)}{s^3 + 2s + 25}$ \hfill (5-296)

(c) $F(s) = \dfrac{s + 5}{4(s^2 + 6)}$ \hfill (5-297)

(d) $F(s) = \dfrac{2s + 1}{s^2(s + 10)}$ \hfill (5-298)

(e) $F(s) = \dfrac{5s^2 + 3s + 8}{2s^3 + 5s^2 + s + 1}$ \hfill (5-299)

5-17. Use Eq. (5-63) to help find the Laplace transforms of the functions (a) $f(t) = t\sin(\omega t)$, (b) $f(t) = t^2$, and (c) $f(t) = t^2 e^{-at}$.

The Laplace Transform

5-18. Find the Laplace transforms of the following differential equations.

(a) $\dot{y} + 5y = 0$ $\qquad y(0) = 2$ (5-300)

(b) $\ddot{y} + 2\zeta\omega_n\dot{y} + \omega_n^2 y = 0$ $\qquad y(0)=A, \dot{y}(0)=B$ (5-301)

(c) $A\ddot{y} + B\dot{y} + Cy = 0$ $\qquad y(0)=D, \dot{y}(0)=0$ (5-302)

5-19. Laplace transform the following differential equations and solve for $Y(s)$. Assume all initial conditions are zero.

(a) $\ddot{y} + 2\zeta\omega_n\dot{y} + \omega_n^2 y = u(t)$ (5-303)

(b) $\dot{y} + 4y = tu(t) + e^{-t}u(t)$ (5-304)

(c) $\ddot{y} + 16y = \cos(4t)$ (5-305)

(d) $\dddot{y} + 4\ddot{y} + 5\dot{y} = \delta(t) - u(t-2)$ (5-306)

5-20. Use partial fractions to find the inverse Laplace transforms of the following functions:

(a) $F(s) = \dfrac{s^2}{s(s+2)}$ (5-307)

(b) $F(s) = \dfrac{1}{(s+3)(s+6)}$ (5-308)

(c) $F(s) = \dfrac{s^2}{s^2 + 5s + 6}$ (5-309)

(d) $F(s) = \dfrac{s+2}{(s+3)(s^2+5)}$ (5-310)

(e) $F(s) = \dfrac{s+1}{s^2(s+5)}$ (5-311)

(f) $F(s) = \dfrac{1}{(s^2+a^2)(s^2+b^2)}$ (5-312)

(g) $F(s) = \dfrac{s}{(s^2+1)(s^2+4)}$ (5-313)

(h) $F(s) = \dfrac{5s-2}{s^2(s+2)(s+1)}$ (5-314)

(i) $F(s) = \dfrac{1}{s(s+2)^2(s+4)}$ \hfill (5-315)

5-21. Use the frequency translation theorem, Eq. (5-51), to find the inverse transforms of

(a) $F(s) = \dfrac{1}{(s+a)^2 + b^2}$ \hfill (5-316)

(b) $F(s) = \dfrac{s}{(s+a)^2 + b^2}$ \hfill (5-317)

5-22. Solve the following differential equations using Laplace transforms:

(a) $\dot{y} + 10y = 0$ \qquad $y(0) = 4$ \hfill (5-318)

(b) $\dot{v}(t) = -3v(t)$ \qquad $v(0) = -5$ \hfill (5-319)

(c) $\dfrac{\dot{q}}{q} + 4 = 2$ \qquad $q(0) = 5$ \hfill (5-320)

(d) $\dot{y} = 4$ \qquad $y(0) = 5$ \hfill (5-321)

5-23. Solve Problem 4-5 using Laplace transforms.
5-24. Solve Problem 4-6 using Laplace transforms.
5-25. Solve Problem 4-7 using Laplace transforms.
5-26. Solve Problem 4-8 using Laplace transforms.
5-27. Solve Problem 4-9 using Laplace transforms.
5-28. Solve Problem 4-11 using Laplace transforms.
5-29. Solve the following differential equations using Laplace transforms:

(a) $\dot{y} + 3y = 2u(t)$ \qquad $y(0) = 0$ \hfill (5-322)

(b) $\dot{\theta} + \theta = e^{-2t}$ \qquad $\theta(0) = 2$ \hfill (5-323)

(c) $\dot{q} + 2q = tu(t)$ \qquad $q(0) = -5$ \hfill (5-324)

(d) $\ddot{\theta} + 5\dot{\theta} + 125\theta = 500u(t)$ \qquad $\theta(0) = \dot{\theta}(0) = 0$ \hfill (5-325)

5-30. Use Laplace transforms to find differential equations which have the following solutions:

(a) $y(t) = 2 + e^{-3t} + e^{-2t}$ \hfill (5-326)

(b) $q(t) = t^2 e^{-5t}$ \hfill (5-327)

The Laplace Transform

(c) $h(t) = e^{-5t}\cos(3t) - 2$ (5-328)

(d) $x(t) = 1 + e^{-t} + e^{-2t} + e^{-3t} + e^{-4t}$ (5-329)

5-31. In each of the networks of Fig. 5-22 switch S moves from position A to position B at t=0. Find the Laplace-transformed network in each case. The switch has been in position A long enough for steady-state conditions to be achieved.

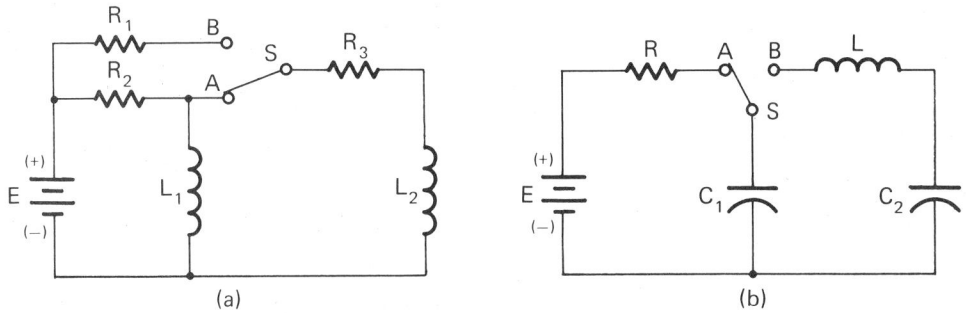

Figure 5-22. Problem 5-31.

5-32. Find the input impedance $Z(s)$ for each of the networks of Fig. 5-23. Express each result as a rational function.

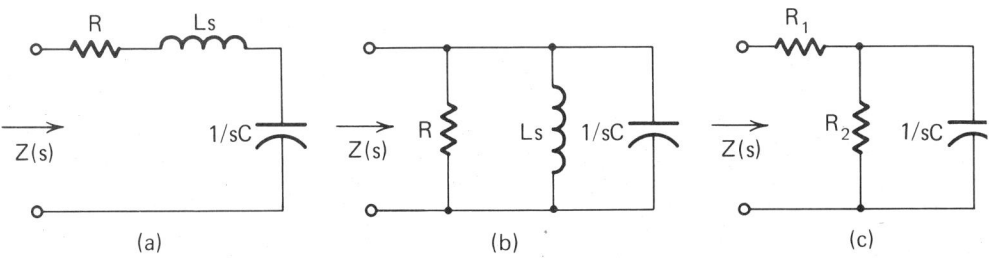

Figure 5-23. Problem 5-32.

5-33. Find the output voltage $v_o(t)$ in the networks of Fig. 5-24 using Laplace methods if necessary.

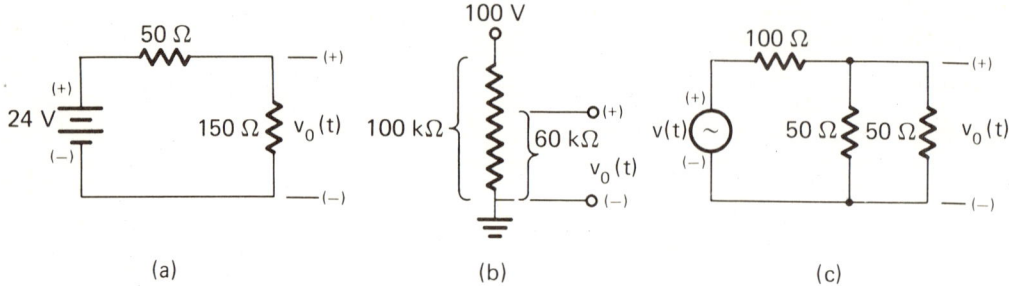

Figure 5-24. Problem 5-33. (c) $v(t) = 50 \sin (377t)$.

5-34. Use the voltage-divider principle to find expressions for $V_o(s)$ in each of the networks of Fig. 5-25.

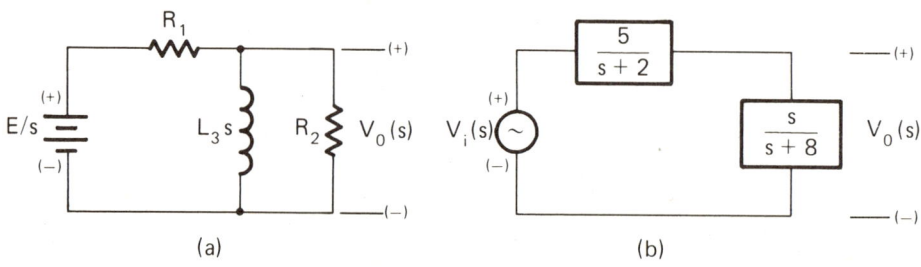

Figure 5-25. Problem 5-34.

5-35. Find Thevenin equivalent circuits to the left of terminals a-a' for the networks of Fig. 5-26.

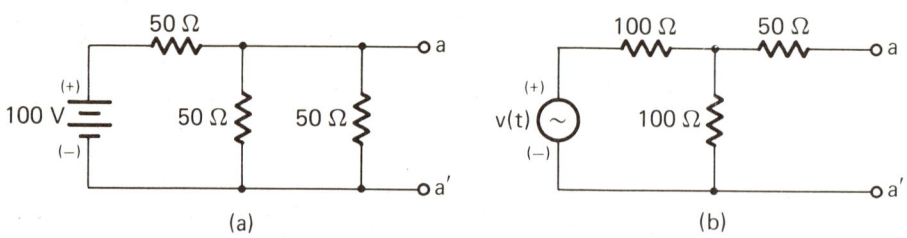

Figure 5-26. Problem 5-35. (c) $v(t) = 110 \sin (377t)$.

5-36. Find s-domain Thevenin equivalent circuits to the left of terminals a-a' for the networks of Fig. 5-27.

The Laplace Transform

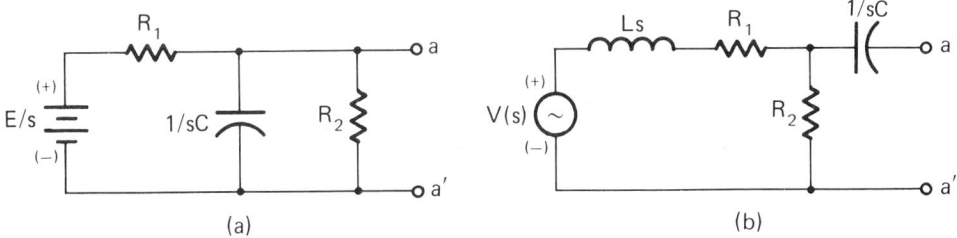

Figure 5-27. Problem 5-36.

5-37. Solve for $V_1(s)$ and $V_2(s)$ in terms of E for the following coupled system:

$$\dot{v}_1 + 2v_1 - v_2 = Eu(t) \qquad (5\text{-}330)$$

$$\dot{v}_2 + 8v_2 - 5v_1 = 0 \qquad (5\text{-}331)$$

Assume zero initial conditions.

5-38. Solve for $V_o(s)$ after the switch closes at t=0 in the network of Fig. 5-28.

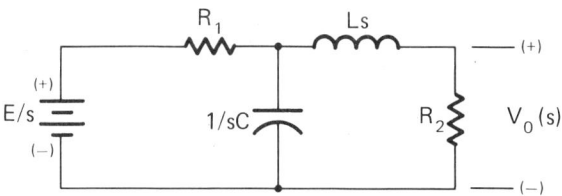

Figure 5-28. Problem 5-38.

5-39. Solve for r(t) using Laplace transforms if r(0)=0 and $\dot{r}(0)=0$.

$$\ddot{r} + 8\dot{r} + 15r = 2\delta(t) \qquad (5\text{-}332)$$

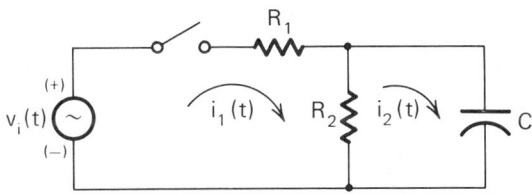

Figure 5-29. Problem 5-40.

5-40. The switch in Fig. 5-29 is closed at t=0. Solve for $i_1(t)$ and $i_2(t)$ if R_1=5.0 kΩ, R_2=9.0 kΩ, C=250 μF, and $v_i(t)$=115cos(377t).

5-41. Prove that an impulse input is equivalent to an initial condition on the next to the highest order derivative for the nth-order differential equation

$$\sum_{i=0}^{n} \frac{d^i y(t)}{dt^i} a_i = 0 \qquad (5\text{-}333)$$

5-42. Convolve each of the following pairs of functions. Use Laplace transforms: (a) f(t)=exp(-2t), g(t)=t+1; (b) f(t)=sin(t), g(t)=u(t); (c) $f(t)=e^{-t}$, $g(t)=e^{-t}$.

5-43. Derive an expression for the inverse Laplace transform of the function $F(s)=(s+a)^{-n}$ where n is an integer.

5-44. Expand the following functions by partial fractions:

(a) $F(s) = \dfrac{s^2 + 3s + 4}{(s+5)(s+4)}$

(b) $Y(s) = \dfrac{1}{s(s^2+4)^2}$

REFERENCES FOR CHAPTER 5

(1) Rainville, E. D.: <u>The Laplace Transform: An Introduction</u>, The Macmillan Company, New York, 1963.
(2) Kaplan, W.: <u>Operational Methods For Linear Systems</u>, Addison-Wesley Publishing Co., Inc., Reading, Massachusetts, 1962.
(3) Goldman, S.: <u>Transform Calculus and Electrical Transients</u>, Prentice-Hall, Inc., New York, 1949.
(4) Aseltine, J. A.: <u>Transform Methods in Linear System Analysis</u>, McGraw-Hill Book Co., Inc., New York, 1958.
(5) Melsa, J. L. and D. G. Schlutz: <u>Linear Control Systems</u>, McGraw-Hill Book Co., Inc., New York, 1969.

(6) Agnew, R. P.: Differential Equations, McGraw-Hill Book Co., Inc., New York, 1960.

(7) Churchill, R. V.: Operational Mathematics, McGraw-Hill Book Co., Inc., New York, 1958.

(8) Widder, D. V.: The Laplace Transform, Princeton University Press, Princeton, New Jersey, 1946.

(9) Hall, D. L., Maple, C. G., and B. Vinograde: Introduction to the Laplace Transform, Appleton-Century-Crofts, Inc., New York, 1959.

(10) Doetsch, G.: Guide to the Applications of Laplace Transforms, D. Van Nostrand Co., Inc., Princeton, New Jersey, 1961.

(11) Bohn, E. V.: The Transform Analysis of Linear Systems, Addison-Wesley Publishing Co., Inc., Reading, Massachusetts, 1963.

(12) Scott, G. J.: Transform Calculus, Harper Brothers Publishers, Inc., New York, 1955.

(13) Flanders, H., Korfhage, R. R. and J. J. Price: Calculus, Academic Press Inc., New York, 1970.

(14) Gibson, J. E. and F. B. Tuteur: Control System Components, McGraw-Hill Book Co., Inc., New York, 1958.

(15) Del Toro, V. and S. R. Parker: Principles of Control Systems Engineering, McGraw-Hill Book Co., Inc., New York, 1960.

(16) Huelsman, L. P.: Basic Circuit Theory With Digital Computation, Prentice-Hall, Inc., Englewood Cliffs, New Jersey, 1972.

(17) Lewis, L. J., Reynolds, D. K., Bergreth, F. R., and F. J. Alexandro, Jr.,: Linear Systems Analysis, McGraw-Hill Book Co., Inc., New York, 1969.

(18) Sabbagh, E. M.: Circuit Analysis, The Ronald Press Company, Inc., New York, 1961.

(19) Kuo, F. F.: Network Analysis and Synthesis, John Wiley and Sons, Inc., New York, 1962.

(20) Golomb, M. and M. Shanks: Elements of Ordinary Differential Equations, McGraw-Hill Book Co., Inc., New York, 1965.

(21) Auslander, D. M., Takahashi, Y. and M. J. Rabins: Introducing Systems and Control, McGraw-Hill Book Co., Inc., New York, 1974.

(22) Zadeh, L. A. and C. A. Desoer: Linear System Theory, McGraw-Hill Book Co., Inc., New York, 1963.

(23) Guillemin, E. A.: Theory of Linear Systems, John Wiley and Sons, Inc., New York, 1963.

CHAPTER 6

The Complex Frequency Plane

The Laplace transformation is a straightforward, unified approach to solving differential equations. In many instances Laplace transforms involve less computational effort than classical time-domain methods, but they are even more useful as the key to frequency-domain representation of linear systems. We shall see in this chapter that by working in the complex frequency domain we can gain a great deal of insight into the behavior of a system without having to solve its describing differential equations. Indeed, much modern system design is done entirely in the frequency domain, without the designer ever seeing a time-domain representation of the predicted system outputs.[1,2,3] As a consequence, in spite of the recent trend toward state variables, modern control theory, and digital computer analysis (all of which are oriented toward the time domain) the frequency domain has stubbornly held its own in the realm of systems analysis and design.

This chapter introduces the complex frequency plane, which we call for short the s-plane. The s-plane is at the end of a progression which starts with a differential equation in the time domain, proceeds through an s-domain algebraic equation and leads finally to a graphical representation of system equations. Among the topics covered are the s-plane itself, how to make judgements about system performance by inspecting s-plane representations of system functions, and how to evaluate partial-fraction residues graphically. An auxiliary but important topic also treated is stability, viewed from the s-plane and verified by the Routh criterion.

6.1 THE s-PLANE

As introduced in Chapter 5, the Laplace transform variable s takes the form $s=\sigma+j\omega$, where σ and ω are real variables. Because s is an independent variable, so are σ and ω. We can plot values of s on what is called the complex frequency or s-plane.[4] The s-plane has two coordinate axes, a horizontal or real axis for σ, and a vertical or imaginary axis for $j\omega$. Points in the plane correspond to constant values of s, while lines or curves are the loci of variable values of s. Figure 6-1 shows the s-plane with five points plotted and indicated by crosses or zeros.

The Pole-Zero Plot [5]

If, as is the usual case in linear systems work, a Laplace transformed function F(s) is a rational function in s, it can be written in terms of the roots or factors of its numerator and denominator polynomials. The general form is

$$F(s) = \frac{K(s - z_1)(s - z_2)\ldots(s - z_m)}{(s - p_1)(s - p_2)\ldots(s - p_n)} \qquad (6-1)$$

The Complex Frequency Plane

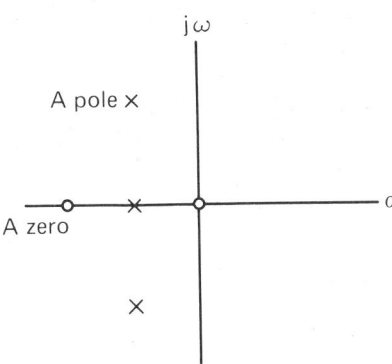

Figure 6-1. Values of the complex variable $s = \sigma + j\omega$ can be plotted on the complex s-plane (also known as the complex frequency plane).

where K is a gain or amplitude factor, and the z_i and p_i are the <u>zeros</u> and <u>poles</u> of F(s). A zero of a rational function is a value of the independent variable for which the numerator polynomial becomes zero; a pole is a value for which the denominator polynomial becomes zero. Both poles and zeros may be either real or complex. We can plot the poles and zeros of a function as points on the complex s-plane. Poles and zeros are conventionally represented by crosses and small circles, respectively. Figure 6-1 shows a constellation of three poles and two zeros.

The pole-zero plot is <u>not</u> a representation of the function F(s). To plot the complex function F(s) itself requires two sets of three-dimensional axes. Because F(s) is a complex function it may have both a real and an imaginary part, each of which is a function of a complex variable. Figure 6-2 shows the coordinate systems which would be needed to plot the real and imaginary parts, denoted by Re{F(s)} and Im{F(s)}, of a complex function F(s).

It is generally more informative to plot the magnitude and angle of a complex function rather than its real and imaginary parts. These quantities are given by

$$|F(s)| = \sqrt{\text{Re}\{F(s)\}^2 + \text{Im}\{F(s)\}^2} \qquad (6\text{-}2)$$

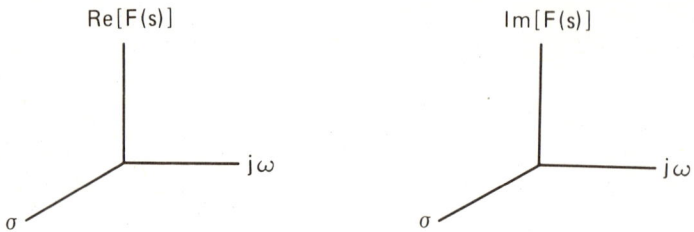

Figure 6-2. Two sets of coordinate axes are needed to plot the real and imaginary parts of a complex function F(s).

and

$$\emptyset(s) = \text{Arg}\{F(s)\} = \tan^{-1}\left\{\frac{\text{Im}\{F(s)\}}{\text{Re}\{F(s)\}}\right\} \quad (6-3)$$

Figure 6-3 shows the coordinate axes needed to plot the magnitude and angle of F(s).

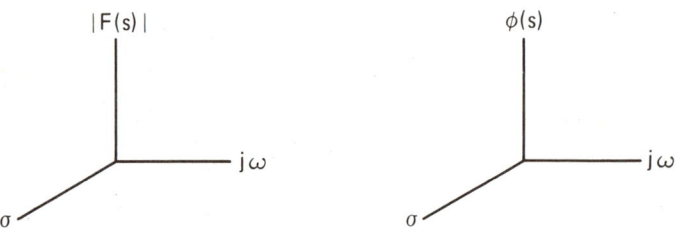

Figure 6-3. Two sets of coordinate axes are needed to plot the magnitude and phase functions associated with a complex function F(s).

Example 6-1. Investigate the real and imaginary parts and the magnitude and angle of the function F(s)=s+2.
Solution. Let s=σ+jω, then

$$F(s) = (\sigma + 2) + j\omega \quad (6-4)$$

from which Re{F(s)}=σ+2, Im{F(s)}=ω,

$$|F(s)| = \sqrt{(\sigma + 2)^2 + \omega^2} \quad (6-5)$$

The Complex Frequency Plane

and

$$\phi(\omega) = \tan^{-1}\left\{\frac{\omega}{\sigma+2}\right\} \qquad (6\text{-}6)$$

These functions could be plotted on coordinate axes similar to those shown in Figs. 6-2 and 6-3.

Example 6-2. Plot the magnitude function for the function $F(s)=1/(s+a)$.

Solution. Again, let $s=\sigma+j\omega$; then

$$F(s) = \frac{1}{\sigma + a + j\omega} \qquad (6\text{-}7)$$

and

$$|F(s)| = \frac{1}{\sqrt{(\sigma + a)^2 + \omega^2}} \qquad (6\text{-}8)$$

Figure 6-4 is a plot of Eq. (6-8).

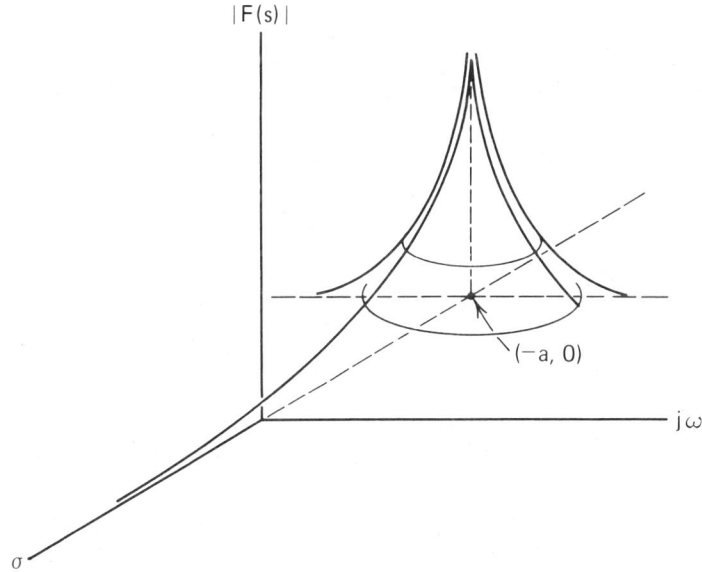

Figure 6-4. The magnitude plot for $F(s) = 1/(s + a)$ becomes unbounded at the point $(-a, 0)$. From the nature of this discontinuity we can see why the value $s = -a$ is called a "pole" of the function.

Example 6-2 illustrates the difficulty of plotting even the simplest of complex functions; it also shows why we use the term pole to describe the behavior at s=-a. Here the magnitude of F(s) tends toward infinity with a pole-like graphical representation. The magnitude of F(s) at an arbitrary point, say $\sigma_1+j\omega_1$, can be found by reading the value of $|F(\sigma_1+j\omega_1)|$ from the vertical axis.

Example 6-3. Plot the magnitude of the function F(s)=4/(s+2) as: (a) a function of ω for $\sigma=0$, and (b) as a function of σ for $\omega=0$.

Solution. (a) For $\sigma=0$ we have

$$|F(j\omega)| = \frac{4}{\sqrt{\omega^2 + 4}} \qquad (6-9)$$

Figure 6-5(a) is a sketch of this function. (b) If $\omega=0$, then

$$|F(\sigma)| = \frac{4}{|\sigma + 2|} \qquad (6-10)$$

with the graph of Fig. 6-5(b). In both figures we have assumed that the variables σ and ω take on both positive and negative values. If ω and σ both vary we need a three-dimensional plot as in Fig. 6-4.

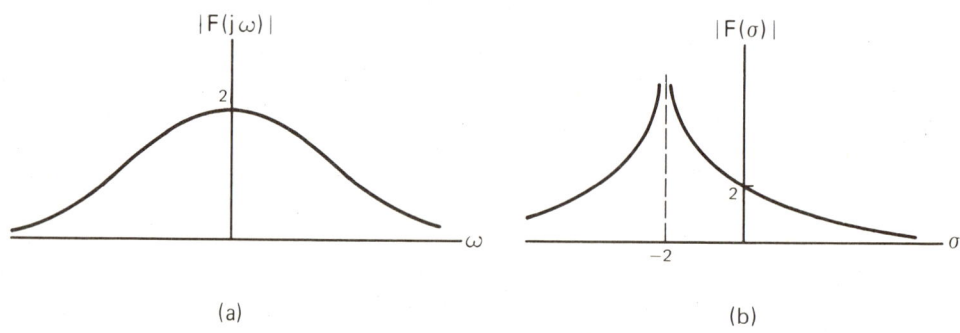

Figure 6-5. Magnitude plots for F(s) = 4/(s + 2). (a) The magnitude of F(s) for s = 0 + jω; (b) The magnitude of F(s) for s = σ + j0.

If it were necessary to resort to the cumbersome procedure of the previous examples to exploit the s-domain,

The Complex Frequency Plane

Chapter 6 would end right here. Fortunately, we can study the properties of F(s) from the simpler pole-zero diagram, such as the one in Fig. 6-1, without having to resort to three-dimensional functional plots. Moreover, as we shall see in a later chapter, much useful information can be gleaned by letting the variable σ be identically zero. This eliminates the need for one coordinate axis and lets us represent $|F(j\omega)|$ with a single two-dimensional plot.

Example 6-4. Identify the poles and zeros for the gust function, $f(t) = At\exp(-bt)$.

Solution. F(s) is found from Appendix D:

$$F(s) = \frac{A}{(s+b)^2} \qquad (6-11)$$

The pole-zero plot consists of a double pole at s=-b. There are no finite zeros, but some theoreticians find it useful to interpret Eq. (6-11) as having two zeros, both at $s \to \infty$. In the remaining material we ignore zeros or poles at infinity and consider only those which are finite.

Example 6-5. Plot the poles and zeros of the function

$$F(s) = \frac{25(s+2)(s-5)}{s(s+6)(s+3\pm j5)} \qquad (6-12)$$

Solution. Because F(s) is given in factored form, its poles and zeros are found by inspection. The pole-zero plot is shown in Fig. 6-6. The zeros are at s=5 and s=-2, while the poles are at s=0, s=-6, and s=-3±j5.

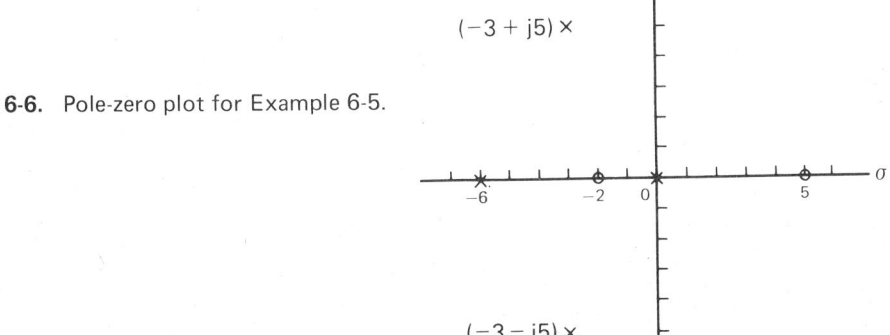

Figure 6-6. Pole-zero plot for Example 6-5.

This example should make it clear that plotting poles and zeros is an easier job than plotting complex functions. The real difficulty arises when the numerator or the denominator must be factored to find the poles and zeros.

Transform Pairs In The s-Plane

To gain an insight into the meaning of pole-zero plots in the s-plane, we will explore the relationship between pole locations in the s-plane and time-domain behavior for several commonly encountered Laplace transform pairs.[6]

<u>Pure integration or differentiation</u>. Recall that multiplication by s in the frequency domain corresponds to differentiation in the time domain and that division by s corresponds to integration. Figure 6-7 shows the s-plane representation of these operations.

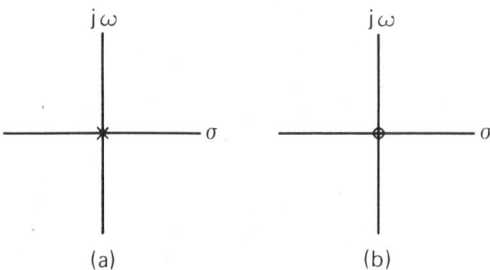

Figure 6-7. s-domain representation of integration and differentiation. (a) Integration, which corresponds to division by s, produces a pole at the origin; (b) Differentiation, which corresponds to multiplication by s, produces a zero at the origin.

<u>The first-order lag</u>. If $f(t)=\exp(-at)$ then

$$F(s) = \frac{1}{s+a} \qquad (6\text{-}13)$$

Figure 6-8 shows the s-plane and time-domain representations for this function, often called the first-order lag, for two positive values of the parameter a.

The Complex Frequency Plane

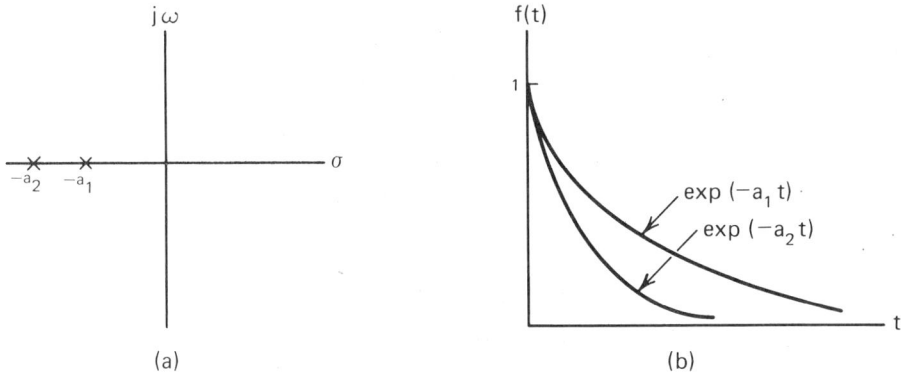

Figure 6-8. The first-order lag, $F(s) = 1/(s + a)$, for $a = a_1$ and $a = a_2$ with $a_2 > a_1$. (a) Pole-zero plots for the first-order lag; (b) Time responses for the first-order lag.

Larger values of the constant give transients which subside more rapidly and a pole which is further to the left in the s-plane. This situation is general: the further a pole is to the left in the s-plane, the more rapidly the time-domain transient decays.

<u>Pure oscillations</u>. The sinusoidal function $f(t) = \sin(\omega_o t)$ has the Laplace transform

$$F(s) = \frac{\omega_o}{s^2 + \omega_o^2} \tag{6-14}$$

The poles are at $s = \pm j\omega_o$ on the imaginary axis as shown in Fig. 6-9(a).

The cosine function $f(t) = \cos(\omega_o t)$ has

$$F(s) = \frac{s}{s^2 + \omega_o^2} \tag{6-15}$$

with poles at $s = \pm j\omega_o$ and a zero at $s=0$. These results have special significance for our study of the sinusoidal steady state in Chapter 8.

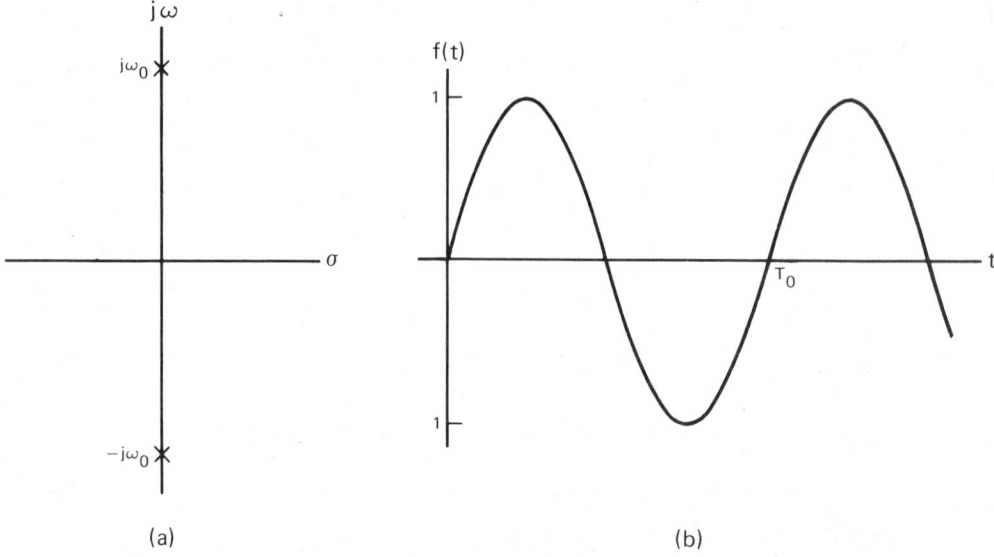

Figure 6-9. The s-domain function $F(s) = \omega_0/(s^2 + \omega_0^2)$ produces the sinusoidal response $f(t) = \sin(\omega_0 t)$. Note that $T_0 = 2\pi/\omega_0$. (a) Pole-zero plot; (b) Transient response.

The damped sinusoid. The s-domain function

$$F(s) = \frac{1}{s^2 + 2\zeta\omega_n s + \omega_n^2} \qquad (6-16)$$

generates the damped sinusoidal response

$$f(t) = \frac{1}{\omega_d} e^{-\sigma t} \sin(\omega_d t) \qquad (6-17)$$

where $\sigma = \zeta\omega_n$ and

$$\omega_d = \omega_n \sqrt{1-\zeta^2} \qquad (6-18)$$

This response form is valid for $\zeta < 1$. If $\zeta > 1$ the oscillations are overdamped and are represented by two real exponentials rather than a damped sinusoid. The pole locations for $\zeta < 1$ are found by factoring the denominator of Eq. (6-16). Apply-

The Complex Frequency Plane

ing the quadratic formula and designating the poles as p_1 and p_2 yields

$$p_1 = -\sigma + j\omega_d \quad \text{and} \quad p_2 = -\sigma - j\omega_d \quad (6\text{-}19)$$

Figure 6-10 shows the pole locations and a typical time response. Because poles and zeros always come in complex-conjugate pairs, pole-zero plots are symmetric about the real axis. The angle θ in Fig. 6-10(a) is given by $\theta = \sin^{-1}(\sigma/\omega_n)$ or $\theta = \sin^{-1}(\zeta)$.

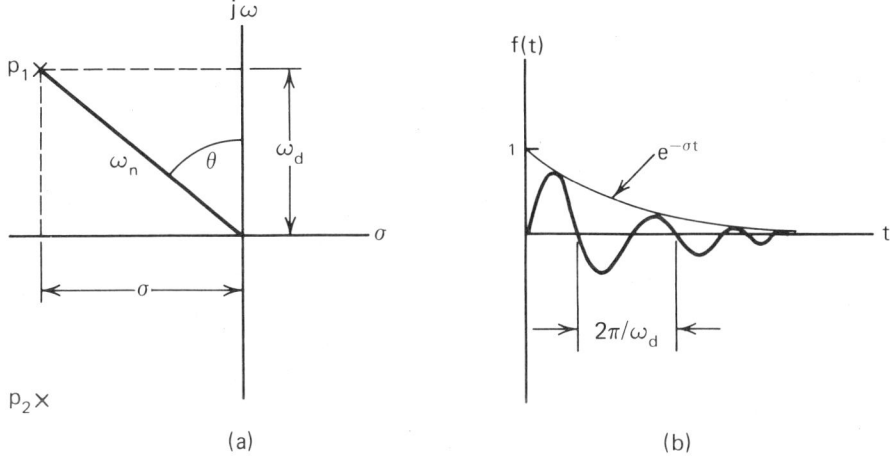

Figure 6-10. The s-domain function of Eq. (6-16) produces the damped sinusoidal time response $f(t) = e^{-\sigma t}\sin(\omega_d t)/\omega_d$. (a) Pole-zero plot; (b) Transient response.

Example 6-6. Investigate the locus of the poles of Eq. (6-16) (a) as ζ varies from 0 to 1 with ω_n constant, (b) as ω_n varies from 0 to infinity with constant ζ, and (c) as $\omega_n \to \infty$ with σ a constant.

Solution. (a) As ω_n is the radial distance from the origin to either of the poles, the root locus is a semicircle in the left half of the s-plane, starting at $\zeta = 0$ with the roots at $\pm j\omega_n$ on the imaginary axis and terminating on the $-\sigma$ axis where $\zeta = 1$ and the roots are real and equal. Figure 6-11(a) shows this behavior.

Glass—Linear Systems—19

(b) If ζ is constant the roots lie on a ray from the origin because $\theta = \sin^{-1}(\zeta)$ and for constant ζ, θ is also constant. As ω_n increases from 0, the roots move outward from the origin as shown in Fig. 6-11(b).

(c) Because σ is the distance from the imaginary axis to the roots, if σ is constant the roots lie on a vertical line σ units to the left of the imaginary axis. As ω_n increases the roots move further from the real axis as shown in Fig. 6-11(c). This behavior can be achieved only by simultaneously varying ζ and ω_n.

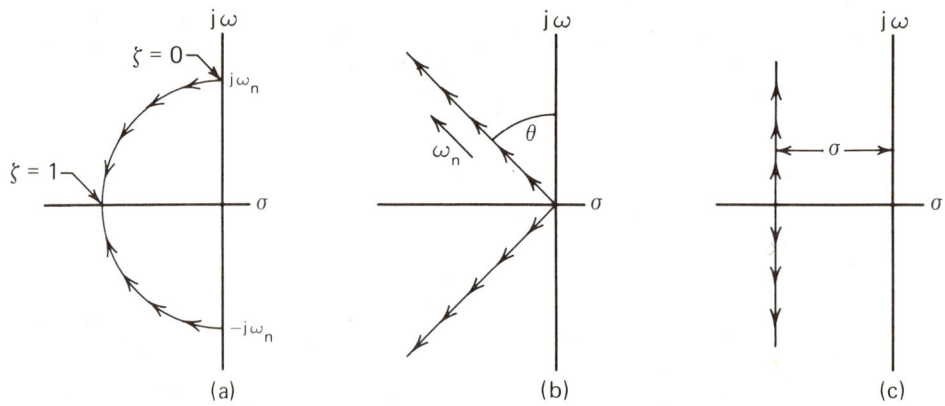

Figure 6-11. Paths of the roots of Eq. (6-16) for Example 6-6. A plot of the paths of the roots of an s-domain function as some parameter varies is called a *root locus*. (a) ζ varies from 0 to 1 with ω_n constant; (b) ω_n varies from 0 to ∞ with ζ constant; (c) ω_n varies from 0 to ∞ with σ constant.

Example 6-7. The path that the poles of a system follow as some system parameter changes is called a <u>root locus</u>. Sketch the root locus of the function

$$F(s) = \frac{K}{s(s + K)} \qquad (6\text{-}20)$$

as K varies from 0 to ∞.

Solution. One pole remains at the origin, s=0. When K=0, both poles are at the origin. As K increases, the pole at s=-K moves to the left along the negative real axis. Figure 6-12 is the root locus.

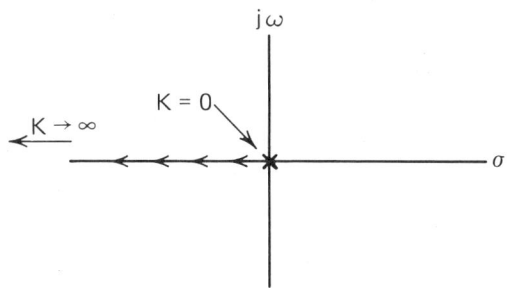

Figure 6-12. Root locus for Example 6-7.

The root-locus method is important in the design of automatic control systems, where the problem is to place the poles and zeros of a system to achieve a desired transient response. In system analysis on the other hand, the roots are fixed and the problem is to study the behavior of a given system--the poles cannot be moved.

Combined Response

If $F(s)$ is the product of two functions, say

$$F(s) = G(s)H(s) \qquad (6-21)$$

then the pole-zero plot for $F(s)$ is the superposition of the plots for $G(s)$ and $H(s)$ (except in the situation where poles and zeros cancel each other). The transient response $f(t)$, however, is not the sum of the transient response terms due to $G(s)$ and $H(s)$; the forms of the terms are the same but the magnitudes differ. These magnitudes can be found from $F(s)$ by a partial-fraction expansion, or they can be found graphically from the pole-zero diagram by the method explained in the next section. A final possibility is to realize that $f(t)$, given by

$$f(t) = L^{-1}\{G(s)H(s)\} \qquad (6-22)$$

is the convolution of the functions $g(t)=L^{-1}\{G(s)\}$ and $h(t)=L^{-1}\{H(s)\}$ as derived in Chapter 5. As a practical matter we prefer to evaluate f(t) by Laplace methods rather than by time-domain convolution.

Example 6-8. A composite system has $F(s)=G(s)H(s)$ where

$$G(s) = \frac{25(s + 2)}{s(s + 25)} \tag{6-23}$$

and

$$H(s) = \frac{s(s + 1.5)}{(s + 2)(s^2 + 16)} \tag{6-24}$$

Identify the poles and zeros of F(s).

Solution. F(s) is given by

$$F(s) = \left\{\frac{25(s + 2)}{s(s + 25)}\right\} \frac{s(s + 1.5)}{(s + 2)(s^2 + 16)} \tag{6-25}$$

After taking cancellations into account, F(s) has a zero at s=-1.5 and poles at s=-25 and s=±j4.

6.2 EVALUATING RESIDUES IN THE s-PLANE

If F(s) is a proper rational function with no mulitiple poles, it can be written in the partial-fraction form

$$F(s) = \frac{A_1}{s - p_1} + \frac{A_2}{s - p_2} + \ldots \frac{A_n}{s - p_n} \tag{6-26}$$

with the inverse transform

$$f(t) = A_1 e^{p_1 t} + A_2 e^{p_2 t} + \ldots A_n e^{p_n t} \tag{6-27}$$

where constants A_i are the residues of the poles p_i. Our problem is to determine the A_i from the pole-zero diagram.[7,8] Suppose F(s) has the form

$$F(s) = K\frac{(s - z_1)(s - z_2)\ldots(s - z_m)}{(s - p_1)(s - p_2)\ldots(s - p_n)} \tag{6-28}$$

The Complex Frequency Plane

where m<n and all poles are assumed to be simple (non-repeated). As shown in Chapter 5, the residues A_i for Eq. (6-28) are determined from

$$A_i = (s - p_i)F(s)\Big|_{s=p_i} \tag{6-29}$$

or, using Eq. (6-28) for $F(s)$,

$$A_i = \frac{K(p_i-z_1)(p_i-z_2)\ldots(p_i-z_m)}{(p_i-p_1)(p_i-p_2)\ldots(p_i-p_{i-1})(p_i-p_{i+1})\ldots(p_i-p_n)} \tag{6-30}$$

Note the skip from (p_i-p_{i-1}) to (p_i-p_{i+1}) in the denominator which indicates the cancelled term (p_i-p_i). We can interpret the terms in Eq. (6-30) on the complex s-plane using vector algebra (see Fig. 6-13).

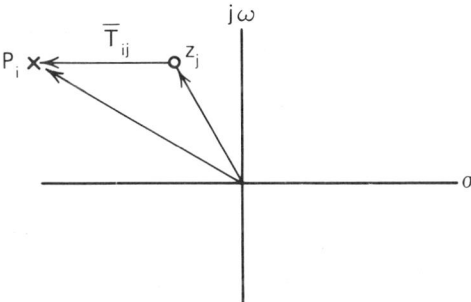

Figure 6-13. The quantity $\overline{T}_{ij} = p_i - z_j$ can be interpreted as a vector from the point z_j to the point p_i in the s-plane.

The quantities p_i and z_j are vectors from the origin to the points p_i and z_j in the s-plane. A vector $\overline{T}_{ij}$ from z_j to p_i satisfies the head-to-tail vector addition rule,

$$\overline{T}_{ij} + z_j = p_i \tag{6-31}$$

or

$$\overline{T}_{ij} = p_i - z_j \tag{6-32}$$

As a result, the terms in Eq. (6-30) represent vectors from the poles and zeros to each other. Define

$$T_{ij} = |p_i - z_j| \qquad (6\text{-}33)$$

$$B_{ij} = |p_i - p_j| \qquad (6\text{-}34)$$

$$\theta_{ij} = \text{Arg}\{p_i - z_j\} \qquad (6\text{-}35)$$

and

$$\phi_{ij} = \text{Arg}\{p_i - p_j\} \qquad (6\text{-}36)$$

where the vertical bars mean magnitude of the complex quantity and Arg{ } means take angle of the bracketed vector quantity. Using these definitions, we write Eq. (6-30) as

$$|A_i| = \frac{K(T_{i1})(T_{i2})(T_{i3})\cdots(T_{im})}{(B_{i1})(B_{i2})\cdots(B_{i,i-1})(B_{i,i+n})\cdots B_{in}} \qquad (6\text{-}37)$$

with

$$\text{Arg}\{A_i\} = \theta_{i1} + \theta_{i2} + \cdots \theta_{im} - (\phi_{i1} + \phi_{i2} + \cdots \phi_{i,i-1} + \phi_{i,i+1} + \cdots \phi_{in}) \qquad (6\text{-}38)$$

All of the magnitudes and angles can be measured on the pole-zero diagram with a ruler and a protractor. Figure 6-14 shows the geometry for a pole-zero plot containing three poles and two zeros.

The figure shows the magnitudes and angles for computing A_3, the residue for p_3. The result is

$$|A_3| = \frac{(T_{31})(T_{32})}{(B_{31})(B_{32})} \qquad (6\text{-}39)$$

with $\text{Arg}\{A_3\} = \theta_{31} + \theta_{32} - \phi_{31} - \phi_{32}$. The vectors are drawn from the other poles and zeros <u>toward</u> the pole whose residue is being found, and the angles θ and ϕ are measured from a line parallel to the positive real axis counterclockwise around to the appropriate vector.

The Complex Frequency Plane

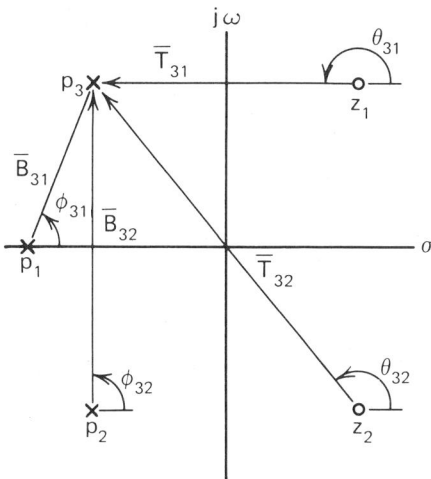

Figure 6-14. Residues can be evaluated by using angles and distances between the s-domain poles and zeros. This pole-zero plot and Eq. (6-39) illustrate the procedure for a system with 3 poles and 2 zeros.

The gain K in Eq. (6-37) does not appear on the pole-zero diagram. It is determined by rearranging F(s) as

$$F(s) = K\frac{(s^m + \ldots)}{(s^n + \ldots)} \qquad (6\text{-}40)$$

where the leading coefficients of the polynomials are 1's. The residue for the pole with the positive imaginary part can be used with Eq. (5-149) to find the time response corresponding to a complex pair. It is never necessary to compute the residue for the other pole.

Example 6-9. Do a graphical partial-fraction expansion for the function $F(s)=(10s-20)/(2s^2+16s)$.

Solution. Arrange F(s) in the standard form

$$F(s) = \frac{5(s-2)}{s(s+8)} = \frac{A_1}{s} + \frac{A_2}{s+8} \qquad (6\text{-}41)$$

Figure 6-15 is the pole-zero plot.

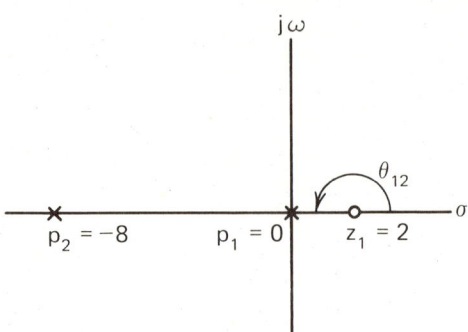

Figure 6-15. Pole-zero plot for Example 6-9.

Let $p_1=0$, $p_2=-8$, $z_1=2$, and $K=5$. Then the magnitude of residue A_1 is $|A_1|=KT_{11}/B_{12}=5(2)/(8)=5/4$. The angle or argument of A_1 is found from $\text{Arg}\{A_1\}=\theta_{12}-\phi_{12}=180°-0°$. As an angle of $180°$ corresponds to a minus sign, we can express A_1 as

$$A_1 = \frac{5}{4}\underline{/180°} = -\frac{5}{4} \tag{6-42}$$

Similarly, $|A_2|=KT_{21}/B_{21}=5(10)/(8)=25/4$ with $\text{Arg}\{A_2\}=0°$. The partial-fraction expansion is then

$$F(s) = \frac{-5}{4s} + \frac{25}{4(s+8)} \tag{6-43}$$

as can be verified by applying the classical partial-fraction method.

Example 6-10. Use graphical methods to find $f(t)$ for the function $F(s)$ whose pole-zero plot is shown in Fig. 6-16. Assume $K=25$.

Solution. The only residue we need is A_1, which corresponds to the pole at p_1. From the geometry of the figure $|A_1|=(25)(6\sqrt{2})/(12)=17.68$ with $\text{Arg}\{A_1\}=45°$. Using A_1 and Eq. (5-149) we have the inverse transform

$$f(t) = 35.36e^{-5t}\cos(6t + \pi/4) \tag{6-44}$$

The Complex Frequency Plane

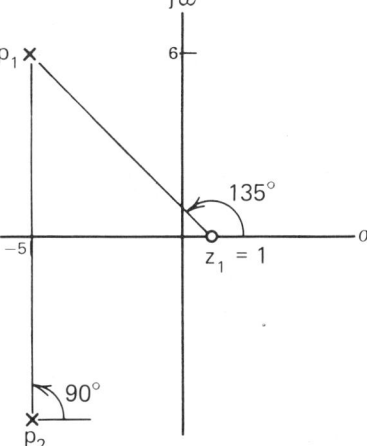

Figure 6-16. Pole-zero plot for Example 6-10.

Dominant Poles

Many practical dynamic systems contain transient phenomena which persist for such a brief time that they can be neglected in the analysis. The ability to identify such transients and to neglect those parts of the system model which cause them, and to do this without compromising the remainder of the model, is one of the marks of a good systems engineer. Although it is generally impossible to decide what to neglect by inspecting the differential equations, the information is readily available from the pole-zero diagram.

Remote poles. We have seen that a pole-zero plot similar to that in Fig. 6-17(a) gives rise in the time domain to the function

$$y(t) = K_1 e^{-a_1 t} + K_2 e^{-a_2 t} \qquad (6-45)$$

If $a_2 \gg a_1$, the second term subsides much more rapidly than the first (see Fig. 6-17(b)). As a consequence, for values of time greater than approximately five time constants, in this case $5/a_2$, only the effects of the $\exp(-a_1 t)$ transient term will be visible in the output. For this reason we say that the pole at $-a_1$ is <u>dominant</u> over the one at $-a_2$ when $a_2 \gg a_1$.

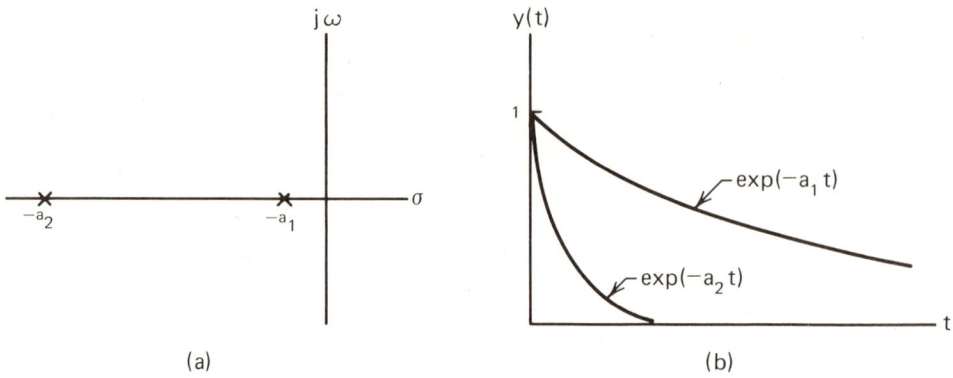

Figure 6-17. Poles near the jω-axis are said to be dominant over those remote from the jω-axis because the transient response terms due to the latter subside more quickly. In this system the pole at $s = a_2$ might be neglected if $a_2 > 10a_1$.
(a) Pole-zero plot; (b) Time response terms.

As a rule of thumb, control system engineers often take the "much, much greater" symbol to mean a factor of at least 10. In this instance we would neglect the pole at $s=-a_2$ provided $a_2>10a_1$ and we are interested in the behavior for $t>5/a_2$. If, on the other hand, we are concerned with the behavior for $t<5/a_2$, then the $\exp(-a_1 t)$ term is relatively constant and the $\exp(-a_2 t)$ term determines the transient behavior. In most practical dynamic systems work we are more concerned with the former situation, neglecting fast transients (poles remote from the jω-axis).

Finally, the concepts of dominant and remote poles also apply when the poles are complex. Although the pole-zero plot of Fig. 6-18 shows a pair of complex-conjugate poles, which would indicate a damped sinusoid in the transient response, for $t>5\sigma_1$ the behavior can be modeled quite well by the real exponential $f(t)=K\exp(-\sigma_1 t/10)$ because the complex pole-pair is remote compared to the real pole at $s=-\sigma_1/10$.

<u>The effect of zeros.</u> The decision as to which poles are dominant is complicated by the presence of zeros. In the diagram of Fig. 6-19 for example the pole at p_1 appears to dominate the complex pair at p_2-p_2^* as the latter are more than 10 times further from the jω-axis. Observe, however, the zero

The Complex Frequency Plane

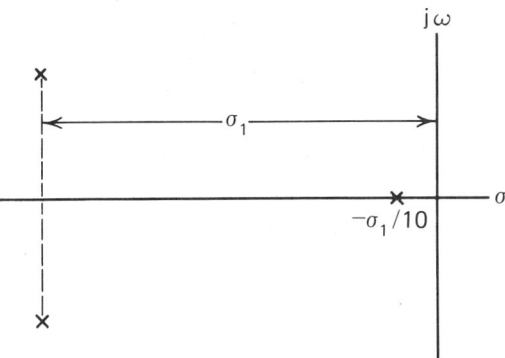

Figure 6-18. The idea of dominant poles also applies if some of the poles are complex. In this instance the complex poles are remote compared to the real pole at $s = -\sigma/10$.

near p_1 and recall from the previous section (see Eq. (6-37)) that the residue for p_1 has in its numerator the product of the distances from zeros to p_1. As the distance from z_1 to p_1 is quite small, we can expect the residue for p_1 to be correspondingly small. Hence, even though the transient due to p_1 persists for a long time, its magnitude is small. The pole pair at p_2-p_2^*, with a large residue due to their distance from z_1, may actually dominate. No general rules exist for establishing which poles are dominant in systems containing zeros: it is a question of engineering judgement.

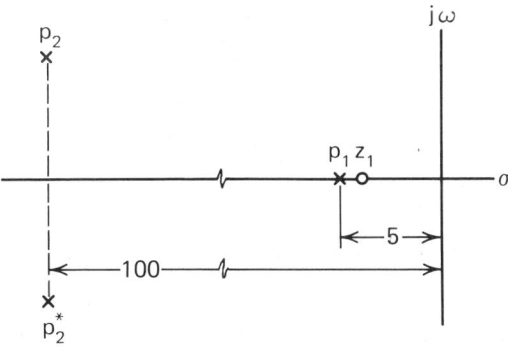

Figure 6-19. A nearby zero can offset the effects of a pole which would otherwise be dominant.

How to discard a pole.

When discarding a remote pole we must be careful to avoid distorting the residues of the remaining poles. As an illustration of this idea, consider the function

$$F(s) = \frac{1}{(s+1)(s+1000)}$$

which has a remote pole at s=-1000. We should be able to disregard this pole if we are interested in the solution for t>5/1000. The exact inverse transform of F(s) is

$$f(t) = \frac{e^{-t}}{999} - \frac{e^{-1000t}}{999}$$

For large t, f(t) is approximately $f(t) \simeq \exp(-t)/999$. If, however, we just discard the term (s+1000) in F(s) the remaining function $F_A(s)$ is

$$F_A(s) = \frac{1}{s+1} \qquad (6\text{-}46)$$

with the inverse transform $f_A(t)=\exp(-t)$. This is **not** the desired result because the coefficient is 1 rather than the correct value of 1/999. In general, to obtain the correct signal level when removing a remote pole, we must express the transfer function in "time-constant" form, or with the zeroth powers of s all 1. For the F(s) given above we write

$$F(s) = \frac{1}{1000(s+1)(.001s+1)} \qquad (6\text{-}47)$$

and when the term containing the remote pole is removed, the approximating function $F_A(s)$ is

$$F_A(s) = \frac{1}{1000(s+1)} \qquad (6\text{-}48)$$

The inverse transform of this expression, $f_A \simeq \exp(-t)/1000$, is the correct approximation.

Example 6-11. Find an approximate second-order transfer function for F(s) by neglecting the remote poles if

$$F(s) = \frac{1}{s(s+1)(s+10^3)(s^2+100s+10^4)} \qquad (6\text{-}49)$$

The Complex Frequency Plane

Solution. Rearrange F(s) in time-constant form:

$$F(s) = \frac{10^{-7}}{s(s+1)(10^{-3}s+1)(10^{-4}s^2+10^{-2}s+1)} \quad (6\text{-}50)$$

Neglecting the pole at s=-1000 and the complex pair whose real part is $-\zeta\omega_n=-50$ leads to the approximate transfer function

$$F_A(s) = \frac{10^{-7}}{s(s+1)} \quad (6\text{-}51)$$

Clearly, it is easier to find f(t) from Eq. (6-51) than from Eq. (6-49), and because the transient behavior for t>5(50)=250 is approximately the same for both functions, a worthwhile simplification has been achieved.

6.3 STABILITY

Most treatments of linear-system theory devote a great deal of space to the topic of stability, probably because it is easy to develop a variety of elegant theorems and proofs on the subject. In the world of real systems stability is often a binary proposition in that a system is either stable or it isn't. If one is concerned with analyzing proposed system designs, then it certainly makes sense first to check to see if the criteria for stability are met. If not, the job of the analyst is ended, and it is back to the drawing board. Our approach to stability theory is pragmatic: we will develop only enough analytical tools to establish quickly whether or not a given system is stable.[9,10]

But just exactly what do we mean by the term stability? Although a number of sophisticated definitions exist for several types of stability, we will be content with the following proposition:

> A linear system is <u>stable</u> if the natural response for finite initial conditions remains bounded as t→∞.

By this definition a purely oscillatory system, one with poles on the jω-axis, is stable because a pure sinusoid remains

bounded even though it does not decay to zero. Many authors
call such oscillatory systems <u>conditionally stable</u>, and we
will follow their lead.

The definition of stability applies only to the natural
or force-free response. We would not be surprised if an
unbounded input produced an unbounded output even for a stable
system. For this reason, we must exclude forcing function inputs from the study of stability. The definition also makes
sense from an energy viewpoint. Initial conditions correspond
to energy stored in the system, and we would expect a finite
amount of stored energy to produce a transient output which
would remain bounded. An unbounded output would require an
infinite quantity of stored energy. A forcing function, on
the other hand, could deliver an arbitrarily large quantity of
energy to the system over a long enough period of time.

Stability From Pole Locations

Linear systems give rise to s-domain output functions
which can always (except for the multiple pole case) be
written as

$$Y(s) = \frac{A_1}{s - p_1} + \frac{A_2}{s - p_2} + \cdots \frac{A_n}{s - p_n} \qquad (6\text{-}52)$$

with time response

$$y(t) = A_1 e^{p_1 t} + A_2 e^{p_2 t} + \cdots A_n e^{p_n t} \qquad (6\text{-}53)$$

According to the definition of stability, $y(t)$ is the response
of a stable system if

$$\lim_{t \to \infty} \{y(t)\} = \lim_{t \to \infty} \{A_1 e^{p_1 t} + A_2 e^{p_2 t} + \cdots A_n e^{p_n t}\} \qquad (6\text{-}54)$$

remains bounded. This implies that every term of the form

$$K_i = \lim_{t \to \infty} \{\exp(p_i t)\} \qquad (6\text{-}55)$$

The Complex Frequency Plane

produces a finite value for K_i. In general the p_i may be complex and of the form $p_i = -\sigma_i \pm j\omega_i$ so that Eq. (6-55) becomes

$$K_i = \lim_{t \to \infty}\left\{\exp\{(-\sigma_i \pm j\omega_i)t\}\right\} \qquad (6\text{-}56)$$

$$= \lim_{t \to \infty}\left\{\exp(-\sigma_i t)\{\cos(\omega_i t) \pm j\sin(\omega_i t)\}\right\} \qquad (6\text{-}57)$$

Because the sinusoidal terms in Eq. (6-57) are bounded with maximum amplitudes of 1, the condition for stability is that the exponential term decays with time. This in turn requires that each σ_i be a positive number, or that the real part of every pole must be negative (or 0 for conditional stability). The left half of the s-plane contains poles whose real parts are negative, hence we have the condition that:

> A linear system is stable if all of its poles lie in the left half of the complex s-plane (or on the $j\omega$-axis for conditional stability).

The zeros of $Y(s)$ have no affect on stability--only the poles are significant. Further, a system need have only one "bad" pole to be unstable. The function

$$Y(s) = \frac{5s(s-10)(s+2)}{(s+50)(s+5)(s+1)(s-3)} \qquad (6\text{-}58)$$

is unstable because the pole at s=3 is in the right half of the s-plane and gives rise to the growing exponential exp(+3t) in the inverse transform. The zero at s=10 in the right-half plane does not effect stability.

Stabilization by pole cancellation. A plausible approach to stabilizing an unstable system--an approach which has often been tried with sad results--is to try to cancel an unstable pole with a zero. If we had the system function

$$G(s) = \frac{50}{(s-5)(s+1)} \qquad (6\text{-}59)$$

we might be tempted to modify the system to include a zero at s=5, cancelling the bad pole and stablizing the system. The resultant transfer function is

$$G(s) = \frac{50(s-5)}{(s-5)(s+1)} = \frac{50}{s+1} \quad (6\text{-}60)$$

The difficulty is that with real-world system components--resistors, capacitors, inductors, masses, springs, dampers, etc.--we can never achieve exact cancellation. The result of an inexact cancellation is in this case the time-domain term $\varepsilon\exp(5t)$ where ε is a small constant which would be zero for exact cancellation. If ε is nonzero the growing exponential would eventually achieve a value large enough to destroy the system or to drive it into a nonlinear mode of operation (saturation).

Two practical ways to stabilize an unstable system are (a) to redesign the system to eliminate the cause of the instability, or (b) to incorporate feedback around the system to drive the unstable poles into the left half of the s-plane. This latter option is explored in Chapter 7.

Relative stability. If we interpret light damping as relative instability, then the further the poles are from the $j\omega$-axis the more relatively stable the system; and the nearer they are to the $j\omega$-axis the more relatively unstable the system. The system of Fig. 6-20(a) is relatively stable compared to that of Fig. 6-20(b) because its dominant poles are further from the $j\omega$-axis. The zeros do not affect stability.

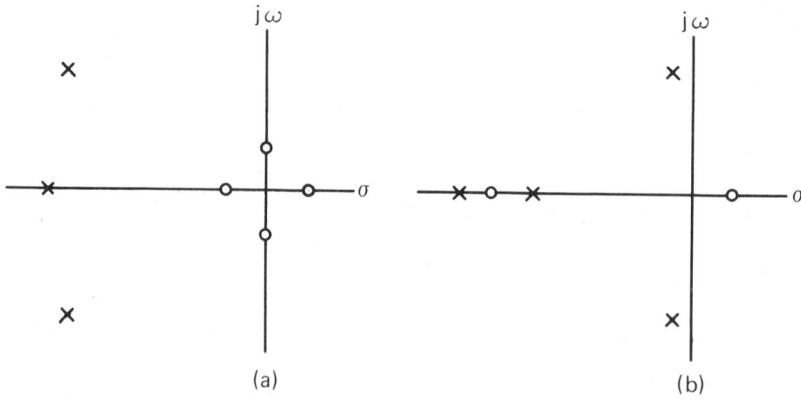

Figure 6-20. Systems with complex poles near the $j\omega$-axis are said to be relatively unstable compared to those whose complex poles are further from the axis. In this instance system (b) is relatively unstable compared to system (a).

The Complex Frequency Plane

The Routh Criterion

The difficulty with using pole-zero plots to determine stability is that one must first find the poles. This involves finding the roots of the characteristic equation, a burdensome task for higher-order systems. The Routh criterion is a convenient method for determining stability without factoring the characteristic equation.[11,12] This section shows how to use the Routh criterion but does not prove its validity. Proofs are found in books on advanced linear-system theory.

Suppose the s-domain polynomial $D(s)$ given by

$$D(s) = a_n s^n + a_{n-1} s^{n-1} + \ldots a_1 s + a_0 \qquad (6-61)$$

is the denominator polynomial of a transfer function whose stability we seek to determine. The first thing to realize is that all of the a_i must have the same algebraic signs. If even one of the a_i differs in sign from the rest, then the polynomial has at least one root in the right half of the s-plane and the system is unstable. In addition, if even one of the a_i, other than a_n or a_0, is 0, then the system can be at best conditionally stable. That is, it will have poles in the right half of the s-plane or on the $j\omega$-axis. If the polynomial passes these inspection tests we are ready to form the Routh array.

The Routh array. Starting with the coefficients a_i of Eq. (6-61), form the array

s^n	a_n	a_{n-2}	a_{n-4}	a_{n-6}
s^{n-1}	a_{n-1}	a_{n-3}	a_{n-5}	a_{n-7}
s^{n-2}	b_1	b_2	b_3	
s^{n-3}	c_1	c_2	c_3	
$\vdots$				
s^1	K_1			
s^0	K_0			

where

$$b_1 = (a_{n-1}a_{n-2} - a_n a_{n-3})/a_{n-1} \qquad (6\text{-}62)$$

$$b_2 = (a_{n-1}a_{n-4} - a_n a_{n-5})/a_{n-1} \qquad (6\text{-}63)$$

$$b_3 = (a_{n-1}a_{n-6} - a_n a_{n-7})/a_{n-1} \qquad (6\text{-}64)$$

and

$$c_1 = (b_1 a_{n-3} - a_{n-1} b_2)/b_1 \qquad (6\text{-}65)$$

$$c_2 = (b_1 a_{n-5} - a_{n-1} b_3)/b_1 \qquad (6\text{-}66)$$

$$c_3 = (b_1 a_{n-7} - a_{n-1} b_4)/b_1 \qquad (6\text{-}67)$$

This pattern is continued until all terms are calculated. The Routh criterion then states that the number of roots of the polynomial D(s) with positive real parts equals the number of sign changes in the first column of the Routh array. If there are no sign changes in the first column the system is stable.

Example 6-12. Investigate the stability of the system whose characteristic equation is

$$D(s) = s^4 - 3s^3 + s^2 + 5s + 2 \qquad (6\text{-}68)$$

Solution. The system is unstable because all of the coefficients of D(s) do not have the same algebraic signs.

Example 6-13. Investigate the stability of

$$D(s) = s^4 + s^2 + s + 1 \qquad (6\text{-}69)$$

Solution. The s^3 term is missing so the system is either unstable or oscillatory.

Example 6-14. Use the Routh criterion to investigate the stability of the system whose characteristic equation is

$$D(s) = s^4 + 2s^3 + 3s^2 + 5s + 10 \qquad (6\text{-}70)$$

The Complex Frequency Plane

Solution. The polynomial passes the inspection tests. The Routh array is

$$
\begin{array}{c|ccc}
s^4 & 1 & 3 & 10 \\
s^3 & 2 & 5 & \\
s^2 & 1/2 & 10 & \\
s^1 & -35 & & \\
s^0 & 10 & &
\end{array}
$$

The first column of the array has two sign changes, hence the characteristic equation has two roots in the right-half plane and the system is unstable.

The Routh criterion breaks down in two special cases:

A zero row. If all of the coefficients in a row are 0, it is impossible to complete the Routh array as this would require division by 0. If this happens we form an <u>auxiliary polynomial</u> (see Example 6-15) from the coefficients of the preceeding row. The auxiliary polynomial is a factor of the original polynomial, so we use long division to find two polynomials whose product is the original function. Now examine the roots of the two polynomials separately using the Routh criterion.

An alternate approach, after finding the auxiliary polynomial, is to differentiate it and use the coefficients of the resulting polynomial in the row which was formerly zero. Then continue with the Routh array.

Example 6-15. Investigate the stability of the system whose characteristic equation is

$$D(s) = s^4 + s^3 + 5s^2 + 4s + 4 \qquad (6\text{-}71)$$

Solution. The Routh array is

$$
\begin{array}{c|ccc}
s^4 & 1 & 5 & 4 \\
s^3 & 1 & 4 & \\
s^2 & 1 & 4 & \\
s^1 & 0 & & \\
s^0 & & &
\end{array}
$$

The s^1 row has all 0 coefficients. The auxiliary polynomial A(s) formed from the s^2 row is $A(s)=s^2+4$. The auxiliary equation is a factor of the original equation. By long division we find that

$$s^4 + s^3 + 5s^2 + 4s + 4 = (s^2 + 4)(s^2 + s + 1) \quad (6\text{-}72)$$

A quadratic characteristic equation is stable if all of its coefficients are positive (see Example 6-16). Hence the system is conditionally stable because of the poles at $s=\pm j2$.

Using the alternate approach we would differentiate A(s) to obtain $A'(s)=2s+0$ and replace the zero row in the Routh array with the coefficients of A'(s). The final two rows of the array would then be

$$\begin{array}{ccc} s^1 & 2 & 0 \\ s^0 & 4 & \end{array}$$

and, as the first column contains no sign changes, the system is either stable or conditionally stable.

A Zero Coefficient In The First Column

If the Routh array contains a 0 element in the first column we cannot complete the array as this would involve division by 0. If this situation occurs we can use either of the following methods:

1. Substitute a small positive quantity ε for the 0 and proceed to evaluate the array. Then let $\varepsilon \to 0$ and see what signs the associated terms have in the limit.

2. Substitute $s=1/g$ in the original equation. Then form the Routh array for the g-domain polynomial. If the g-polynomial is stable so is the s-polynomial.

Example 6-16. Prove that a second-order equation is stable if all of its coefficients are positive.

Solution. Let the general second-order characteristic equation be $D(s)=s^2+as+b$. The Routh array is

$$\begin{array}{ccc} s^2 & 1 & b \\ s^1 & a & \\ s^0 & b & \end{array}$$

The Complex Frequency Plane

The first column of the Routh array has no sign changes if the coefficients a and b are both positive.

Example 6-17. Derive the conditions for stability of the general third-order equation

$$D(s) = s^3 + as^2 + bs + c \qquad (6\text{-}73)$$

Solution. The Routh array is

$$\begin{array}{ccc} s^3 & 1 & b \\ s^2 & a & c \\ s^1 & \dfrac{ab-c}{a} & \\ s^0 & c & \end{array}$$

The conditions for stability are a>0, c>0, and ab>c. The latter condition, along with a>0, also implies that b>0. Hence if all coefficients are positive and ab>c the third-order equation is stable.

PROBLEMS FOR CHAPTER 6

6-1. Plot the poles and zeros of the following s-domain functions:

(a) $F(s) = \dfrac{s(s+1)}{s^2 + 16}$ \qquad (6-74)

(b) $F(s) = \dfrac{s(s^2+4)}{(s+2)(s-3)}$ \qquad (6-75)

(c) $Y(s) = \dfrac{s^2(s+2)}{s^2 + 4s + 16}$ \qquad (6-76)

(d) $G(s) = \dfrac{K}{s^2(s^2+9)(s^2+2s+8)}$ \qquad (6-77)

6-2. List the poles and zeros of the functions which have the following time-domain representations: (a) $f(t) = u(t) - e^{-3t}$, (b) $y(t) = \cos(4t) - \sin(4t)$, (c) $g(t) = te^{-3t} + t$, and (d) $h(t) = e^{-t} + e^{-2t} - 5e^{2t}$.

6-3. Evaluate the magnitude and angle of the function $F(s)=s(s+2)/(s^2+5)$ when: (a) $s=-1$, (b) $s=j4$, and (c) $s=1+j2$.

6-4. Sketch the magnitude and phase of the following functions for $s=j\omega$: (a) $F(s) = 1/s$, (b) $F(s)=1/(s^2+a^2)$, (c) $F(s)=1/\{s(s+2)\}$, and (d) $Y(s)=(s+a)/(s+10a)$.

6-5. The second-order system in the standard form below is often encountered in systems work. Find the value of ω for which $|F(j\omega)|$ is a maximum.

$$F(s) = \frac{1}{s^2 + 2\zeta\omega_n s + \omega_n^2} \qquad (6\text{-}82)$$

6-6. Repeat Problem 6-5 if

$$F(s) = \frac{s}{s^2 + 2\zeta\omega_n s + \omega_n^2} \qquad (6\text{-}83)$$

6-7. Plot the locus of the roots of the following functions as K varies from 0 to ∞.

(a) $Y(s) = \dfrac{1}{s + K}$ (6-84)

(b) $F(s) = \dfrac{K}{s^2(s + K)}$ (6-85)

(c) $H(s) = \dfrac{K}{s^2 + 2as + K}$ (6-86)

(d) $F(s) = K + \dfrac{1}{s}$ (6-87)

(e) $Y(s) = \dfrac{1}{s^2 + 2Ks + K^2}$ (6-88)

(f) $Y(s) = \dfrac{K}{s^2 + 2Ks + 4}$ (6-89)

6-8. Find the s-domain functions which have the pole-zero plots shown. Assume $K=1$.

6-9. Use graphical methods to find partial-fraction expansions for the following functions:

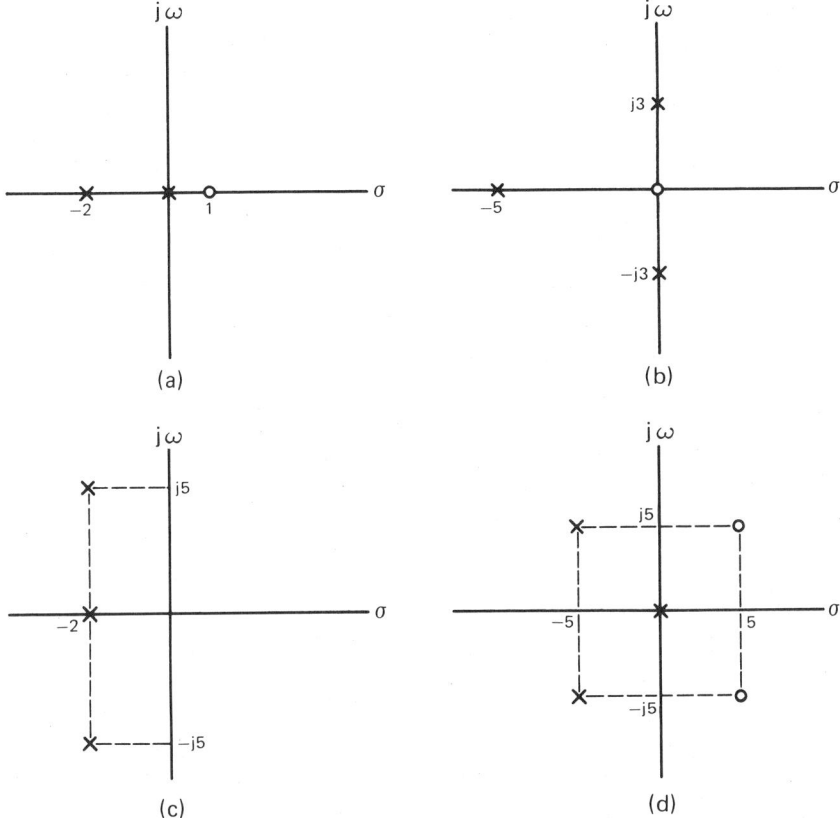

Figure 6-21. Problem 6-8.

(a) $G(s) = \dfrac{5}{s(s+2)(s+10)}$ (6-90)

(b) $G(s) = \dfrac{s+2}{s(s^2+16)}$ (6-91)

(c) $G(s) = \dfrac{s+1}{s(s^2+4s+16)}$ (6-92)

(d) $G(s) = \dfrac{s^2+4}{s(s+5)(s^2+25)}$ (6-93)

6-10. Derive a function which represents the absolute value of the error as a function of time which results from neglecting the remote pole in the function

$$F(s) = \frac{1}{(s + a)(s + 10a)} \quad (6\text{-}94)$$

6-11. Using the one-tenth rule for remoteness, simplify the following transfer functions by neglecting remote poles:

(a) $\quad G(s) = \dfrac{5}{s(s + 2)(s + 50)} \quad (6\text{-}95)$

(b) $\quad G(s) = \dfrac{2(s + 10)}{s(s + 5)(s^2 + 100s + 10000)} \quad (6\text{-}96)$

(c) $\quad G(s) = \dfrac{25}{s(0.01s + 1)(2s + 1)} \quad (6\text{-}97)$

6-12. Identify the following functions as being stable, conditionally stable, or unstable:

(a) $\quad F(s) = \dfrac{s}{(s + 1)(s + 2)} \quad (6\text{-}98)$

(b) $\quad F(s) = \dfrac{s - 3}{s(s^2 + 4)} \quad (6\text{-}99)$

(c) $\quad F(s) = \dfrac{s + 5}{s(s - 8)} \quad (6\text{-}100)$

(d) $\quad F(s) = \dfrac{s(s - 2)}{(s - 3)(s^2 + 4)} \quad (6\text{-}101)$

(e) $\quad F(s) = \dfrac{s^2 + 10}{s^2 + 8s + 36} \quad (6\text{-}102)$

(f) $\quad F(s) = \dfrac{1}{s + 2} - \dfrac{1}{s^2 + 16} \quad (6\text{-}103)$

(g) $\quad H(s) = \dfrac{1}{s^2} \quad (6\text{-}104)$

(h) $\quad F(s) = \dfrac{1}{s^4 + 16} \quad (6\text{-}105)$

(i) $\quad F(s) = \dfrac{s(s + 4)}{s^2 - 3s + 24} \quad (6\text{-}106)$

6-13. Use the Routh criterion to investigate the stabil-

The Complex Frequency Plane

ity of the systems having the following characteristic equations:

(a) $D(s) = s^5 + 4s^4 + 6s^3 + 6s^2 + 6s + 3$ (6-107)

(b) $D(s) = s^3 + 8s^2 + 3s + 5$ (6-108)

(c) $D(s) = s^5 + s^4 + 2s^3 + 4s + 15$ (6-109)

(d) $D(s) = s^5 + s^4 + s^3 + s^2 + 2s$ (6-110)

(e) $D(s) = s^4 + 2s^3 + 11s^2 + 18s + 18$ (6-111)

6-14. Use the Routh criterion to derive the conditions for stability for the general fourth-order function:

$$D(s) = s^4 + as^3 + bs^2 + cs + d \qquad (6-112)$$

REFERENCES FOR CHAPTER 6

(1) Gille, J. C., Pelegrin, M. J. and P. Decaulne: <u>Feedback Control Systems, Analysis, Synthesis, and Design</u>, McGraw-Hill Book Co., Inc., New York, 1959.

(2) Barbe, E. C.,: <u>Linear Control Systems</u>, International Textbook Company, Scranton, Pennsylvania, 1963.

(3) Harris, L. D.: <u>Introduction to Feedback Systems</u>, John Wiley and Sons, Inc., New York, 1961.

(4) Murphy, G. J.: <u>Control Engineering</u>, D. Van Nonstrand Company, Inc., Princeton, New Jersey, 1959.

(5) Clark, R. N.: <u>Introduction to Automatic Control Systems</u>, John Wiley and Sons, Inc., New York, 1962.

(6) Cannon, R. H., Jr.: <u>Dynamics of Physical Systems</u>, McGraw-Hill Book Co., Inc., New York, 1967.

(7) Evans, W. R.: <u>Control System Dynamics</u>, McGraw-Hill Book Co., Inc., New York, 1954.

(8) D'Azzo, J. J. and C. H. Houpis: <u>Feedback Control Systems Analysis and Synthesis</u>, McGraw-Hill Book Co., Inc., New York, 1960.

(9) Wilts, C. H.: <u>Principles of Feedback Control</u>, Addison-Wesley Publishing Co., Inc., Reading, Massachusetts, 1960.

(10) Truxal, J. G.: <u>Automatic Feedback Control System Synthesis</u>, McGraw-Hill Book Co., Inc., New York, 1955.

(11) Kuo, B. C.: <u>Automatic Control Systems</u>, Prentice-Hall, Inc., Englewood Cliffs, New Jersey, 1962.

(12) Savant, C. J., Jr.: <u>Control System Design</u>, McGraw-Hill Book Co., Inc., New York, 1958.

CHAPTER 7

Systems Analysis

7.1 PRELIMINARIES

Up to this point our intent has been to develop the conceptual tools needed to analyze continuous dynamic systems. This chapter ties many of these concepts together in a unified approach to systems analysis. We use differential equations, system modeling, Laplace transforms, and other techniques from the previous chapters, but we also introduce three new topics: transfer functions, block diagrams and feedback.[1] But the emphasis is still on analysis as opposed to design. In addition, our analysis methods are oriented mainly toward the frequency domain and toward systems with a single input function and a single output function. Multi-input-multi-output systems are considered separately in Chapter 10 in conjunction with state-variable methods. Finally, we limit our treatment to linear systems, but this is not a serious restriction as many inherently nonlinear systems can be linearized about an operating point.

The System Transfer Function

Most large-scale engineering systems--from telephones to space vehicles--are designed, not by a single engineer, but by engineering groups, often from a number of subcontractors. Various teams of engineers design and test different subassemblies or subsystems. These are then combined or interfaced in the hope that the overall system will work. A difficulty with this approach is that while for maximum efficiency each team should be concerned only with its own subsystem, it must also take into account what the other teams are doing. The output of one subsystem is the input to the next.

The transfer-function concept is an attempt to circumvent this difficulty. A linear system can be described, at least insofar as its inputs and outputs are concerned, by an s-domain function called a transfer function. If transfer functions are used, each engineering team need provide only its transfer function (and the limitations on its applicability) to the others. The internal details of the subsystems are taken into account by the transfer functions.

Block Diagrams

A large system such as an airplane may contain literally hundreds of subsystems, each described mathematically by algebraic equations, differential equations, or transfer functions. As an aid to the analyst or designer, a complicated system can be represented as an interconnection of blocks,

Systems Analysis

each of which contains a transfer function, with the pathways between blocks representing the flow of system variables.

The block diagram then becomes a universal systems language, allowing an engineer from one discipline to understand systems from a variety of others without getting bogged down in physics and fundamental laws. If, for example, we are told that the transient model of a certain heat exchanger is $G(s)=1/(s+a)$, then we can analyze its behavior, even though we are ignorant of the principles of heat transfer.

Computer Simulation

A final advantage of the transfer-function viewpoint is that it leads to efficient ways to simulate dynamic systems on a digital or an analog computer. By decomposing the overall system into a number of simpler transfer function blocks, we can simulate it as an interconnection of subprograms in the digital computer. Each of these can be debugged and tested separately before the overall simulation is attempted.

7.2 TRANSFER FUNCTIONS

A transfer function is an s-domain function by which the Laplace transform of an input function is multiplied to obtain the transform of the output function.[2,3] If $F(s)$ is the transform of the system input and $Y(s)$ is its output, then

$$Y(s) = G(s)F(s) \qquad (7-1)$$

where $G(s)$ is the transfer function. Transfer functions for linear time-invariant systems have the following properties:

(1) The transfer function is a rational function in the Laplace variable s. The two general forms are the <u>polynomial form</u>

$$G(s) = \frac{K(s^m + b_{m-1}s^{m-1} + \ldots b_1 s + b_0)}{s^n + a_{n-1}s^{n-1} + \ldots a_1 s + a_0} \qquad (7-2)$$

and the pole-zero form

$$G(s) = \frac{K(s - z_1)(s - z_2)\ldots(s - z_m)}{(s - p_1)(s - p_2)\ldots(s - p_n)} \qquad (7\text{-}3)$$

For most physical systems the order of the numerator polynomial is less than that of the denominator polynomial, or m<n. For large n, converting from the polynomial to the pole-zero form can be quite difficult, often requiring a digital computer and a root-finding program.

(2) Transfer functions are defined only for cases where the system has no stored energy: all initial conditions are zero. If the initial conditions are nonzero, then Eq. (7-1) is no longer valid.

(3) Transfer functions exist, in general, only for linear time-invariant systems. Nonlinear or time-varying systems do not have transfer functions in the conventional sense.

(4) The inverse Laplace transform g(t) of a transfer function G(s) is the response of the system to a unit impulse function applied at t=0. The response to any input function f(t) can be found by convolving f(t) and g(t) by the methods of Chapter 5. Convolution is generally not a practical way to find system response, however.

Transfer Functions From Differential Equations

A pertinent first question is how do we derive transfer functions? Where do they come from? To answer this, consider the nth-order constant-coefficient differential equation

$$\frac{d^n y}{dt^n} + a_{n-1}\frac{d^{n-1} y}{dt^{n-1}} + \ldots a_1\frac{dy}{dt} + a_0 y = f(t) \qquad (7\text{-}4)$$

which, when all initial conditions are zero, is Laplace transformed to

$$\{s^n + a_{n-1}s^{n-1} + \ldots a_1 s + a_0\}Y(s) = F(s) \qquad (7\text{-}5)$$

Systems Analysis   301

The solution for Y(s) is

$$Y(s) = \frac{1}{s^n + a_{n-1}s^{n-1} + \ldots + a_1 s + a_0} F(s) \qquad (7\text{-}6)$$

Comparing Eqns. (7-1) and (7-6) reveals that

$$G(s) = \frac{1}{s^n + a_{n-1}s^{n-1} + \ldots + a_1 s + a_0} \qquad (7\text{-}7)$$

for the class of systems described by the given differential equation. It is clear then that if we can write the differential equation which represents a system, we can find the transfer function with little extra work. Further, for systems which can be modeled in the s-domain, such as the Laplace-transformed networks of Chapter 5, we can often write G(s) directly, without passing through the differential-equation stage. For the reader who wonders what effect the numerator of G(s) has in the time domain we offer the following example.

Example 7-1. Investigate the time-domain significance of the transfer function

$$G(s) = \frac{Y(s)}{R(s)} = \frac{s^2 + Cs + 1}{s^2 + As + B} \qquad (7\text{-}8)$$

Solution. Cross multiply Eq. (7-8) to obtain

$$(s^2 + As + B)Y(s) = (s^2 + Cs + 1)R(s) \qquad (7\text{-}9)$$

Take inverse Laplace transforms to obtain

$$\ddot{y}(t) + A\dot{y}(t) + By(t) = \ddot{r}(t) + C\dot{r}(t) + r(t) \qquad (7\text{-}10)$$

The effect of the numerator of G(s) has been to produce differentiation of the input function r(t). Many physical systems exhibit this phenomenon.

Example 7-2. Find the transfer function $G(s) = V_o(s)/V_i(s)$ for the network of Fig. 7-1.

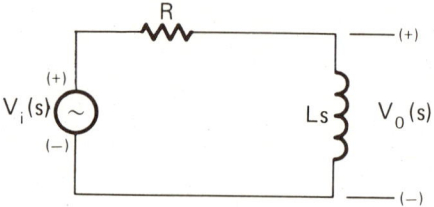

Figure 7-1. Network for Example 7-2. The transfer function between $V_i(s)$ and $V_o(s)$ is $G(s) = s/(s + R/L)$.

Solution. Using the voltage-divider principle for impedances, we have

$$V_o(s) = V_i(s)\frac{Ls}{Ls + R} \tag{7-11}$$

so that

$$G(s) = \frac{s}{s + R/L} \tag{7-12}$$

Example 7-3. Find the transfer function $G(s)=Y(s)/F(s)$ for the mass-spring system of Fig. 7-2 if the differential equation of motion is

$$m\ddot{y} + ky = f(t) \tag{7-13}$$

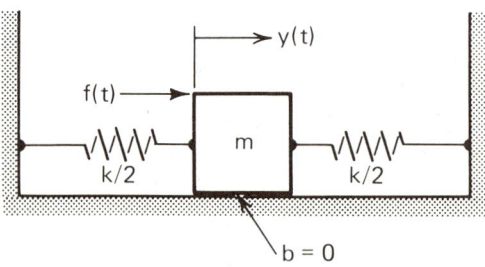

Figure 7-2. Mechanical system for Example 7-3. See Appendix H for a treatment of mechanical systems.

Solution. Taking Laplace transforms for zero initial conditions and solving for $Y(s)$ yields

$$Y(s) = \frac{1}{ms^2 + k}F(s) \tag{7-14}$$

Systems Analysis

The transfer function $G(s)$ is therefore

$$G(s) = \frac{1/m}{s^2 + k/m} \tag{7-15}$$

Transfer Functions of Coupled Systems

The transfer functions of coupled systems can be found by algebra. For example, the coupled equations

$$\dot{y}_1 + ay_1 - k_2 y_2 = f(t) \tag{7-16}$$

$$\dot{y}_2 + by_2 - k_1 y_1 = 0 \tag{7-17}$$

are transformed to

$$(s + a)Y_1(s) - k_2 Y_2(s) = F(s) \tag{7-18}$$

$$(s + b)Y_2(s) - k_1 Y_1(s) = 0 \tag{7-19}$$

provided the initial conditions are zero. Equation (7-19) is solved for

$$Y_2(s) = \frac{k_1 Y_1(s)}{s + b} \tag{7-20}$$

and the solution substituted into Eq. (7-18) to give

$$(s + a)Y_1(s) - \frac{k_1 k_2 Y_1(s)}{s + b} = F(s) \tag{7-21}$$

from which

$$G_1(s) = \frac{Y_1(s)}{F(s)} = \frac{s + b}{(s + a)(s + b) - k_1 k_2} \tag{7-22}$$

Also,

$$G_2(s) = \frac{Y_2(s)}{F(s)} = \frac{k_1}{(s + a)(s + b) - k_1 k_2} \tag{7-23}$$

The transfer functions $G_1(s)$ and $G_2(s)$ are the transfer functions between $F(s)$ and $Y_1(s)$ and $Y_2(s)$ respectively. Once f(t) is known Eqns. (7-22) and (7-23) can be solved for $y_1(t)$ and $y_2(t)$. In theory, any number of coupled equations can be Laplace transformed and reduced by algebra to find the associated transfer functions. In practice, for higher-order systems we must turn to more formalized methods which can be implemented algorithmically on a digital computer[4] (or the system can be simulated directly on an analog machine.)

Example 7-4. Find the transfer function $G(s)=V_2(s)/V_1(s)$ for the network of Fig. 7-3.

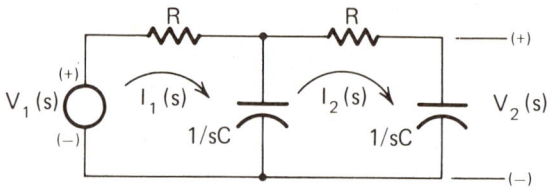

Figure 7-3. Network for Example 7-4. Equation (7-27) is the transfer function between $V_1(s)$ and $V_2(s)$.

Solution. Writing KVL for the s-domain loop currents yields

$$V_1(s) = I_1(R + \frac{1}{sC}) - \frac{1}{sC}I_2 \qquad (7\text{-}24)$$

$$0 = -\frac{1}{sC}I_1 + I_2(R + \frac{2}{sC}) \qquad (7\text{-}25)$$

These equations are solved for $I_2(s)$,

$$I_2(s) = V_1(s) \frac{sC}{(RCs)^2 + 3RCs + 1} \qquad (7\text{-}26)$$

and, because $V_2 = I_2/sC$, the transfer function is

$$G(s) = \frac{V_2(s)}{V_1(s)} = \frac{1}{(RCs)^2 + 3RCs + 1} \qquad (7\text{-}27)$$

Example 7-5. Find the transfer function $G(s)=V(s)/I(s)$ for the Laplace-transformed network of Fig. 7-4.

Systems Analysis

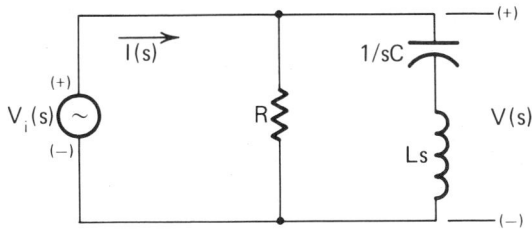

Figure 7-4. Network for Example 7-5. Equation (7-30) is the transfer function $G(s) = V(s)/I(s)$.

Solution. The voltage $V(s)$ is given by $V(s)=Z(s)I(s)$ where $Z(s)$ is found by the rules for combining impedances:

$$Z(s) = \frac{R(Ls + 1/sC)}{R + Ls + 1/sC} \tag{7-28}$$

Multiplying $Z(s)$ by $I(s)$ and rearranging slightly yields

$$V(s) = \frac{R(LCs^2 + 1)}{LCs^2 + RCs + 1} I(s) \tag{7-29}$$

from which

$$G(s) = \frac{V(s)}{I(s)} = \frac{R(s^2 + 1/LC)}{s^2 + Rs/L + 1/LC} \tag{7-30}$$

Stability From Transfer Functions

A system is stable if the poles of the transfer function have negative real parts. This follows from the result that

$$y(t) = L^{-1}\{Y(s)\} = L^{-1}\{F(s)G(s)\} \tag{7-31}$$

and, using Eq. (7-3) for $G(s)$,

$$y(t) = L^{-1}\left\{\frac{KF(s)(s-z_1)(s-z_2)\ldots(s-z_m)}{(s-p_1)(s-p_2)\ldots(s-p_n)}\right\} \tag{7-32}$$

A partial-fraction expansion of Eq. (7-32) has the form

$$y(t) = L^{-1}\left\{\frac{A_1}{s-p_1} + \frac{A_2}{s-p_1} + \ldots + \frac{A_n}{s-p_n} + \text{terms due to } F(s)\right\} \tag{7-33}$$

The terms due to the input F(s) do not affect stability.
Stability is determined entirely by the poles of G(s). As
shown in Chapter 6, these must be in the left half of the s-
plane, which in turn means that they must have negative real
parts. Stability can thus be tested by examining the pole
locations of G(s) if it is given in factored form, or by
applying the Routh criterion to the denominator of G(s) if it
is given in polynomial form. This stability criterion applies
only to systems consisting of a single transfer function.
The overall stability of a system containing many transfer
functions cannot, in general, be determined by examining the
individual transfer functions, particularly if the system has
feedback paths.

Example 7-6. Comment on the stability of the systems containing the following transfer functions:

(a) $G(s) = \dfrac{5(s - 2)}{(s + 1)(s + 3)}$ (7-34)

(b) $G(s) = \dfrac{25(s + 5)}{(s - 4)(s + 8)}$ (7-35)

(c) $G(s) = \dfrac{1}{s^2 + 3s + 4}$ (7-36)

(d) $G(s) = \dfrac{100}{s^2 + 5s - 2}$ (7-37)

Solution. System (a) is stable because both poles have
negative real parts. The location of the zero does not affect
stability. System (b) is unstable due to the pole at s=4.
This pole is located in the right half of the s-plane and
gives rise to a growing exponential Aexp(4t) in the transient
response. System (c) is stable; the denominator is a quadratic with positive coefficients, which implies that the poles
are in the left-half plane. System (d) is unstable because
the denominator contains a negative coefficient. Recall that
having all coefficients positive is a <u>necessary</u> condition for
stability. (It is necessary <u>and</u> <u>sufficient</u> only for polynomials of the first or second order.)

Systems Analysis

The Impulse Response [5]

For a system described by $Y(s)=G(s)F(s)$, if $f(t)=\delta(t)$, the unit impulse function, then $F(s)=1$ and

$$Y(t) = L^{-1}\{G(s)\} = g(t) \qquad (7\text{-}38)$$

The inverse Laplace transform $g(t)$ of the transfer function $G(s)$ is thus identical to the response of the system to a unit impulse function.

Example 7-7. Find the impulse response of the system whose transfer function is

$$G(s) = \frac{Y(s)}{F(s)} = \frac{K}{s+a} \qquad (7\text{-}39)$$

Solution. The impulse response $y(t)$, when the input is $f(t)=\delta(t)$, is the inverse transform of $G(s)$, or

$$y(t) = g(t) = Ke^{-at} \qquad (7\text{-}40)$$

Example 7-8. Show that the response of a system to a unit step function input can be obtained by integrating the impulse response function between limits of 0 and t.

Solution. Let $f(t)=u(t)$ so that $F(s)=1/s$; hence, the step response $Y(s)$ is

$$Y(s) = \frac{G(s)}{s} \qquad (7\text{-}41)$$

Taking inverse transforms using Entry 4 of Appendix E gives the desired result

$$y(t) = \int_0^t g(t)\,dt \qquad (7\text{-}42)$$

(Do you suppose that the response to $f(t)=tu(t)$ is the integral of the step response?)

7.3 BLOCK DIAGRAMS

One advantage of using transfer functions is that they let us represent a system as an interconnection of functional blocks, each of which contains the s-domain transfer function of a part of the system. In this way complex systems can be represented as interconnections of simpler subsystems. Moreover, the system representation in terms of block diagrams is divorced from the original physical model; the representation does not depend on whether the parent system was electrical, mechanical, economic, or whatever. This property of a block diagram makes the study of dynamic systems truly general. The informed systems analyst can handle problems from a wide variety of engineering disciplines without being a specialist in any of them.

If $G(s)$ is the transfer function from $F(s)$ to $Y(s)$, the associated block diagram is that of Fig. 7-5.

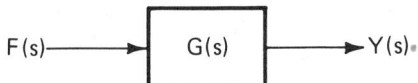

Figure 7-5. This block diagram satisfies the relationship $Y(s) = F(s)G(s)$.

The block-diagram notation implies that we multiply the input function by the contents of the block to get the output function. Some writers use time functions on block diagrams with the understanding that the analysis **must be done with the transformed versions of these functions**.

Block-Diagram Algebra

To analyze complex systems represented by interconnections of blocks, we must show how to combine certain standard configurations. These are explained in the following paragraphs.[6]

<u>The series connection</u>. Blocks in series can be combined into a single block by multiplying together the transfer functions contained in the blocks. The two blocks in Fig. 7-6 are replaced by a single block with the transfer function

$$G(s) = G_1(s)G_2(s) \qquad (7\text{-}43)$$

Systems Analysis

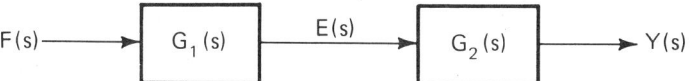

Figure 7-6. Blocks in series (or cascade) combine by multiplication. These two blocks can be replaced by a single block whose transfer function is $G(s) = G_1(s)G_2(s)$.

The proof of Eq. (7-42) is almost trivial. In Fig. 7-6, the intermediate variable $E(s)$ is found from

$$E(s) = F(s)G_1(s) \tag{7-44}$$

and

$$Y(s) = E(s)G_2(s) \tag{7-45}$$

Combining Eqns. (7-44) and (7-45) yields

$$Y(s) = F(s)\{G_1(s)G_2(s)\} \tag{7-46}$$

and $Y(s)=F(s)G(s)$ where $G(s)=G_1(s)G_2(s)$.

The parallel connection. Blocks in parallel are combined by adding the transfer functions contained in the blocks. The three blocks in Fig. 7-7 can be replaced by a single block whose transfer function is

$$G(s) = G_1(s) + G_2(s) + G_3(s) \tag{7-47}$$

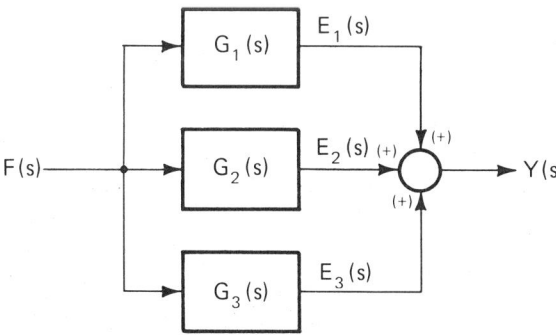

Figure 7-7. Blocks in parallel combine by addition. These three blocks can be replaced by a single block whose transfer function is $G(s) = G_1(s) + G_2(s) + G_3(s)$.

Observe the notation used for a summing point in Fig. 7-7. The summing point is described by the relationship

$$Y(s) = E_1(s) + E_2(s) + E_3(s) \qquad (7\text{-}48)$$

Equation (7-48) is proved by noting that $E_1(s)=F(s)G_1(s)$, $E_2(s)=F(s)G_2(s)$, and $E_3(s)=F(s)G_3(s)$. Substituting these relationships into Eq. (7-48) yields

$$Y(s) = F(s)G_1(s) + F(s)G_2(s) + F(s)G_3(s) \qquad (7\text{-}49)$$

$$= F(s)\{G_1(s) + G_2(s) + G_3(s)\} \qquad (7\text{-}50)$$

Equation (7-50) has the form $Y(s)=F(s)G(s)$ provided $G(s)= G_1(s)+G_2(s)+G_3(s)$.

The feedback connection. Consider the feedback connection of Fig. 7-8, where the input to the transfer function depends on the output $Y(s)$ as well as the input function $F(s)$. The many and varied implications of feedback in dynamic systems are the basis for automatic control theory and the physical devices--from record changers to airplanes--that are based on this theory.

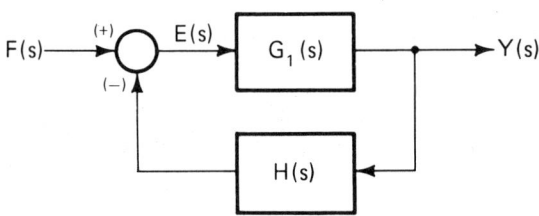

Figure 7-8. Blocks arranged in the feedback configuration can be replaced by a single block whose transfer function is given by Eq. (7-55).

To reduce Fig. 7-8 to a single block which relates $Y(s)$ to $F(s)$ we proceed as follows:
At the summing point

$$E(s) = F(s) - H(s)Y(s) \qquad (7\text{-}51)$$

Systems Analysis

and at the output of the $G_1(s)$ block

$$Y(s) = G_1(s)E(s) \tag{7-52}$$

Combining Eqns. (7-51) and (7-52) yields

$$Y(s) = G_1(s)\{F(s)-H(s)Y(s)\} \tag{7-53}$$

Solving Eq. (7-53) for $Y(s)$ results in

$$Y(s) = \frac{G_1(s)}{1 + G_1(s)H(s)} F(s) \tag{7-54}$$

Equation (7-54) has the form $Y(s)=G(s)F(s)$ provided we define the new transfer function $G(s)$ as

$$G(s) = \frac{G_1(s)}{1 + G_1(s)H(s)} \tag{7-55}$$

Example 7-9. Reduce the block diagram of Fig. 7-9 to a single block.

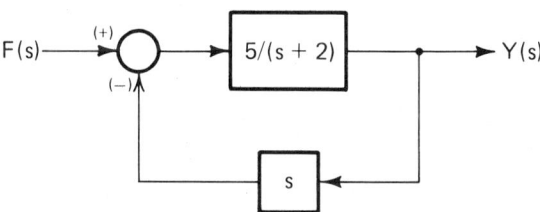

Figure 7-9. Block diagram for Example 7-9. Equation (7-55) can be used to reduce this system to the single transfer function $G(s) = 5/(6s + 2)$.

Solution. Comparing Figs. 7-8 and 7-9 reveals that Eq. (7-55) applies. The equivalent transfer function is

$$G(s) = \frac{5/(s+2)}{1 + 5s/(s+2)} = \frac{5}{6s+2} \tag{7-56}$$

Equivalent Block Diagrams

The systems analyst is often faced with the problem of finding the response of a system consisting of a large number of blocks, each containing an s-domain transfer function or some other operation. Although we have not emphasized the point, blocks in system diagrams may be used for purposes other than to represent transfer functions. Figure 7-10 for example shows blocks used for (a) multiplication by the constant K, Fig. 7-10(a), (b) squaring, Fig. 7-10(b), and (c) application of a nonlinear transfer characteristic, Fig. 7-10(c). A transfer-function block which indicates multiplication by a constant applies equally well to s-domain or time-domain variables. Blocks which represent functional operations, such as the squaring block of Fig. 7-10(b), apply only to the time-domain variables. Thus, in Fig. 7-10(b) the output of the block is $f(t)^2$, <u>not</u> $F(s)^2$. These rules hold regardless of whether the block diagram is labeled with time-domain or s-domain variables.

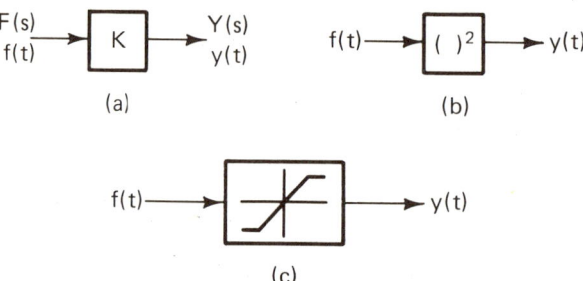

Figure 7-10. Blocks in system diagrams often represent operations other than multiplication by a transfer function. The system variables may be specified in either the time domain or the s-domain. (a) Multiplication by a constant; y(t) = Kf(t); (b) A squaring block for which y(t) = f(t)²; (c) A nonlinear transfer characteristic.

The algebra of block diagrams as outlined in the above paragraphs can be used to reduce a complicated block diagram to a single block, provided we augment the procedure with the following proposition. Two block diagrams are equivalent

Systems Analysis

if the overall transfer functions along each straight-through path and around every closed-loop path are identical. By overall transfer function we mean the product of the contents of each transfer function block along the path. This definition of equality of block diagrams gives us a powerful tool for rearranging a system diagram to give greater insight into the behavior of the overall system.(7,8) Consider, for example, the problem of removing the differentiation block from the feedback path of the system shown in Fig. 7-11(a). The block containing the transfer function Ks represents differentiation because, as shown in Chapter 5, multiplying a Laplace transformed function by s is equivalent to differentiating that function in the time domain. The reason that one might want to remove the differentiation operation is as a preparatory step to simulating the system on a digital computer where numerical differentiation can introduce errors.

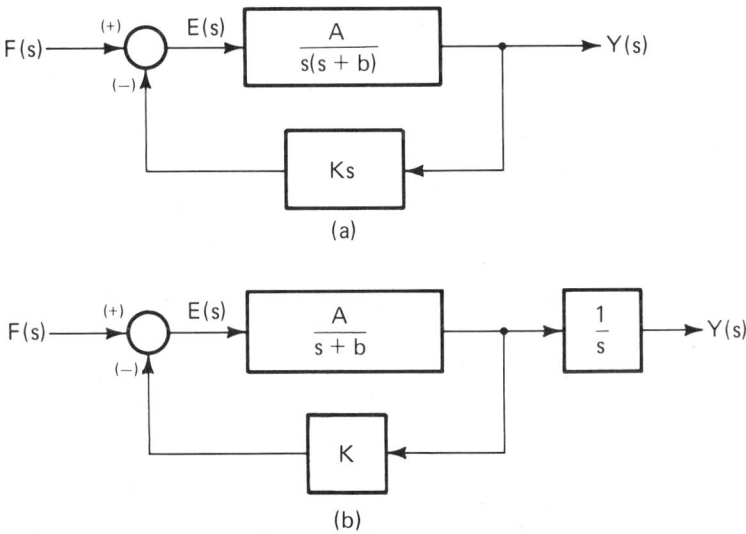

Figure 7-11. Two feedback systems are equivalent if they have identical loop gains and straight-through gains. (a) A system with differentiation in the feedback path; (b) An equivalent system without differentiation in the feedback path.

The system of Fig. 7-11(b) is equivalent to that of Fig. 7-11(a) because the straight-through transfer function,

$A/\{s(s+b)\}$, and the transfer function around the feedback loop, $AK/(s+b)$, are identical for both diagrams. Differentiation has been eliminated in Fig. 7-11(b).

Throughout the remainder of the chapter we occasionally omit the explicit dependence of functions and transfer functions on s. This simplifies equations and block diagrams and is done only where there is no chance of misunderstanding. We may, for example, write G_1 rather than $G_1(s)$.

Example 7-10. For the system whose block diagram is shown in Fig. 7-12, find the transfer function between the unwanted disturbance $D(s)$ and the output $Y(s)$.

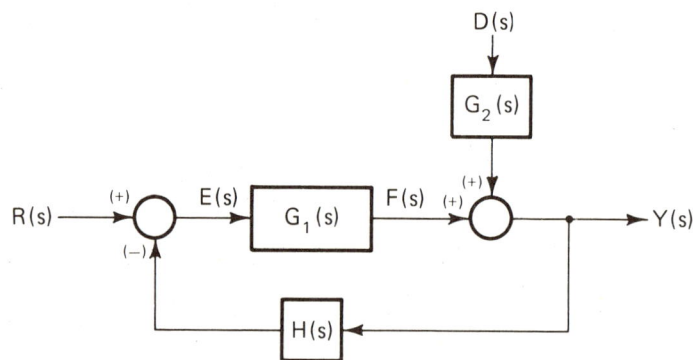

Figure 7-12. Block diagram for Example 7-10. This system has an unwanted disturbance $D(s)$ applied through the transfer function $G_2(s)$.

Solution. As we are concerned only with the response to $D(s)$, we can safely let $R(s)=0$. This procedure is valid because the system is linear and the responses due to different inputs are independent of each other. The output $Y(s)$ is given by

$$Y(s) = F(s) + D(s)G_2(s) \qquad (7\text{-}57)$$

where $F(s)=G_1(s)E(s)$ or

$$F(s) = G_1(s)\{-H(s)Y(s)\} \qquad (7\text{-}58)$$

Systems Analysis

Combining these expressions yields

$$Y(s) = -Y(s)H(s)G_1(s) + D(s)G_2(s) \qquad (7\text{-}59)$$

and, solving for $Y(s)$, we have

$$Y(s) = \frac{D(s)G_2(s)}{1 + G_1(s)H(s)} \qquad (7\text{-}60)$$

The transfer function $G_3(s)$ between $D(s)$ and $Y(s)$ is then

$$G_3(s) = \frac{Y(s)}{D(s)} = \frac{G_2(s)}{1 + G_1(s)H(s)} \qquad (7\text{-}61)$$

Example 7-11. Reduce the system of Fig. 7-13 to a single block.

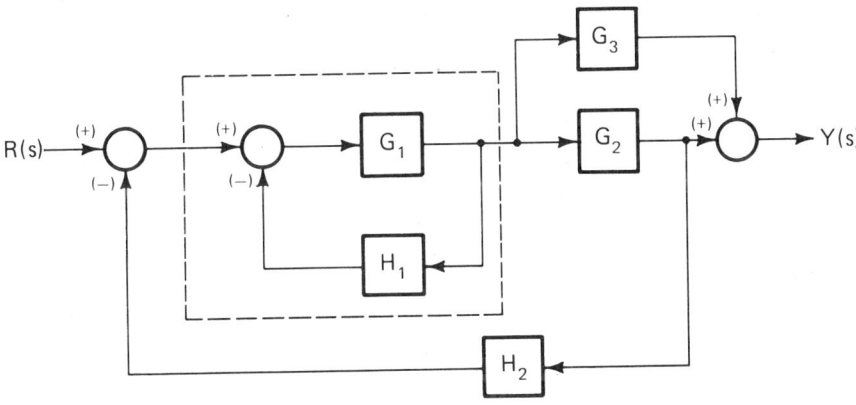

Figure 7-13. Block diagram for Example 7-11. A complicated configuration such as this can be reduced to a single block by using the algebra of block diagrams.

Solution. The G_1-H_1 subsystem enclosed in dashed lines is in the feedback form and Eq. (7-55) applies. Let G_4 be defined as

$$G_4 = \frac{G_1}{1 + G_1 H_1} \qquad (7\text{-}62)$$

The system now appears as in Fig. 7-14.

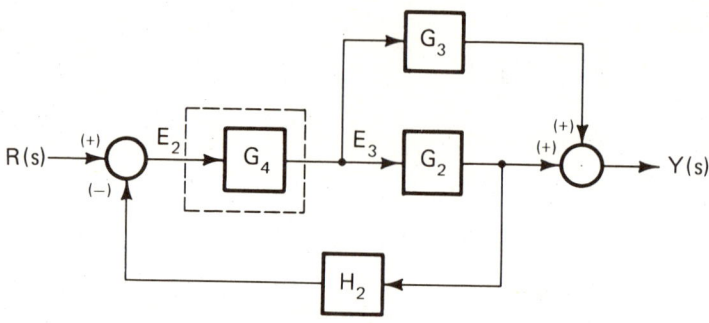

Figure 7-14. Intermediate result for Example 7-11. This is the system of Fig. 7-13 with the subsystem enclosed in dashed lines reduced to G_4 by the relationship $G_4 = G_1/(1 + G_1 H_1)$.

From Fig. 7-14, $E_3 = G_4 E_2$,

$$Y = E_3(G_2 + G_3) \qquad (7\text{-}63)$$

and

$$E_2 = R - E_3 G_2 H_2 \qquad (7\text{-}64)$$

Using Eq. (7-64) in $E_3 = G_4 E_2$ yields

$$E_3 = G_4(R - E_3 G_2 H_2) \qquad (7\text{-}65)$$

and, solving for E_3, we have

$$E_3 = \frac{G_4 R}{1 + G_4 G_2 H_2} \qquad (7\text{-}66)$$

Substituting Eq. (7-66) into Eq. (7-63) yields

$$Y = \frac{(G_2 + G_3) G_4 R}{1 + G_4 G_2 H_2} \qquad (7\text{-}67)$$

The overall transfer function is then

$$G = \frac{Y}{R} = \frac{G_4(G_2 + G_3)}{1 + G_4 G_2 H_2} \qquad (7\text{-}68)$$

Systems Analysis 317

7.4 FEEDBACK SYSTEMS

Feedback, an all-important topic for anyone who seeks an understanding of modern dynamic systems,[9,10] arises in two important contexts. First, the performance of many systems can be best understood if they are interpreted in terms of feedback. Second, the designer often introduces feedback into proposed systems (1) to offset his ignorance of various parameters, (2) to increase the linearity of non-linear devices or subsystems, (3) to reduce the sensitivity of the output to parameter variations, (4) to stabilize an unstable system, (5) to tailor the response to meet such design criteria as relative damping, and (6) to provide automatic control, eliminating the need for a human operator. Feedback is not a theoretical abstraction: it is ubiquitous in practical systems and is found everywhere from missile guidance to toilet-flushing mechanisms.

Analysis In Terms Of Feedback

We can often gain a greater insight into the role played by parameters in a complex system if we can represent the system in terms of elementary transfer functions connected by feedback paths. The second-order differential equation

$$\ddot{y} + 2\zeta\omega_n \dot{y} + \omega_n^2 y = f(t) \qquad (7\text{-}69)$$

with transfer function

$$G(s) = \frac{Y(s)}{F(s)} = \frac{1}{s^2 + 2\zeta\omega_n s + \omega_n^2} \qquad (7\text{-}70)$$

can be represented as a feedback system according to the block diagram of Fig. 7-15. (Check this using Eq. (7-55).) Using a second level of feedback, the second-order system can be decomposed even further as in Fig. 7-16.

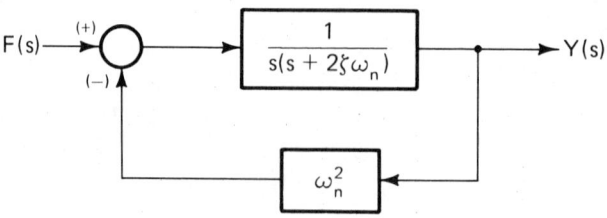

Figure 7-15. This block diagram shows how Eq. (7-69) can be interpreted as a feedback system.

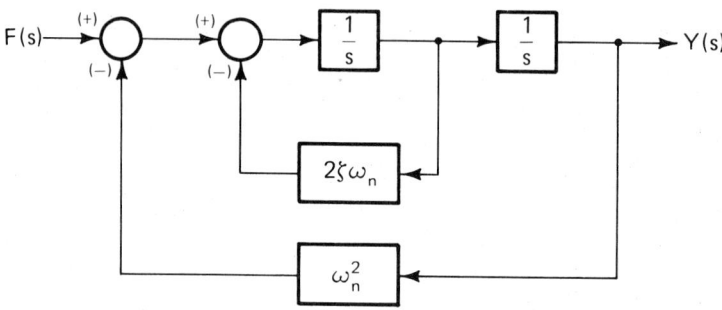

Figure 7-16. This block diagram represents Eq. (7-70) by gains and pure integrations.

This decomposition reduces the system to an interconnection of gains, integrations and summations, showing the roles ζ and ω_n play in the overall system behavior. This kind of decomposition is what we must do indirectly when simulating a dynamic system on a digital or an analog computer.

Feedback As a Design Tool

A more important reason for studying feedback is that the designer often purposely introduces it into a system to overcome his ignorance of the exact behavior of some subsystems or parameters.[11] We saw in Chapter 3 that feedback around an operational amplifier can be used to give a precise overall gain whose value is determined by the input and feedback elements. The operational amplifier network of Fig. 7-17 has

Systems Analysis

the input-output relationship $v_o(t)=-\{R_f/R_i\}v_i(t)$ or, in the s-domain,

$$v_o(s) = -\frac{R_f}{R_i}V_i(s) \qquad (7-71)$$

The transfer function is independent of the gain K of the operational amplifier as long as K is large, say in the range $10^4 < K < 10^8$.

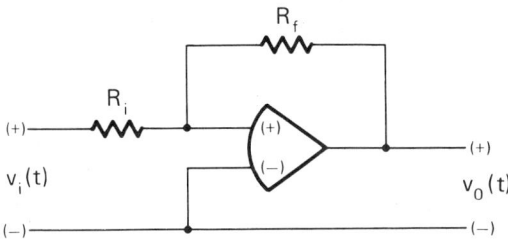

Figure 7-17. An operational amplifier with a feedback resistor is a feedback system.

In this way we use the feedback element R_i whose value can be precisely controlled--and resistors with one-percent tolerances can be readily obtained--to offset the effects of a system parameter (in this case the amplifier gain K) whose value is either unknown or difficult to control accurately. It is not unusual for the gains of operational amplifiers to vary by 50% from the advertised value. Further, the value of K may change during prolonged operation without materially degrading the performance of the feedback amplifier.

Finally, feedback is used to provide automatic control--to eliminate the need for a human observer or operator. Consider the watering trough for livestock shown in Fig. 7-18. In Fig. 7-18(a) the tank is supplied by a manually operated valve. Someone must check the water level periodically and refill the tank by hand when it is low. This is an open-loop system. Figure 7-18(b) shows the same system redesigned to include feedback. When the water level is low, the float drops, opening the valve to admit water; when the water level is high, the float keeps the valve closed. Thus, information about the output variable (the height of the water) is used to control the value of the input (the input flow rate). Such systems are called closed-loop control systems.

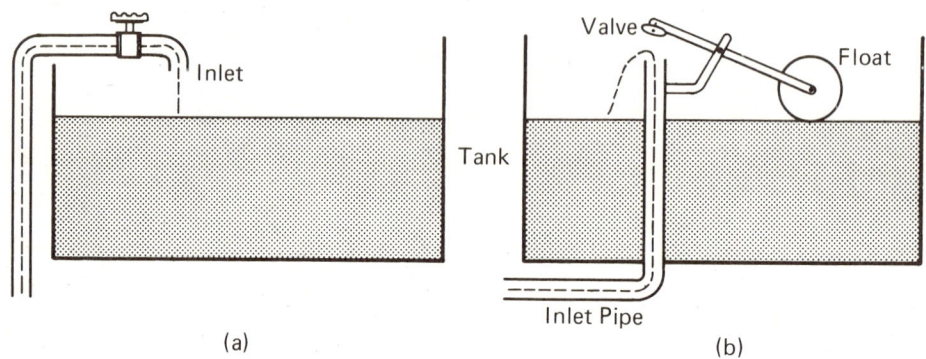

Figure 7-18. Open- and closed-loop control systems. In a closed-loop system the output affects the input via feedback.

Since World War II with its requirements for sophisticated weaponry, a new engineering discipline has grown up around the subject matter area of automatic control Automatic control has found continued application not only in weapons systems but also in the control of industrial processes and many other civil applications. One can now be a full-fledged control systems engineer without being committed to applications from a specific engineering discipline.

The Standard System

Figure 7-19 shows a configuration which is typical of control systems from a variety of practical applications.[12,13] The system consists of a plant, which is the system to be controlled, and feedback and compensator elements which the designer selects to modify the behavior of the plant. The reference input r(t) is the forcing function which drives the system. In many systems there are other inputs, such as random noise, which must be controlled or suppressed but whose study is beyond the scope of our treatment.[14] The

Systems Analysis

controlled output y(t) is the final product of the system. It is also used as feedback information to modify the performance of the system. The modified output b(t) is an altered version of the output which is used in conjunction with the reference input to drive the plant.

The feedback elements are often needed to change the energy form of the output--y(t) may be mechanical movement of a shaft, for example, while the reference input is electrical. The error signal e(t), as the name implies, is a measure of the error between the desired value of the output and the actual value. As long as the error is nonzero, there is an input to the plant and the output continues to change. When the error is zero, the output of the plant is correct and there is no input to the plant which would cause its output to change.

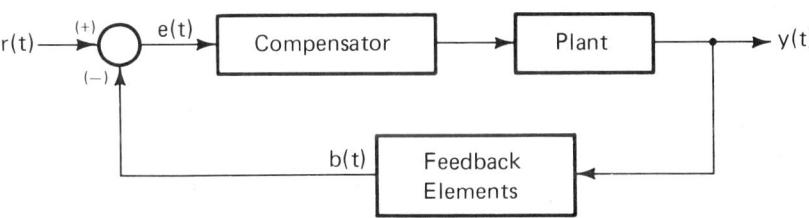

Figure 7-19. The standard feedback control system has a reference input, r(t), a controlled output y(t), an error signal e(t), and a modified output b(t).

A number of analytical techniques applicable to control system analysis and design have been perfected in the last few decades. Some of these techniques are:

1. Bode Plots and The Nichols Chart[15]
2. Nyquist or Polar Plots[15]
3. Root Locus[1]
4. Modern Control Theory[16]
5. Analog Simulation[17]
6. Digital Computation[18]

All of these methods are essentially design by repeated analysis. The designer analyzes the system, either in the time or the frequency domain, then selects a compensator and/or feedback elements according to some rule of thumb or other approximation technique. He then reanalyzes the system to see if it now meets the design criteria. The process is repeated until satisfactory performance is achieved.

The first three methods, Bode plots, polar plots, and the root locus, are frequency-domain methods. They take advantage of the relationships poles, damping ratio, natural frequency, and gain have to time-domain performance. The frequency-domain methods let one design a system without solving its differential equations. Computers, on the other hand, can be used to find the actual time response, which can be examined and modified as necessary by changing system parameters.

Frequency-domain methods are still important conceptual tools for understanding systems, but most design, particularly for complex systems, is now done with the aid of a computer. Before computers, all engineering work was done with pencil-and-paper methods, but these fail for many of today's complex systems. In the educational environment, however, we still must start at the beginning and learn the fundamental ideas behind control systems before proceeding to computer procedures. For this reason we continue to study the classical techniques, not just because they are useful as design tools, but because they give us insight as to why systems behave as they do.

Sensitivity of Feedback Systems

To better understand how using feedback reduces the sensitivity of the output variable to variations in plant parameters, we can define a sensitivity index and compute its value for both open- and closed-loop systems.[19] Define the sensitivity of a variable with respect to a parameter as the percent change in the variable divided by the percent change in the parameter. Consider the problem of determining the

Systems Analysis

sensitivity of a Laplace-transformed function Y(s) with respect to another function G(s). According to the definition just given we can write

$$S_G^Y = \frac{\Delta Y(s)}{Y(s)} \div \frac{\Delta G(s)}{G(s)} \qquad (7-72)$$

where S_G^Y is read "the sensitivity of Y with respect to G", and the notation $\Delta Y(s)$ refers to an incremental change in Y(s). Equation (7-72) can be rearranged as

$$S_G^Y = \frac{\Delta Y(s)}{\Delta G(s)} \frac{G(s)}{Y(s)} \qquad (7-73)$$

and, in the limit as the incremental changes become smaller, we pass from increments to differentials with the result

$$S_G^Y = \frac{dY(s)}{dG(s)} \frac{G(s)}{Y(s)} \qquad (7-74)$$

For an open-loop control system where Y(s) is given by Y(s)= F(s)G(s) we have

$$\frac{dY(s)}{dG(s)} = F(s) \qquad (7-75)$$

and, from Eq. (7-74),

$$S_G^Y = F(s)\frac{G(s)}{Y(s)} = \frac{F(s)G(s)}{F(s)G(s)} = 1 \qquad (7-76)$$

The unit sensitivity means that changes in G(s) appear as changes in Y(s) on a one-to-one basis: if G(s) doubles so does Y(s). For the feedback configuration of Fig. 7-8 we have

$$Y(s) = \frac{F(s)G_1(s)}{1 + G_1(s)H(s)} \qquad (7-77)$$

with

$$\frac{dY(s)}{dG_1(s)} = \frac{F(s)}{\{1+G_1(s)H(s)\}^2} \qquad (7-78)$$

The sensitivity is

$$S_{G_1}^Y = \frac{F(s)}{\{1+G_1(s)H(s)\}^2} \left\{ \frac{G_1(s)}{Y(s)} \right\} \qquad (7\text{-}79)$$

and, again using Eq. (7-77),

$$S_{G_1}^Y = \frac{1}{1 + G_1(s)H(s)} \qquad (7\text{-}80)$$

If $|G_1(s)H(s)| \gg 1$ for all values of s of interest, then

$$S_{G_1}^Y = \frac{1}{G_1(s)H(s)} \qquad (7\text{-}81)$$

If the $G_1(s)H(s)$ product is large, the sensitivity to plant variations is small. This result demonstrates the improved sensitivity of the feedback system over the open-loop system where the sensitivity was 1. The sensitivity of Y(s) to H(s) for the system of Eq. (7-77) is

$$S_H^Y = \frac{-H(s)G_1(s)}{1+G_1(s)H(s)} \qquad (7\text{-}82)$$

or $S_H^Y \approx -1$ for large $G_1(s)H(s)$ as was the case for the open-loop system. Thus adding feedback has exchanged sensitivity to plant variations for sensitivity to feedback variations. This is a profitable trade if we can closely control the parameters in H(s), as is usually the case.

7.5 TRANSFER FUNCTION SIMULATION

The systems analyst must often analyze a system containing transfer functions. He may decide to simulate his transfer functions on an analog computer or convert them to differential equations for solution by numerical methods on a digital computer. In either situation he must address the problem of how to handle the numerator polynomial N(s) in a

Systems Analysis

transfer function of the form

$$G(s) = \frac{Y(s)}{F(s)} = \frac{N(s)}{D(s)} \qquad (7\text{-}83)$$

Recall that $N(s)$ results in differentiation of the input function $f(t)$ when we attempt to convert Eq. (7-83) to a time-domain differential equation. If $f(t)$ is a step function its derivative is an impulse function at the origin, a function which is difficult to simulate on either an analog or a digital computer. In any event, it is advantageous to be able to treat the system represented by Eq. (7-83) as a differential equation whose input does not need to be differentiated. To do this we use **feed-forward programming**.[20,21]

Define an arbitrary variable $z(t)$ with transform $Z(s)$ such that

$$G(s) = \frac{Y(s)}{F(s)} = \frac{Y(s)}{Z(s)} \frac{Z(s)}{F(s)} = N(s)\frac{1}{D(s)} \qquad (7\text{-}84)$$

where

$$\frac{Z(s)}{F(s)} = \frac{1}{D(s)} \qquad (7\text{-}85)$$

and

$$\frac{Y(s)}{Z(s)} = N(s) \qquad (7\text{-}86)$$

Equation (7-85) is converted to the time domain--without differentiating the input because the numerator is constant--and programmed as a differential equation. Suppose $G(s)$ has the form

$$G(s) = \frac{b_m s^m + b_{m-1} s^{m-1} + \ldots b_1 s + b_0}{s^n + a_{n-1} s^{n-1} + \ldots a_1 s + a_0} \qquad (7\text{-}87)$$

The denominator of Eq. (7-87) is identified as $D(s)$; therefore Eq. (7-85), when transformed back to the time domain, becomes

$$\frac{d^n z}{dt^n} + a_{n-1}\frac{d^{n-1} z}{dt^{n-1}} + \ldots a_1\frac{dz}{dt} + a_0 z = f(t) \qquad (7\text{-}88)$$

Equation (7-88) is a constant-coefficient differential equation with a conventional input function. Equation (7-86) then gives y(t) in terms of the coefficients of N(s) and the derivatives of z(t), which are available from the solution of Eq. (7-88). Using the numerator of Eq. (7-87) for N(s) and taking inverse transforms, Eq. (7-86) becomes

$$y(t) = b_m \frac{d^m z}{dt^m} + b_{m-1} \frac{d^{m-1} z}{dt^{m-1}} + \ldots b_1 \frac{dz}{dt} + b_0 \qquad (7\text{-}89)$$

As m<n for physically realizable transfer functions, we are guaranteed that Eq. (7-89) will always work.

Example 7-12. Show how we could determine the unit step response of the system for which

$$\frac{Y(s)}{F(s)} = \frac{s^2 + 2s + 3}{s^2 + 10s + 50} \qquad (7\text{-}90)$$

The response is to be found in the time domain without recourse to the first and second derivatives of the unit step function.

Solution. Following the feed-forward method we define the auxiliary variable z(t) for which

$$\ddot{z} + 10\dot{z} + 50z = u(t) \qquad (7\text{-}91)$$

The output y(t) is then given by

$$y(t) = \ddot{z} + 2\dot{z} + 3 \qquad (7\text{-}92)$$

Equations (7-91) and (7-92) could be solved on a digital computer by numerical methods or simulated on an analog computer without the need to differentiate u(t).

This concludes our introduction to systems analysis. The interested student who wants to pursue the subject in greater depth is directed to any of the references at the end of this chapter. Chapters 8, 9, and 10 introduce more topics which are needed for an understanding of continuous dynamic systems, of which automatic control systems are an important subset.

PROBLEMS FOR CHAPTER 7

7-1. Rearrange the following transfer function into one of the standard forms represented by Eqns. (7-2) and (7-3):

(a) $G(s) = \dfrac{5s + 20}{s(2s + 8)}$ \hfill (7-93)

(b) $G(s) = 1 - \dfrac{4s + 16}{s^2 + 5s + 6}$ \hfill (7-94)

(c) $G(s) = \dfrac{as^2 + bs + c}{ds^3 + es^2 + fs + g}$ \hfill (7-95)

(d) $G(s) = \dfrac{1}{s+1} + \dfrac{2}{s+2} + \dfrac{3}{s+3}$ \hfill (7-96)

7-2. Assuming differentiation of the input function $f(t)$ is permissible, derive the differential equations which are represented by the transfer functions of Problem 7-1. Assume $G(s)=Y(s)/F(s)$ and that all initial conditions are zero.

7-3. Find the transfer function $G(s)=Y(s)/F(s)$ for the following differential equations. Assume zero initial conditions.

(a) $\ddot{y} + 4\dot{y} + 18y = f(t)$ \hfill (7-97)

(b) $\dddot{y} + 4\ddot{y} = f(t)$ \hfill (7-98)

(c) $\dot{y} + 4y - 8z = f(t) \qquad \dot{z} + 2z + 3y = 0$ \hfill (7-99)

7-4. Find the transfer function $G(s)=V_o(s)/V_i(s)$ for each of the networks of Fig. 7-20.

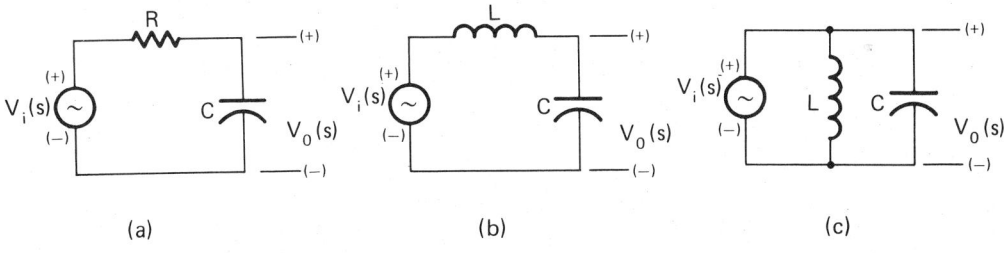

(a) \qquad\qquad (b) \qquad\qquad (c)

Figure 7-20. Problem 7-4.

7-5. Show that the transfer function $G(s)=V_o(s)/V_i(s)$ of the operational amplifier network shown in Fig. 7-21 is $G(s)=-Z_f(s)/Z_i(s)$.

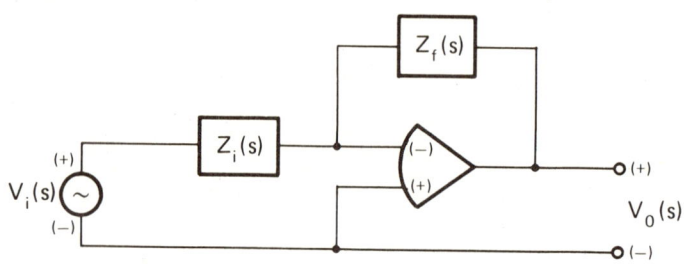

Figure 7-21. Problem 7-5.

7-6. Find the transfer function $G(s)=V_o(s)/V_i(s)$ for each of the networks shown in Fig. 7-22.

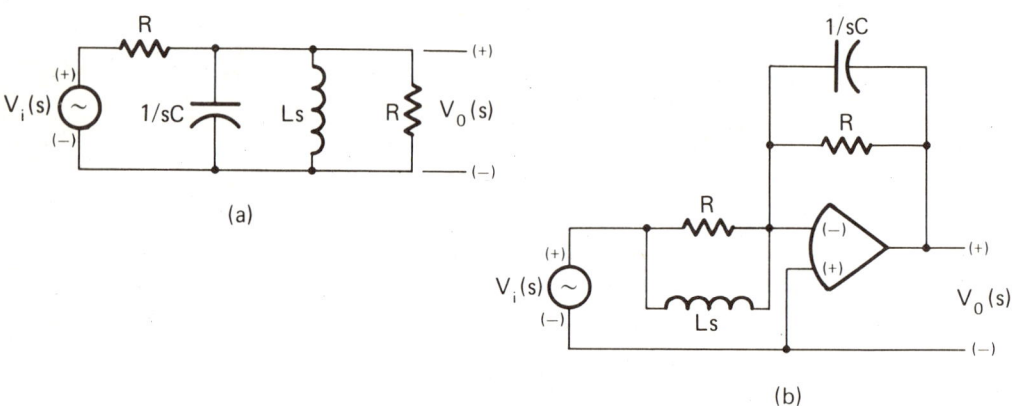

Figure 7-22. Problem 7-6.

7-7. Represent the second-order differential equation

$$\ddot{y} + 2\zeta\omega_n\dot{y} + \omega_n^2 y = \omega_n^2 f(t) \qquad (7\text{-}100)$$

as a feedback system with unity gain in the feedback path.

Systems Analysis

7-8. Generalize Eq. (7-43) to a series connection of N blocks, $G_1(s)$, $G_2(s)$,...$G_N(s)$.

7-9. Show the equivalence of the block diagrams of Fig. 7-23.

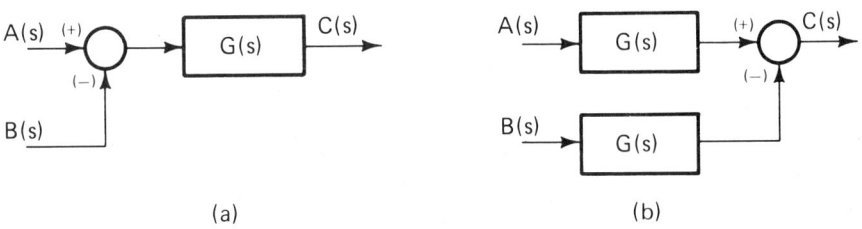

Figure 7-23. Problem 7-9.

7-10. Show the equivalence of the block diagrams of Fig. 7-24.

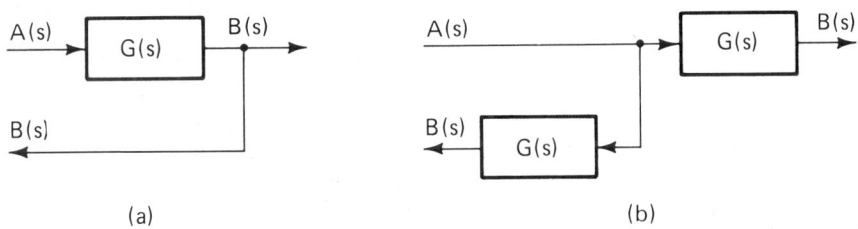

Figure 7-24. Problem 7-10.

7-11. Show the equivalence of the block diagrams of Fig. 7-25.

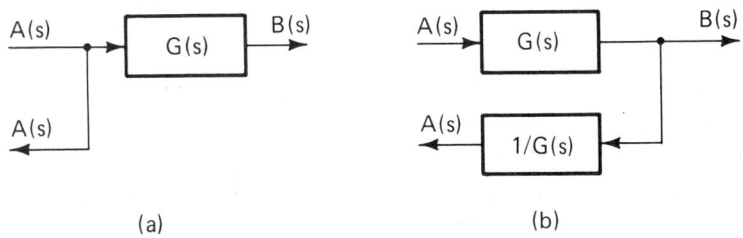

Figure 7-25. Problem 7-11.

7-12. Show the equivalence of the block diagrams of Fig. 7-26.

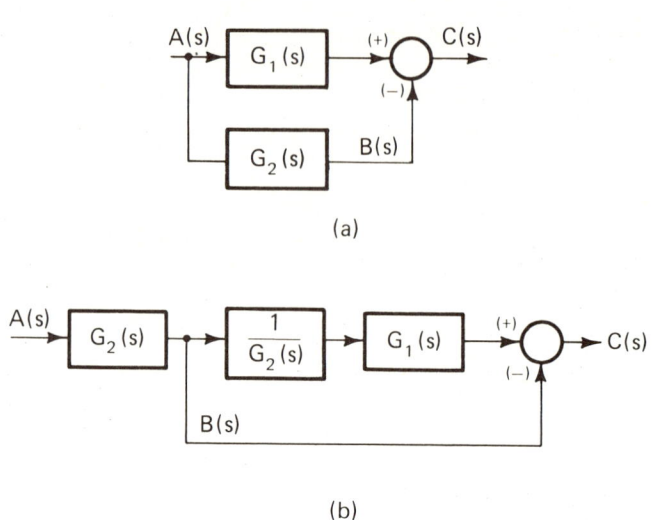

Figure 7-26. Problem 7-12.

7-13. Reduce the feedback systems shown in Fig. 7-27 to a single block:

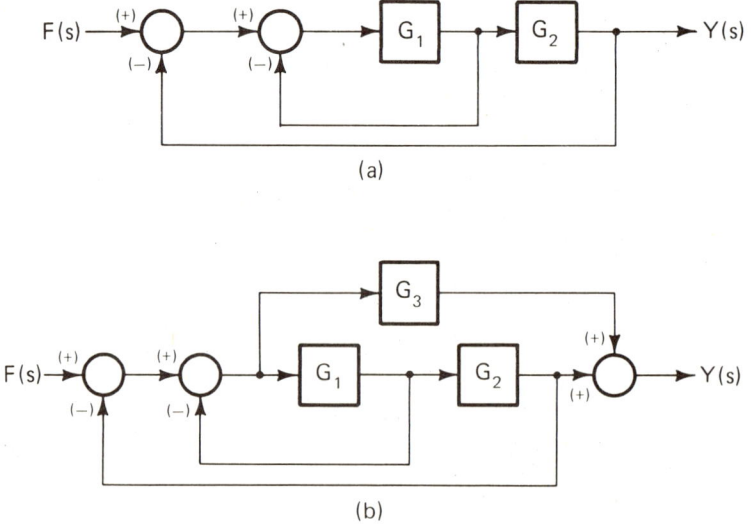

Systems Analysis

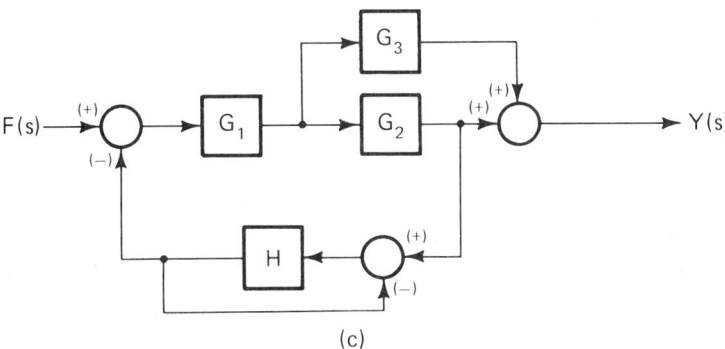

Figure 7-27. Problem 7-13.

7-14. Calculate the value of feedback gain K needed to cause the overall system of Fig. 7-28 to have a damping coefficient of $\zeta=0.6$.

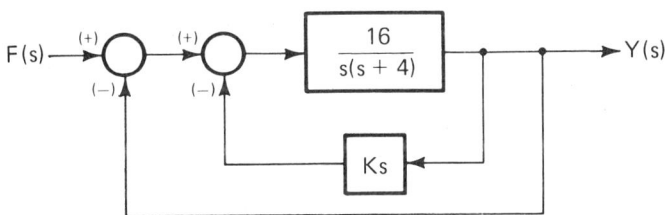

Figure 7-28. Problem 7-14.

7-15. Choose a value of K to cause the time constant of the system of Fig. 7-29 to be $T=1/\zeta\omega_n=0.001$.

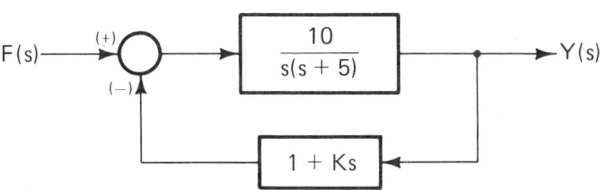

Figure 7-29. Problem 7-15.

7-16. Choose a value of K to make the natural frequency $\omega_n = 5000$ rad/s in the system of Fig. 7-30. What is the value of ζ in this instance?

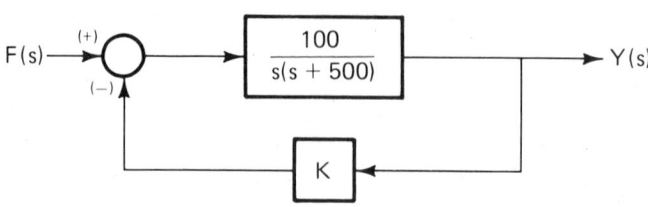

Figure 7-30. Problem 7-16.

7-17. Derive an expression for S_K^Y, the sensitivity of the output Y(s) to the parameter K, for the system of Problem 7-16.

7-18. Replace the constant 4 by a parameter b in the system of Fig. 7-28. Then calculate S_b^Y, the sensitivity of Y(s) to the parameter b.

7-19. Investigate the stability of the systems with the following transfer functions:

(a) $G(s) = \dfrac{20(s - 3)}{s(s + 5)}$ (7-101)

(b) $G(s) = \dfrac{100}{s^2 + 4s - 5}$ (7-102)

(c) $G(s) = \dfrac{s^2 + 10s + 20}{-s(s + 3)^2}$ (7-103)

(d) $G(s) = \dfrac{s - 1}{s + 2} + \dfrac{s}{s^2 + 4s + 5} - \dfrac{3}{s - 1}$ (7-104)

7-20. Find the impulse response functions for the systems with the following transfer functions:

(a) $G(s) = \dfrac{5(s + 2)}{s(s + 10)}$ (7-105)

(b) $G(s) = \dfrac{1}{s + 5} - \dfrac{2}{s + 24}$ (7-106)

(c) $\quad G(s) = \dfrac{s + 4}{s + 8}$ \hfill (7-107)

(d) $\quad G(s) = \dfrac{s + a}{(s + b)^2 + c^2}$ \hfill (7-108)

REFERENCES FOR CHAPTER 7

(1) Evans, W. R.: *Control System Dynamics*, McGraw-Hill Book Co., Inc., New York, 1954.
(2) Auslander, D. M., Takahashi, Y., and M. J. Rabins: *Introducing Systems and Control*, McGraw-Hill Book Co., Inc., New York, 1974.
(3) Schwarz, R. J. and B. Friedland: *Linear Systems*, McGraw-Hill Book Co., Inc., New York, 1965.
(4) Shinners, S. M.: *Modern Control System Theory and Application*, Addison-Wesley Publishing Co., Inc., Reading, Massachusetts, 1972.
(5) Cheng, D. K.: *Analysis of Linear Systems*, Addison-Wesley Publishing Co., Inc., Reading, Massachusetts, 1959.
(6) Towill, D. R.: *Transfer Function Techniques For Control Engineers*, Daniel Davey and Co., Inc., Hartford, Connecticut, 1970.
(7) Shinners, S. M.: *Control System Design*, John Wiley and Sons, Inc., New York, 1964.
(8) Dorf, R. C.: *Modern Control Systems*, Addison-Wesley Publishing Co., Inc., Reading, Massachusetts, 1967.
(9) Melsa, J. L. and D. G. Schultz: *Linear Control Systems*, McGraw-Hill Book Co., Inc., New York, 1969.
(10) Ahrendt, W. R. and J. F. Taplin: *Automatic Feedback Control*, McGraw-Hill Book Co., Inc., New York, 1951.
(11) Smith, O. J. M.: *Feedback Control Systems*, McGraw-Hill Book Co., Inc., New York, 1958.
(12) D'Azzo, J. J. and C. H. Houpis: *Feedback Control System Analysis and Synthesis*, McGraw-Hill Book Co., Inc., New York, 1960.

(13) Fett, G. H.: Feedback Control Systems, Prentice-Hall Inc., Englewood Cliffs, New Jersey, 1954.

(14) Chang, S. S. L.: Synthesis of Optimum Control Systems, McGraw-Hill Book Co., Inc., New York, 1961.

(15) Thaler, G. J.: Design of Feedback Systems, Dowden, Hutchinson, and Ross, Inc., Stroudsburg, Pennsylvania, 1973.

(16) Tou, J. J.: Modern Control Theory, McGraw-Hill Book Co., Inc., New York, 1964.

(17) Jackson, A. S.: Analog Computation, McGraw-Hill Book Co., Inc., New York, 1960.

(18) Dorf, R. C.: Time-Domain Analysis and Design of Control Systems, Addison-Wesley Publishing Co., Inc., Reading, Massachusetts, 1965.

(19) Elgerd, O. I.: Control Systems Theory, McGraw-Hill Book Co., Inc., New York, 1967.

(20) Hausner, A.: Analog and Analog/Hybrid Computer Programming, Prentice-Hall, Inc., Englewood Cliffs, New Jersey, 1971.

(21) Bowers, J. C. and S. R. Sedore: SCEPTRE: A Computer Program For Circuit and System Analysis, Prentice-Hall Inc., Englewood Cliffs, New Jersey, 1971.

CHAPTER 8

The Sinusoidal Steady State

When we switch on an electric fan a brief switching transient occurs, but after the transient dies out (in at most a few seconds) the device runs for a relatively long period of time in what is called the <u>steady state</u>. Many dynamic systems operate much of the time under steady-state as opposed to transient conditions. From the standpoint of differential equations, steady-state operation corresponds to the forced response while the transient response corresponds somewhat to the natural response. The parallelism is not exact as we will see later.

The most typical input function for steady-state operation of electrical systems is the sinusoid. Commercial electrical power is provided to the consumer as sinusoidally varying voltages and currents, not because some engineer has a penchant for sinusoids, but because the rotating electrical machines which generate power produce sinusoidal voltages due to the nature of their construction.[1] Many electronic communication and control systems also use sinusoidal voltages as carrier waveforms or as control signals, again because of the ease with which sinusoids can be generated, processed and otherwise manipulated.[2] A final reason for studying the sinusoidal steady state is that it provides a natural introduction to the more general topic of frequency response.

The widespread use of sinusoids in engineering systems thus justifies the in-depth treatment of the sinusoidal steady state contained in this chapter. The analysis techniques we will develop apply to dynamic systems from any engineering discipline, but the examples given are confined mainly to electrical networks. We shall see that one can find the sinusoidal steady state response of a system without solving any differential equations. Because one of the main purposes of driving a system with a forcing function is to deliver selected amounts of energy to the system components, the topics of energy and power are treated in conjunction with the sinusoidal steady state. In addition, we introduce the subject of phasors, an important classical tool for steady-state analysis.[3,4]

8.1 TRANSFER FUNCTIONS IN THE SINUSOIDAL STEADY STATE

If a system is stable and has no poles on the $j\omega$-axis, the transient terms due to the initial conditions eventually subside to zero. Equivalently, we can assume that the initial conditions are in fact zero, in which case we can describe our system by an s-domain transfer function. This makes for neat, tidy derivations and allows us to consider complicated systems in a straightforward manner.[5,6] Consider the system of Fig. 8-1 where $Y(s)=G(s)F(s)$ with

$$G(s) = \frac{K(s - z_1)(s - z_2)\ldots(s - z_m)}{(s - p_1)(s - p_2)\ldots(s - p_n)} \qquad (8\text{-}1)$$

The Sinusoidal Steady State

and m<n. Assume further that none of the poles of G(s) lie on the jω-axis in the s-plane.

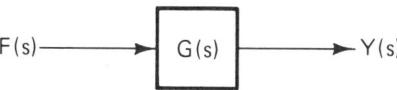

Figure 8-1. The system transfer function relates the input F(s) to the output Y(s) by the expression Y(s) = F(s)G(s).

If the input is $f(t) = A\cos(\omega_o t)$, so that

$$F(s) = \frac{As}{s^2 + \omega_o^2} \qquad (8\text{-}2)$$

then

$$Y(s) = \frac{AsG(s)}{s^2 + \omega_o^2} \qquad (8\text{-}3)$$

Assuming a partial-fraction expansion with no repeated poles in the denominator of G(s), Eq. (8-3) becomes

$$Y(s) = \frac{K_1}{s - j\omega_o} + \frac{K_1^*}{s + j\omega_o} + \frac{A_1}{s - p_1} + \frac{A_2}{s - p_2} + \cdots \frac{A_n}{s - p_n} \qquad (8\text{-}4)$$

Using the shortcut method for evaluating the residues for non-repeated poles,

$$K_1 = A \left. \frac{s(s - j\omega_o)}{s^2 + \omega_o^2} G(s) \right|_{s = j\omega_o} = \frac{A}{2} G(j\omega_o) \qquad (8\text{-}5)$$

From Eq. (8-5) we see that K_1 is determined by the value of $G(j\omega_o)$; the term $s/(s^2 + \omega_o^2)$ contributes only the factor $1/2$ to the value of K_1. If we define a new constant $R = G(j\omega_o)$ then the steady-state solution becomes

$$y_{ss}(t) = \frac{AR}{2} e^{j\omega_o t} + \frac{AR^*}{2} e^{-j\omega_o t} \qquad (8\text{-}6)$$

Expand the complex exponentials in Eq. (8-6) to get

$$y_{ss}(t) = \frac{A}{2}\cos(\omega_o t)(R+R^*) + j\frac{A}{2}\sin(\omega_o t)(R-R^*) \qquad (8-7)$$

Now let $R=C+jD$, in which case $R^*=C-jD$, $(R+R^*)/2=C$ and $(R-R^*)/2=jD$. Using these values, Eq. (8-7) becomes

$$y_{ss}(t) = A\{C\cos(\omega_o t) - D\sin(\omega_o t)\} \qquad (8-8)$$

and

$$y_{ss}(t) = A\sqrt{C^2 + D^2}\cos\{\omega_o t + \tan^{-1}(D/C)\} \qquad (8-9)$$

Relating Eq. (8-9) back to the original function $G(s)$ reveals that if $R=G(j\omega_o)=C+jD$ then

$$\sqrt{C^2 + D^2} = |R| = |G(j\omega_o)| \qquad (8-10)$$

and

$$\tan^{-1}(D/C) = \text{Arg}\{R\} = \underline{/G(j\omega_o)} = \emptyset \qquad (8-11)$$

Substituting Eqns. (8-10) and (8-11) into Eq. (8-9) yields the final expression

$$y_{ss}(t) = A|G(j\omega_o)|\cos\{\omega_o t + \emptyset\} \qquad (8-12)$$

Equation (8-12) is what we have been seeking.[7] It says that if we apply a sinusoidal function (either sine or cosine) of amplitude A and frequency ω_o to the input of a linear system with transfer function $G(s)$, then the steady-state output is also a sinusoidal function with frequency ω_o but with amplitude $A|G(j\omega_o)|$ and phase angle $\text{Arg}\{G(j\omega_o)\}$. Figure 8-2 emphasizes this point. If the input cosine has a phase angle it is added to the phase angle of Eq. (8-12).

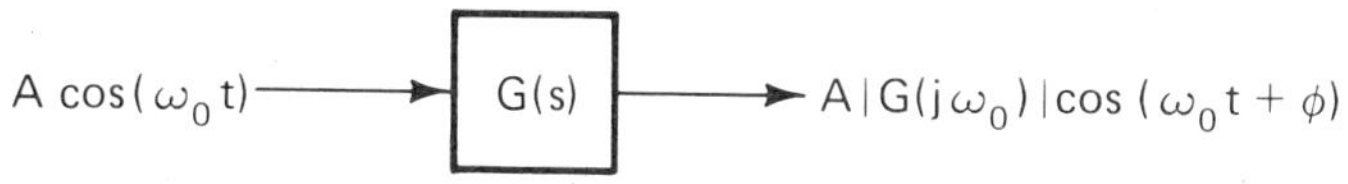

The Sinusoidal Steady State 339

$$A\cos(\omega_0 t) \longrightarrow \boxed{G(s)} \longrightarrow A|G(j\omega_0)|\cos(\omega_0 t + \phi)$$

Figure 8-2. If the input to a linear system is a sinusoid, the steady-state output is a sinusoid with the same frequency but with a different magnitude and phase angle.

Example 8-1. A system with output $y(t)$ has the transfer function

$$G(s) = \frac{1}{s^2 + 2s + 16} \qquad (8-13)$$

If the input function is $f(t)=8\cos(5t-30°)$ find the steady-state output $y_{ss}(t)$.

Solution. By inspecting $f(t)$ we see that $\omega_0=5$; hence

$$G(j5) = \frac{1}{(j5)^2 + 2(j5) + 16} \qquad (8-14)$$

and

$$|G(j5)| = \frac{1}{\sqrt{(16-25)^2 + (10)^2}} = 0.0743 \qquad (8-15)$$

The angle of $G(j5)$ is

$$\underline{/G(j5)} = -\{\pi - \tan^{-1}(10/9)\} = -132° \qquad (8-16)$$

Using Eq. (8-12), $y_{ss}(t)$ is

$$y_{ss}(t) = 0.595\cos\{5t-162.0°\} \qquad (8-17)$$

Example 8-2. The high-pass filter of Fig. 8-3(a) is excited by a voltage source with $v(t)=V\cos(\omega_0 t)$. Determine the steady-state value of the output voltage $v_o(t)$ in terms of V, R, C, and ω_0. Investigate the response for different values of ω_0.

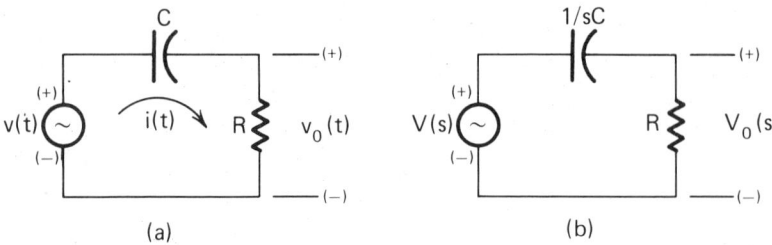

Figure 8-3. A high-pass filter network. A high-pass filter attenuates low-frequency inputs and emphasizes high-frequency inputs. (a) Time-domain network; (b) Laplace-transformed network.

Solution. Figure 8-3(b) is the Laplace transformed network. Using the voltage-divider principle for s-domain networks yields

$$V_o(s) = V(s) \frac{R}{R + 1/sC} = \frac{V(s)RCs}{RCs + 1} \qquad (8\text{-}18)$$

and the transfer function is

$$G(s) = \frac{RCs}{RCs + 1} \qquad (8\text{-}19)$$

Evaluating $G(s)$ for $s = j\omega_o$ gives

$$G(j\omega_o) = \frac{j\omega_o RC}{j\omega_o RC + 1} \qquad (8\text{-}20)$$

with magnitude

$$|G(j\omega_o)| = \frac{\omega_o RC}{\sqrt{(\omega_o RC)^2 + 1}} \qquad (8\text{-}21)$$

and phase angle

$$\angle G(j\omega_o) = 90° - \tan^{-1}(\omega_o RC) \qquad (8\text{-}22)$$

Equation (8-12) yields the result

$$v_o(t) = A\cos\{\omega_o t + 90° - \tan^{-1}(\omega_o RC)\} \qquad (8\text{-}23)$$

The Sinusoidal Steady State

where

$$A = \frac{\omega_o RCV}{\sqrt{(\omega_o RC)^2 + 1}} \qquad (8\text{-}24)$$

As $\omega_o \to 0$, the amplitude also approaches 0, and as $\omega_o \to \infty$, the amplitude approaches the limiting value of $A=V$. The higher the input frequency the greater the amplitude of the steady state output sinusoid; hence the name "high-pass" filter for the network. The filter action is emphasized in the following example.

Example 8-3. The product RC in the network of Example 8-2 has the value RC=1.0. Find the steady-state output $v_o(t)$ if the input voltage is $v(t)=100\cos(0.1t)+100\cos(10t)$.

Solution. Because the system is linear, we can consider each input separately and sum the results. For $v_1(t)=100\cos(0.1t)$ we have

$$|G(j0.1)| = \frac{0.1}{\sqrt{(0.1)^2 + 1}} = 0.0995 \qquad (8\text{-}25)$$

$$\underline{/G(j0.1)} = 90° - \tan^{-1}(0.1) = 84.3° \qquad (8\text{-}26)$$

and the partial output

$$v_{o1}(t) = 9.95\cos(0.1t + 84.3°) \qquad (8\text{-}27)$$

For $v_2(t)=100\cos(10t)$,

$$|G(j10)| = \frac{10}{\sqrt{(10)^2 + 1}} = 0.995 \qquad (8\text{-}28)$$

$$\underline{/G(j10)} = 90° - \tan^{-1}(10) = 5.7° \qquad (8\text{-}29)$$

and

$$v_{o2}(t) = 99.5\cos(10t+5.7°) \qquad (8\text{-}30)$$

The complete output is then $v_o(t) = v_{o1}(t) + v_{o2}(t)$ or

$$v_o(t) = 9.95\cos(0.1t+84.3°) + 99.5\cos(10t+5.7°)$$
(8-31)

The high-frequency input has been passed through the network with a greater magnitude and less phase shift than the low-frequency input.

The method presented above for calculating the sinusoidal steady state response is easier than (a) applying the method of undetermined coefficients to the original differential equation, or (b) using Laplace transforms to find the complete solution and then examining it for large t to identify the steady-state response. We have achieved this simplification by exploiting the functional form of the sinusoid and by avoiding the necessity of calculating the transient response.

8.2 PHASORS

The phasor method presented in the following paragraphs was historically popular because it involved a less sophisticated mathematical background on the part of the engineer. Finding the steady-state response of a linear system by phasor methods requires only a knowledge of the algebra of complex numbers—no calculus, differential equations, or Laplace transforms. In the early days of electrical engineering, before electronics and control applications had gained their present stature, electrical engineers were more oriented toward the generation, conversion, and transmission of energy, all at a single frequency. For this reason, phasor analysis was predominant,[7,8] although it did not always go by that name. Even though modern electrical engineering has expanded to include a broader selection of topics, phasors are still important (particularly due to the recent upsurge of interest in energy and energy conversion).

The Sinusoidal Steady State

Complex Notation For Time Functions

The first step on the road to understanding phasors is to learn how to represent real time-domain functions in terms of complex functions. Recalling that $\exp(jx)=\cos(x)+j\sin(x)$, the real time function $f(t)=A\cos(\omega_o t-\emptyset)$ can be written as the real part of a complex time function, or

$$f(t) = \text{Re}\{Ae^{j(\omega_o t-\emptyset)}\} \qquad (8-32)$$

where Re{ } means take the real part of the bracketed complex quantity. The quantity

$$\bar{F}(j\omega_o) = Ae^{j(\omega_o t-\emptyset)} \qquad (8-33)$$

is a complex function from which we can recover the time function $f(t)$ by the operation $f(t)=\text{Re}\{\bar{F}(j\omega_o)\}$. Because the operations of differentiation and taking the real part are interchangeable, or

$$\frac{d}{dt}\text{Re}\{\bar{F}(j\omega_o)\} = \text{Re}\left\{\frac{d}{dt}\bar{F}(j\omega_o)\right\} \qquad (8-34)$$

we can use a complex function as the input to a differential equation and then take the real part of the complex solution to find the real solution. Although it may seem that this procedure artificially complicates the problem, such is not the case; the "nice" properties of exponentials come to the rescue.

<u>Solutions in terms of complex functions</u>. In the following derivation we use the second-order differential equation

$$\ddot{y} + a_1\dot{y} + a_0 y = f(t) \qquad (8-35)$$

although the reader should realize that the results apply to higher-order equations as well. If the input to Eq. (8-35) is a sinusoid, the steady-state output is also a sinusoid but with different amplitude and phase. Write $f(t)$ and $y(t)$ as the real parts of complex quantities, or $f(t)=\text{Re}\{\bar{F}(j\omega_o)\}$ and

$y(t) = \text{Re}\{\bar{Y}(j\omega_o)\}$. Substitute these expressions into Eq. (8-35) to obtain

$$\frac{d^2}{dt^2}\text{Re}\{\bar{Y}(j\omega_o)\} + a_1\frac{d}{dt}\text{Re}\{\bar{Y}(j\omega_o)\} + a_0\text{Re}\{\bar{Y}(j\omega_o)\} = \text{Re}\{\bar{F}(j\omega_o)\}$$

(8-36)

Interchanging the operations of differentiation and taking the real part results in

$$\text{Re}\left\{\frac{d^2}{dt^2}\bar{Y}(j\omega_o) + a_1\frac{d}{dt}\bar{Y}(j\omega_o) + a_0\bar{Y}(j\omega_o)\right\} = \text{Re}\{\bar{F}(j\omega_o)\}$$

(8-37)

Now consider the operation of differentiating a complex function. Starting with

$$\bar{Y}(j\omega_o) = Be^{j(\omega_o t - \theta)}$$

(8-38)

where B is a real constant, we differentiate to obtain

$$\frac{d}{dt}\bar{Y}(j\omega_o) = j\omega_o Be^{j(\omega_o t - \theta)} = j\omega_o \bar{Y}(j\omega_o)$$

(8-39)

and

$$\frac{d^2}{dt^2}\bar{Y}(j\omega_o) = (j\omega_o)^2 \bar{Y}(j\omega_o)$$

(8-40)

In general, differentiating the complex exponential is equivalent to multiplying the function by $j\omega_o$; thus, for the nth-order derivative

$$\frac{d^{(n)}}{dt^n}\bar{Y}(j\omega_o) = (j\omega_o)^n \bar{Y}(j\omega_o)$$

(8-41)

Applying Eqns. (8-39) and (8-40), and dropping the Re{ } notation, Eq. (8-37) becomes

$$\{(j\omega_o)^2 + a_1(j\omega_o) + a_0\}\bar{Y}(j\omega_o) = \bar{F}(j\omega_o)$$

(8-42)

Substituting Eqns. (8-33) and (8-38) into Eq. (8-42) gives

$$\{(j\omega_o)^2 + a_1(j\omega_o) + a_0\}Be^{j(\omega_o t - \theta)} = Ae^{j(\omega_o t - \emptyset)}$$

(8-43)

The Sinusoidal Steady State

The bracketed quantity on the left is a complex number which can be written in magnitude-angle form,

$$(j\omega_o)^2 + a_1(j\omega_o) + a_0 = Ce^{j\alpha} \tag{8-44}$$

and Eq. (8-43) becomes

$$BCe^{j\alpha}e^{j\omega_o t}e^{-j\theta} = Ae^{j\omega_o t}e^{-j\phi} \tag{8-45}$$

Equation (8-45) is solved for $B=A/C$ and $\theta=\phi+\alpha$. The steady-state solution $y_{ss}(t)$ is then $y_{ss}(t)=\text{Re}\{\bar{Y}(j\omega_o)\}$, or

$$y_{ss}(t) = \text{Re}\{Be^{j(\omega_o t - \theta)}\} = B\cos(\omega_o t - \theta) \tag{8-46}$$

Substituting for B and θ yields

$$y_{ss}(t) = \frac{A}{C}\cos(\omega_o t - \phi - \alpha) \tag{8-47}$$

This result agrees with Eq. (8-12) if we identify $|G(j\omega_o)|=1/C$ and $\alpha=\text{Arg}\{G(j\omega_o)\}$.

Thus far it appears that we have merely confirmed that the magnitude and phase of the input sinusoid are modified by the magnitude and phase of the system transfer function to produce the steady-state output. But this is only the beginning.

Example 8-4. A system has as its input $f(t)=25\cos(35t-\pi/4)$. What is the complex function $\bar{F}(j\omega)$ associated with $f(t)$?

Solution. Referring to Eqns. (8-32) and (8-33), we have $\bar{F}(j\omega)=25\exp\{j(35t-\pi/4)\}$.

Example 8-5. The general second-order system with a sinusoidal input is

$$\ddot{y} + 2\zeta\omega_n\dot{y} + \omega_n^2 y = A\cos(\omega_o t - \phi) \tag{8-48}$$

Derive expressions for (a) the complex output $\bar{Y}(j\omega_o)$, and (b) the steady-state solution $y_{ss}(t)$.

Solution. The complex input function, $\bar{F}(j\omega_o)$, is $\bar{F}(j\omega_o) = A\exp\{j(\omega_o t - \emptyset)\}$ and the transfer function is

$$G(s) = \frac{Y(s)}{F(s)} = \frac{1}{s^2 + 2\zeta\omega_n s + \omega_n^2} \qquad (8\text{-}49)$$

Replacing s by $j\omega_o$ yields

$$G(j\omega_o) = \frac{1}{(j\omega_o)^2 + 2j\omega_o\zeta\omega_n + \omega_n^2} \qquad (8\text{-}50)$$

from which

$$|G(j\omega_o)|^2 = \frac{1}{(\omega_n^2 - \omega_o^2)^2 + (2\zeta\omega_o\omega_n)^2} \qquad (8\text{-}51)$$

and

$$\underline{/G(j\omega_o)} = -\tan^{-1}\left\{\frac{2\zeta\omega_o\omega_n}{\omega_n^2 - \omega_o^2}\right\} \qquad (8\text{-}52)$$

The complex output $\bar{Y}(j\omega_o)$ is then

$$\bar{Y}(j\omega_o) = A|G(j\omega_o)|e^{j\{\omega_o t - \emptyset + \underline{/G(j\omega_o)}\}} \qquad (8\text{-}53)$$

and the time-domain solution is the real part of the complex solution, or

$$y_{ss}(t) = A|G(j\omega_o)|\cos\{\omega_o t - \emptyset + \underline{/G(j\omega_o)}\} \qquad (8\text{-}54)$$

The Sinusoidal Steady State Response Via Phasors

As a consequence of the previous derivations, we see that the only pieces of information needed to determine the sinusoidal steady state response of a linear system are (1) the magnitude of the input sinusoid, (2) the phase angle of the input sinusoid, and (3) $G(j\omega_o)$, the system transfer function evaluated at the frequency of the input sinusoid. The procedure is straightforward, but we are faced with the problem of determining the system transfer function $G(s)$ from

The Sinusoidal Steady State

the system model. We can avoid even this difficulty by introducing the concepts of phasor variables and phasor system models.$^{(9,10)}$

Phasor variables. Suppose we have a system with input $f(t)$ and dependent variable $y(t)$, where $f(t)=A\cos(\omega_o t-\emptyset)$ and $y(t)=B\cos(\omega_o t-\theta)$. The associated complex functions are

$$\bar{F}(j\omega_o) = Ae^{j(\omega_o t-\emptyset)} = Ae^{-j\emptyset}e^{j\omega_o t} \qquad (8\text{-}55)$$

and

$$\bar{Y}(j\omega_o) = Be^{j(\omega_o t-\theta)} = Be^{-j\theta}e^{j\omega_o t} \qquad (8\text{-}56)$$

But the terms $\exp(j\omega_o t)$, common to both Eqns. (8-55) and (8-56), convey no information; they just remind us of the frequency ω_o of the input sinusoid. Recall that the $\exp(j\omega_o t)$ terms cancelled in Eq. (8-45) and hence were not really needed in the derivation. For this reason we can, with no loss of generality, ignore the $\exp(j\omega_o t)$ terms and introduce the complex __phasor__ quantities $\bar{F}=A\exp(-j\emptyset)$ and $\bar{Y}=B\exp(-j\theta)$, where the absence of arguments on $\bar{F}$ and $\bar{Y}$ indicates the deletion of the $\exp(j\omega_o t)$ factors. We will use this notation in the remainder of the chapter. Capital letters with overbars denote phasor quantities. A phasor then is nothing more than a complex constant which tells us the amplitude and phase of a sinusoidal variable. The frequency information has been lost and must be noted separately.

__Example 8-6.__ What is the phasor voltage associated with $v(t)=50\cos(33t-25°)$?

__Solution.__ When the phase angle is given in degrees, we avoid the exponential notation. In this instance the phasor voltage is given either by $\bar{V}=50\underline{/-25°}$ or by $\bar{V}=50\exp(-j0.4363)$ where 0.4363 is $25°$ expressed in radians.

Finally, the phasor does not tell us whether the time domain function is a sine or a cosine; valid analyses can be made with either. In the remaining material we assume that phasors always represent cosine functions in the time domain. This involves no loss in generality because $\sin(x)=\cos(\pi/2-x)$.

The phasor network model. We have seen how the essence of a time function can be contained in a complex number or phasor, eliminating the need either for time-domain or s-domain functions. Next we must develop a means of by-passing the need for determining G(s) if it is not readily available. To do this we introduce still another network model--the phasor model. This procedure does not further complicate life as the phasor network model is just the s-domain model with s replaced by $j\omega$. The impedances of the network components become

$$Z_R(j\omega) = R \qquad Z_L(j\omega) = j\omega L \qquad Z_C(j\omega) = \frac{1}{j\omega C} \qquad (8\text{-}57)$$

Figure 8-4 shows the circuit diagram notation for the network impedances.

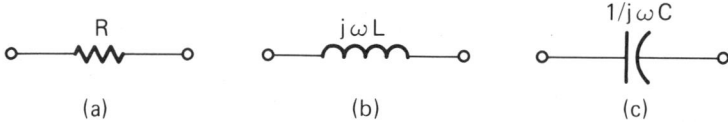

(a)   (b)   (c)

Figure 8-4. The impedances of the network elements are found by replacing s by $j\omega$. (a) Resistor; (b) Inductor; (c) Capacitor.

Sources are replaced by their phasor equivalents. A voltage source which produces $v(t)=A\cos(\omega_o t-\emptyset)$ is replaced on the phasor diagram by a source which produces the phasor voltage $\bar{V}=A\exp(-j\emptyset)$ or $V=A\underline{/-\emptyset}$. Note the different way we treat dependent variables as opposed to network elements. Impedances are _not_ phasor quantities; they are s-domain functions with s replaced by $j\omega$. But dependent variables are not s-domain quantities with s replaced by $j\omega$. If $y(t)=\cos(\omega_o t)$ then

$$Y(s) = \frac{s}{s^2 + \omega_o^2} \qquad (8\text{-}58)$$

and

$$Y(j\omega_o) = Y(s)\Big]_{s=j\omega_o} \quad \to\infty \qquad (8\text{-}59)$$

The Sinusoidal Steady State

because $j\omega_o$ is a pole of $Y(s)$. $Y(j\omega_o)$ is a perfectly valid expression, it is just not helpful in determining the sinusoidal steady state. The phasor $\bar{Y}$ for $y(t)=\cos(\omega_o t)$ is the well-behaved function $\bar{Y}=1\exp(j0)$.

Example 8-7. The electrical network of Fig. 8-5(a) is to be analyzed in the sinusoidal steady state. (a) Redraw the network as a phasor network. (b) Solve for the phasor output voltage $\bar{V}_o$. (c) Solve for the steady state output voltage in the time domain if $v(t)=V_m\cos(\omega t)$. Use C=2.5 µF, R=100 Ω, ω =5000 rad/s, and V_m=100 V.

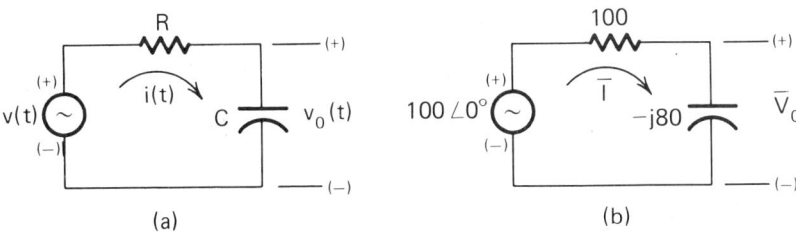

Figure 8-5. Networks for Example 8-7. (a) Time-domain network; (b) Phasor network.

Solution. The complex impedances needed are Z_R=R=100 and Z_C=1/jωC=-j80.

Write KVL around the loop in terms of phasors, or

$$100\underline{/0°} = 100\bar{I} - j80\bar{I} \quad (8\text{-}60)$$

Solve for $\bar{I}$:

$$\bar{I} = \frac{100 + j0}{100 - j80} \quad (8\text{-}61)$$

or

$$\bar{I} = \frac{100}{\sqrt{(100)^2 + (80)^2}} \underline{/\tan^{-1}(80/100)} \quad (8\text{-}62)$$

and $\bar{I}$=0.78$\underline{/38.6°}$. Because $\bar{V}_o$=-j80$\bar{I}$, we have $\bar{V}_o$=62.4$\underline{/-51.4°}$. The time-domain steady-state output voltage is then

$$v_o(t) = 62.4\cos(5000t-51.4°) \quad (8\text{-}63)$$

Example 8-8. Derive an expression for the phasor output voltage $\bar{V}_o$ for the network of Fig. 8-6(a). Assume $v(t) = V_m \cos(\omega t)$.

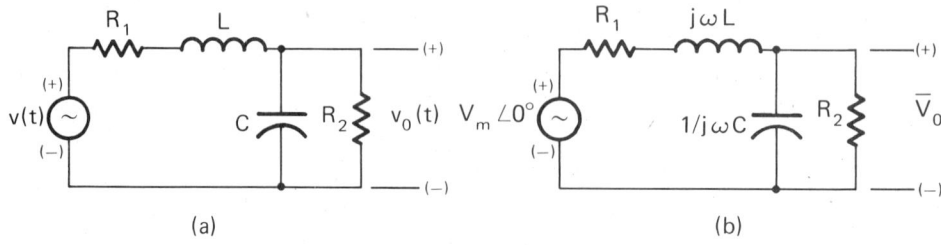

Figure 8-6. Network for Example 8-8. (a) Time-domain network; (b) Phasor network.

Solution. (a) Replacing the source by a phasor source and the elements by their complex impedances gives the phasor network of Fig. 8-6(b). Rather than write loop equations in terms of phasor loop currents (which is a perfectly valid approach), we can combine R_2 and $1/j\omega C$ in parallel, and then use the voltage-divider principle to find $\bar{V}_o$. Thus,

$$\bar{V}_o = \frac{V_m \angle 0° \; R_2/(1 + j\omega R_2 C)}{R_2/(1 + j\omega R_2 C) + R_1 + j\omega L} \qquad (8\text{-}64)$$

$$= \frac{R_2 V_m \angle 0°}{R_1 + R_2 - \omega^2 L C R_2 + j(\omega L + \omega R_1 R_2 C)} \qquad (8\text{-}65)$$

or $\bar{V}_o = B\angle{-\beta}$ where

$$B^2 = \frac{(R_2 V_m)^2}{(R_1 + R_2 - \omega^2 L C R_2)^2 + (\omega L + \omega R_1 R_2 C)^2} \qquad (8\text{-}66)$$

and

$$\beta = \tan^{-1}\left\{\frac{\omega L + \omega R_1 R_2 C}{R_1 + R_2 - \omega^2 L C R_2}\right\} \qquad (8\text{-}67)$$

The Sinusoidal Steady State

The time-domain output voltage is then $v_o(t) = B\cos(\omega t - \beta)$.

Example 8-9. Calculate the steady state output voltage $v_o(t)$ for the network of Fig. 8-7 if $\bar{V}_i = 110\underline{/0°}$ and the frequency of the source voltage is $\omega = 377$ rad/s.

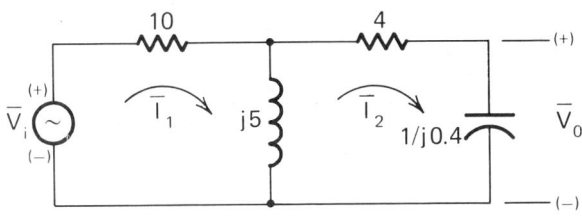

Figure 8-7. Network for Example 8-9.

Solution. Write loop equations for the phasor currents $\bar{I}_1$ and $\bar{I}_2$:

$$\bar{V}_i = 10\bar{I}_1 + j5(\bar{I}_1 - \bar{I}_2) \qquad (8\text{-}68)$$

$$0 = j5(\bar{I}_2 - \bar{I}_1) + \bar{I}_2(4 + (-j2.5)) \qquad (8\text{-}69)$$

or, after rearranging,

$$(10 + j5)\bar{I}_1 - j5\bar{I}_2 = \bar{V}_i \qquad (8\text{-}70)$$

$$-j5\bar{I}_1 + (4 + j2.5)\bar{I}_2 = 0 \qquad (8\text{-}71)$$

Using Cramer's rule and determinants, the solution for $\bar{I}_2$ is

$$\bar{I}_2 = \frac{\begin{vmatrix} 10 + j5 & \bar{V}_i \\ -j5 & 0 \end{vmatrix}}{\begin{vmatrix} 10 + j5 & -j5 \\ -j5 & 4 + j2.5 \end{vmatrix}} \qquad (8\text{-}72)$$

$$= \frac{j5\bar{V}_i}{(10 + j5)(4 + j2.5) + -25} \qquad (8\text{-}73)$$

from which $\bar{I}_2 = 12.2\underline{/3.18°}$. The phasor output voltage is $\bar{V}_o =$

$-j2.5\bar{I}_2 = 30.5\underline{/-86.8°}$. The steady-state time-domain output voltage is then

$$v_o(t) = 30.5\cos(377t - 86.8°) \qquad (8\text{-}74)$$

Phasor analysis applies to linear systems from any engineering discipline; it is not limited to electrical networks. However, it is mainly in the electrical world that we are interested in the steady-state response of a system to a sinusoidal input function. In mechanical systems we are often more concerned with transient analysis, in which case phasor methods do not apply.

Phasor Diagrams

Just as the s-plane leads to greater insight into Laplace transformed functions, a graphical interpretation of phasor quantities can be a useful aid to the systems analyst.[11] A phasor diagram is a two-dimensional plot of complex phasors. Figure 8-8 shows a typical phasor diagram.

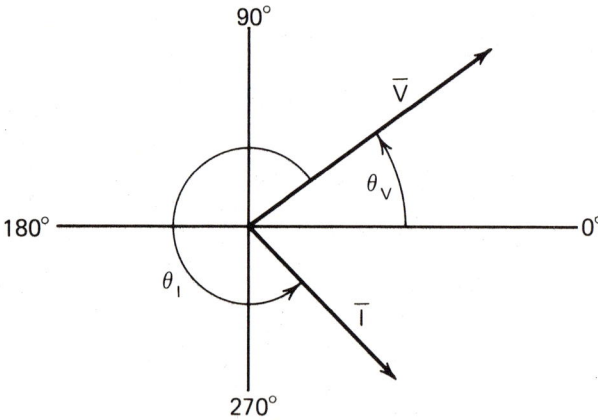

Figure 8-8. Phasor voltages and currents are represented geometrically on the phasor diagram.

Phasor voltages and currents become vectors on the phasor diagram, each with its appropriate magnitude and

The Sinusoidal Steady State

phase. The length of each arrow corresponds to the magnitude of the phasor and the angle from the $+0°$ axis is the argument.

It is useful to sketch phasor diagrams for the network elements. For the resistor $\bar{V}_R = \bar{I}_R R$ and the voltage and current are in phase. Figure 8-9(a) is the associated phasor diagram. For the inductor, $\bar{V}_L = j\omega L \bar{I}_L = \omega L I_L \underline{/90°}$ so that if $\bar{V}_L = V_L \underline{/0°}$ then

$$\bar{I}_L = \frac{V_L}{\omega L} \underline{/-90°} \tag{8-75}$$

(A phasor symbol without the overbar denotes the magnitude of the phasor; thus $V = |\bar{V}|$.) Figure 8-9(b) is the phasor diagram. Finally, for the capacitor we have

$$\bar{V}_C = \frac{\bar{I}_C}{j\omega C} = \frac{I_C}{\omega C} \underline{/-90°} \tag{8-76}$$

If $\bar{V}_C = V_C \underline{/0°}$, then

$$\bar{I}_C = V_C \omega C \underline{/90°} \tag{8-77}$$

Figure 8-9(c) is the phasor diagram.

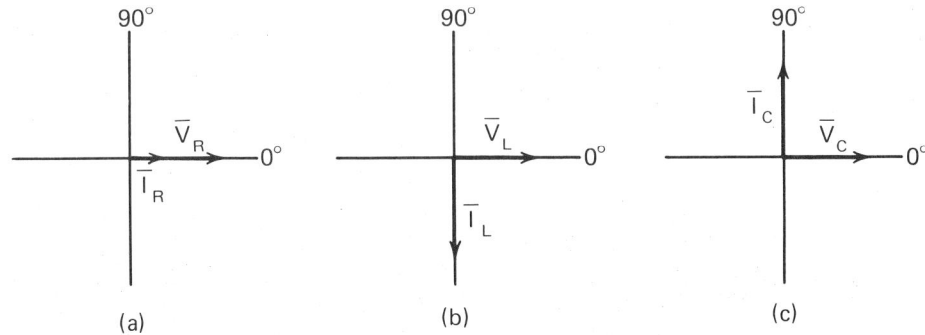

Figure 8-9. Phasor diagrams for the network elements. Voltage and current are in phase for the resistor. Inductor voltage leads inductor current, but capacitor voltage lags capacitor current. (a) Resistor; (b) Inductor; (c) Capacitor.

8.3 RESONANCE

A system is said to be in <u>resonance</u> if it is driven with a sinusoidal input whose frequency is equal to one of the natural frequencies of the system. An equivalent statement for electrical networks is that the phasor voltage and current at a particular terminal are in phase. This condition also implies that the impedance is a real rather than a complex quantity at the resonant frequency. In the following paragraphs we investigate the phenomenon of **resonance** and show the equivalence of the above statements.[12,13]

From the definitions above, resonance may seem to be a benign mathematical abstraction, but this is far from the case. Resonance can be a desirable **phenomenon, as** when one attempts to deliver the maximum amount of power from a source to a load, or it can bring disaster, as when soldiers marching in step across a bridge excite one of the natural frequencies and cause the structure to collapse, or when a properly pitched musical note shatters a glass.

The Series RLC Network

Consider the series RLC circuit of Fig. 8-10.

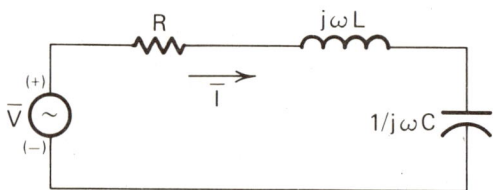

Figure 8-10. A series RLC network.

The phasor input voltage $\bar{V}$ is a sinusoid with a frequency of ω, or $v(t) = V\cos(\omega t)$. The current $\bar{I}$ has the form $\bar{I} = I\,\underline{/\theta^\circ}$ where I is a real constant. The network elements are written as impedances, and the total impedance is

$$Z(j\omega) = R + j\omega L + \frac{1}{j\omega C} \tag{8-78}$$

Using Ohm's law for impedances yields $\bar{V} = \bar{I}Z(j\omega)$ or

$$\bar{I} = \frac{\bar{V}}{R + j(\omega L - 1/\omega C)} \tag{8-79}$$

The condition for resonance, or for $\bar{V}$ and $\bar{I}$ to have the same phase angle, is that the angle of $Z(j\omega_o)$ be 0 where ω_o is the <u>resonant frequency</u> of the network. This in turn requires that the imaginary part of the impedance be 0, or

$$j(\omega_o L - \frac{1}{\omega_o C}) = 0 \tag{8-80}$$

Solving for ω_o yields $\omega_o = 1/\sqrt{LC}$.

If we use s-domain impedances, the equation for the Laplace transform of the current is

$$I(s) = \frac{V(s)}{R + Ls + 1/sC} \tag{8-81}$$

or

$$I(s) = \frac{sV(s)/L}{s^2 + Rs/L + 1/LC} \tag{8-82}$$

Comparing the denominator of Eq. (8-82) with $s^2 + 2\zeta\omega_n s + \omega_n^2$ reveals that $\omega_n = 1/\sqrt{LC}$. This verifies that the resonant frequency corresponds to the natural frequency of the network. If a higher-order system has several natural frequencies, it will resonate at any one of them.

Another property of the series resonant circuit is that at resonance $\bar{I}$ assumes its maximum value. From Eq. (8-79) the magnitude I of the phasor current $\bar{I}$ at resonance is

$$I = \frac{V}{\sqrt{R^2 + (\omega_o L - 1/\omega_o C)^2}} \tag{8-83}$$

At resonance, the second term in the denominator becomes 0, and $I = V/R$. For any value of frequency other than ω_o the magnitude of the current is less than V/R.

Example 8-10. A series RLC network (Fig. 8-10) is driven by a sinusoidal voltage source whose frequency is f=60 Hz and whose peak value is 115 V. If the inductance is 2.0 H and the resistance is 50 Ω, determine (a) the value of C for resonance, and (b) the expression for i(t), the current in the circuit.

Solution. (a) The expression $\omega_o = 1/\sqrt{LC}$ can be solved for the value of capacitance needed for resonance, $C = 1/\omega_o^2 L$, from which C=3.52 μF. (b) We can assume that the phase of v(t) is $0°$ and represent the phasor voltage as $\bar{V} = V\underline{/0°} = 115\underline{/0°}$. Similarly, we write $\bar{I} = I\underline{/\theta}$. At the resonant frequency, I=V/R is the maximum value of $\bar{I}$; hence I=115/50=2.3 A. Finally, at resonance i(t) is in phase with v(t) so that $\theta = 0°$; the expression for i(t) is therefore

$$i(t) = (2.3)\cos(2\pi 60 t) \qquad (8\text{-}84)$$

Once the capacitor value is set at 3.52 μF, if voltage at any frequency other than f=60 Hz were applied to the network, the current amplitude would be less than 2.3 A and the phase shift would be nonzero.

The Parallel RLC Network

Figure 8-11 is a parallel RLC circuit driven by a phasor current source.

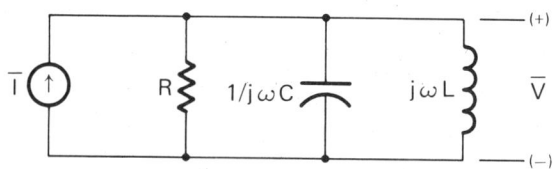

Figure 8-11. A parallel RLC network.

The total admittance is

$$Y(j\omega) = \frac{1}{R} + j\omega C + \frac{1}{j\omega L} \qquad (8\text{-}85)$$

The Sinusoidal Steady State 357

and the output voltage is found from $\bar{V}=\bar{I}/Y(j\omega)$ or

$$\bar{V} = \frac{\bar{I}}{1/R + j(\omega C - 1/\omega L)} \tag{8-86}$$

Again, resonance occurs if the input phasor has the frequency $\omega_0 = 1/\sqrt{LC}$ to make the current and voltage in phase, but here the admittance is a minimum so the impedance is at a maximum. This condition is often called <u>anti-resonance</u>. In this case the output voltage is at a maximum.

Example 8-11. Use differential calculus to show that V is a maximum at resonance in the parallel resonant circuit of Fig. 8-11.

Solution. The expression for V is

$$V = \frac{I}{\sqrt{(1/R)^2 + (\omega C - 1/\omega L)^2}} \tag{8-87}$$

Taking the derivative of V with respect to ω yields

$$\frac{d}{d\omega}V = \frac{-(\omega C - 1/\omega L)(C + 1/\omega^2 L)I}{\{(1/R)^2 + (\omega C - 1/\omega L)^2\}^{3/2}} \tag{8-88}$$

Setting the derivative equal to 0 results in

$$(\omega C - 1/\omega L)(C + 1/\omega^2 L) = 0 \tag{8-89}$$

and the only positive solution for ω is $\omega = \omega_0$ where $\omega_0^2 = 1/LC$.

8.4 POWER IN THE SINUSOIDAL STEADY STATE

When a dynamic system undergoes a brief period of operation, we are not normally interested in the power developed or the energy consumed during the transient. But when such a system operates for long periods in the steady state, we are vitally concerned with the amount of energy required to operate the system or the amount it can deliver to another system. In this section we derive relationships for the power developed in an electrical network in the sinusoidal steady state.[14,15]

Power And Energy

Energy, defined as the ability to do work, is a fundamental quantity in the study of natural phenomena. The dimensions of energy are force times distance. Power is the time rate of change of energy; if W(t) stands for energy and p(t) for power, then p(t)=dW(t)/dt and

$$W(t) = \int_0^t p(t)dt + W(0) \qquad (8\text{-}90)$$

To obtain the amount of energy consumed, we sum (integrate) the power over a period of time. Note too that the energy associated with a system is a function of time and is measured with respect to a reference level. For electrical variables, we showed in Section 3.1 that the instantaneous power associated with an electrical component is p(t)=v(t)i(t) where v(t) and i(t) are the instantaneous voltage and current.

Average Power

The instantaneous power can be positive, negative, or zero depending on the values of v(t) and i(t) at a given instant. For this reason instantaneous power is not a particularly useful concept. In Chapter 3 we defined the <u>average power</u> by

$$P_{ave} = \frac{1}{t_2 - t_1} \int_{t_1}^{t_2} p(t)dt \qquad (8\text{-}91)$$

where t_2-t_1 is the interval of time over which we want the average taken.

Let us consider the power delivered to an impedance by a sinusoidal voltage source in steady-state operation. We can, with no loss of generality, take the phase reference for v(t) as 0° and assume that the voltage source produces v(t)= $V_m \cos(\omega t)$. In this case the current through the impedance

The Sinusoidal Steady State

takes the form $i(t)=I_m\cos(\omega t-\emptyset)$ where the magnitude and phase of $i(t)$ depend on the complex impedance (which, of course, need not be a pure resistance, inductance, or capacitance but can be the equivalent impedance of a complicated network). Taking the product of $v(t)$ and $i(t)$ yields the instantaneous power

$$p(t) = V_m I_m \cos(\omega t)\cos(\omega t-\emptyset) \qquad (8\text{-}92)$$

Substituting Eq. (8-92) into Eq. (8-91) results in

$$P_{ave} = \frac{V_m I_m}{t_2-t_1} \int_{t_1}^{t_2} \cos(\omega t)\cos(\omega t-\emptyset)dt \qquad (8\text{-}93)$$

Because $v(t)$ and $i(t)$ are both sinusoids with the same frequency, we need to compute the average over only one period. By symmetry, the average power must be the same over every period provided we are in steady state operation. Consequently, in Eq. (8-93) we choose $t_1=0$ and $t_2=2\pi/\omega$ where ω is the frequency of the sinusoids. Using the trigonometric identity $\cos(x-y)=\cos(x)\cos(y)+\sin(x)\sin(y)$, Eq. (8-93) becomes

$$P_{ave} = \frac{V_m I_m}{T} \int_0^T \cos(\omega t)\{\cos(\omega t)\cos(\emptyset) + \sin(\omega t)\sin(\emptyset)\}dt \qquad (8\text{-}94)$$

$$= \frac{V_m I_m}{T} \left\{ \cos(\emptyset) \int_0^T \cos^2(\omega t)dt + \sin(\emptyset) \int_0^T \cos(\omega t)\sin(\omega t)dt \right\} \qquad (8\text{-}95)$$

Using the identities $\cos^2(\omega t)=\{1+\cos(2\omega t)\}/2$ and $\sin(\omega t)\cos(\omega t)=\sin(2\omega t)/2$, Eq. (8-95) becomes

$$P_{ave} = \frac{V_m I_m}{2T} \left\{ \cos(\emptyset) \int_0^T (1)dt + \cos(\emptyset) \int_0^T \cos(2\omega t)dt \right.$$

$$\left. + \sin(\emptyset) \int_0^T \sin(2\omega t)dt \right\} \qquad (8\text{-}96)$$

Because the integral of a sine or a cosine over an integral number of its periods is 0, Eq. (8-96) reduces to

$$P_{ave} = \frac{V_m I_m \cos(\emptyset)}{2T} \int_0^T (1) dt \qquad (8-97)$$

and

$$P_{ave} = \frac{V_m I_m \cos(\emptyset)}{2} \qquad (8-98)$$

Equation (8-98) is packed with significance. It implies that an impedance consumes a maximum amount of energy if the voltage and current are in phase, $\emptyset=0$, in which case $\cos(\emptyset)$ has its maximum value of 1. This in turn means that power transfer is maximized in resonant circuits. One practical result of this phenomena is that if we wish to transfer a maximum amount of energy to a system in a minimum time span, then we should design the system so that its input impedance resonates at the frequency of the input voltage. This is exactly what is done in practice.

At the other end of the spectrum, if $\emptyset=90°$, or if the voltage and current are in <u>quadrature</u> ($90°$ out of phase), as is the case with a pure inductor or a pure capacitor, then $\cos(\emptyset)=0$ and no energy is consumed by the impedance! There is an oscillatory ebb and flow of energy from the source to the impedance and then back from the impedance to the source. Because no energy is consumed in the impedance, it might seem that this situation is not deleterious, but remember that in the real world wires have resistance, and energy will be consumed in the conductors which connect the source to the impedance.

<u>Example 8-12</u>. The voltage $v(t)=120\cos(\omega t-\pi/2)$ is applied to an unknown impedance and a steady-state sinusoidal current $i(t)=5.0\cos(\omega t-\pi/12)$ is observed. Calculate the average power developed in the impedance.

The Sinusoidal Steady State

Solution. Equation (8-98) applies and

$$P_{ave} = \frac{V_m I_m}{2} \cos(\emptyset) = \frac{(120)(5)}{2} \cos(\pi/2 - \pi/12) \quad (8-99)$$

and $P_{ave} = 77.6$ W.

Example 8-13. The voltage waveform of Fig. 8-12 is applied to a 10-ohm resistor. Calculate the average power developed across the resistor.

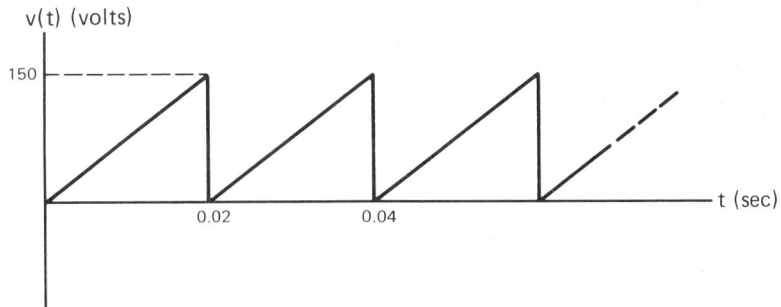

Figure 8-12. Voltage waveform for Example 8-13.

Solution. Equation (8-98), developed for sinusoidal voltages and currents, does <u>not</u> apply to this problem. We must use the definition of average power, Eq. (8-91). The equation for the voltage in the first period is

$$v(t) = \left\{\frac{150}{0.02}\right\} t = 7500t \quad (8-100)$$

for $0 < t < 0.02$. The current, according to Ohm's law, is $i(t) = v(t)/R = 750t$, also for $0 < t < 0.02$. The instantaneous power is

$$p(t) = v(t)i(t) = 5.625 \times 10^6 t^2 \quad (8-101)$$

in the first period. Using Eq. (8-91),

$$P_{ave} = \frac{1}{T} \int_0^T p(t) dt = \frac{1}{0.02} \int_0^{0.02} (5.625 \times 10^6) t^2 dt \quad (8-102)$$

$$= 2.8125 \times 10^8 \left. \frac{t^3}{3} \right]_0^{0.02} \tag{8-103}$$

from which $P_{ave} = 750$ W.

Root-Mean-Square Variables

Equation (8-98) reveals that the power in the sinusoidal steady state can be computed from a knowledge of the maximum values of the voltage and current and the phase difference between them. But this is exactly the information contained in the phasor voltage and current. If, for example, the voltage and current associated with an impedance are $v(t) = V_m \cos(\omega t - \alpha)$ and $i(t) = I_m \cos(\omega t - \beta)$, then $\bar{V} = V_m \underline{/\alpha}$, $\bar{I} = I_m \underline{/\beta}$, and

$$P_{ave} = \frac{V_m I_m}{2} \cos(\alpha - \beta) \tag{8-104}$$

For this reason, information about average power consumption is readily available if we work with phasors. We can, however, go one step further and introduce the so-called root-mean-square (RMS) values for voltage and current.

Assume sinusoidal variables and use Ohm's law, $v(t) = i(t)R$. With the equation for instantaneous power, we find that the power across a resistor is $p(t) = v(t)i(t) = i(t)^2 R$ and the average power is

$$P_{ave} = \frac{1}{T} \int_0^T i(t)^2 R \, dt \tag{8-105}$$

where $T = 2\pi/\omega$. We now define the root-mean-square or <u>effective</u> current as the value of a direct current, say I, which will produce the same average power as the sinusoidal current: or

$$P_{ave} = \frac{1}{T} \int_0^T I^2 R \, dt \tag{8-106}$$

The Sinusoidal Steady State

Equating Eqns. (8-105) and (8-106) results in

$$RI^2 \int_0^T dt = R \int_0^T i(t)^2 dt \qquad (8-107)$$

from which

$$I = \sqrt{\frac{1}{T} \int_0^T i(t)^2 dt} \qquad (8-108)$$

Equation (8-108) shows the reason for calling the effective value the root-mean-square current. I is the square-root of the mean of the squared current. If $i(t)$ is a sinusoid, say $i(t) = I_m \cos(\omega t)$, then

$$I^2 = \frac{1}{T} \int_0^T I_m^2 \cos^2(\omega t) dt \qquad (8-109)$$

$$= \frac{I_m^2}{2T} \int_0^T \{1 + \cos(2\omega t)\} dt \qquad (8-110)$$

and

$$I^2 = \frac{I_m^2}{2T} \left\{ \int_0^T (1) dt + \int_0^T \cos(2\omega t) dt \right\} \qquad (8-111)$$

Recognizing that the integral of $\cos(2\omega t)$ over the interval $0 < t < T$ is 0, the RMS value is $I = I_m/\sqrt{2}$.

Example 8-14. Compute the RMS value of the waveform of Fig. 8-12.

Solution. The RMS value of a sinusoid is its peak value divided by $\sqrt{2}$, but for the waveform of Fig. 8-12 we must return to the definition of RMS value, Eq. (8-108). In general,

$$V_{RMS}^2 = \frac{1}{T}\int_0^T v(t)^2 dt \qquad (8\text{-}112)$$

For this example waveform

$$V_{RMS}^2 = \frac{1}{0.02}\int_0^{0.02} (7500)^2 t^2 dt \qquad (8\text{-}113)$$

and $V_{RMS}=86.6$ V.

Example 8-15. Develop a general expression for the RMS value of the waveform of Fig. 8-13.

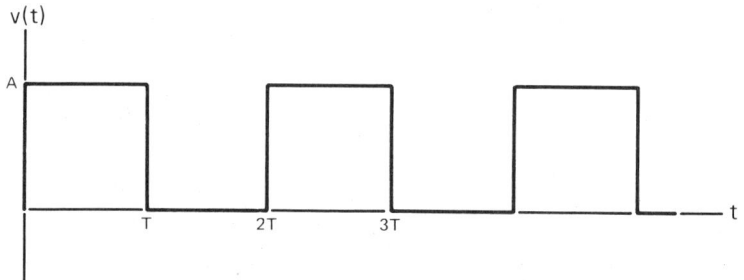

Figure 8-13. Example 8-15. Application of Eq. (8-112) shows that $V_{RMS} = A/\sqrt{2}$ for this waveform.

Solution. Using Eq. (8-108), and observing that the period is 2T, we have

$$V_{RMS}^2 = \frac{1}{2T}\left[\int_0^T A^2 dt + \int_T^{2T} 0^2 dt = \frac{A^2}{2T}t\right]_0^T \qquad (8\text{-}114)$$

and $V_{RMS}=A/\sqrt{2}$. That the result is identical to that for the sine wave is just a happy coincidence.

Up to this point we have used the notation $\bar{V}$ to represent the phasor voltage, $\bar{V}=V_m\underline{/\theta^\circ}$ or $\bar{V}=V_m\exp(j\theta)$, associated with the time-domain voltage $v(t)=V_m\cos(\omega t-\theta)$. In a similar fashion we can define a root-mean-square phasor as $V_m/\sqrt{2}\underline{/\theta^\circ}$.

The Sinusoidal Steady State

Using RMS values, we can write the expression for average power, Eq. (8-98), as

$$P_{ave} = \frac{V_m I_m}{2} \cos(\emptyset) = \frac{V_m}{\sqrt{2}} \frac{I_m}{\sqrt{2}} \cos(\emptyset) \qquad (8\text{-}115)$$

or

$$P_{ave} = VI\cos(\emptyset) \qquad (8\text{-}116)$$

where V and I are the magnitudes of the RMS voltage and current. The angle $\emptyset$ is called the <u>power factor angle</u> and the <u>power factor</u>, pf, of the impedance associated with $\bar{V}$ and $\bar{I}$ is defined by pf=cos($\emptyset$).

Other Power Relationships[16]

Consider the network of Fig. 8-14 where a sinusoidal source described by the RMS phasor $\bar{V}$ is applied to a complex load impedance $Z_o(j\omega)$ to produce the RMS phasor current $\bar{I}$.

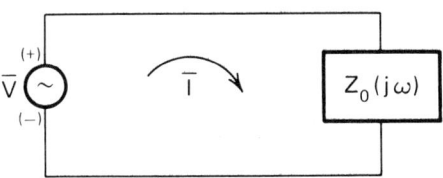

Figure 8-14. Power relationships are based on this network wherein a voltage source delivers energy to a complex load impedance $Z_o(j\omega)$.

Let $\bar{V}$ and $\bar{I}$ have the forms $\bar{V}$=Vexp(jα) and $\bar{I}$=Iexp(jβ) where V=$V_m/\sqrt{2}$ and I=$I_m/\sqrt{2}$. Further, let the load impedance $Z_o(j\omega)$ be written

$$Z_o(j\omega) = R(\omega) + jX(\omega) \qquad (8\text{-}117)$$

where R(ω)=Re{$Z_o(j\omega)$} and X(ω)=Im{$Z_o(j\omega)$}. Although R(ω) is the real part of the impedance, it need not be a constant resistance; it may be a function of ω (see Example 8-18).

The imaginary part $X(\omega)$ is called the <u>reactance</u> of the impedance. The impedance of a capacitor is $Z_C(j\omega)=1/j\omega C$; hence, its reactance is $X_C'=-1/\omega C$. The impedance of an inductor is $Z_L(j\omega)=j\omega L$ and its reactance is $X_L=\omega L$. The reactance of a resistor is 0.

If the load impedance of Fig. 8-14 is written as $Z_o(j\omega)=|Z_o|\exp(j\emptyset)$ then, using Ohm's law for impedance, we have $Z_o(j\omega)=\bar{V}/\bar{I}$ and

$$|Z_o|e^{j\emptyset} = \frac{Ve^{j\alpha}}{Ie^{j\beta}} \qquad (8\text{-}118)$$

from which $\emptyset=\alpha-\beta$. Recall that the power factor of a load is the cosine of the angle between the voltage and current associated with that load. As that phase difference is $\alpha-\beta$, we see that the power factor is the cosine of the angle of the load impedance. This in turn means that we can determine the power factor by examining the load alone, without assuming voltages and currents.

Example 8-16. Compute the power factor of a load consisting of (a) $Z_o=R$, (b) $Z_o=j\omega L$ and (c) $Z_o=1/j\omega C$.

<u>Solution</u>. (a) The angle of an impedance consisting of a pure resistance is $0°$; hence, the power factor is $\cos(0)=1$. (b) and (c) For the pure inductor or the pure capacitor, the impedance angle is either $+90°$ or $-90°$; in either case the power factor is $\cos(90°)=0$.

But let us return to our task of developing new power relationships. Using the notation of Eq. (8-116), the power developed at the load is $P_{ave}=VI\cos(\alpha-\beta)$. But

$$\cos(\alpha - \beta) = \text{Re}\{e^{j(\alpha-\beta)}\} \qquad (8\text{-}119)$$

and

$$P_{ave} = VI\text{Re}\{e^{j(\alpha-\beta)}\} = \text{Re}\{VIe^{j(\alpha-\beta)}\} \qquad (8\text{-}120)$$

$$= \text{Re}\{Ve^{j\alpha}Ie^{-j\beta}\} \qquad (8\text{-}121)$$

The Sinusoidal Steady State

Finally, as $\bar{V}=V\exp(j\alpha)$ and $\bar{I}=I\exp(j\beta)$,

$$P_{ave} = \text{Re}\{\bar{V}\bar{I}^*\} \qquad (8\text{-}122)$$

where $\bar{I}^*$ is the complex conjugate of $\bar{I}$ to account for $\exp(-j\beta)$ in Eq. (8-121). Using $\bar{V}=\bar{I}Z(j\omega)$ in Eq. (8-122), we obtain

$$P_{ave} = \text{Re}\{Z(j\omega)\bar{I}\bar{I}^*\} \qquad (8\text{-}123)$$

and, as $\bar{I}\bar{I}^*=I^2$, we have the alternate power formula $P_{ave}=I^2\text{Re}\{Z(j\omega)\}$ or

$$P_{ave} = I^2 R(\omega) \qquad (8\text{-}124)$$

where, as before, $R(\omega)$ is the real part of $Z(j\omega)$. If we use admittance rather than impedance and write

$$Y(j\omega) = \frac{1}{Z(j\omega)} = G(\omega) + jB(\omega) \qquad (8\text{-}125)$$

where $G(\omega)=\text{Re}\{Y(j\omega)\}$ is the <u>conductance</u> of the load, and $B(\omega)=\text{Im}\{Y(j\omega)\}$ is its <u>susceptance</u>, then it is easy to derive the additional power expression

$$P_{ave} = V^2 \text{Re}\{Y(j\omega)\} \qquad (8\text{-}126)$$

or, using Eq. (8-125),

$$P_{ave} = V^2 G(\omega) \qquad (8\text{-}127)$$

Conductance and susceptance have the same units as admittance (mhos) while reactance, like impedance, is expressed in ohms.

The power relationships are summarized in Table 8-1. The network is that of Fig. 8-14.

Example 8-17. Compute the following for the series RLC network of Fig. 8-15: (a) $Z(j\omega)$, (b) $R(\omega)$, (c) $X(\omega)$, (d) pf (e) $Y(j\omega)$, (f) $G(\omega)$, (h) $\bar{I}$, and (i) P_{ave}.

Table 8-1.

Average Power Relationships In The Sinusoidal Steady State

Voltage	$\bar{V} = V\exp(j\alpha)$ $v(t) = (V\sqrt{2})\cos(\omega t + \alpha)$
Current	$\bar{I} = I\exp(j\beta)$ $i(t) = (I\sqrt{2})\cos(\omega t + \beta)$
Network	$\bar{V} = \bar{I}Z(j\omega)$
Impedance	$Z(j\omega) = |Z|\exp(j\phi)$ $= R(\omega) + jX(\omega)$
Phase Relationship	$\phi = \alpha - \beta$
Admittance	$Y(j\omega) = |Y|\exp(-j\phi)$ $= G(\omega) + jB(\omega)$
Average Power Relationships	$P_{ave} = VI\cos(\phi) = I^2 R(\omega)$ $= V^2 G(\omega)$

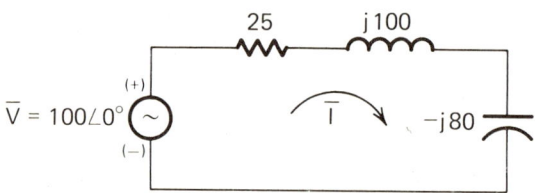

Figure 8-15. Network for Example 8-17.

Solution. The components are in series so we have

$$Z(j\omega) = 25 + j100 - j80 \qquad (8\text{-}128)$$

or (a) $Z(j\omega) = 25 + j20$. By inspecting this expression we see that (b) $R(\omega) = 25$, and (c) $X(\omega) = 20$. (We have preserved the argument ω for these quantities even though in this example a specific frequency has already been used.)

The Sinusoidal Steady State

To compute the power factor we have

$$Z(j\omega) = \sqrt{(25)^2 + (20)^2} \; \underline{/\tan^{-1}(20/25)} \quad (8\text{-}129)$$

and $Z(j\omega) = 32.02 \underline{/38.66°}$ Ω. Hence, (d) pf=cos(38.66°)=0.781. Admittance is the reciprocal of impedance, so

$$Y(j\omega) = \frac{1}{Z(j\omega)} = \frac{1}{25 + j20} \quad (8\text{-}130)$$

$$= \frac{25 - j20}{(25 + j20)(25 - j20)} \quad (8\text{-}131)$$

and (e) $Y(j\omega) = 0.0244 - j0.195$. From this expression we have (f) $G(\omega) = 0.0244$ mho and, (g) $B(\omega) = -0.195$ mho. (Note that $G(\omega)$ is not the reciprocal of $R(\omega)$.) (h) The current is found using Ohm's law for impedance:

$$\bar{I} = \frac{\bar{V}}{Z(j\omega)} = \frac{100\underline{/0°}}{32.02 \underline{/38.66°}} = 3.123 \; \underline{/-38.66°} \text{ A} \quad (8\text{-}132)$$

(i) Finally, assuming RMS phasors,

$$P_{ave} = VI\cos(\theta) = (100)(3.123)(0.781) \quad (8\text{-}133)$$

and $P_{ave} = 243.9$ W. We can confirm this result by using the alternate expression

$$P_{ave} = I^2 R(\omega) = (3.123)^2(25) = 243.9 \text{ W} \quad (8\text{-}134)$$

Example 8-18. Derive expressions for (a) $Y(j\omega)$, (b) $G(\omega)$, (c) $B(\omega)$, (d) $Z(j\omega)$, (e) $R(\omega)$, and (f) $X(\omega)$ for the network of Fig. 8-16.

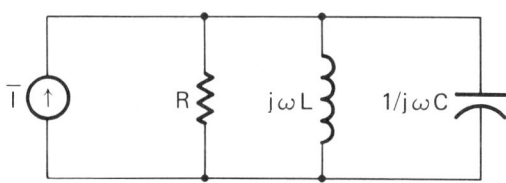

Figure 8-16. Network for Example 8-18.

Solution. As admittances in parallel combine by addition we have (a)

$$Y(j\omega) = \frac{1}{R} + \frac{1}{j\omega L} + j\omega C = \frac{1}{R} + j\{\omega C - \frac{1}{\omega L}\} \quad (8\text{-}135)$$

from which (b) $G(\omega)=1/R$ and (c) $B(\omega)=\omega C-1/\omega L$. Taking the reciprocal of $Y(j\omega)$ yields (d)

$$Z(j\omega) = \frac{1}{Y(j\omega)} = \frac{1}{G(\omega) + jB(\omega)} \quad (8\text{-}136)$$

$$= \frac{G(\omega)}{G(\omega)^2 + B(\omega)^2} + j\frac{-B(\omega)}{G(\omega)^2 + B(\omega)^2} \quad (8\text{-}137)$$

and (e)

$$R(\omega) = \frac{G(\omega)}{G(\omega)^2 + B(\omega)^2} = \frac{1/R}{1/R^2 + (\omega C - 1/\omega L)^2} \quad (8\text{-}138)$$

Also, (f)

$$X(\omega) = \frac{-B(\omega)}{G(\omega)^2 + B(\omega)^2} = \frac{-(\omega C - 1/\omega L)}{1/R^2 + (\omega C - 1/\omega L)^2} \quad (8\text{-}139)$$

Again, $R(\omega)$ is not a constant resistance but depends on ω, C and L as well.

Power In The Network Elements

It is useful to consider the power situation for loads consisting of the pure network elements.

<u>Resistor power</u>. For the resistor, $v(t)=Ri(t)$, and the voltage and current are in phase. Hence, Eq. (8-116) gives the average power as

$$P_{ave} = VI\cos(0^\circ) = VI = I^2 R \quad (8\text{-}140)$$

<u>Reactive power</u>. For the inductor,

$$\bar{V}_L = Z_L(j\omega)\bar{I}_L = j\omega L \bar{I}_L \quad (8\text{-}141)$$

The Sinusoidal Steady State

or, as multiplication by j is equivalent to a $90°$ phase shift, $\bar{V}_L = \bar{I}_L \omega L \underline{/90°}$. Because $\bar{V}_L$ and $\bar{I}_L$ differ in phase by $90°$, Eq. (8-116) gives $P_{ave}=0$. This result agrees with reality since the ideal inductor <u>stores</u> energy in a magnetic field. It does not consume any energy. Similarly, for the capacitor,

$$\bar{V}_C = Z_C(j\omega)\bar{I}_C = \frac{\bar{I}_C}{j\omega C} \tag{8-142}$$

or $\bar{V}_C = \bar{I}_C/\omega C \underline{/-90°}$. Again, $\bar{V}_C$ and $\bar{I}_C$ are $90°$ out of phase and Eq. (8-116) indicates that $P_{ave}=0$. This result follows because capacitors store charge on their plates; they do not consume energy.

The Power Triangle

Electrical engineers who work extensively with power find it useful to define some additional power-related quantities and then to represent some of these graphically by what is called the power triangle.[17] Define

$$\bar{W} = \bar{V}\bar{I}* \tag{8-143}$$

as the <u>complex power</u>, also called the <u>apparent power</u>, where $\bar{V}$ and $\bar{I}$ are RMS phasors. We can then write $\bar{W}=P+jQ$ where

$$P = \text{Re}\{\bar{W}\} = \text{Re}\{\bar{V}\bar{I}*\} \tag{8-144}$$

and P is seen to be the average or <u>real power</u>. The quantity Q, given by

$$Q = \text{Im}\{\bar{W}\} = \text{Im}\{\bar{V}\bar{I}*\} \tag{8-145}$$

is called the <u>reactive power</u>. If $\bar{V}=V\exp(j\alpha)$ and $\bar{I}=I\exp(j\beta)$, then

$$\bar{W} = VI e^{j(\alpha-\beta)} \tag{8-146}$$

and we see that the phase angle of the complex power $\bar{W}$ is identical to the power factor angle ϕ since $\phi=\alpha-\beta$. Thus,

$$\phi = \underline{/\bar{W}} = \tan^{-1}\frac{Q}{P} \tag{8-147}$$

The power triangle is a graphical representation of $\bar{W}$ (see

Fig. 8-17). Further, as a consequence of Eq. (8-147) we have
P=VIcos(∅) and Q=VIsin(∅).

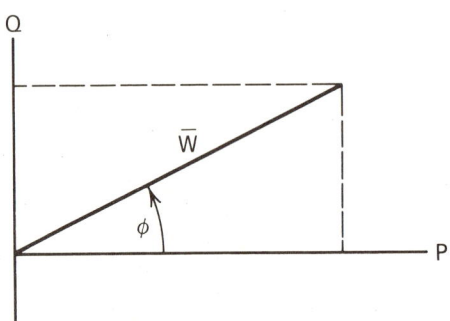

Figure 8-17. A typical power triangle. Real power is the P component of W; ϕ is the power factor angle.

In Fig. 8-17 the vertical axis represents Q, the reactive power and the horizontal axis P, the real or active power. $\bar{W}$, the complex power, is a vector with length VI and angle ∅, the power factor angle. If ∅ is positive, we say that the power factor is <u>lagging</u>. If ∅ is negative the power factor is said to be <u>leading</u>. The power triangle should not be confused with the phasor diagram such as that of Fig. 8-8.

Maximum Power Transfer[18,19]

In many electrical applications the goal is to transfer maximum power from one network to another during sinusoidal steady state operation. The power source may be modeled as an ideal sinusoidal voltage source $\bar{V}_S$ in series with a complex source impedance,

$$Z_S = R_S + jX_S \tag{8-148}$$

We can always use Thevenin's theorem to reduce a more complicated network to this form, although the actual voltage source might be an electrical generating station and the load a

The Sinusoidal Steady State

factory. The load impedance is, in general, complex and of the form

$$Z_L = R_L + jX_L \tag{8-149}$$

The following paragraphs treat the problem of maximizing the power transfered from the source to the load under three conditions: (1) $X_L = 0$ with R_L variable, (2) X_L and R_L both variable, and (3) X_L fixed but R_L variable.

Variable resistance load. Figure 8-18(a) shows the network for this case. Let V_S and I be the magnitudes of the RMS phasors $\bar{V}_S$ and $\bar{I}$.

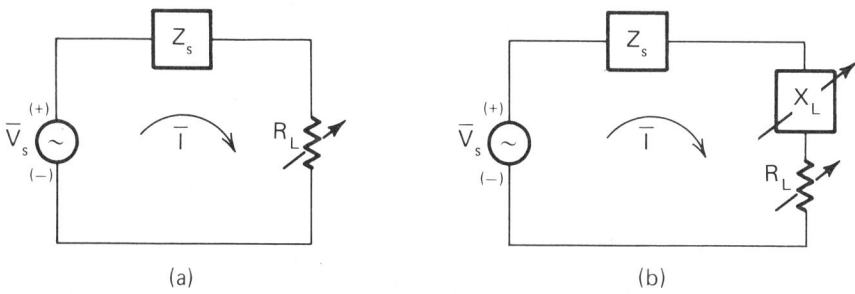

Figure 8-18. Networks to illustrate maximum power transfer. An arrow through the symbol for a network element indicates that its value is adjustable. (a) The load is purely resistive and variable; (b) The load has both resistance and reactance, both variable.

The current $\bar{I}$ is

$$\bar{I} = \frac{\bar{V}_S}{R_L + R_S + jX_S} \tag{8-150}$$

with magnitude squared

$$I^2 = \frac{V_S^2}{(R_L + R_S)^2 + (X_S)^2} \tag{8-151}$$

The power in R_L is $P=I^2 \text{Re}\{Z_L(j\omega)\}=I^2 R_L$ and

$$P = \frac{V_S^2 R_L}{(R_L + R_S)^2 + (X_S)^2} \quad (8-152)$$

Taking the derivative of P with respect to R_L gives

$$\frac{dP}{dR_L} = \frac{\{(R_L + R_S)^2 + (X_S)^2\} - 2(R_L + R_S)R_L}{\{(R_L + R_S)^2 + (X_S)^2\}^2} V_S^2 \quad (8-153)$$

Setting the derivative equal to 0 and solving for the required value of R_L leads to

$$R_L^2 + 2R_L R_S + R_S^2 + X_S^2 - 2R_L^2 - 2R_L R_S = 0 \quad (8-154)$$

and

$$R_S^2 + X_S^2 = R_L^2 \quad (8-155)$$

or

$$R_L = \sqrt{R_S^2 + X_S^2} = |Z_S| \quad (8-156)$$

Thus power transfer is maximized when the load resistance just equals the <u>magnitude</u> of the source impedance. If the source is resistive ($X_S=0$), then Eq. (8-156) reduces to $R_L = R_S$.

Example 8-19. A source with impedance $Z_S = 3+j4$ Ω delivers power to a resistive load R_L. Calculate the value of R_L for maximum power transfer.

Solution. Using Eq. (8-156),

$$R_L = \sqrt{R_S^2 + X_S^2} = \sqrt{3^2 + 4^2} = 5 \text{ Ω} \quad (8-157)$$

<u>Variable resistance-variable reactance load</u>. If a source with impedance $Z_S = R_S + jX_S$ feeds a load impedance $Z_L = R_L + jX_L$ as

shown in Fig. 8-18(b), the magnitude squared of the current is

$$I^2 = \frac{V_S^2}{(R_L + R_S)^2 + (X_L + X_S)^2} \tag{8-158}$$

and the load power is

$$P = I^2 R_L = \frac{V_S^2 R_L}{(R_L + R_S)^2 + (X_L + X_S)^2} \tag{8-159}$$

If R_L is held fixed in Eq. (8-159), then the power is a maximum if $X_L = -X_S$; this reduces Eq. (8-159) to

$$P = \frac{V_S^2 R_L}{(R_L + R_S)^2} \tag{8-160}$$

If R_L is now variable, we can compute dP/dR_L and show that maximum power transfer occurs if $R_L = R_S$. Combining these results reveals that maximum power is transferred if $Z_L = Z_S^*$, or if the load impedance is the complex conjugate of the source impedance. The complex conjugate is needed because X_L must equal $-X_S$ to cancel the reactance of the source.

Example 8-20. For the network of Fig. 8-19, compute the values of R_L and C_L required for maximum power transfer if the frequency of the source voltage is 60 Hz, $L_S = 5.0$ H and $R_S = 500$ Ω.

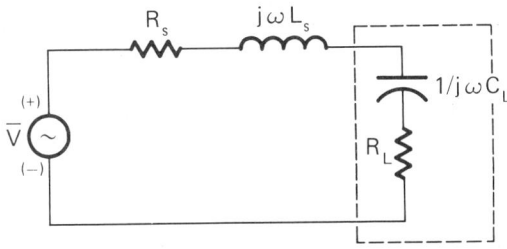

Figure 8-19. Network for Example 8-20. The load impedance is enclosed in dashed lines.

Solution. The source impedance is $R_s + j\omega L_s$. To maximize power transfer the load impedance must satisfy

$$R_L + \frac{1}{j\omega C_L} = R_s - j\omega L_s \qquad (8\text{-}161)$$

Equating real and imaginary parts gives $R_L = R_s = 500\ \Omega$ and $1/\omega C_L = \omega L_s$ or

$$C_L = \frac{1}{\omega^2 L_s} = \frac{1}{(2\pi 60)^2 (5)} = 1.41\ \mu F \qquad (8\text{-}162)$$

From this example we see that maximum power transfer occurs when the series RLC circuit is in resonance. This result is logical, since under resonant conditions the current is a maximum and $P = I^2 R(\omega)$.

<u>Variable resistance-fixed reactance load</u>. If $Z_s = R_s + jX_s$ and $Z_L = R_L + jX_L$, where R_L is variable but X_L is fixed, then X_L can be lumped with X_s to give a net source reactance of $X_L + X_s$. The results of the resistive-load case now apply and

$$R_L = \sqrt{R_s^2 + (X_L + X_s)^2} \qquad (8\text{-}163)$$

for maximum power transfer.

PROBLEMS FOR CHAPTER 8

8-1. If $A = 2 + j3$, $B = 1 + j5$, and $C = 4 - j8$, calculate (a) $A + B - C$, (b) AB, (c) $|A|$, (d) $\underline{/A}$, (e) A/C, (f) B^2, and (g) $\sqrt{A}$.

8-2. Find $v_4(t)$, the sum of the three voltages $v_1(t) = \sin(5t - 40°)$, $v_2(t) = 50\cos(5t)$, and $v_3(t) = 75\sin(5t + 60°)$. Express the result as (a) a sum of sines and cosines, (b) a single sine function with a phase angle, and (c) a single cosine with a phase angle.

8-3. If $V_1 = 10\exp(j2)$ and $V_2 = 5\underline{/30°}$, calculate (a) V_1/V_2, (b) $V_1 + V_2$, (c) $V_1 V_2$, and (d) $\sqrt{V_1}/V_2^2$.

The Sinusoidal Steady State

8-4. Show that $y(t)=A\sin(\omega_o t)$ is a solution of the differential equation $\ddot{y}+2\zeta\omega_n\dot{y}+\omega_n^2 y=K\sin(\omega_o t)$ for some choice of A.

8-5. Show that the system with transfer function

$$G(s) = \frac{Y(s)}{F(s)} = \frac{1}{(s+5)(s^2+4)} \qquad (8\text{-}164)$$

has a sinusoidal steady-state response even though the input is the unit step function $f(t)=u(t)$.

8-6. Find the sinusoidal steady-state response for each of the differential equations (a) $\dot{y}+2y=\cos(5t)$, (b) $\dot{y}+3y=\sin(t)+\cos(t)$, (c) $\ddot{q}+3\dot{q}+2q=5\sin(2t)$.

8-7. The input function $f(t)=8\cos(2t)$ is applied to systems with the following s-domain transfer functions. Find the sinusoidal steady-state response $y_{ss}(t)$ for each: (a) $G(s)=10/(s+2)$, (b) $G(s)=(s+2)/\{s(s+10)\}$, (c) $G(s)=8/(s^2+2s+8)$, and (d) $G(s)=(s^2-2s-1)/(s^2+2s+1)$.

8-8. Express Eq. (8-12) in the form $y_{ss}(t)=B\cos(\omega_o t)+C\sin(\omega_o t)$.

8-9. For the network of Fig. 8-20: (a) derive an expression for $G(s)=V_o(s)/V_i(s)$, (b) evaluate $|G(j\omega)|$ and $\underline{/G(j\omega)}$, (c) find the steady-state output $v_{oss}(t)$ if $v_i(t)=K\sin(\omega_o t)$, and (d) investigate the behavior of the magnitude of $v_{oss}(t)$ as $\omega_o \to \infty$.

Figure 8-20. Problem 8-9.

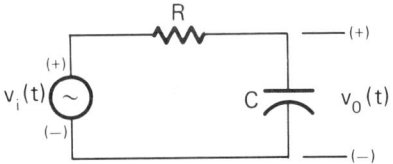

8-10. Represent each of the following quantities as the real part of a complex function: (a) $f(t)=5\cos(10t-25°)$, (b) $f(t)=25\sin(3t)$, (c) $g(t)=\cos(t)+\sin(t)$, and (d) $y(t)=50u(t)$.

8-11. Find the complex output $\bar{Y}(j2)$ for each of the systems of Problem 8-7.

8-12. The networks of Fig. 8-21 are excited with 60-Hz sinusoidal input voltages. Find the equivalent phasor network for each.

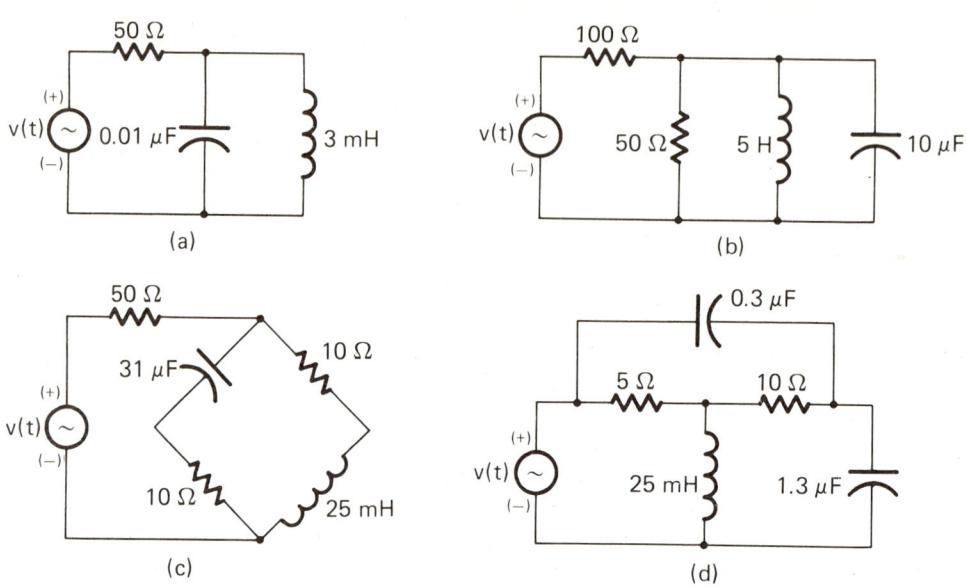

Figure 8-21. Problem 8-12.

8-13. Derive expressions for the phasor output voltage $\bar{V}_o$ for each of the networks of Fig. 8-22:

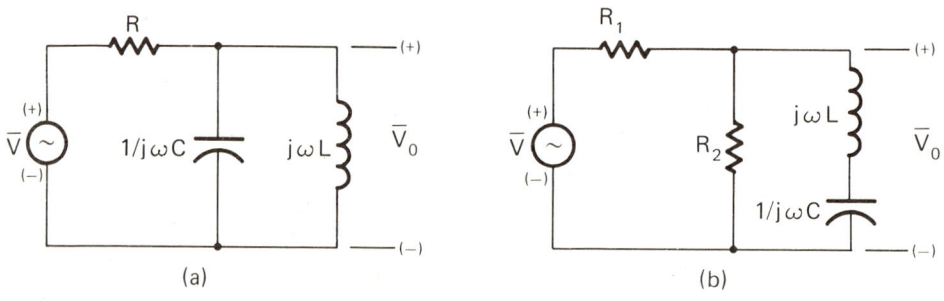

Figure 8-22. Problem 8-13.

8-14. The network of Fig. 8-23 arises in the study of 3-phase power distribution systems. Find the phasor line currents $\bar{I}_A$, $\bar{I}_B$, and $\bar{I}_C$.

Figure 8-23. Problem 8-14.

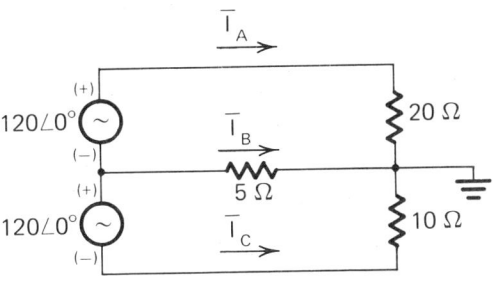

8-15. Solve for $\bar{V}_o$ in the network of Fig. 8-24 if $\bar{V}_i = 25\exp(j0)$.

Figure 8-24. Problem 8-15.

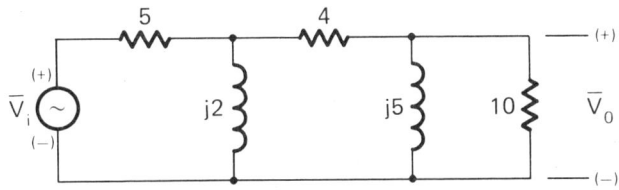

8-16. Reduce each of the networks of Fig. 8-25 to a single impedance expressed in the form $Z=R+jX$.

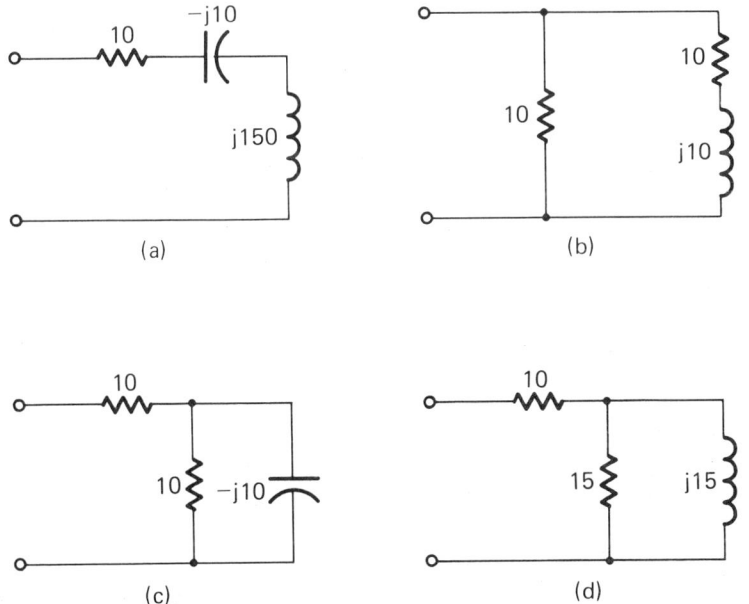

Figure 8-25. Problem 8-16.

8-17 Sketch $|Z|$ as a function of ω for a series RLC network where $Z=R+j\omega L+1/j\omega C$. Then show that $|Z|$ has its minimum value at the resonant frequency $\omega_o^2=1/LC$.

8-18. Derive a phasor Thevenin equivalent network for the network to the left of terminals a-a'.

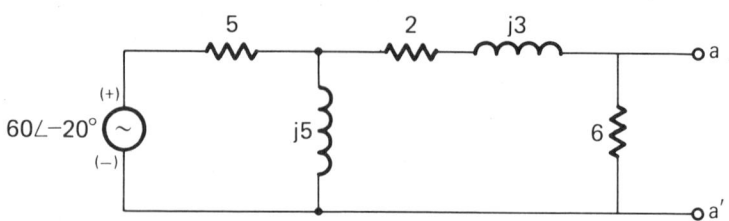

Figure 8-26. Problem 8-18.

8-19. Solve for the phasor quantities $\bar{I}_1$, $\bar{I}_2$, and $\bar{V}_o$ if $\bar{V}_i=100\exp(j0)$. Use loop analysis.

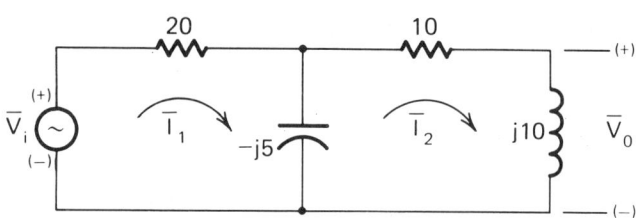

Figure 8-27. Problem 8-19.

8-20. If the time-domain input voltage in Problem 8-19 is $v_i(t)=100\sin(377t+0^\circ)$, find $i_1(t)$, $i_2(t)$, and $v_o(t)$ in the sinusoidal steady state.

8-21. Find the input impedance Z_i for the network shown if $Z_1=1+j2$, $Z_2=5-j8$, and $Z_3=-j12$.

Figure 8-28. Problem 8-21.

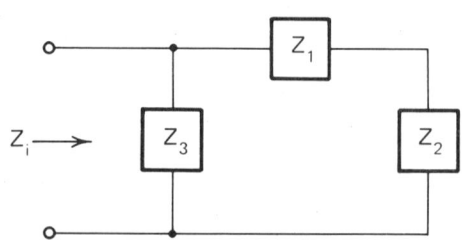

The Sinusoidal Steady State

8-22. The following voltage-current pairs are associated with an impedance in the sinusoidal steady state. Determine (a) the phasor voltages and currents, and (b) the magnitude and angle of the impedance.

(1) $v(t) = 400\sin(200t + 150°)$ (8-165)

$i(t) = 5.0\sin(200t + 100°)$

(2) $v(t) = 120\cos(377t - 30°)$ (8-166)

$i(t) = 2.0\cos(377t + 5°)$

(3) $v(t) = 50\cos(4t)$ (8-167)

$i(t) = 10\sin(4t)$

(4) $v(t) = 2000\sin(t - 45°)$ (8-168)

$i(t) = 2.5\cos(t - 45°)$

8-23. Construct phasor diagrams for the networks implied by the voltage-current pairs of problem 8-22.

8-24. A series RC circuit has R=20 Ω, C=50 μF, and an applied sinusoidal voltage such that the steady-state current leads the applied voltage by 60°. Calculate (a) the frequency of the input voltage and (b) the frequency which would be required to cause the current to lead the voltage by 80°.

8-25. Find the resonant frequencies of the series RLC circuits with the following parameters: (a) R=100 Ω, L=3 mH, C=5 μF; (b) R=1.0 Ω, L=1.0 H, C=1.0 F; and (c) R=500 kΩ, L=0.5 mH, C=0.002 μF.

8-26. It is required to make the input impedance of each of the following networks appear purely resistive to an input sinusoid whose frequency is 400 Hz. Find the type and value of the required element and the resulting input impedance. If it is impossible to make the input impedance resistive with an element placed as shown, show why.

8-27. Show that the input impedance Z_i=R if both of the resistors in the following network have the value R=$\sqrt{L/C}$. What is the resonant frequency?

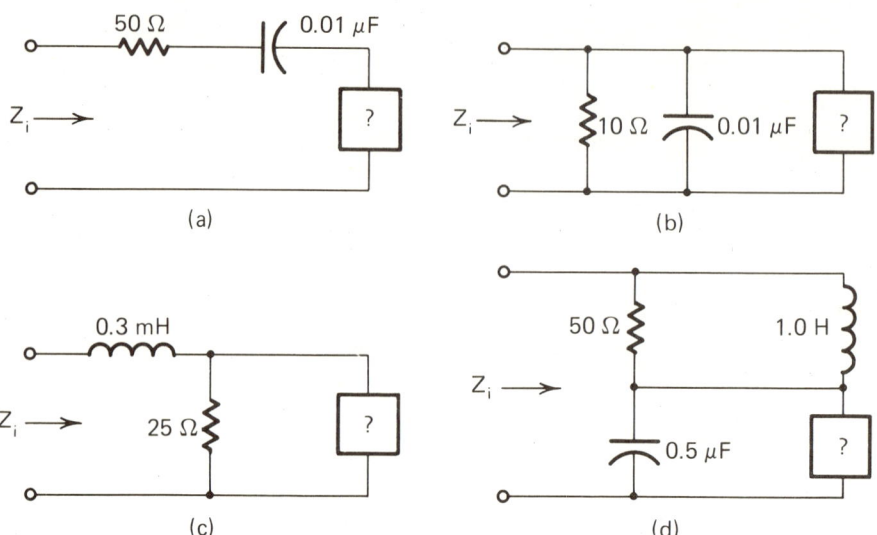

Figure 8-29. Problem 8-26.

Figure 8-30. Problem 8-27.

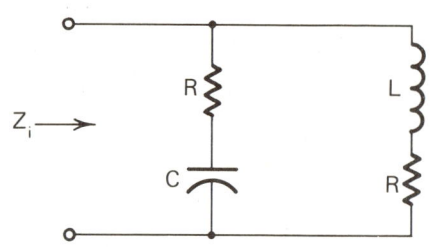

8-28. A voltage v(t)=100cos(377t) is applied to an unknown impedance. A steady-state current of i(t)=2.0cos(377t-30°) is observed to flow. Calculate the average power developed in the impedance.

8-29. Find the average power developed in each of the impedances of Problem 8-22.

8-30. An RMS phasor voltage $\bar{V}$=220exp(j0) is applied to each of the impedances given. In each case find the phasor current $\bar{I}$ which flows and the average power developed: (a) Z=25+j0, (b) Z=0.5+j0.85, (c) Z=-j30, (d) Z=1/(5+j4), (e) Z=exp(j0.2), and (f) Z=100exp(jπ/4).

8-31. Derive an expression for the resonant frequency of the general two-branch network shown. (Hint: Resonance occurs when the complex admittance is a real number.)

Figure 8-31. Problem 8-31.

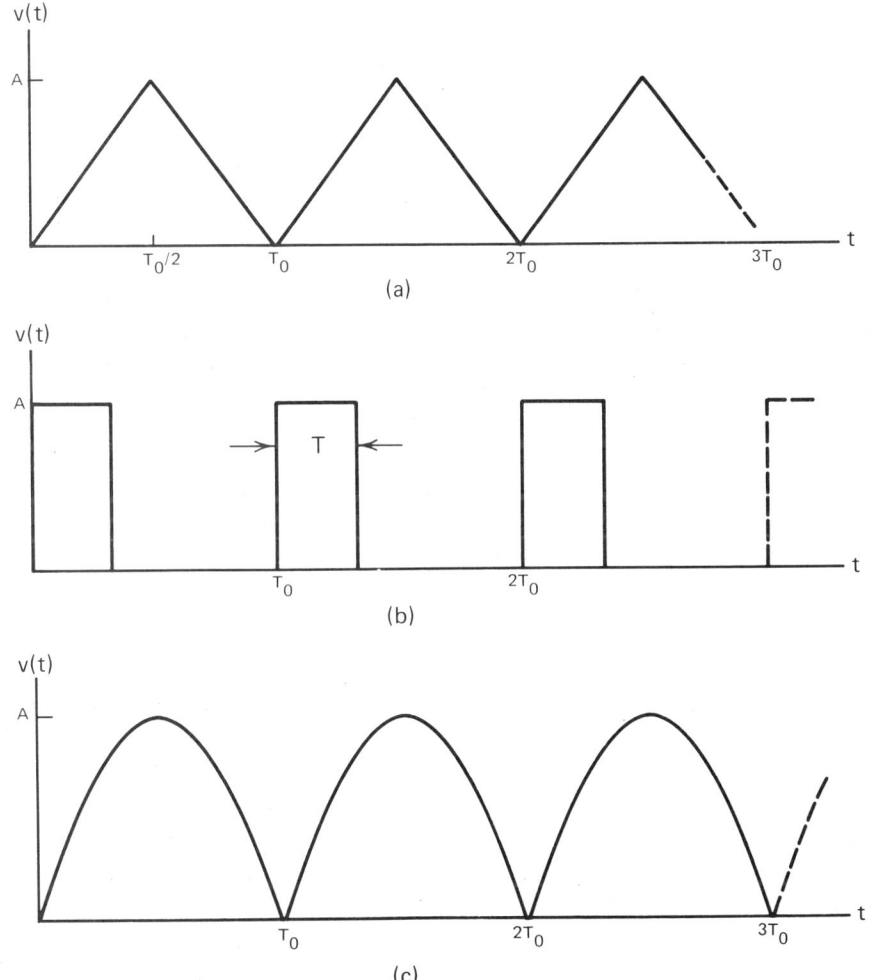

8-32. Show that when $R_L = R_C = \sqrt{L/C}$ the network of Problem 8-31 is resonant at <u>all</u> frequencies.

8-33. Derive an expression for the RMS value of each of the waveforms shown:

Figure 8-32. Problem 8-33.

Glass–Linear Systems—26

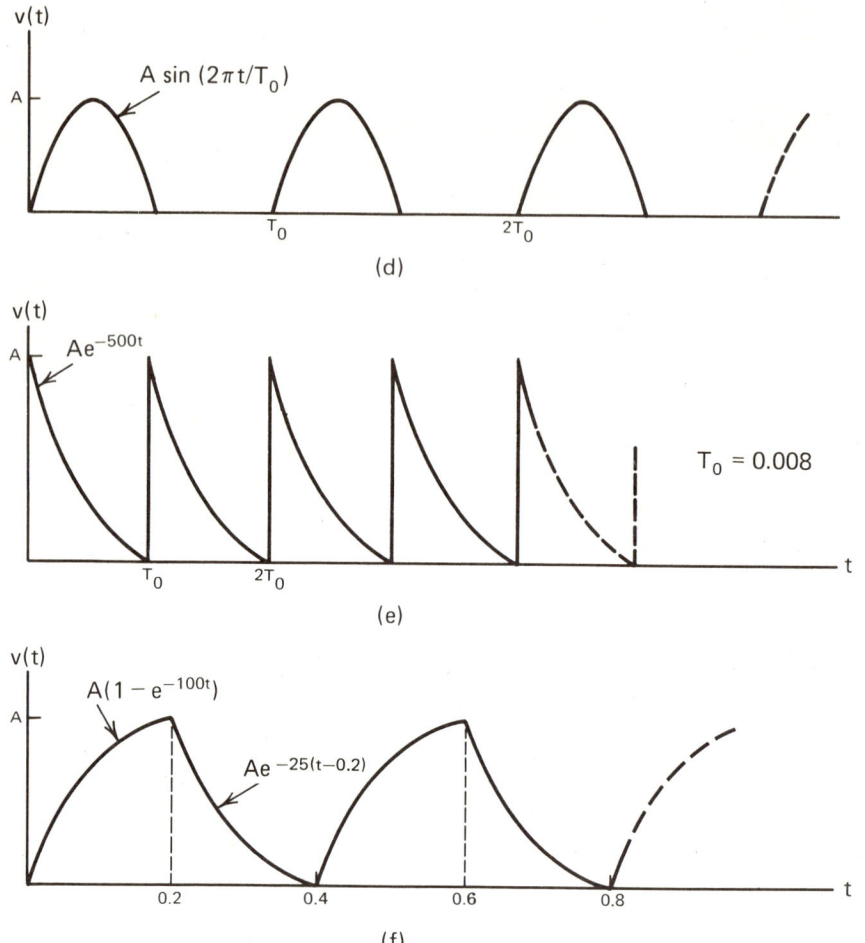

Figure 8-32. Problem 8-33 continued.

8-34. Find the average power developed across a one-ohm resistor by voltages whose waveforms are those given in Problem 8-33.

8-35. Derive an expression for the RMS value of the function $v(t)=A+B\sin(\omega t)$.

8-36. Compute the power factor of each of the impedances of Problem 8-22. Be sure to indicate whether the power factor is leading or lagging.

8-37. Compute the power factor of each of the impedances of Problem 8-30.

8-38. An industrial plant has an "input" impedance of $Z=5+j20$ Ω at 60 Hz. It has been decided for reasons of economy to correct the power factor to 0.85 lagging by inserting capacitor banks. Find the amount of capacitance which must be added if (a) the capacitance is inserted in series with the input impedance and (b) the capacitance is inserted in parallel.

8-39. Compute the power factor of the input impedance of each of the networks of Fig. 8-33.

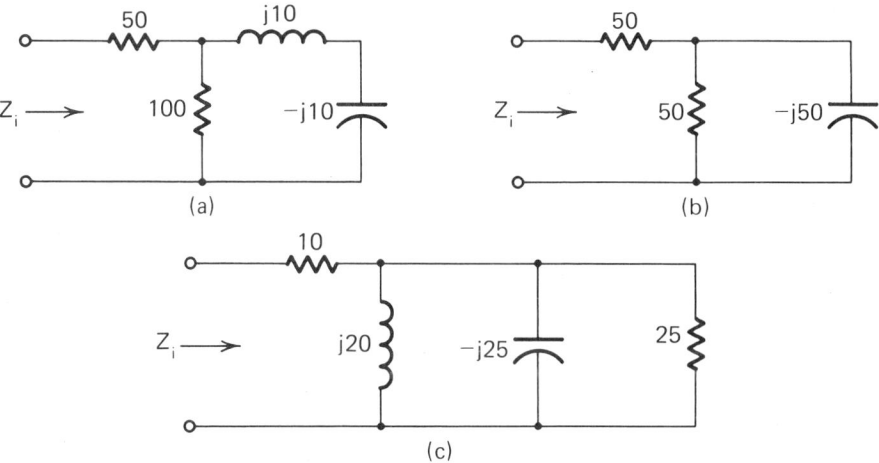

Figure 8-33. Problem 8-39.

8-40. Find the values of the elements in the networks of Problem 8-39 if the frequency of operation is 60 Hz.

8-41. What value of capacitance must be incorporated in series with the inductance in the circuit of Fig. 8-34 to correct the power factor to (a) 1.0, lagging (b) 0.95 lagging. The frequency of operation is 50 Hz.

Figure 8-34. Problem 8-41.

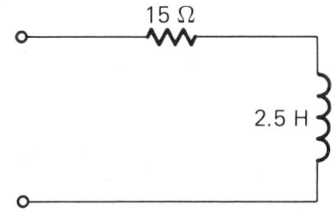

8-42. Determine the power triangle for the network of Fig. 8-35.

Figure 8-35. Problem 8-42.

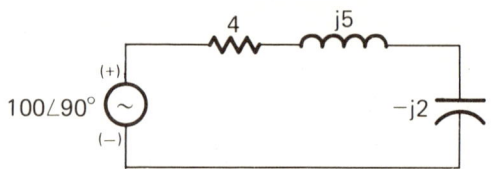

8-43. Determine the power triangle for each branch of the parallel network of Fig. 8-36. Then add to obtain the power triangle for the overall network. Use $Z_1 = 2\exp(j\pi/6)$ and $Z_2 = 5\exp(j\pi/3)$.

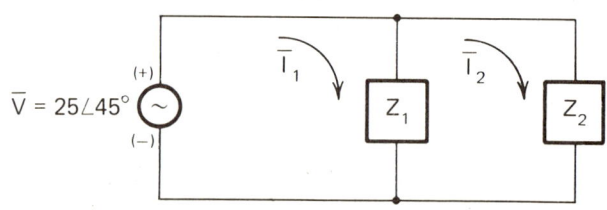

Figure 8-36. Problem 8-43.

8-44. A certain electrical motor has a 2-horsepower (hp) output with an efficiency of 85%. At this load the power factor is 0.90 lagging. Determine the input power, the input volt-amps, and the input reactive power. (Hint: 1 hp = 746 W.)

8-45. Find the value of R_L for maximum power transfer in each of the networks below:

8-46. Find the value of Z_L for maximum power transfer.

8-47. Find the resistance, reactance, conductance, and susceptance of the input impedance for each of the networks given in Problem 8-39.

8-48. Repeat Problem 8-47 for the impedances given in Problem 8-16.

The Sinusoidal Steady State

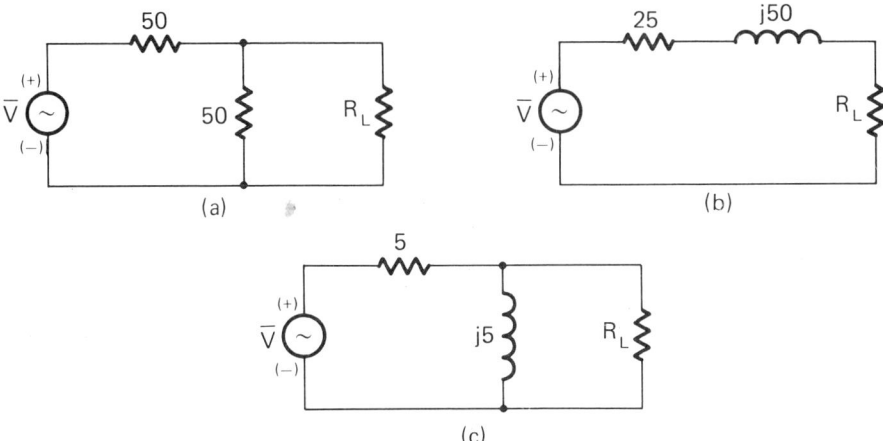

Figure 8-37. Problem 8-45.

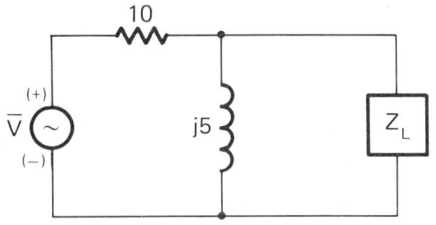

Figure 8-38. Problem 8-46.

8-49. Shown below are the classical Δ and Y connections. Use phasor loop and nodal analysis to derive the Δ-Y and Y-Δ transformations:

$$Z_A = \frac{Z_P}{Z_3} \qquad Z_B = \frac{Z_P}{Z_2} \qquad Z_C = \frac{Z_P}{Z_1} \qquad (8\text{-}169)$$

and

$$Z_1 = \frac{Z_A Z_B}{Z_T} \qquad Z_2 = \frac{Z_A Z_C}{Z_T} \qquad Z_3 = \frac{Z_B Z_C}{Z_T} \qquad (8\text{-}170)$$

where $Z_P = Z_1 Z_2 + Z_1 Z_3 + Z_2 Z_3$ and $Z_T = Z_A + Z_B + Z_C$. (Hint: Equate input, output, and transfer impedances, where $Z_{trans} = \bar{V}_i / \bar{I}_o = \bar{V}_o / \bar{I}_i$.)

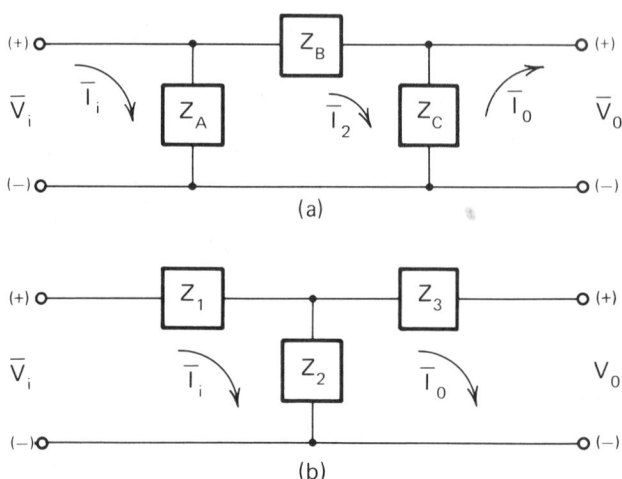

Figure 8-39. Problem 8-49.

8-50. Transform a Y-connected network with $Z_1 = Z_2 = Z_3 = 10\exp(j\pi/6)$ to an equivalent Δ-connected network.

8-51. Transform a Δ-connected network with $Z_A = Z_B = Z_C = 5\exp(-j\pi/4)$ to an equivalent Y-connected network.

REFERENCES FOR CHAPTER 8

(1) Dawes, C. L.: Industrial Electricity, McGraw-Hill Book Co., Inc., New York, 1960.
(2) Kingsley, C., Jr., and A. Kusko: Electric Machinery, McGraw-Hill Book Co., Inc., New York, 1971.
(3) Steinmetz, C. P.: Lectures on Electrical Engineering (vol. I), Dover Publications, Inc., New York, 1971.
(4) Chirlian, P. M.: Basic Network Theory, McGraw-Hill Book Co., Inc., New York, 1969.
(5) Desoer, C. A. and E. S. Kuh: Basic Circuit Theory, McGraw-Hill Book Co., Inc., New York, 1969.
(6) Craig, E. J.: Laplace and Fourier Transforms For Electrical Engineers, Holt, Rinehart and Winston, Inc., New York, 1964.

(7) Blalock, G. C.: *Principles of Electrical Engineering*, McGraw-Hill Book Co., Inc., New York, 1950.

(8) Cook, A. L. and C. C. Carr: *Elements of Electrical Engineering*, John Wiley and Sons, Inc., New York, 1974.

(9) Oppenhimer, S. L., Hess, R. F., Jr., and J. P. Borchers: *Direct and Alternating Currents*, McGraw-Hill Book Co., Inc., New York, 1973.

(10) Del Toro, V.: *Electrical Engineering Fundamentals*, Prentice-Hall, Inc., Englewood Cliffs, New Jersey, 1972.

(11) Van Valkenburg, M. E.: *Network Analysis*, Prentice-Hall Inc., Englewood Cliffs, New Jersey, 1974.

(12) Driscoll, F. F.: *Analysis of Electric Circuits*, Prentice-Hall, Inc., Englewood Cliffs, New Jersey, 1973.

(13) Hayt, W. H. Jr., and J. E. Kemmerly: *Engineering Circuit Analysis*, McGraw-Hill Book Co., Inc., New York, 1971.

(14) Johnson, J. R.: *Electric Circuits: Part 2, Alternating Current*, Holt, Rinehart and Winston, Inc., New York, 1970.

(15) Erickson, W. H. and N. H. Bryant: *Electrical Engineering Theory and Practice*, John Wiley and Sons, Inc., New York, 1952.

(16) Calahan, D. A., Macnee, A. B. and E. L. McMahon: *Introduction to Modern Circuit Analysis*, Holt, Rinehart and Winston, Inc., New York, 1974.

(17) Clement, P. R. and W. C. Johnson: *Electrical Engineering Science*, McGraw-Hill Book Co., Inc., New York, 1960.

(18) Cox, C. W. and W. L. Reuter: *Circuits, Signals, and Networks*, The MacMillian Company, New York, 1969.

(19) Kerchner, R. M. and G. F. Corcoran: *Alternating-Current Circuits*, John Wiley and Sons, Inc., New York, 1943.

*

CHAPTER 9

Frequency Response of Linear Systems

Frequency response is a way of interpreting the sinusoidal steady-state response of a system. We again restrict the system inputs to sinusoids and wait until steady state is achieved, but rather than investigating the response at a single input frequency, we consider the amplitude and phase of the output as <u>functions</u> of the frequency of the input sinusoid. This method of analysis can give us additional insight into the behavior of a linear system without our having to solve any differential equations.

Frequency response methods are used in studying mechanical vibrations,[1] designing automatic control systems,[2] analyzing electronic devices,[3] and in a variety of other practical engineering applications.

9.1 FREQUENCY RESPONSE OF FIRST–ORDER SYSTEMS

As preliminaries to the general case, we apply frequency-response methods to first- and second-order systems; first, because such systems are important special cases in their own right, and second, because studying these simpler applications avoids obscuring the fundamental issues with mathematical details.

The Frequency Response Function

As we showed in Chapter 8, if the sinusoidal input function $f(t)=A\cos(\omega t-\theta)$ is applied to a system with transfer function $G(s)$, then the sinusoidal steady-state output $y(t)$ is[4]

$$y(t) = A|G(j\omega)|\cos\{\omega t - \theta + \underline{/G(j\omega)}\} \qquad (9-1)$$

That is, the magnitude of the input is multiplied by the magnitude of $G(s)$ evaluated for $s=j\omega$, and the phase is shifted by the angle of $G(j\omega)$. In general, $|G(j\omega)|$ and $\underline{/G(j\omega)}$ depend on the value of ω. We can, in fact, interpret them as continuous functions of ω. To simplify notation we define the <u>magnitude function</u> $M(\omega)$ by

$$M(\omega) = |G(j\omega)| \qquad (9-2)$$

and the <u>phase function</u> $\emptyset(\omega)$ by

$$\emptyset(\omega) = \underline{/G(j\omega)} = \text{Arg}\{G(j\omega)\} \qquad (9-3)$$

Plots of $M(\omega)$ and $\emptyset(\omega)$ versus ω are the key to analysis by frequency response. Using Eqns. (9-2) and (9-3), Eq. (9-1) becomes

$$y(t) = M(\omega)A\cos\{\omega t - \theta + \emptyset(\omega)\} \qquad (9-4)$$

If the input is a sine rather than a cosine function, Eq. (9-4) is still valid if we replace cosine by sine.

In practice, we seldom need to find the actual expression for $y(t)$ but are able to derive much information about the

Frequency Response of Linear Systems

behavior of a system by examining $M(\omega)$ and $\emptyset(\omega)$. Consider the first-order system described by the transfer function $G(s) = K/(s+a)$. Evaluating for $s = j\omega$ gives $G(j\omega) = K/(j\omega+a)$ from which

$$M(\omega) = |G(j\omega)| = \frac{K}{\sqrt{\omega^2 + a^2}} \qquad (9-5)$$

The phase function is $\emptyset(\omega) = \text{Arg}\{K\} - \text{Arg}\{j\omega+a\}$ or

$$\emptyset(\omega) = -\tan^{-1}(\omega/a) \qquad (9-6)$$

Figure 9-1 shows plots of $M(\omega)$ and $\emptyset(\omega)$ versus ω for the first-order system.

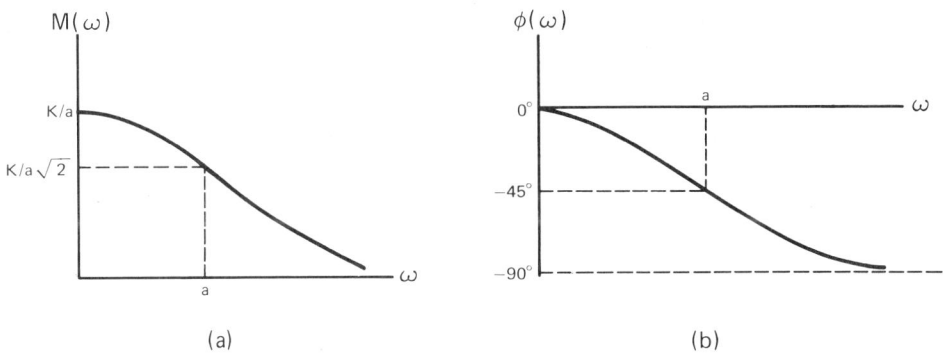

Figure 9-1. Frequency response functions for the first-order system with transfer function $G(s) = K/(s+a)$. (a) Magnitude response; (b) Phase response.

Figure 9-1 shows that the magnitude or "peak" response M_p of the first-order system is $M_p = K/a$ at $\omega = 0$ and that the phase shift of the output is $0°$ at $\omega = 0$. The steady-state transient response is readily found from the frequency response plots. If, for example, the input function is $f(t) = A\cos(at - \emptyset)$, so that $\omega = a$, then we can use the plots of Fig. 9-1 to write the steady-state output

$$y(t) = \frac{AK}{a\sqrt{2}} \cos\{at - \theta - 45°\} \qquad (9-7)$$

Example 9-1. A first-order system has the transfer function $G(s)=50/(0.2s+1)$. Calculate the steady-state response to the input $f(t)=5\sin(2t+30°)$.

Solution. Rearrange $G(s)$ as $G(s)=250/(s+5)$ from which $G(j\omega)=250/(j\omega+5)$ and

$$M(\omega) = |G(j\omega)| = \frac{250}{\sqrt{\omega^2 + 25}} \tag{9-8}$$

The phase function is $\phi(\omega)=\text{Arg}\{G(j\omega)\}=-\tan^{-1}(1/5)$. Evaluating $M(\omega)$ and $\phi(\omega)$ for $\omega=2$, the frequency of the input sinusoid, results in

$$M(2) = \frac{250}{\sqrt{4 + 25}} = 46.42 \tag{9-9}$$

and $\phi(2)=-\tan^{-1}(2/5)=-21.8°$. Equation (9-4) gives the steady-state response

$$y(t) = 232.1\sin\{2t + 8.2°\} \tag{9-10}$$

Bandwidth

An important quantitative descriptor of the magnitude response is the <u>bandwidth</u>,[5] which is the range of frequencies outside of which the magnitude response remains less than $1/\sqrt{2}$ or 0.707 of the maximum value M_p. For the first-order system with magnitude function given by Eq. (9-5), let ω_U be the value of frequency for which the magnitude is 0.707 times the maximum value of K/a. Thus we write

$$\frac{M_p}{\sqrt{2}} = \frac{K}{a\sqrt{2}} = \frac{K}{\sqrt{\omega_U^2 + a^2}} \tag{9-11}$$

and solve for $\omega_U=a$. This frequency is variously called the <u>cutoff frequency</u> of the low-pass characteristic or the <u>half-power</u> frequency. In general, we may have two cutoff frequencies, an upper one ω_U and a lower one ω_L. The bandwidth BW of the system is then defined as

$$BW = \omega_U - \omega_L \qquad (9-12)$$

For the first-order system discussed above, the lower cutoff frequency is $\omega_L=0$ and BW=a-0=a rad/s.

Ideal Filter Characteristics

Electrical filters[6,7] are devices whose purpose is to suppress or emphasize inputs on the basis of frequency. Many systems, though not designed to be filters, have magnitude responses which can be interpreted with reference to certain ideal filter properties. Many mechanical systems, for example, because of the masses of their components, act as if they were low-pass filters. The four standard filter characteristics are the low-pass, the high-pass, the band-pass, and the band-reject. These are shown in Fig. 9-2.

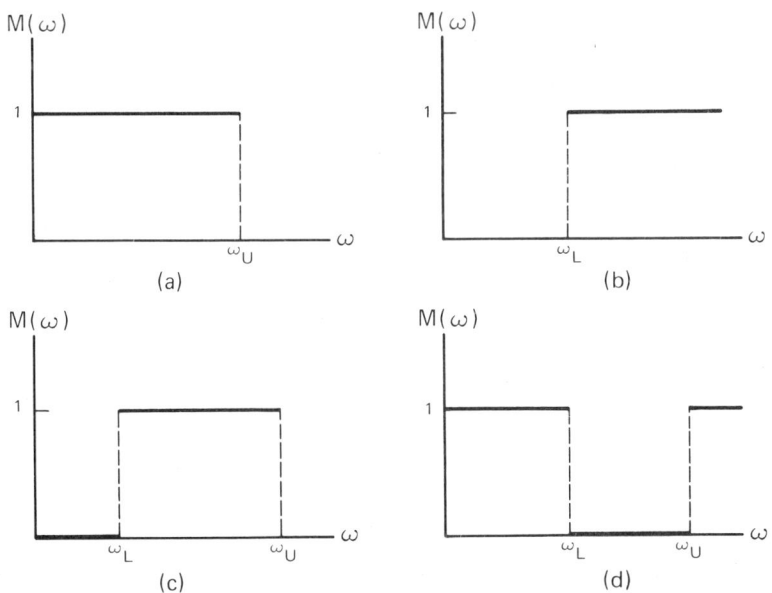

Figure 9-2. Magnitude response characteristics for the four types of ideal filters. (a) Low pass; (b) High pass; (c) Band pass; (d) Band reject.

The ideal low-pass filter passes without attenuation inputs with frequencies less than ω_U and attenuates those with frequencies above ω_U. The high-pass filter rejects inputs with frequencies below ω_L and passes those above. The ideal band-pass filter passes frequencies in the band $\omega_U - \omega_L$ and rejects all others, while the band-reject filter rejects frequencies in the band $\omega_U - \omega_L$ and passes all others. Practical filters only approximate these ideal characteristics.

Filter Characteristics Of The First-Order System

An interesting interpretation of the first-order system with transfer function $G(s)=K/(s+a)$ is that it behaves as a low-pass filter.[8] Sinusoidal inputs with frequencies substantially higher than the cutoff frequency $\omega_U = a$ are greatly attenuated or reduced, while those with frequencies much lower than ω_U are passed with little reduction in amplitude. Comparing Figs. 9-1(a) and 9-2(a) shows that the first-order magnitude response is a crude approximation to the ideal low-pass characteristic. Higher-order transfer functions give better approximations to the ideal case, but the actual characteristic will always have "rounded" corners.

Example 9-2. Design a first-order low-pass filter with $M(0)=1$ and cutoff frequency at $f_U = 60$ Hz.

Solution. The desired transfer function is $G(s)=K/(s+a)$ with $\omega_U = a$, where $\omega_U = 2\pi f_U = 120\pi$ rad/s. To achieve gain 1 for $\omega=0$, Eq. (9-5) shows that we must choose $K=a$. The required transfer function is then $G(s)=120\pi/(s+120\pi)$.

Example 9-3. The input function $f(t)=\cos(12\pi t) + \cos(1200\pi t)$ is applied to the filter of Example 9-2. Compute the steady-state output $y(t)$.

Solution. From Eq. (9-5) $M(\omega)$ is

$$M(\omega) = \frac{120\pi}{\sqrt{\omega^2 + (120\pi)^2}} \tag{9-13}$$

Frequency Response of Linear Systems 397

For ω=12π we have M(12π)=1/√1.01=0.995, and for ω=1200π, M(1200π)=1/√101=0.10. The phase shifts are ∅(12π)= -tan⁻¹(0.1)=-5.7° and ∅(1200π)=-tan⁻¹(10)=-84.3°. The steady-state response is then

$$y(t) = 0.995\cos\{12\pi t - 5.7°\} + 0.10\cos\{1200\pi t - 84.3°\}$$

(9-14)

Notice the degree to which the low-pass property of the magnitude response has decreased the response to the 1200π **rad/s** input compared to that for the 12π **rad/s** input.

Example 9-4. Investigate the magnitude response, phase response, and filter characteristics of the class of first-order systems with the transfer function G(s)=Ks/(s+a).

Solution. The magnitude and phase functions are found from G(jω)=jωK/(jω+a); thus,

$$M(\omega) = \frac{K\omega}{\sqrt{\omega^2 + a^2}} \qquad (9\text{-}15)$$

and ∅(ω)=90°-tan⁻¹(ω/a). These functions are sketched in Fig. 9-3.

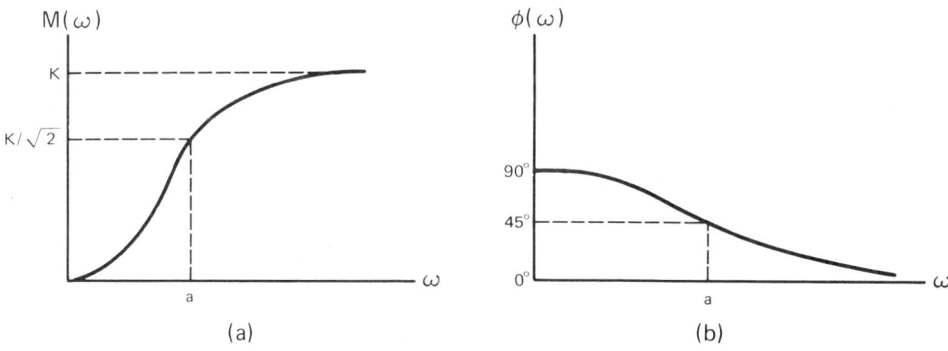

Figure 9-3. Magnitude and phase plots for G(s) = Ks/(s + a). (a) Magnitude response; (b) Phase response.

Comparing Fig. 9-3(a) with Fig. 9-2(b) shows that this response approximates that of the ideal high-pass filter. The system rejects input sinusoids with low frequencies and passes those with high frequencies.

Example 9-5. Find the magnitude and phase responses of the system with the transfer function $G(s)=10(s+1)/(s+10)$.
Solution. From $G(j\omega)=10(j\omega+1)/(j\omega+10)$ we have

$$M(\omega) = 10 \sqrt{\frac{\omega^2 + 1}{\omega^2 + 100}} \qquad (9\text{-}16)$$

and $\phi(\omega)=\tan^{-1}(\omega)-\tan^{-1}(\omega/10)$. In a later section we show how to sketch such functions without laboriously plotting points.

9.2 FREQUENCY RESPONSE OF SECOND–ORDER SYSTEMS

We turn next to the class of second-order systems described by the transfer function

$$G(s) = \frac{K}{s^2 + 2\zeta\omega_n s + \omega_n^2} \qquad (9\text{-}17)$$

where $\zeta<1$. (If $\zeta>1$, the poles of $G(s)$ are real and the system can be decomposed into a series connection of two first-order systems and the results of the previous section can be applied to each subsystem.) If the input to the system is $f(t)=A\cos(\omega t-\theta)$, then the steady-state output $y(t)$ is again given by Eq. (9-4). Evaluating Eq. (9-17) for $s=j\omega$ gives

$$G(j\omega) = \frac{K}{(\omega_n^2 - \omega^2) + 2j\zeta\omega\omega_n} \qquad (9\text{-}18)$$

As $M(\omega)=|G(j\omega)|$, we have, after some rearrangement,

$$M(\omega) = \frac{K/\omega_n^2}{\sqrt{\{1 - (\omega/\omega_n)^2\}^2 + 4\zeta^2(\omega/\omega_n)^2}} \qquad (9\text{-}19)$$

Frequency Response of Linear Systems

The phase function, $\phi(\omega)=\text{Arg}\{G(j\omega)\}$, is

$$\phi(\omega) = -\tan^{-1}\left\{\frac{2\zeta\omega/\omega_n}{1-(\omega/\omega_n)^2}\right\} \quad (9\text{-}20)$$

Figure 9-4 is a plot of Eq. (9-19) for two values of ζ. The horizontal axis is plotted in units of the ratio ω/ω_n.

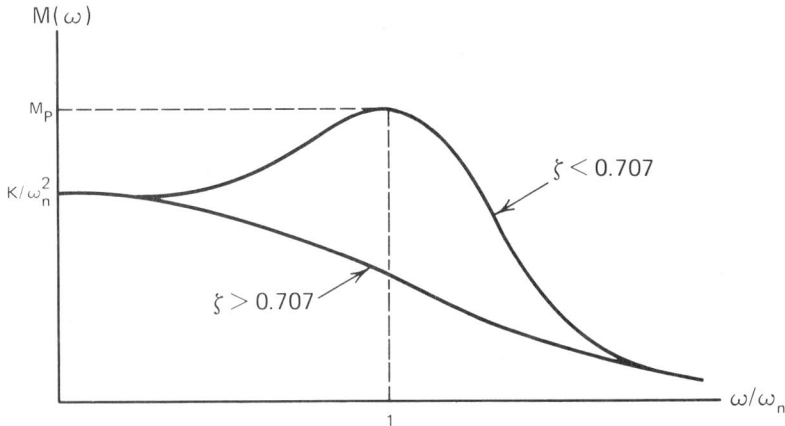

Figure 9-4. Magnitude response for the second-order system with transfer function $G(s) = K/(s^2 + 2\zeta\omega_n s + \omega_n^2)$.

Figure 9-5 shows the phase as a function of ω/ω_n. The phase lag has its maximum value of $0°$ at $\omega=0$ and decreases to $-180°$ as $\omega \to \infty$. Equation (9-20) shows that the phase is $-90°$ at $\omega=\omega_n$.

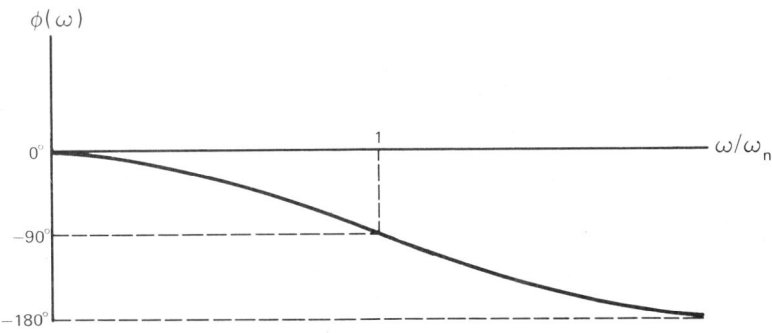

Figure 9-5. Phase response for the second-order system with transfer function $G(s) = K/(s^2 + 2\zeta\omega_n s + \omega_n^2)$.

Peak response. To understand the behavior near the peak M_p of the magnitude response, we must investigate Eq. (9-19) in some detail.[11] Define a new variable $q=\omega/\omega_n$; then Eq. (9-19) becomes

$$M(q) = \frac{K/\omega_n^2}{\sqrt{(1-q^2) + 4\zeta^2 q^2}} \qquad (9-21)$$

To find the value of q for which M(q) is a maximum, differentiate Eq. (9-21) with respect to q, set dM/dq=0, and solve for q. This results in

$$q = \sqrt{1 - 2\zeta^2} \qquad (9-22)$$

Equation (9-22) gives a real solution only if ζ is limited to positive values in the range $0<\zeta<1/\sqrt{2}=0.707$. Substituting Eq. (9-22) into Eq. (9-21) yields an expression for the peak value M_p of $M(\omega)$:

$$M_p = \frac{K/\omega_n^2}{2\zeta\sqrt{1-\zeta^2}} \qquad (9-23)$$

At the limiting value of $\zeta=0.707$, the "hump" disappears and Eq. (9-23) yields $M_p=K/\omega_n^2=M(0)$ as expected. Observe that M_p does not occur at $\omega=\omega_n$, or q=1, as one's intuition might indicate. On the contrary, Eq. (9-22), written in terms of ω, shows that the frequency ω_p which corresponds to M_p is

$$\omega_p = \omega_n\sqrt{1 - 2\zeta^2} \qquad (9-24)$$

The maximum occurs at ω_n only when $\zeta=0$; as ζ increases from zero, the maximum moves to the left. Note too that when $\zeta=0$ in Eq. (9-23) $M_p \to \infty$. This is the destructive resonance phenomenon which has been popularized in science demonstrations. As the peak response cannot become infinite, the system either destroys itself or goes into some other less drastic nonlinear mode of operation--saturation, for example.

Frequency Response of Linear Systems

Resonance.(10,12) As defined in Chapter 8, a system is in resonance if it is driven at one of its natural frequencies. For second-order systems, the resonant frequency is $\omega=\omega_n$. In the case of the constant-numerator system of Eq. (9-17), the resonant frequency neither causes a maximum response, nor does it cause $G(j\omega)$ to be a real function, nor does it correspond to zero phase shift as was the case for the series and parallel resonant electrical networks. When $\omega=\omega_n$ in Eq. (9-19) we have $M(\omega_n)=K/2\zeta\omega_n^2$, or, using Eq. (9-23) $M(\omega_n)=M_p\sqrt{1-\zeta^2}$. Clearly, $M(\omega_n)$ is less than M_p for permissible values of ζ, those in the range $0<\zeta<0.707$.

Example 9-6. Derive expressions for the magnitude and phase functions associated with the class of second-order systems having the transfer function

$$G(s) = \frac{Ks}{s^2 + 2\zeta\omega_n s + \omega_n^2} \tag{9-25}$$

Also investigate the behavior of $M(\omega)$ at the resonant frequency $\omega=\omega_n$.

Solution. Let $s=j\omega$ so that

$$G(j\omega) = \frac{j\omega K}{\omega_n^2 - \omega^2 + j2\zeta\omega\omega_n} \tag{9-26}$$

and

$$M(\omega) = \frac{K\omega/\omega_n^2}{\sqrt{\{1 - (\omega/\omega_n)^2\}^2 + 4\zeta^2(\omega/\omega_n)^2}} \tag{9-27}$$

The phase function is

$$\phi(\omega) = 90° - \tan^{-1}\{2\zeta\omega\omega_n/(\omega_n^2 - \omega^2)\} \tag{9-28}$$

To find the peak magnitude M_p we compute dM/dq, where $q=\omega/\omega_n$. If

$$M(q) = \frac{Kq/\omega_n}{\sqrt{(1 - q^2)^2 + 4\zeta^2 q^2}} \tag{9-29}$$

then setting $dM/dq=0$ leads to $0=(1-q^2)(1+q^2)$. The only realistic value is $q=1$, for which $\omega=\omega_n$. The peak value is then $M_p = K/2\zeta\omega_n$. For this class of systems, if the input sinusoid is at the resonant frequency ω_n, then (1) $M(\omega_n)$ is the maximum value of the magnitude response, (2) $G(j\omega_n)$ is real, and (3) the input and output responses are in phase. The point is that resonance means different things for different types of transfer functions. The situation is further confused by the fact that different writers define resonance in different ways.

Filter Characteristics Of The Second-Order System

If $\zeta >> 0.707$, the magnitude response given by Eq. (9-19) has no overshoot and somewhat resembles that of a low-pass filter (see $\zeta > 0.707$ in Fig. 9-4) whose upper cutoff frequency ω_U is found by setting Eq. (9-21) equal to $0.707M(0) = 0.707K/\omega_n^2$. The interested reader can verify that this leads to

$$\omega_U = \omega_n \sqrt{(1 - 2\zeta^2) + \sqrt{\zeta^4 - \zeta^2 + 1/2}} \qquad (9-30)$$

The more interesting situation occurs when $\zeta < 0.707$. In this case we can obtain band-pass characteristics from the second-order system. Two nonzero half-power points, ω_U and ω_L, may exist and lead to a bandwidth of $BW = \omega_U - \omega_L$. To solve for these cutoff frequencies, note that the peak value for $\zeta < 0.707$ is M_p as given in Eq. (9-23). To get the cutoff frequencies we set $M(\omega) = 0.707 M_p$ in Eq. (9-21) and solve for q. This gives

$$q = \sqrt{1 - 2\zeta^2 \pm 2\zeta\sqrt{1 - \zeta^2}} \qquad (9-31)$$

At the limiting value of $\zeta = 0.707$, $q=1$ as expected, provided the positive root is taken. This single value for q implies that there is a single nonzero cutoff frequency corresponding

Frequency Response of Linear Systems

to $\omega_U = \omega_n$, and we again have a low-pass filter. For smaller values of ζ, the amount of overshoot in the magnitude response increases, but we do not get two nonzero cutoff frequencies until ζ is small enough for the quantity under the outer radical to be zero for the negative root. When this occurs,

$$1 - 2\zeta^2 - 2\zeta\sqrt{1 - \zeta^2} = 0 \qquad (9\text{-}32)$$

which can be solved for $\zeta = 0.383$. For $\zeta < 0.383$, Eq. (9-31) has two real nonzero solutions and a band-pass magnitude characteristic results. Figure 9-6 illustrates the band-pass case.

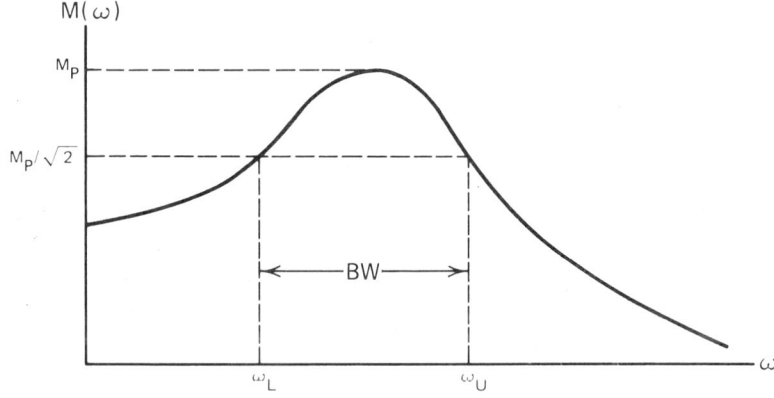

Figure 9-6. This is the type of band-pass response that can be achieved with a second-order system.

Example 9-7. A second-order system has $G(s) = 1/(s^2 + 0.2s + 1)$. Calculate the upper and lower cutoff frequencies and the bandwidth.

Solution. The magnitude function is

$$M(\omega) = \frac{1}{\sqrt{(1 - \omega^2)^2 + (0.2\omega)^2}} \qquad (9\text{-}33)$$

The given $G(s)$ has $\omega_n = 1$ and $\zeta = 0.1$ and, according to Eq. (9-23),

$$M_p = \frac{1}{0.2\sqrt{1 - .01}} = 5.025 \qquad (9\text{-}34)$$

To find the cutoff frequencies we set $M(\omega)=0.707\ M_p$ and solve for ω. Using this procedure, we have

$$3.553 = \frac{1}{\sqrt{(1-\omega^2)^2 + 0.04\omega^2}} \qquad (9-35)$$

Solving for ω by the quadratic formula yields $\omega_U=1.09$ and $\omega_L=0.884$. The bandwidth is then $BW=\omega_U-\omega_L=0.206$ rad/s.

9.3 MAGNITUDE AND PHASE FROM THE POLE-ZERO DIAGRAM

For transfer functions of order greater than the second, direct evaluation of the magnitude and phase functions becomes tedious. One can write digital computer programs to do the job, but if a computer is not available, we can use the s-plane pole-zero diagram.[13,14] Consider the third-order transfer function

$$G(s) = \frac{K(s-z_1)(s-z_2)}{(s-p_1)(s-p_2)(s-p_3)} \qquad (9-36)$$

For $s=j\omega$, $G(s)$ can be written as

$$G(j\omega) = \frac{K(j\omega-z_1)(j\omega-z_2)}{(j\omega-p_1)(j\omega-p_2)(j\omega-p_3)} \qquad (9-37)$$

Because each factor corresponds to a vector from the pole or zero to the point $j\omega$ on the imaginary axis, we can evaluate $G(j\omega)$ using graphical quantities from the pole-zero plot. Figure 9-7 shows the geometry for Eq. (9-37).
For a particular value of ω, say $\omega=\omega_1$ as shown in the figure,

$$G(j\omega_1) = \frac{KT_1 T_2}{B_1 B_2 B_3} \underline{/\alpha_1 + \alpha_2 - \beta_1 - \beta_2 - \beta_3} \qquad (9-38)$$

where T_1 and T_2 are the distances from the zeros of $G(s)$ to the point $j\omega_1$; B_1, B_2, and B_3 are the distances from the poles of $G(s)$ to the point $j\omega_1$; α_1 and α_2 are the angles from

Frequency Response of Linear Systems

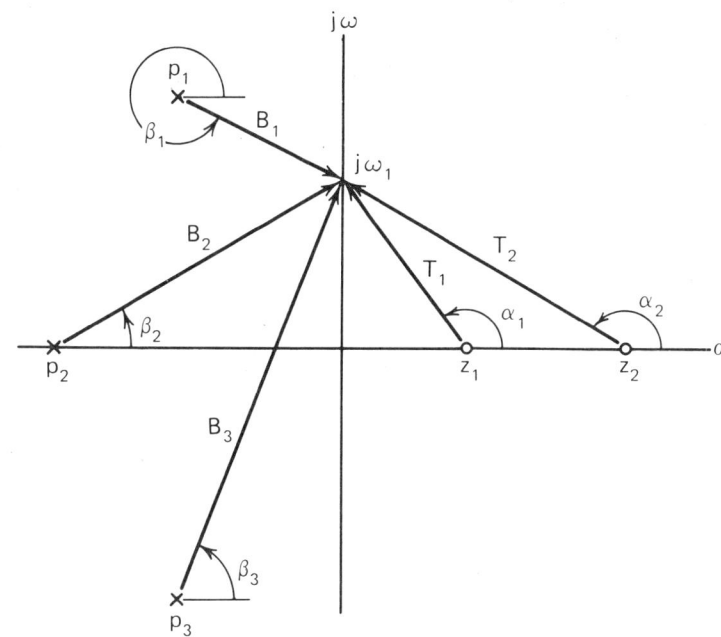

Figure 9-7. Geometrical relationships for determining $M(\omega_1)$ and $\phi(\omega_1)$ for a system with three poles and two zeros.

the zeros; and β_1, β_2 and β_3 are the angles from the poles. From Eq. (9-38) we have $M(\omega_1)=|G(j\omega_1)|$, or $M(\omega_1)=KT_1T_2/B_1B_2B_3$ and $\phi(\omega_1)=\alpha_1+\alpha_2-\beta_1-\beta_2-\beta_3$. For an nth-order system with transfer function

$$G(s) = \frac{K(s-z_1)(s-z_2)\ldots(s-z_m)}{(s-p_1)(s-p_2)\ldots(s-p_n)} \quad (9\text{-}39)$$

the magnitude and phase responses are

$$M(\omega_1) = \frac{KT_1T_2\ldots T_m}{B_1B_2\ldots B_n} \quad (9\text{-}40)$$

and

$$\phi(\omega) = \alpha_1 + \alpha_2 + \ldots \alpha_m - \{\beta_1 + \beta_2 + \ldots \beta_n\} \quad (9\text{-}41)$$

The pole-zero method lends itself to evaluation on a digital computer, provided the transfer function is given in factored form, or if a root-finding program is available.

Example 9-8. Compute M(5) and ∅(5) for the transfer function whose poles and zeros are shown in Fig. 9-8(a). Assume K=1.

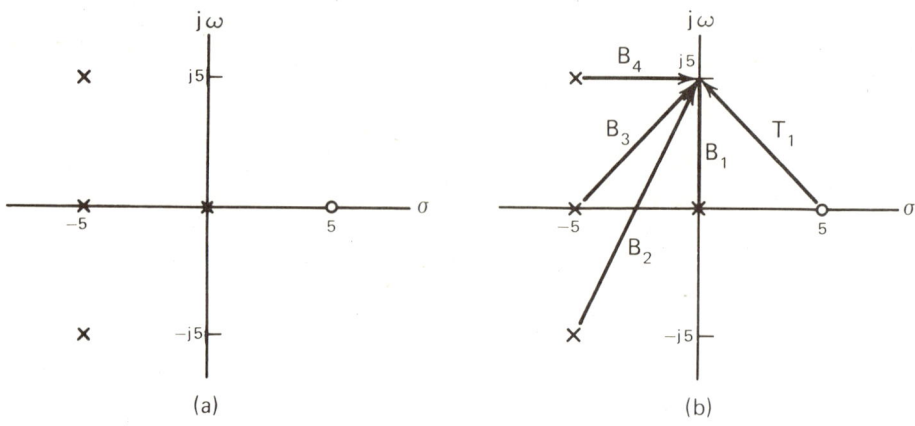

Figure 9-8. Example 9-8. From figure (b), M(5) = $T_1/(B_1 B_2 B_3 B_4)$. (a) Pole-zero plot for Example 9-8; (b) Geometrical relationships for determining M(5).

Solution. Figure 9-8(b) shows the pole-zero plot with vectors drawn from the poles and zeros to the point j5. The magnitudes and angles are evaluated by elementary geometry: $T_1 = 5\sqrt{2}$, $\alpha_1 = 135°$; $B_1 = 5$, $\beta_1 = 90°$; $B_2 = \sqrt{125}$, $\beta_2 = \tan^{-1}(10/5) = 63.44°$; $B_3 = 5\sqrt{2}$, $\beta_3 = 45°$; and $B_4 = 5$, $\beta_4 = 0°$. From Eq. (9-40) the magnitude function is $M(5) = KT_1/B_1 B_2 B_3 B_4$, or $M(5) = 0.00358$. The phase function is $\emptyset(5) = \alpha_1 - \beta_1 - \beta_2 - \beta_3 - \beta_4$ or $\emptyset(5) = -63.44°$.

9.4 DISTORTION IN THE SINUSOIDAL STEADY STATE

Suppose the input to a system is a signal which can be represented as the sum of a number of sinusoids at different frequencies--and virtually all signals of interest in dynamic analysis can be so represented using the Fourier series. We may intentionally want to suppress high frequencies, low frequencies, or frequencies inside or outside of a known band of frequencies. Examples are removing high-frequency noise or

Frequency Response of Linear Systems

60-cycle hum from an audio system. In such cases we use the filter properties of systems by appropriately shaping the magnitude response. On the other hand, in many applications--communications systems, for example--the objective is to reproduce faithfully the input signal at the output without rejecting, suppressing, or otherwise altering any of the input frequencies. This brings us to the subject of <u>distortion</u>. Distortion may occur in the time domain, as when a sinusoidal signal is flattened on top, but it may also occur in the frequency domain. The three kinds of frequency-domain distortion are <u>magnitude distortion</u>, <u>phase distortion</u>, and <u>delay distortion</u>. These are discussed in the following paragraphs. (15)

Magnitude Distortion

Magnitude or amplitude distortion occurs if the amplitudes of some frequency components are changed more than those of others. For the system whose magnitude function is shown in Fig. 9-9, all desired input frequencies should be well above ω_{min} to avoid magnitude distortion.

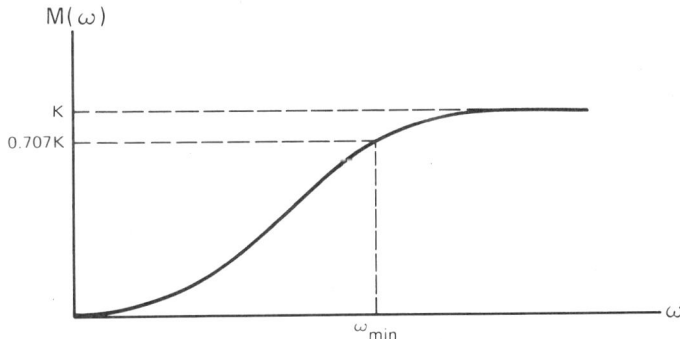

Figure 9-9. The system with this magnitude characteristic has no magnitude distortion for input frequencies substantially greater than ω_{min}.

Frequencies substantially below ω_{min} would be suppressed by this system. (All filters purposely introduce magnitude distortion into a system.) To avoid magnitude distortion, the magnitude function must be "flat" over the range of frequencies of interest. In practice, we generally do not require absolute flatness but are given an allowable deviation.

Phase Distortion

We have seen that the effect of a linear system is to alter the magnitude and shift the phase of a sinusoidal input as it passes through the system. If the phases of some frequencies are altered more than those of others, then the system has <u>phase distortion</u>. If all frequency components of the input are to have the same phase shift at the output, then the phase response $\phi(\omega)$ must be flat, or a horizontal straight line. The phase response shown in Fig. 9-10 indicates a system which would have no phase distortion for input frequencies above ω_{min}. For frequencies below ω_{min} phase distortion would occur because $\phi(\omega)$ is a variable function of frequency. Most practical systems have phase distortion; the best we can do is design so that the amount of distortion is within allowable limits for some band of frequencies of interest.

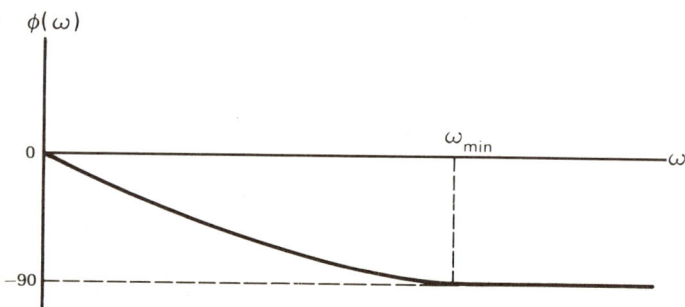

Figure 9-10. The system with this phase characteristic has no phase distortion for input frequencies greater than ω_{min}.

Delay Distortion

If we are willing to accept an output signal whcih is a faithful reproduction of the input but delayed in time, the restrictions on the phase response are not so severe.

Frequency Response of Linear Systems

We showed earlier that the Lapace transform of a signal delayed in time by T seconds is found by multiplying the undelayed transform by exp(-sT); that is, if L{f(t)}=F(s), then

$$L\{f(t-T)\} = e^{-sT}F(s) \qquad (9\text{-}42)$$

The delay term exp(-sT) has the magnitude response $M(\omega)=|\exp(-j\omega T)|=1$ and the phase response $\phi(\omega)=\text{Arg}\{\exp(-j\omega T)\}=-\omega T$. From this expression we can determine the delay T by $T=-\phi(\omega)/\omega$. If we allow for the possibility that the delay is a function of ω, then we can define a <u>delay function</u> $T(\omega)$ as

$$T(\omega) = \frac{-d\phi(\omega)}{d\omega} \qquad (9\text{-}43)$$

If we now interpret a linear system as having delay, the value of the delay is the negative derivative of the phase response. In order for all frequencies to be delayed the same amount, thus avoiding delay distortion in the output, then the delay function must be a constant. This in turn implies that the <u>slope</u> of the phase response must be constant over the frequency range of interest, a less restrictive condition than for the nondelayed case which required that the phase itself remain constant. The deviation of $T(\omega)$ from a constant produces <u>delay distortion</u>.

9.5 INTRODUCTION TO BODE PLOTS

Plots of magnitude and phase versus frequency on semilog paper are called log plots, or, more commonly, <u>Bode plots</u>.[16] Bode plots can be used (1) as an aid in sketching complicated magnitude and phase functions without laboriously plotting points; (2) in identifying unknown transfer functions, and (3) in designing compensators for automatic control systems[17,18]

The Log-Magnitude Function [10]

Recall that for the general transfer function in factored form

$$G(s) = \frac{K(s + z_1)(s + z_2)\ldots(s + z_m)}{(s + p_1)(s + p_2)\ldots(s + p_n)} \quad (9\text{-}44)$$

the magnitude function is

$$M(\omega) = \frac{K|j\omega + z_1||j\omega + z_2|\ldots|j\omega + z_m|}{|j\omega + p_1||j\omega + p_2|\ldots|j\omega + p_n|} \quad (9\text{-}45)$$

where we have written the factors with positive rather than negative signs. The phase function is

$$\phi(\omega) = \alpha_1 + \alpha_2 + \ldots \alpha_m - \{\beta_1 + \beta_2 + \ldots \beta_n\} \quad (9\text{-}46)$$

where $\alpha_i = \text{Arg}\{j\omega + z_i\}$ and $\beta_i = \text{Arg}\{j\omega + p_i\}$. Define the log-magnitude, $LM\{\ \}$, of a function as

$$LM\{M(\omega)\} = 20\text{Log}\{M(\omega)\} \quad (9\text{-}47)$$

where Log means take logarithms to the base 10. Log-magnitudes are given in units of <u>decibels</u>, abbreviated "db". We can take log-magnitudes of either real functions, complex functions, or constants. If a quantity is real its log magnitude is 20 times the logarithm of its absolute value. If a quantity is complex, we must take its magnitude and then compute 20 times the logarithm of the magnitude. Thus if $A = 3 + j4$, then $LM\{A\} = 20\text{Log}(5)$.

For the magnitude function of Eq. (9-45),

$$LM\{M(\omega)\} = 20\text{Log}(K) + \sum_{i=1}^{m} 20\text{Log}|j\omega + z_i| - \sum_{j=1}^{n} 20\text{Log}|j\omega + p_j|$$

$$(9\text{-}48)$$

where we have used the results that $\text{Log}(xy) = \text{Log}(x) + \text{Log}(y)$ and $\text{Log}(x/y) = \text{Log}(x) - \text{Log}(y)$. The interesting feature of Eq. (9-48) is that the magnitude is now represented by a <u>sum</u> rather than a product of factors. We are therefore able to examine the effects of each pole and zero separately and sum the results.

Frequency Response of Linear Systems   411

A Bode magnitude plot is a plot of LM{M(ω)} versus ω on semi-log paper, again with the logarithmic axis used for ω.

Analysis Of Standard Factors [17, 19]

If G(s), as given by Eq. (9-44), is rearranged in time-constant form (with the zeroth powers of s forced to 1 by factoring out suitable constants), then G(s) can be expressed as a product of partial transfer functions, G(s)= $G_1(s)G_2(s)\ldots G_N$. The $G_i(s)$ may have any of the forms K, s/ω_1, s/ω_2+1, $(s/\omega_n)^2+2\zeta s/\omega_n+1$, $1/(s/\omega_1)$, $1/(s/\omega_2+1)$, or $\{(s/\omega_n)^2+2\zeta(s/\omega_n)+1\}^{-1}$. In addition, we might have repeated factors of any of these types--$(s/\omega_1)^n$, for example. When evaluated for s=jω, the variable terms become $j\omega/\omega_1$, $j\omega/\omega_2+1$, and $1-(\omega/\omega_n)^2+j2\zeta\omega/\omega_n$ along with the reciprocals of these expressions. In the following paragraphs we examine the Bode plots for these classes of factors.

The constant K. For the constant K, the decibel (or "db") value is

$$LM\{K\} = 20Log(K) \qquad (9-49)$$

LM(K) is positive if K>1, 0 if K=1, and negative if K<1. The phase response is a constant $0°$ if K is positive and $180°$ if K is negative (we do not differentiate between plus or minus $180°$). Figure 9-11 shows Bode plots for the constant K.

The terms jω and 1/jω. The log-magnitude of the term jω is LM(jω)=20Log($|j\omega|$) or

$$LM(j\omega) = 20Log(\omega) \qquad (9-50)$$

Equation (9-50) has the form y=mx and hence is that of a straight line with slope 20 on semilog paper. Since $Log(10^n)$= n, we see that LM(j1)=0, LM(j10)=20, LM(j100)=40 and so on. When ω changes by a "decade" (a factor of ten) LM(ω) changes by 20 db. Finally, from LM(j1)=0 it is clear that the plot passes through 0 db when ω=1. Figure 9-12(a) is the Bode magnitude plot.

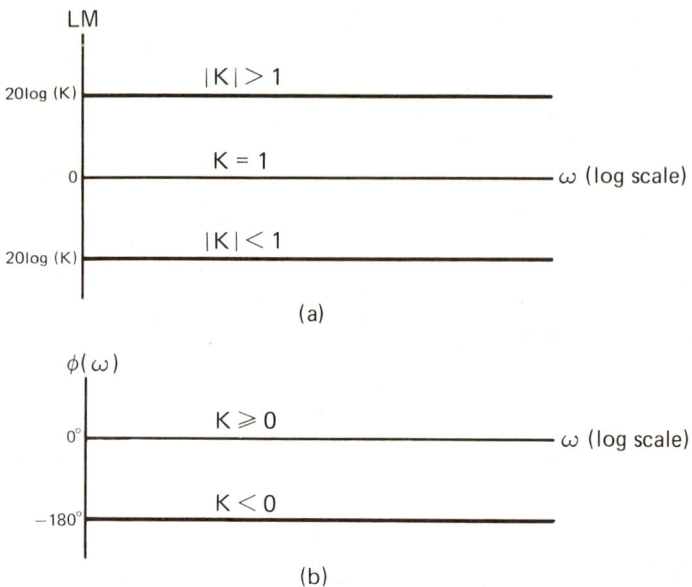

Figure 9-11. Bode plots for a constant K.

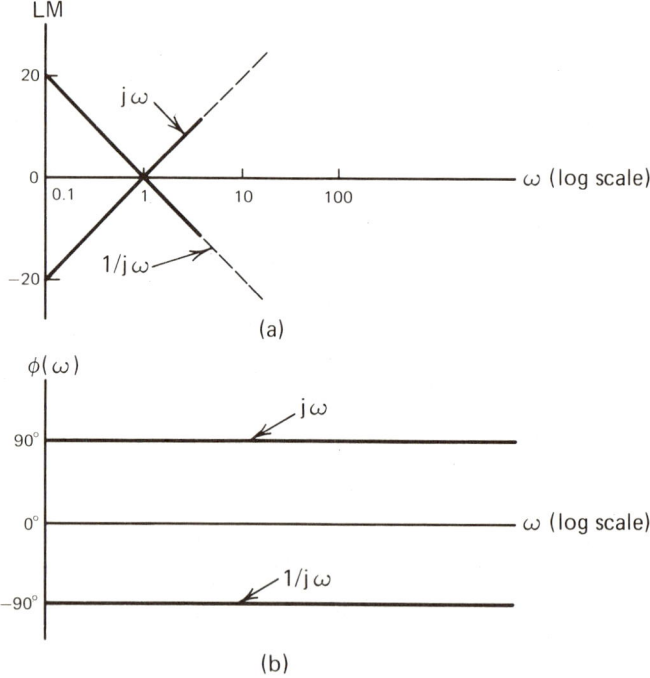

Figure 9-12. Bode magnitude and phase plots for $G(j\omega) = j\omega$ and $G(j\omega) = 1/j\omega$. (a) Magnitude; (b) Phase.

Frequency Response of Linear Systems 413

The phase of the term $j\omega$ has the constant value $\phi(\omega)=\text{Arg}\{j\omega\}=90°$ as shown in Fig. 9-12(b).

For the term $1/j\omega$, the log-magnitude and phase are

$$LM(j\omega) = 20\text{Log}|1/j\omega| = -20\text{Log}(\omega) \qquad (9\text{-}51)$$

and $\phi(\omega)=\text{Arg}\{1/j\omega\}=-90°$. The Bode plots are similar to those for the term $j\omega$ except that the phase and the slope of the magnitude curve are both negative as shown in Fig. 9-12.

A more general form. A more general form in which we may encounter the pole or zero at the origin is

$$G(j\omega) = (j\omega/\omega_1)^{\pm n} \qquad (9\text{-}52)$$

where we have written $G(j\omega)$ with the understanding that this may be only one factor of a more complicated transfer function. The log-magnitude function corresponding to Eq. (9-52) is

$$LM\{G(j\omega)\} = 20\text{Log}\{(\omega/\omega_1)^{\pm n}\} \qquad (9\text{-}53)$$

or

$$LM\{G(j\omega)\} = \pm 20n\text{Log}(\omega/\omega_1) \qquad (9\text{-}54)$$

The Bode magnitude plot thus has a slope of $\pm 20n$ db/decade and crosses the 0-db axis when $\omega=\omega_1$. The phase is $\phi(\omega)=\pm 90n°$. The Bode plots for Eqns. (9-53) and (9-54) are shown in Appendix G.

Example 9-9. Sketch the Bode magnitude plot for the transfer function $G(s)=5/s$.

Solution. The given transfer function has the form $G(j\omega)=(j\omega/\omega_1)^{-1}$ where $\omega_1=5$; hence Eq. (9-53) applies with the minus sign and $n=1$. Figure 9-13 is the magnitude plot.

The term $(j\omega/\omega_2+1)$. This is the case where we have a non-repeated real pole or zero in the transfer function of Eq. (9-44). To simplify the notation in the following derivations we can, with no loss of generality, assume a partial transfer function of the form $G(j\omega)=(j\omega/\omega_2+1)$. A term can always be rearranged in this form by factoring out a constant which is incorporated in an overall constant K. The magnitude function is

$$M(\omega) = \sqrt{(\omega/\omega_2)^2 + 1} \qquad (9\text{-}55)$$

with log-magnitude

$$LM\{M(\omega)\} = 20\text{Log}\left\{\sqrt{(\omega/\omega_2)^2 + 1}\right\} \qquad (9\text{-}56)$$

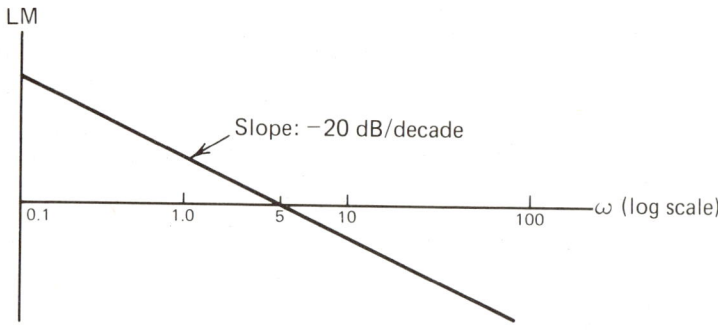

Figure 9-13. Bode magnitude plot for Example 9-9.

We will now begin to see the power of Bode plots in approximating frequency-response functions. Suppose $\omega \ll \omega_2$ in Eq. (9-56), then $\omega/\omega_2 \ll 1$ and $LM\{M(\omega)\} \approx 20\text{Log}(1) = 0$. The asymptotic behavior, the behavior for $\omega \ll \omega_2$, is a horizontal straight line at 0 db, the ω-axis. If, on the other hand, $\omega/\omega_2 \gg 1$, then,

$$LM\{M(\omega)\} \approx \pm 20\text{Log}\sqrt{(\omega/\omega_2)^2} = 20\text{Log}(\omega/\omega_2) \qquad (9\text{-}57)$$

Equation (9-57) is a straight line which passes through 0 db when $\omega = \omega_2$. Its slope is 20 db/decade. Figure 9-14(a) shows these two asymptotes and the actual curve, which must be plotted point-by-point from Eq. (9-56).

Frequency Response of Linear Systems

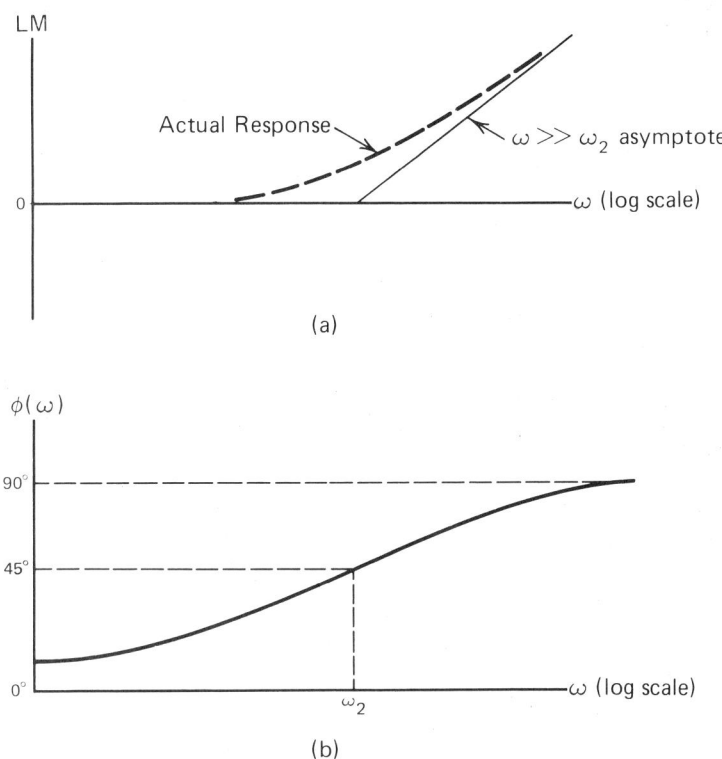

Figure 9-14. Bode plots for $G(j\omega) = j\omega/\omega_2 + 1$. (a) Magnitude; (b) Phase.

The point $\omega=\omega_2$, where the high-frequency and low-frequency asymptotes merge, is called a **break-point** for the asymptotic approximation to the Bode magnitude plot. In fact, the asymptotic approximation is often called a **slope-breakpoint** approximation, because only slopes and breakpoints are needed to sketch the Bode magnitude plot. The maximum error in using asymptotes rather than the actual curve occurs at the break-point. We can calculate this error by evaluating Eq. (9-56) for $\omega=\omega_2$; the result is

$$LM\{M(\omega_2)\} = 20\text{Log}\left\{\sqrt{(\omega_2/\omega_2)^2 + 1}\right\} = 3 \text{ db} \qquad (9\text{-}58)$$

Define an **octave** as a factor of two. By this definition $\omega=2\omega_2$ is one octave above $\omega=\omega_2$ while $\omega=\omega_2/4$ is two octaves below $\omega=\omega_2$. Table 9-1 lists the deviations between the actual curve and the slope-breakpoint approximation for the first-order factor $(j\omega/\omega_2+1)^{\pm 1}$. The negative errors in the table are for the term $(j\omega/\omega_2+1)^{-1}$.

Table 9-1.

Error in the Asymptotic Approximation For $(j\omega/\omega_2+1)^{\pm 1}$

Frequency	Error (db)
Breakpoint ($\omega=\omega_2$)	±3
One Octave above ($\omega=2\omega_2$)	±1
One Octave below ($\omega=\omega_2/2$)	±1
Two Octaves above ($\omega=4\omega_2$)	±0.3
Two Octaves below ($\omega=\omega_2/4$)	±0.3

The phase of the term $G(j\omega)=(j\omega/\omega_2+1)$ is $\phi(\omega)=\tan^{-1}(\omega/\omega_2)$. When $\omega=\omega_2$, $\phi(\omega_2)=45°$; for $\omega=0$ and $\omega\to\infty$, the asymptotic values are $\phi(0)=0°$ and $\phi(\infty)=90°$, respectively (see Fig. 9-14(b)).

The term $\overline{1/\{j\omega/\omega_2+1\}}$. When the first-order linear factor is a pole, we have the partial transfer function $G(j\omega)=1/(j\omega/\omega_2+1)$ with log-magnitude

$$LM\{G(j\omega)\} = -20\text{Log}\sqrt{(\omega/\omega_2)^2 + 1} \qquad (9-59)$$

which is the negative of the log-magnitude for the case above. For this reason, the Bode plots are identical, except that for the pole case the slope is -20 db/decade. Figure 9-15(a) is the Bode magnitude plot.

The phase for $G(j\omega)=1/(j\omega/\omega_2+1)$ is $\phi(\omega)=-\tan^{-1}(\omega/\omega_2)$. Clearly, $\phi(0)=0°$, $\phi(\omega_2)=-45°$, and $\phi(\infty)=-90°$. The phase plot is shown in Fig. 9-15(b).

Frequency Response of Linear Systems 417

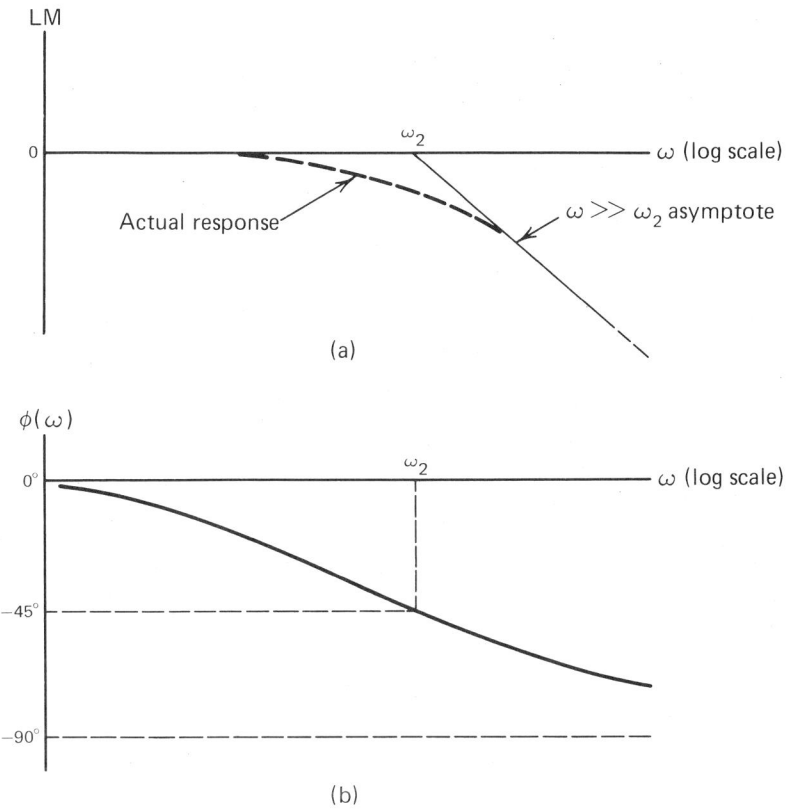

Figure 9-15. Bode plots for $G(j\omega) = 1/(j\omega/\omega_2 + 1)$. (a) Magnitude; (b) Phase.

Example 9-10. Derive the relationship needed to sketch Bode magnitude and phase plots for the repeated first-order factor

$$G(j\omega) = \{j\omega/\omega_2 + 1\}^{\pm n} \qquad (9\text{-}60)$$

Solution. The log-magnitude function is

$$LM\{G(j\omega)\} = \pm 20n\text{Log}\left\{\sqrt{(\omega/\omega_2)^2 + 1}\right\} \qquad (9\text{-}61)$$

For $\omega \ll \omega_2$, the asymptote is $LM\{G(j\omega)\} \approx 0$, which is the ω-axis. For $\omega \gg \omega_2$

$$LM\{G(j\omega)\} \approx \pm 20n\text{Log}(\omega/\omega_2) \qquad (9\text{-}62)$$

This is a straight line with slope $\pm 20n$ db/decade passing through 0 db when $\omega=\omega_2$. Appendix G shows the appropriate magnitude plot.

The phase is obtained by writing $G(j\omega)$ as

$$G(j\omega) = M(\omega)\left\{\exp\{j\tan^{-1}(\omega/\omega_2)\}\right\}^{\pm n} \qquad (9\text{-}63)$$

$$= M(\omega)\exp\{\pm jn\tan^{-1}(\omega/\omega_2)\} \qquad (9\text{-}64)$$

from which $\phi(\omega)=\pm n\tan^{-1}(\omega/\omega_2)$. If $\omega=0$, $\phi(0)=0°$. If $\omega\gg\omega_2$, $\phi(\omega)=\pm(90n)°$. The phase plot is also shown in Appendix G.

Quadratic factors. If a quadratic factor appears in the denominator, we have a partial transfer function with the form

$$G(s) = \frac{1}{(s/\omega_n)^2 + 2\zeta(s/\omega_n) + 1} \qquad (9\text{-}65)$$

or, for $s=j\omega$,

$$G(j\omega) = \frac{1}{1 - (\omega/\omega_n)^2 + j2\zeta(\omega/\omega_n)} \qquad (9\text{-}66)$$

The log-magnitude function is

$$LM\{G(j\omega)\} = -20\text{Log}\left\{\sqrt{\{1 - (\omega/\omega_n)^2\}^2 + 4\zeta(\omega/\omega_n)^2}\right\} \qquad (9\text{-}67)$$

Again, consider the asymptotic cases. If $\omega\ll\omega_n$, then $LM\{G(j\omega)\}\approx -20\text{Log}(1)=0$, and the asymptote is the ω-axis. For $\omega\gg\omega_n$, Eq. (9-67) becomes

$$LM\{G(j\omega)\} \approx -20\text{Log}\left\{\sqrt{(\omega/\omega_n)^4}\right\} = -40\text{Log}(\omega/\omega_n) \qquad (9\text{-}68)$$

This is a straight line with a slope of -40 db/decade passing through 0 db when $\omega=\omega_n$. The behavior in the neighborhood of the breakpoint is influenced by the value of ζ. For small values of ζ a "hump" appears in the true curve near $\omega=\omega_n$. For large values of ζ the curve is "overdamped". Figure 9-16(a) shows the asymptotic approximation along with the actual response for two values of ζ.

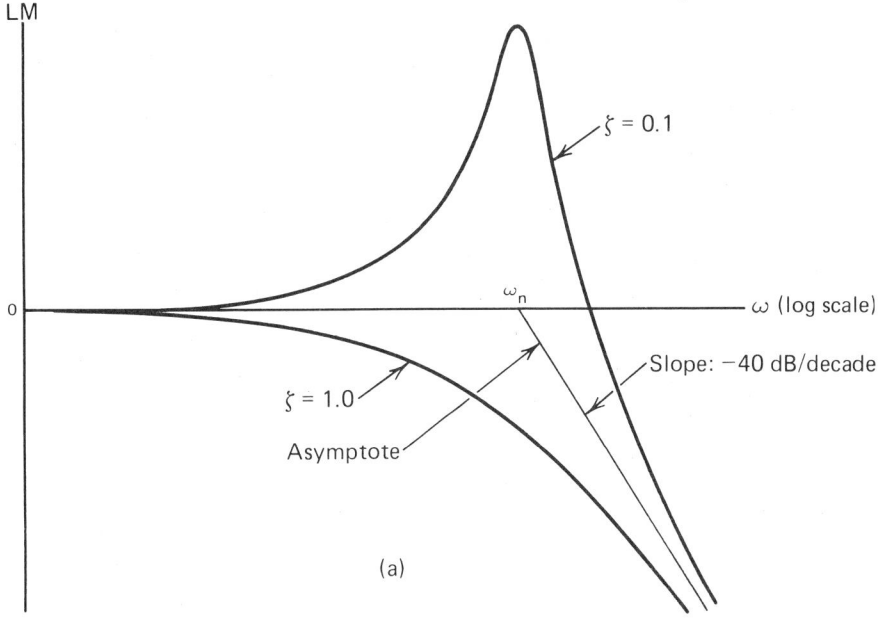

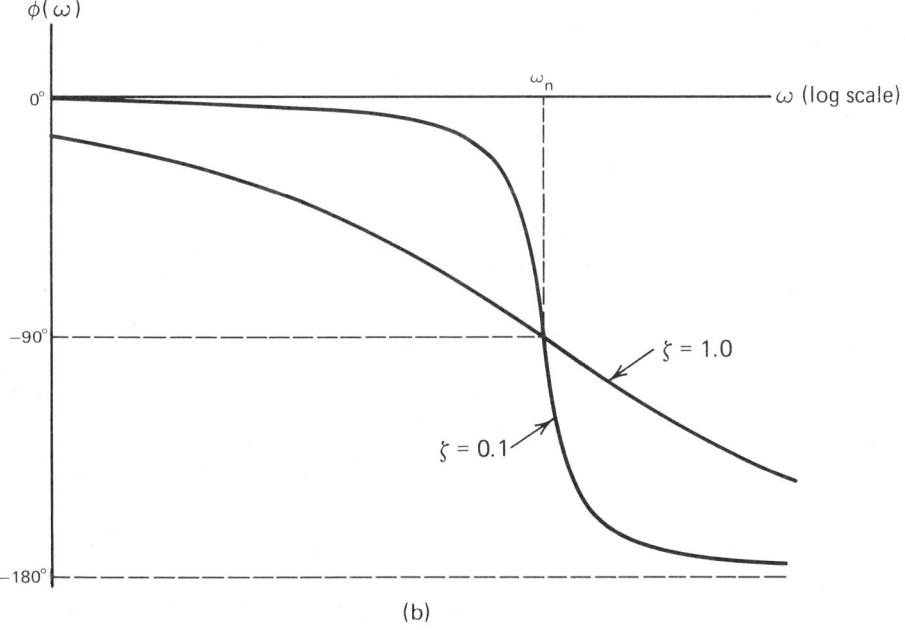

Figure 9-16. Bode plots for the quadratic factor $G(j\omega) = [1 - (\omega/\omega_n)^2 + j2\zeta(\omega/\omega_n)]^{-1}$
(a) Magnitude; (b) Phase.

The phase response which corresponds to Eq. (9-66) is

$$\phi(\omega) = -\tan^{-1}\left\{\frac{2\zeta\omega/\omega_n}{1-(\omega/\omega_n)^2}\right\} \tag{9-69}$$

The critical values are $\phi(0)=0°$, $\phi(\omega_n)=-90°$, and $\phi(\infty)=-180°$. The shape of the phase plot is also sensitive to ζ, as seen from Fig. 9-16(b).

If a quadratic factor appears in the numerator of a transfer function, the partial transfer function is

$$G(s) = (s/\omega_n)^2 + 2\zeta(s/\omega_n) + 1 \tag{9-70}$$

The magnitude and phase responses are identical to those for the pole case except that the high-frequency asymptote slopes upward at +40 db/decade and the phase is always positive.

<u>General comments about Bode plots</u>. Asymptotic, or slope-breakpoint, approximations are also available to simplify plotting the phase response on semilog paper. We have not given these in this preliminary treatment as they are used mostly by system designers.[2,9,10]

The error in the asymptotic approximation for quadratic factors is more complex than for first-order factors because, for the former, the error is a factor of both ζ and ω_n. Fortunately, if one plots the dimensionless ratio ω/ω_n on the frequency axis, then it is possible to design <u>universal</u> magnitude and phase charts for quadratic factors, charts which no longer depend on the value of ω_n for a specific system. Moreover, one can also design a universal error chart which gives the error between the asymptotic approximation and the actual response as a function of ζ and the dimensionless ratio ω/ω_n. Such universal charts are found in standard texts on control system design.[20,21]

<u>Example 9-11</u>. A frequently encountered term in dynamic systems analysis has the transfer function

$$G(s) = \left\{\frac{s}{s+\omega_3}\right\}^{\pm n} \tag{9-71}$$

Derive expressions which can be used to sketch the Bode magnitude and phase plots.

Solution. Dividing numerator and denominator by s and substituting $s=j\omega$ leads to $G(j\omega)=(1-j\omega_3/\omega)^{\pm n}$ with the log-magnitude function

$$LM\{G(j\omega)\} = \pm 20n\text{Log}\left\{\sqrt{1 + (\omega_3/\omega)^2}\right\} \qquad (9\text{-}72)$$

For $\omega \gg \omega_3$ the asymptotic approximation is $LM\{G(j\omega)\} \approx 0$, which is the ω-axis. For $\omega \ll \omega_3$, Eq. (9-72) becomes

$$LM\{G(j\omega)\} \approx \pm 20n\text{Log}(\omega_3/\omega) \qquad (9\text{-}73)$$

Equation (9-73) is a straight line with slope ± 20 db/decade and a breakpoint at $\omega=\omega_3$. This magnitude response is shown in Appendix G. To get the phase response, we write

$$G(j\omega) = M(\omega)\left\{\exp\{-j\tan^{-1}(\omega_3/\omega)\}\right\}^{\pm n} \qquad (9\text{-}74)$$

so that $\phi(\omega)=\pm n\tan^{-1}(\omega_3/\omega)$. The critical points are $\phi(0)= \pm(90n)^\circ$, $\phi(\omega_3)=\pm(45n)^\circ$, and $\phi(\infty)=0^\circ$. This phase response is also shown in Appendix G.

Example 9-12. Sketch Bode magnitude and phase plots for the transfer function

$$G(s) = \frac{50(s + 10)}{s(s + 100)} \qquad (9\text{-}75)$$

Solution. The best procedure is to plot the responses for each individual factor and then to sum these graphically. The factors 50, $s+10$, $1/s$, and $1/(s+100)$ have Bode magnitude responses which can be sketched by inspection using the methods of the preceeding paragraphs. The first step is to rearrange $G(s)$ in time-constant form:

$$G(s) = \frac{5(s/10 + 1)}{s(s/100 + 1)} \qquad (9\text{-}76)$$

Figure 9-17(a) shows the magnitude responses of the individual terms. Figure 9-17(b) shows the overall magnitude response obtained by summing the responses for the individual terms. The responses can be summed accurately if the breakpoints are at least a decade apart. If the breakpoints are close together, the error is considerably greater. In the figure, the solid lines are the asymptotic approximations and the dashed lines are the actual responses. In general, we should start the plot at a frequency at least one decade below the lowest breakpoint.

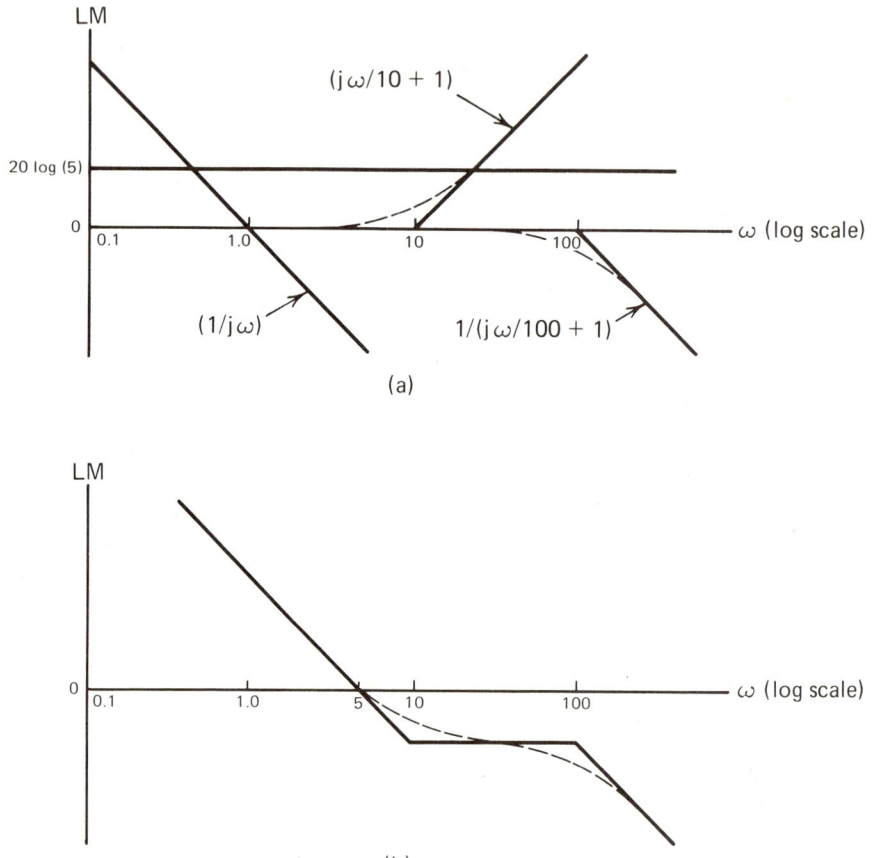

Figure 9-17. Bode magnitude plots for Example 9-12. (All slopes are ± 20 dB/decade.) (a) Magnitude response for the factors of Eq. (9-76); (b) Overall magnitude response.

The phase response is the sum of the phase responses of the individual terms. Figure 9-18(a) shows the individual phase response terms, Fig. 9-18(b) the overall phase response.

Frequency Response of Linear Systems

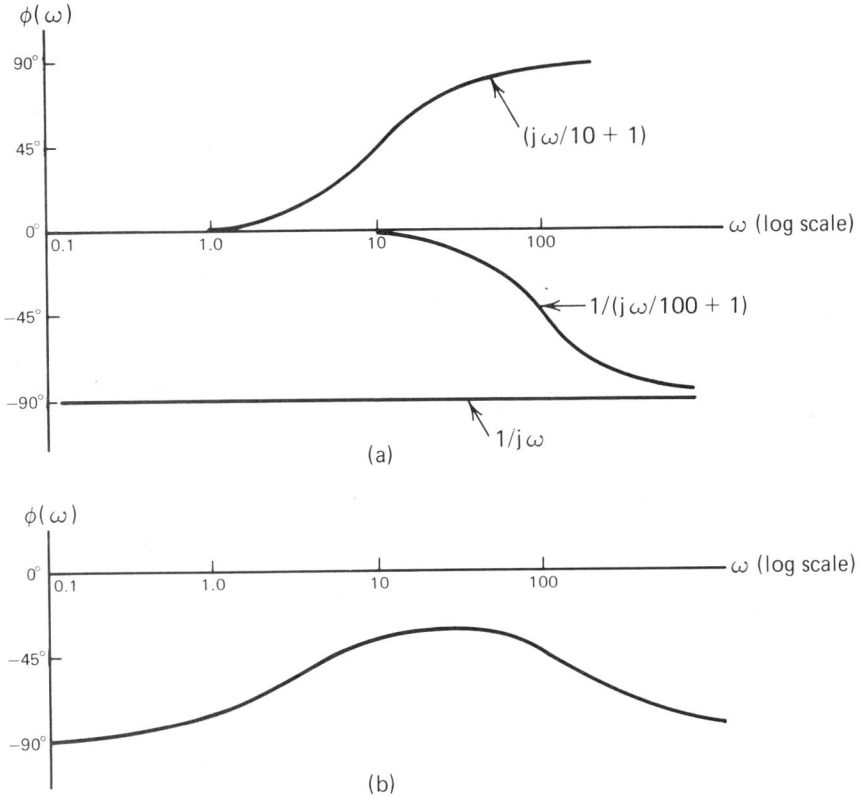

Figure 9-18. Phase response for Examples 9-12. (a) Phase response for the factors of Eq. (9-76); (b) Overall phase response.

9.6 EXPERIMENTAL DETERMINATION OF TRANSFER FUNCTIONS

In many instances the engineer must analyze a system whose transfer function he does not know. He may not even know the order of the system. To develop a mathematical model for such a system so that it can be analyzed by hand or simulated on a computer, we must first evaluate its transfer function. If equipment is available to excite the system with sinusoids of different frequencies, then we can plot the frequency response. From a Bode plot of $LM\{G(j\omega)\}$ we can often approximate the transfer function.

Given the Bode magnitude plot for an unknown system, we can apply the procedure outlined above <u>in reverse</u>. That is, we attempt to fit straight-line asymptotes to the response curve and then identify the slopes and breakpoints. Once the slopes and breakpoints are known, we can guess at a suitable combination of factors. The procedure works well provided the breakpoints are separated by at least a decade of frequency. If the breakpoints are close together, the job becomes more difficult and the approximation becomes less accurate. The phase response is used as an auxiliary check on the approximation. The procedures gives unambiguous results only for so-called <u>minimum-phase</u> systems, or those with no zeros in the right-half of the s-plane.

Example 9-13. Find a transfer function which gives the magnitude response of Fig. 9-19(a).

Solution. A set of approximate asymptotes are shown superimposed on the actual magnitude response. The approximate transfer function is obtained by inspecting the slope-breakpoint approximation. Figure 9-19(b) shows the component slopes and breakpoints which can be summed to give the desired Bode plot. The transfer function is therefore

$$G(s) = \frac{10(s+1)}{(s/50 + 1)^2 (s/10 + 1)} \qquad (9-77)$$

PROBLEMS FOR CHAPTER 9

9-1. Compute the magnitude and phase functions associated with the following differential equations, assuming sinusoidal steady state operation:

(a) $\dot{q} + 5q = f(t)$ \hfill (9-78)

(b) $\ddot{x} + 2\zeta\omega_n \dot{x} + \omega_n^2 = f(t)$ \hfill (9-79)

(c) $\ddot{y} + 10\dot{y} = f(t)$ \hfill (9-80)

(d) $\dddot{w} + \ddot{w} + \dot{w} + w = f(t)$ \hfill (9-81)

9-2. Calculate the magnitude and phase functions, $M(\omega)$ and $\emptyset(\omega)$, for the following transfer functions: (a) $G(s)=K$,

Frequency Response of Linear Systems

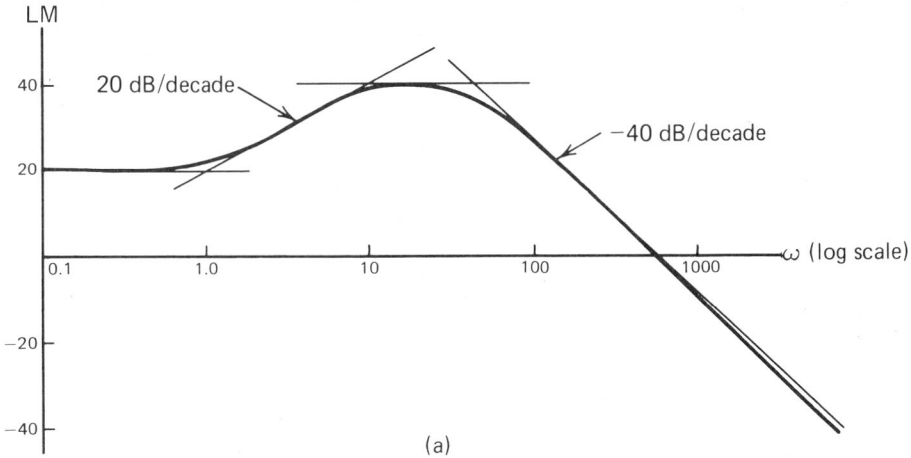

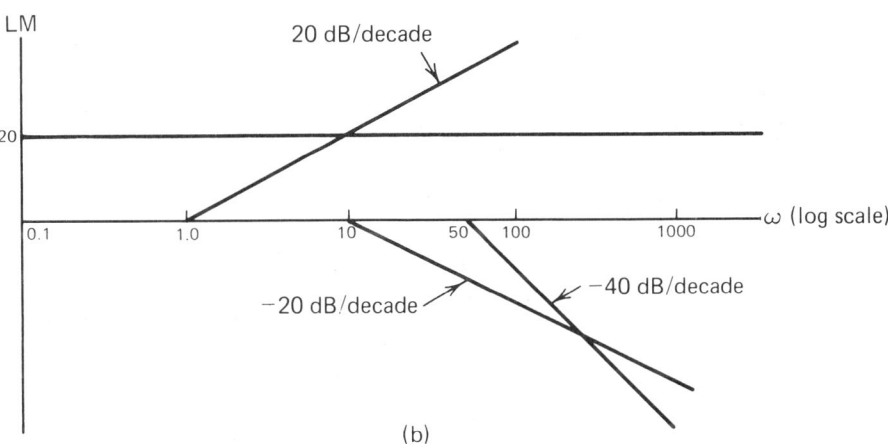

Figure 9-19. Bode magnitude plots for Example 9-13. (a) Given magnitude response with straight-line approximations; (b) Asymptotic approximations for terms which can be summed to give the response of (a).

(b) $G(s)=50s$, (c) $G(s)=K/s$, (d) $G(s)=s+a$, (e) $G(s)=1/(s+a)$, (f) $G(s)=s/(s+5)$, (g) $G(s)=25/\{s(s+5)\}$, (h) $G(s)=(s+1)/(s+2)$, and (i) $G(s)=(s^2-2s+3)/(s^2+2s+3)$.

9-3. Evaluate $M(4)$ and $\emptyset(4)$ if

$$G(s) = \frac{(s+1)(s+2)}{s(s+3)(s^2+25)} \qquad (9\text{-}82)$$

9-4. Calculate the bandwidth of the magnitude characteristic of the following differential equations:

(a) $\dot{x} + 0.001x = f(t)$ (9-83)

(b) $\ddot{y} + 5\dot{y} + 500y = 500f(t)$ (9-84)

(c) $\dddot{w} + w = f(t)$ (9-85)

(d) $\dot{y} + 3y = f(t)$ (9-86)

(e) $\dddot{y} + 5\ddot{y} = f(t)$ (9-87)

9-5. Design a first-order low-pass filter with gain $G(0)=500$ and cutoff frequency at 1000 Hz.

9-6. Design a first-order high-pass filter with $G(\infty)=20$ and cutoff frequency at 500 Hz. (Hint: See Example 9-4.)

9-7. Compute the peak response and cutoff frequencies for the systems with the following transfer functions: (a) $G(s)=25/(s^2+20s+25)$, (b) $G(s)=1/(s^2+0.8s+16)$, and (c) $G(s)=1/(s^2+s+1)$.

9-8. Evaluate $M(\omega_1)$ and $\phi(\omega_1)$ using graphical methods. Assume $K=1$.

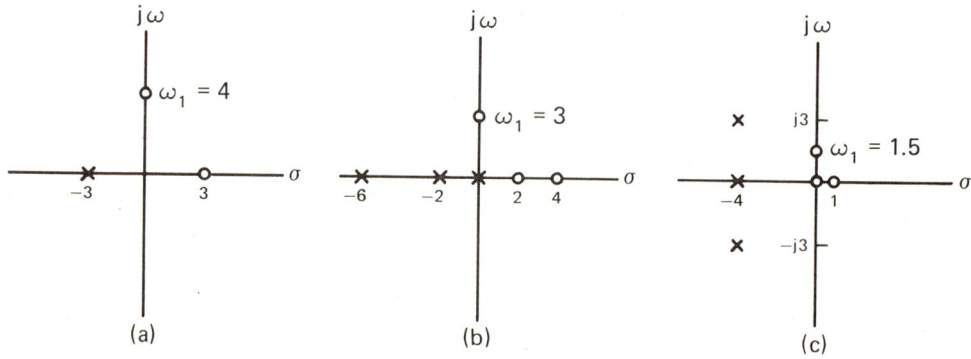

Figure 9-20. Problem 9-8.

9-9. Find the magnitude and phase of the sinusoidal steady state response of the system whose transfer function is $G(s)=25/\{s(s+50)\}$. Assume the input is a sinusoid with a unit amplitude and a frequency (a) one octave below the breakpoint, and (b) one decade below the breakpoint.

Frequency Response of Linear Systems 427

9-10. Calculate the bandwidth of each of the systems described by the following transfer functions: (a) $G(s)=50/(s+50)$, (b) $G(s)=10/\{s(s+4)\}$, (c) $G(s)=K/(s+a)^2$, (d) $G(s)=K/(s+a)^n$, and (e) $G(s)=(s+1)/\{s(s+2)\}$.

9-11. Find the frequency ω_p at which the magnitude response of the transfer function

$$G(s) = \frac{25(s+2)}{s^2 + 0.2s + 25} \qquad (9\text{-}88)$$

assumes its maximum value. Also find the peak response M_p.

9-12. Repeat Problem 9-11 for the transfer function $G(s)=s/(s^2+s+5)$.

9-13. Compute the decibel values for the following constants: (a) 25, (b) 0.001, (c) 857, (d) 0.025, and (e) -5.837.

9-14. It is often convenient to use decibels rather than percentages to express fractions of a given quantity. Compute the decibel equivalents of the following percentages: (a) 10%, (b) 20%, (c) 50%, (d) 75%, (e) 90%, (f) 95%, (g) 99%, and (h) 1.0%.

9-15. Derive the error relationships given in Table 9-1 for the one- and two-octave cases.

9-16. Sketch Bode magnitude and phase plots for the following transfer functions:

(a) $\quad G(s) = \dfrac{10(s/5 + 1)}{(s/10 + 1)(s/100 + 1)} \qquad (9\text{-}89)$

(b) $\quad G(s) = \dfrac{s^2 + 5s + 100}{s(s^2 + s + 10000)} \qquad (9\text{-}90)$

(c) $\quad G(s) = \dfrac{s(s + 1)}{(s + 10)(s + 50)(s + 500)} \qquad (9\text{-}91)$

9-17. Calculate the delay function $T=-d\phi(\omega)/d\omega$ for the systems of Problem 9-2.

9-18. Sketch Bode magnitude plots for the following functions:

(a) $\quad G(s) = \dfrac{100}{s(s + 100)} \qquad (9\text{-}92)$

(b) $\quad G(s) = \dfrac{50s}{(s + 5)(s + 50)} \qquad (9\text{-}93)$

(c) $\quad G(s) = \dfrac{20(s + 5)}{(s + 50)(s + 500)}$ \hfill (9-94)

(d) $\quad G(s) = \dfrac{s + 1}{s^2 + 150s + 1500}$ \hfill (9-95)

REFERENCES FOR CHAPTER 9

(1) Harris, C. M. and C. E. Crede: <u>Shock and Vibration Handbook (Vol. 1)</u>, McGraw-Hill Book Co., Inc., New York, 1961.

(2) Shinners, S. M.: <u>Control System Design</u>, John Wiley and Sons, Inc., New York, 1964.

(3) Ryder, J. D.: <u>Engineering Electronics</u>, McGraw-Hill Book Co., Inc., New York, 1967.

(4) Budak, A.: <u>Passive and Active Network Analysis and Synthesis</u>, Houghton and Mifflin Company, Boston, Massachusetts, 1974.

(5) Friedland, B., Wing, O., and R. Ash: <u>Principles of Linear Networks</u>, McGraw-Hill Book Co., Inc., New York, 1961.

(6) Millman, J. and C. C. Halkias: <u>Integrated Electronics: Analog and Digital Circuits and Systems</u>, McGraw-Hill Book Co., Inc., New York, 1972.

(7) Landee, R. W., Davis, D. C. and A. P. Albrecht: <u>Electronic Designers Handbook</u>, McGraw-Hill Book Co., Inc., New York, 1957.

(8) Cannon, R. H., Jr.: <u>Dynamics of Physical Systems</u>, McGraw-Hill Book Co., Inc., New York, 1967.

(9) Clark, R. N.: <u>Introduction To Automatic Control Systems</u>, John Wiley and Sons, Inc., New York, 1962.

(10) D'Azzo, J. J. and C. H. Houpis: <u>Feedback Control System Analysis and Synthesis</u>, McGraw-Hill Book Co., Inc., New York, 1960.

(11) Thaler, G. J.: <u>Design of Feedback Systems</u>, Dowden, Hutchinson and Ross, Inc., Stroudsburg, Pennsylvania, 1973.

Frequency Response of Linear Systems 429

(12) Dorf, R. C.: <u>Modern Control Systems</u>, Addison-Wesley Publishing Co., Inc., Reading, Massachusetts, 1974.

(13) Angelo, E. J. and A. Papoulis: <u>Pole-Zero Patterns In The Analysis and Design of Low-Order Systems</u>, McGraw-Hill Book Co., Inc., New York, 1964.

(14) Van Valkenburg, M. E.: <u>Modern Network Synthesis</u>, John Wiley and Sons, Inc., New York, 1960.

(15) Kuo, F. F.: <u>Network Analysis and Synthesis</u>, John Wiley and Sons, Inc., New York, 1962.

(16) Bode, H. W.: <u>Network Analysis and Feedback Amplifier Design</u>, D. Van Nostrand Company, Inc., Princeton, New Jersey, 1945.

(17) Chestnut, H. and R. W. Mayer: <u>Servomechanisms and Regulating System Design</u>, John Wiley and Sons, Inc., New York, 1959.

(18) Kuo, B. C.: <u>Automatic Control Systems</u>, Prentice-Hall Inc., Englewood Cliffs, New Jersey, 1962.

(19) DiStefano, J. J. III, Stubberud, A. R., and I. J. Williams: <u>Theory and Problems of Feedback and Control Systems</u>, McGraw-Hill Book Co., Inc., New York, 1967.

(20) Truxal, J. G.: <u>Automatic Feedback Control System Synthesis</u>, McGraw-Hill Book Co., Inc., New York, 1955.

(21) Gille, J. C., Pelegrin, M. J. and P. Decaulne: <u>Feedback Control Systems: Analysis, Synthesis, and Design</u>, McGraw-Hill Book Co., Inc., New York, 1959.

CHAPTER **10**

State Space Analysis

This chapter introduces an alternate method of analyzing dynamic systems from a variety of engineering applications. The method, called <u>state-space analysis</u>, is a body of mathematical tools which systematize many of the techniques given in the preceeding chapters and, in addition, provide the means to treat an important new class of dynamic systems—those with multiple inputs and multiple outputs.[1,2] Moreover, state-space analysis is phrased mathematically in such a way that analysis by digital or analog computer follows naturally from the state model. But possibly the most compelling reason to turn to state-space analysis is that the methods introduced previously, when applied to complicated systems, often result in mathematical models which relate only the system inputs to its outputs, ignoring internal variables which do not appear at the output of the system. The result is that a system may have satisfactory input-output performance but be capable of destroying itself because of some internal instability. State-space models contain all of the information about a system, not just the input-output relationships.

Analysis by state-space methods has increased greatly in popularity in recent years, both because of the increasing use of computers in systems analysis, and because of the increasing complexity of modern engineering systems. State-space methods are widely used in modern control theory.[3,4] A highly theoretical offshoot of classical automatic control technology, modern control theory addresses such topics as controllability, observability and optimization of control systems. Although modern control theory has had limited practical application in the brief time since its inception, when its methods do apply, the payoff is large. We do not delve into modern control theory in this preliminary treatment, but Chapter 10 does provide some of the prerequisites necessary for the student who plans to take up the subject in a later course.

10.1 MATRICES

State-space analysis relies on matrix notation. One must know matrix algebra (addition, subtraction, multiplication, and taking inverses), matrix calculus (how to differentiate, integrate, and Laplace transform matrices whose elements are functions), and functions of a matrix (what is the meaning of sin(A) where A is a matrix, for example). The following paragraphs introduce just enough matrix theory to enable one to follow the derivations in the remainder of the chapter. The more serious student will find entire courses devoted to matrices in the mathematics curricula of most colleges.[5,6,7]

Matrix Algebra

Notation. A matrix is a square or rectangular array of elements, arranged in rows and columns, and enclosed in matrix brackets. The elements may be numbers (either real or complex) or symbols which represent parameters or functions. We differentiate between a matrix, which may have any number of rows or columns, a row vector, which has a single row of elements, and a column vector, which has a single column of elements. Examples of these kinds of matrices are

$$A = \begin{bmatrix} a_{11} & a_{12} \\ a_{21} & a_{22} \end{bmatrix} \quad \underline{b} = \begin{bmatrix} b_1 \\ b_2 \end{bmatrix} \quad (10\text{-}1)$$

$$\underline{c} = \begin{bmatrix} c_1 & c_2 & c_3 \end{bmatrix} \quad (10\text{-}2)$$

A is a 2 by 2 (written in the sequel as (2 x 2)) square matrix, $\underline{b}$ is a (2 x 1) column vector, and $\underline{c}$ is a (1 x 3) row vector. We shall follow the notation that capital letters stand for matrices and small letters with underbars for row or column vectors. The context of the problem will generally indicate whether a row or column vector is intended. Lower case letters with double subscripts designate the elements of a matrix: a_{ij} is the element in the ith row and jth column of matrix A, for example.

The dimensions of a matrix (the numbers of rows or columns) are indicated by writing appropriate subscripts; thus an (m x n) matrix $R = r_{m,n}$ is

$$R = \begin{bmatrix} r_{11} & r_{12} \cdots r_{1n} \\ \vdots & \vdots \\ r_{m,1} & \cdots \cdots r_{m,n} \end{bmatrix} \quad (10\text{-}3)$$

Null, identity, and diagonal matrices. Certain standard matrix forms appear so often that they require special names. A null matrix is a matrix all of whose elements are identically zero. We use the terminology $0_{m,n}$ for an (m x n) null matrix and $\underline{0}_n$ for a null n-element row or column vector. Again, the context will make clear whether a row or column vector is intended. Null matrices correspond to zero in the algebra of real numbers.

The (n x n) <u>identity</u> <u>matrix</u> I_n has 1's on the main diagonal and 0's elsewhere; for example,

$$I_4 = \begin{bmatrix} 1 & 0 & 0 & 0 \\ 0 & 1 & 0 & 0 \\ 0 & 0 & 1 & 0 \\ 0 & 0 & 0 & 1 \end{bmatrix} \qquad (10\text{-}4)$$

The identity matrix corresponds to unity in the algebra of real numbers.

A <u>diagonal</u> matrix is a square matrix which has elements on the main (northwest to southeast) diagonal and zeros elsewhere. The abbreviated notation

$$A = \text{Diag}\begin{bmatrix} a_{11}, a_{22}, \ldots a_{nn} \end{bmatrix} \qquad (10\text{-}5)$$

is used for the diagonal matrix

$$A = \begin{bmatrix} a_{11} & \cdots & 0 \\ \vdots & \ddots & \vdots \\ 0 & \cdots & a_{nn} \end{bmatrix} \qquad (10\text{-}6)$$

<u>Transposition</u>. The <u>transpose</u> A^T of a matrix A is the matrix formed by interchanging the rows and columns of the given matrix. If A is an (n x m) matrix, then A^T has dimensions (m x n). For example, if

$$A = \begin{bmatrix} 1 & 2 & 3 \\ 4 & 5 & 6 \end{bmatrix} \qquad (10\text{-}7)$$

then

$$A^T = \begin{bmatrix} 1 & 4 \\ 2 & 5 \\ 3 & 6 \end{bmatrix} \qquad (10\text{-}8)$$

The transpose of a row vector is a column vector and vise versa. The transpose of a diagonal matrix is itself.

State Space Analysis

Determinants. The determinant of a square matrix A is indicated by the notation Det(A). We assume the reader knows how to take determinants, but as an example, Det(A), where A is given by

$$A = \begin{bmatrix} 1 & 2 & 0 \\ 0 & 1 & 5 \\ 0 & 2 & 4 \end{bmatrix} \qquad (10\text{-}9)$$

is

$$Det(A) = \begin{vmatrix} 1 & 2 & 0 \\ 0 & 1 & 5 \\ 0 & 2 & 4 \end{vmatrix} \qquad (10\text{-}10)$$

and Det(A)=1{(1)(4)-(5)(2)}-2{(0)(4)-(5)(0)} + 0{(0)(2)-(1)(0)}=-6. Note the different brackets used for determinants.

Addition and subtraction. Two matrices or vectors can be added if they have the same numbers of rows and columns. We just add the corresponding elements; thus

$$\begin{bmatrix} 1 & 2 \\ 3 & 4 \end{bmatrix} + \begin{bmatrix} 5 & -6 \\ 7 & 18 \end{bmatrix} = \begin{bmatrix} 6 & -4 \\ 10 & 22 \end{bmatrix} \qquad (10\text{-}11)$$

Subtraction is done by subtracting the corresponding elements.

Multiplication. Two kinds of multiplication are defined: multiplying a matrix by a constant, and multiplying a matrix by another matrix. To multiply a matrix A by a constant c, we multiply each element of A by c. If, for example,

$$A = \begin{bmatrix} 1 & 2 & 3 \\ 4 & 5 & 6 \end{bmatrix} \qquad (10\text{-}12)$$

then the matrix 5A is

$$5A = \begin{bmatrix} 5 & 10 & 15 \\ 20 & 25 & 30 \end{bmatrix} \qquad (10\text{-}13)$$

Two matrices or vectors can be multiplied together if they are conformable; that is, if number of rows of the second matrix equals the number of columns of the first. For example, the matrices $A_{3,2}$ and $B_{2,6}$ are conformable for forming the product C=AB because the number of columns of A (2) equals the number of rows of B. They are not conformable for forming the product of D=BA. Two square matrices with the same dimensions are conformable for multiplication, but even in this case multiplication is not commutative; in general AB≠BA. The product of two matrices is a matrix with the number of rows equal to that of the first matrix and a number of columns equal to that of the second; thus $A_{m,n}$ and $B_{n,k}$ are conformable and produce the matrix

$$C_{m,k} = A_{m,n} B_{n,k} \tag{10-14}$$

Consider the problem of multiplying a row and a column vector, $\underline{a}$ and $\underline{b}$, given by, say,

$$\underline{a} = \begin{bmatrix} 1 & 2 & 3 \end{bmatrix} \qquad \underline{b} = \begin{bmatrix} 4 \\ 5 \\ 6 \end{bmatrix} \tag{10-15}$$
$$(1 \times 3)$$

We can take the product $\underline{a}\,\underline{b}$ since $\underline{a}$ has three columns and $\underline{b}$ three rows. The product is a (1 x 1) matrix $\underline{c}$ where

$$\underline{c} = \underline{a}\,\underline{b} = \begin{bmatrix} 1 & 2 & 3 \end{bmatrix} \begin{bmatrix} 4 \\ 5 \\ 6 \end{bmatrix} \tag{10-16}$$

The product is $\underline{c} = [1(4)+2(5)+3(6)] = [32]$. The product was found by multiplying the first element in $\underline{a}$ by the first element in $\underline{b}$, the second element in $\underline{a}$ by the second element in $\underline{b}$, etc., and then summing the products.

The algorithm for multiplying two general rectangular matrices is to multiply the rows of the first matrix by the columns of the second to obtain the elements of the product

State Space Analysis

matrix. The procedure is better illustrated than explained. Consider the product shown:

$$\begin{bmatrix} 1 & 2 \\ 3 & 4 \\ 5 & 6 \end{bmatrix} \begin{bmatrix} 0 & 4 & 6 & -2 \\ 8 & 1 & 1 & 5 \end{bmatrix} = \begin{bmatrix} 16 & 6 & 8 & 8 \\ 32 & 16 & 22 & 14 \\ 48 & 26 & 36 & 20 \end{bmatrix} \quad (10\text{-}17)$$

(3 x 2)　　　　(2 x 4)　　　　　　　(3 x 4)

The column-one entry whose value is 16 is formed from $(1)(0)+(2)(8)$; the 20 entry in the 3,4 position is formed from $(5)(-2)+(6)(5)$. The multiplication algorithm is stated as follows: if $A_{m,n}$ has elements a_{ij} and $B_{n,k}$ has elements b_{ij}, then the elements c_{ij} of the product matrix $C_{m,k}$, where $C=AB$, are formed by the relationship

$$c_{ij} = \sum_{p=1}^{n} a_{ip} b_{pj} \quad (10\text{-}18)$$

Equation (10-18) is useful for programming matrix multiplication on a digital computer.

Example 10-1. Compute the product, $\underline{c} = \underline{a}\underline{b}$ if

$$\underline{a} = \begin{bmatrix} a_1 \\ a_2 \\ a_3 \end{bmatrix} \quad \underline{b} = \begin{bmatrix} b_1 & b_2 & b_3 \end{bmatrix} \quad (10\text{-}19)$$

Solution. The matrices are conformable since the dimensions are (3 x 1) and (1 x 3). The product is a (3 x 3) matrix:

$$\underline{c} = \begin{bmatrix} a_1 \\ a_2 \\ a_3 \end{bmatrix} \begin{bmatrix} b_1 & b_2 & b_3 \end{bmatrix} \quad (10\text{-}20)$$

$$= \begin{bmatrix} a_1 b_1 & a_1 b_2 & a_1 b_3 \\ a_2 b_1 & a_2 b_2 & a_2 b_3 \\ a_3 b_1 & a_3 b_2 & a_3 b_3 \end{bmatrix} \quad (10\text{-}21)$$

Matrix inversion. Matrix division is not defined, but we do have inversion, where the inverse of a matrix is analogous to its reciprocal. The inverse of an (n x n) square matrix A--and only square matrices have inverses--is a matrix A^{-1} such that the product $AA^{-1}=I_n$ where I_n is the (n x n) identity matrix. This situation is analogous to the result that a number times its reciprocal is 1 in the algebra of real numbers.

If A is a square matrix, its inverse is defined by

$$A^{-1} = \frac{Adj(A)}{Det(A)} \qquad (10-22)$$

where Det(A) is the determinant of A--if Det(A)=0 the matrix is singular and does not have an inverse--and Adj(A) is the adjoint of A, or the transpose of the matrix whose elements are the co-factors of the elements of matrix A. The co-factor of an element is formed by striking out the row and column containing the element and taking the determinant of the remaining matrix. The procedure is illustrated by the following example.

Example 10-2. Find the inverse of the matrix

$$A = \begin{bmatrix} 1 & 3 & 3 \\ 1 & 4 & 3 \\ 1 & 3 & 4 \end{bmatrix} \qquad (10-23)$$

Solution. Det(A)=16+9+9-{12+9+12}=1. To find the co-factor of a_{11}, strike out the first row and the first column,

$$\begin{bmatrix} \cancel{1} & \cancel{3} & \cancel{3} \\ \cancel{1} & 4 & 3 \\ \cancel{1} & 3 & 4 \end{bmatrix} \qquad (10-24)$$

and the remaining determinant is 16-9=7. Proceeding in this manner--and remembering to introduce a minus sign as needed

State Space Analysis

when expanding determinants--the adjoint of A is

$$Adj(A) = \begin{bmatrix} 7 & -1 & -1 \\ -3 & 1 & 0 \\ -3 & 0 & 1 \end{bmatrix}^T = \begin{bmatrix} 7 & -3 & -3 \\ -1 & 1 & 0 \\ -1 & 0 & 1 \end{bmatrix} \qquad (10-25)$$

and

$$A^{-1} = \frac{Adj(A)}{Det(A)} = \frac{1}{(1)} \begin{bmatrix} 7 & -3 & -3 \\ -1 & 1 & 0 \\ -1 & 0 & 1 \end{bmatrix} \qquad (10-26)$$

To verify that we have indeed found the inverse, form the product AA^{-1} to see if it gives the identity matrix:

$$\underbrace{\begin{bmatrix} 1 & 3 & 3 \\ 1 & 4 & 3 \\ 1 & 3 & 4 \end{bmatrix}}_{A} \underbrace{\begin{bmatrix} 7 & -3 & -3 \\ -1 & 1 & 0 \\ -1 & 0 & 1 \end{bmatrix}}_{A^{-1}} = \underbrace{\begin{bmatrix} 1 & 0 & 0 \\ 0 & 1 & 0 \\ 0 & 0 & 1 \end{bmatrix}}_{I_3} \qquad (10-27)$$

It is clear from Example 10-2 that finding the inverse of a matrix is, in general, a laborious task. Fortunately, the procedure is readily implemented on a digital computer. Diagonal matrices are a useful special case; the inverse of a diagonal matrix is a diagonal matrix whose elements are the reciprocals of the elements of the original matrix; if

$$A = \begin{bmatrix} 2 & 0 & 0 \\ 0 & 4 & 0 \\ 0 & 0 & 5 \end{bmatrix} \qquad (10-28)$$

then

$$A^{-1} = \begin{bmatrix} 1/2 & 0 & 0 \\ 0 & 1/4 & 0 \\ 0 & 0 & 1/5 \end{bmatrix} \qquad (10-29)$$

and

$$AA^{-1} = \begin{bmatrix} 1 & 0 & 0 \\ 0 & 1 & 0 \\ 0 & 0 & 1 \end{bmatrix} \qquad (10\text{-}30)$$

as required.

Differentiation And Integration

In state-space analysis one often encounters the problem of differentiating a matrix, some of whose elements are functions. The approach here is straightforward; we differentiate or integrate each element of the matrix. For example,

$$\frac{d}{dt}\begin{bmatrix} t^2 & \sin(t) \\ 5 & t \end{bmatrix} = \begin{bmatrix} 2t & \cos(t) \\ 0 & 1 \end{bmatrix} \qquad (10\text{-}31)$$

while

$$\int_0^t \begin{bmatrix} x & 0 \\ 5 & e^{-x} \end{bmatrix} dx = \begin{bmatrix} x^2/2 & 0 \\ 5x & -e^{-x} \end{bmatrix}_0^t \qquad (10\text{-}32)$$

$$= \begin{bmatrix} t^2/2 & 0 \\ 5t & 1-e^{-t} \end{bmatrix} \qquad (10\text{-}33)$$

The integration rule implies that to take the Laplace transform of a matrix we transform each element.

Functions Of A Real Matrix

State-space analysis requires us to introduce the concept of a function of a matrix. The following paragraphs treat the matrix functions needed in solving the matrix-vector differential equations which arise in state-space analysis.

Powers of a real matrix. A square matrix can be raised to an integer power by repeated multiplication. If, for

State Space Analysis

instance

$$A = \begin{bmatrix} 1 & 2 \\ 3 & 4 \end{bmatrix} \quad (10\text{-}34)$$

then

$$A^2 = (A)(A) = \begin{bmatrix} 1 & 2 \\ 3 & 4 \end{bmatrix}\begin{bmatrix} 1 & 2 \\ 3 & 4 \end{bmatrix} = \begin{bmatrix} 7 & 10 \\ 15 & 22 \end{bmatrix} \quad (10\text{-}35)$$

This procedure applies for any higher power--$A^4=(A)(A)(A)(A)$ and so on.

The matrix exponential. By far the most important matrix function for analyzing state-variable systems is the matrix exponential

$$f(A) = e^{At} \quad (10\text{-}36)$$

where A is a constant square matrix. Having defined powers of matrices in the previous paragraph, we can now interpret Eq. (10-36) in terms of an infinite series. Just as

$$e^{at} = 1 + at + \frac{(at)^2}{2!} + \frac{(at)^3}{3!} + \ldots \quad (10\text{-}37)$$

if a is a real number, we define

$$e^{At} = I_n + At + \frac{A^2 t^2}{2!} + \frac{A^3 t^3}{3!} + \ldots \quad (10\text{-}38)$$

where A is an (n x n) matrix and I_n is the (n x n) identity matrix. Thus exp(At) is itself an (n x n) square matrix. Note too that, as seen from setting t=0 in Eq. (10-28), $\exp(0_{n,n}t)=I_n$.

Example 10-3. Show that

$$\frac{de^{At}}{dt} = Ae^{At} \quad (10\text{-}39)$$

if A is an (n x n) matrix.

Solution. Differentiate Eq. (10-38) to obtain

$$\frac{de^{At}}{dt} = 0_{n,n} + A + A^2 t + \frac{3A^3 t^2}{3!} + \frac{4A^4 t^3}{4!} \qquad (10\text{-}40)$$

$$= A\left\{ I_n + At + \frac{A^2 t^2}{2!} + \frac{A^3 t^3}{3!} + \ldots \right\} \qquad (10\text{-}41)$$

Recognizing the bracketed quantity as exp(At) leads to the desired result.

Example 10-4. At times it is convenient to approximate a matrix exponential by the first few terms of its series expansion. This procedure often gives satisfactory results when t is small. Compute a three-term approximation to exp(At) if

$$A = \begin{bmatrix} 1 & 0 \\ 2 & 1 \end{bmatrix} \qquad (10\text{-}42)$$

Solution. Our approximation is

$$e^{At} \simeq I_2 + At + \frac{A^2 t^2}{2} \qquad (10\text{-}43)$$

Compute A^2, or

$$A^2 = (A)(A) = \begin{bmatrix} 1 & 0 \\ 2 & 1 \end{bmatrix} \begin{bmatrix} 1 & 0 \\ 2 & 1 \end{bmatrix} = \begin{bmatrix} 1 & 0 \\ 4 & 1 \end{bmatrix} \qquad (10\text{-}44)$$

Then

$$e^{At} \simeq \begin{bmatrix} 1 & 0 \\ 0 & 1 \end{bmatrix} + \begin{bmatrix} 1 & 0 \\ 2 & 1 \end{bmatrix} t + \begin{bmatrix} 1 & 0 \\ 4 & 1 \end{bmatrix} \frac{t^2}{2} \qquad (10\text{-}45)$$

and

$$e^{AT} \simeq \begin{bmatrix} 1+t+t^2/2 & 0 \\ 0+2t+4t^2/2 & 1+t+t^2/2 \end{bmatrix} \qquad (10\text{-}46)$$

State Space Analysis

Other matrix functions can be defined by the infinite series method. The sine of a square matrix is

$$\sin(A) = A - \frac{A^3}{3!} + \frac{A^5}{5!} - \frac{A^7}{7!} + \ldots \qquad (10\text{-}47)$$

10.2 STATE VARIABLES

State variables are the basis for state-space analysis.[8,9] If we define a <u>system</u> as a collection of physical components assembled to perform some specified task, or as a <u>process</u> which can be described mathematically, and a <u>dynamic system</u> as a system whose inputs and outputs can be related by sets of differential or difference equations, then we can define the state of a system as the minimum amount of present information about its past history needed to predict its future behavior. The variables needed to describe the state of a system are called the <u>state variables</u>. It should be clear that state variables are dependent variables and that they depend on the independent variable time. Further, state variables are not unique: if one has a satisfactory set of state variables, then he can derive an infinite number of other sets of state variables which convey the same information.

State variables are intimately related to the energy storage elements in the system, a phenomena which often helps us in establishing the minimum number of state variables needed to describe a system. In an electrical network energy is stored in the inductors and capacitors, and the inductor currents and the capacitor voltages are often chosen as the state variables. Life is not so simple, however, as we cannot simply add up the total number of inductors and capacitors in the network to get the required number of state variables. Two inductors in series can be combined into a single equivalent inductor to reduce the complexity of the network. Except in simple cases it is impossible to determine the required number of state variables needed to analyze a dynamic system by inspecting its block or circuit diagram. Rigorous proofs relating the number of state variables to the system diagram are not available, but ways do exist to "squeeze" the order of the mathematical model to a minimum after a nonminimum state model has been developed. Our approach is to establish some simple guidelines for selecting state variables--nontrivial inductor currents and capacitor voltages in electrical networks, for example--and leave the problem of minimization to a more advanced treatment. The remainder of the chapter assumes either that the minimum number of state variables has been used, or that whatever redundancy there is, is unimportant.

The Standard System

A broad class of dynamic systems can be described by a set of input variables, a set of output variables, and a set of internal, or state variables (Fig. 10-1).

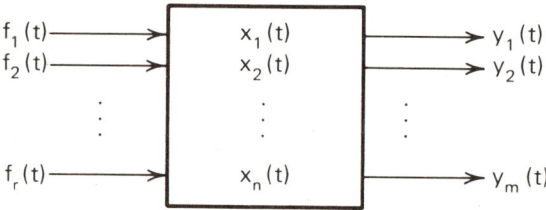

Figure 10-1. This block diagram represents a multi-input multi-output system with r inputs, m outputs, and n state variables.

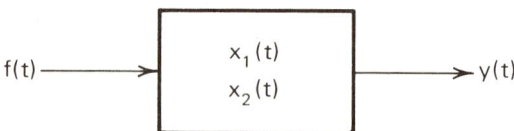

Figure 10-2. This second-order system has input f(t), output y(t) and two state variables, $x_1(t)$ and $x_2(t)$.

State Space Analysis

If the derivatives of the state variables depend only on present state of the system, and not on its past history, we can describe the behavior of the system by ordinary differential equations. For example, a second-order system with one input and one output, such as the one in Fig. 10-2, is described by the equations

$$\dot{x}_1(t) = a_{11}x_1(t) + a_{12}x_2(t) + b_1 f(t)$$

$$\dot{x}_2(t) = a_{21}x_1(t) + a_{22}x_2(t) + b_2 f(t) \quad (10\text{-}48)$$

$$y(t) = c_1 x_1(t) + c_2 x_2(t) + f(t)d$$

with the initial conditions $x_1(0)$ and $x_2(0)$. The quantities a_{11}, a_{12}, a_{21}, a_{22}, b_1, b_2, c_1, c_2, and d in Eqns. (10-48) are parameters of the system. The second-order system is described by two first-order differential equations, and the state variables are the dependent variables. The equations are <u>coupled</u> because $x_2(t)$ appears in the $x_1(t)$ equation and vise versa. An additional equation is required to relate the state variables to the output variable $y(t)$. Finally, the derivatives $\dot{x}_1(t)$ and $\dot{x}_2(t)$ depend only on the present values of $x_1(t)$, $x_2(t)$, and $f(t)$, not on their past histories.

State-variable systems are classified as linear and constant-coefficient, linear and time-varying, nonlinear, or nonlinear and time-varying, depending on whether the coefficients a_{ij} are constants, functions of time, functions of the state variables, or functions of both time and the state variables. For simplicity, we deal only with linear constant-coefficient systems in this book.

Matrix-Vector Differential Equations

The next step is to write Eqns. (10-48), and their more general version for systems with several inputs and outputs, in terms of constant matrices and vector functions.[10] This simplifies the notation, reduces the labor of manipulating repetitious equations, and lets us use matrix methods for analysis. Equations (10-48) can be written

$$\begin{bmatrix} \dot{x}_1(t) \\ \dot{x}_2(t) \end{bmatrix} = \begin{bmatrix} a_{11} & a_{12} \\ a_{21} & a_{22} \end{bmatrix} \begin{bmatrix} x_1(t) \\ x_2(t) \end{bmatrix} + \begin{bmatrix} b_1 \\ b_2 \end{bmatrix} f(t) \tag{10-49}$$

$$y(t) = \begin{bmatrix} c_1 & c_2 \end{bmatrix} \begin{bmatrix} x_1(t) \\ x_2(t) \end{bmatrix} + df(t) \tag{10-50}$$

as can be verified by performing the indicated matrix multiplications. If we introduce the definitions

$$A = \begin{bmatrix} a_{11} & a_{12} \\ a_{21} & a_{22} \end{bmatrix} \qquad \underline{b} = \begin{bmatrix} b_1 \\ b_2 \end{bmatrix} \tag{10-51}$$

$$\underline{c} = \begin{bmatrix} c_1 & c_2 \end{bmatrix}$$

$$\underline{x}(t) = \begin{bmatrix} x_1(t) \\ x_2(t) \end{bmatrix} \qquad \underline{x}(0) = \begin{bmatrix} x_1(0) \\ x_2(0) \end{bmatrix} \tag{10-52}$$

then Eqns. (10-49) and (10-50) can be expressed in the compact form

$$\underline{\dot{x}}(t) = A\underline{x}(t) + \underline{b}f(t)$$
$$y(t) = \underline{c}\,\underline{x}(t) + df(t) \tag{10-53}$$

Equations (10-53) are the most general form for the second-order, single-input-single-output state-variable system. For a system with r inputs, m outputs, and n state variables the formulation is

$$\underline{\dot{x}}(t) = A\underline{x}(t) + B\underline{f}(t)$$
$$\underline{y}(t) = C\underline{x}(t) + D\underline{f}(t) \tag{10-54}$$

where the dimensions of A, B, C, and D are (n x n), (n x r), (m x n), and (m x r) respectively. The vectors $\underline{y}(t)$, $\underline{x}(t)$ and

State Space Analysis

$\underline{f}(t)$, given by

$$\underline{x}(t) = \begin{bmatrix} x_1(t) \\ x_2(t) \\ \vdots \\ x_n(t) \end{bmatrix} \quad \underline{f}(t) = \begin{bmatrix} f_1(t) \\ f_2(t) \\ \vdots \\ f_r(t) \end{bmatrix} \quad \underline{y}(t) = \begin{bmatrix} y_1(t) \\ y_2(t) \\ \vdots \\ y_m(t) \end{bmatrix} \quad (10\text{-}55)$$

are called respectively the state vector, the input vector and the output vector.

Equations (10-54) are the starting point for state-space analysis, either by hand calculation or by computer methods. The direct contribution of $\underline{f}(t)$ to $\underline{y}(t)$ comes by way of the D matrix; this corresponds to a "resistance path" from input to output and does not contribute to the system dynamics. In the following material we often assume $D=0_{m,r}$ with no loss of generality.

Block Diagram Representation[11]

The general state-variable system of Eqns. (10-54) can be represented by the block diagram of Fig. 10-3.

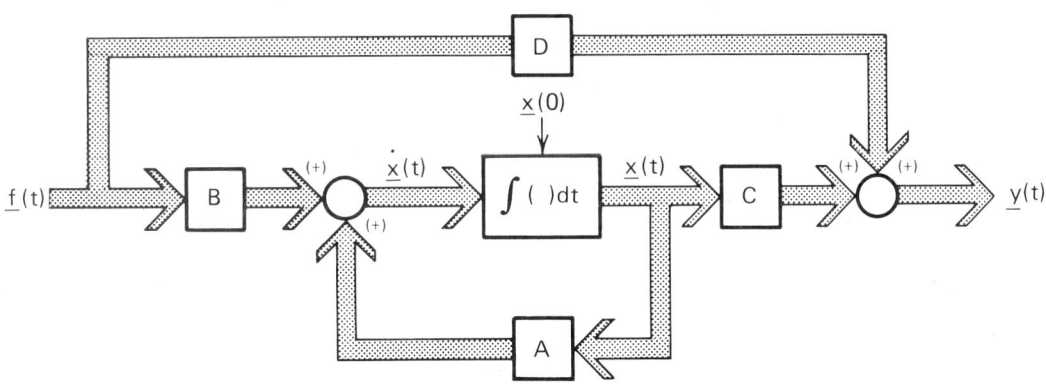

Figure 10-3. The state-variable system of Eqns. (10-54) can be represented by this block diagram. The broad arrows represent flow of vector quantities.

In the figure, broad lines represent vector flow. The integrator block indicates that $\underline{\dot{x}}(t)$ must be integrated to obtain

$\underline{x}(t)$, with the initial conditions vector $\underline{x}(0)$ appropriately applied. The figure also shows how the D matrix provides a bypass path from input to output.

The State-Space Interpretation

The n-dimensional state vector $\underline{x}(t)$ can be interpreted as a point in an n-dimensional state-space. As time passes and $\underline{x}(t)$ changes, the point corresponding to the state vector moves along a trajectory in state space. This geometric interpretation leads to some interesting proofs and theorems in modern control theory, but we will not need any of these.

Conversion To State-Variable Form

A more pertinent question is how to transform higher-order differential equations, such as we have studied in previous chapters, into state-variable form. The procedure is to identify $x_1(t)$ with the dependent variable $y(t)$, $x_2(t)$ with $\dot{y}(t)$ and so on up to $x_n(t) = y(t)^{(n-1)}$ for an nth-order differential equation. The rest is easy, as we demonstrate in the following examples.

Example 10-5. Express the following third-order differential equation in state-variable form:

$$\dddot{y} + 8\ddot{y} + 25\dot{y} + 10y = f(t) \qquad (10\text{-}56)$$

Assume zero initial conditions.

Solution. Define three state variables $x_1 = y$, $x_2 = \dot{y}$ and $x_3 = \ddot{y}$. From these we see that $\dot{x}_1 = x_2$, $\dot{x}_2 = x_3$, and, using the given differential equation along with $\dot{x}_3 = \dddot{y}$,

$$\dot{x}_3 = f(t) - 8x_3 - 25x_2 - 10x_1 \qquad (10\text{-}57)$$

When arranged in matrix form these relationships are

$$\begin{bmatrix} \dot{x}_1 \\ \dot{x}_2 \\ \dot{x}_3 \end{bmatrix} = \begin{bmatrix} 0 & 1 & 0 \\ 0 & 0 & 1 \\ -10 & -25 & -8 \end{bmatrix} \begin{bmatrix} x_1 \\ x_2 \\ x_3 \end{bmatrix} + \begin{bmatrix} 0 \\ 0 \\ 1 \end{bmatrix} f(t) \qquad (10\text{-}58)$$

State Space Analysis

The output equation is $y=x_1$, or

$$y = \begin{bmatrix} 1 & 0 & 0 \end{bmatrix} \begin{bmatrix} x_1 \\ x_2 \\ x_3 \end{bmatrix} \qquad (10\text{-}59)$$

This approach can be used to convert any constant-coefficient differential equation to matrix form.

State Variables For Electrical Networks

Algorithmic procedures are available for writing state-variable equations which describe electrical networks.[12] These can be implemented on digital computers in such a way that the programmer need only specify the element values and the nodes between which they are connected. For simple networks, however, we can use ordinary network analysis. The latter approach is taken in this book because our goals are to understand the principles which underlie state-space analysis rather than to develop polished problem-solving techniques.

One approach to deriving state-variable equations for an electrical network is to use the loop or nodal analysis methods of Chapter 3 and then convert the resulting differential equations to state-variable form using the technique of Example 10-5. This plausible approach works, but it often yields more state-variable equations than there are energy storage elements in the network, a situation which not only increases the labor of solving the equations, but also can lead to unstable simulations if we attempt to use computer assistance. A better approach is to establish the inductor currents and capacitor voltages as state variables and try to derive the state-variable equations directly from the KCL, KVL, and device relationships.[13,14]

The procedure is illustrated by a running example rather than a formal derivation. Consider the network of Fig. 10-4. The switch closes at t=0. The problem is to write state-variable equations which can be solved for the transient output voltage $v_4(t)$. Assume zero initial conditions for the energy storage elements.

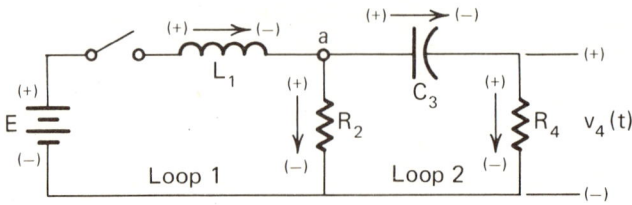

Figure 10-4. A network for state-variable analysis. Inductor currents and capacitor voltages make good choices for the state variables in an electrical network.

<u>The connection equations</u>. The connection equations are Kirchhoff's voltage and current laws written for the loops and nodes of the network in terms of the element variables. For the network of Fig. 10-4 these are

$$\text{Loop 1:} \quad E = v_1 + v_2 \tag{10-60}$$

$$\text{Loop 2:} \quad v_3 + v_4 - v_2 = 0 \tag{10-61}$$

$$\text{Node a:} \quad i_1 = i_2 + i_3 \tag{10-62}$$

$$\text{Node b:} \quad i_3 = i_4 \tag{10-63}$$

Theorems are available to determine exactly how many connections equations are needed and how to select them, but we leave these considerations to a more advanced treatment. For planar networks, we use the meshes for loops and use every node except the ground or reference node.

<u>The device equations</u>. The device equations are nothing more than the v-i relationships for the components in the network. In this instance

$$v_1 = L_1 \frac{di_1}{dt} \tag{10-64}$$

$$v_2 = R_2 i_2 \tag{10-65}$$

$$i_3 = C_3 \frac{dv_3}{dt} \tag{10-66}$$

State Space Analysis

and

$$v_4 = R_4 i_4 \tag{10-67}$$

<u>The state equations</u>. The connection and device equations are a complete description of the network. When judiciously combined they lead either to loop analysis, nodal analysis, or state-variable analysis. Identifying the inductor current i_1 and the capacitor voltage v_3 as the state variables, we start with Eq. (10-64), then substitute Eq. (10-60) to obtain

$$L_1 \frac{di_1}{dt} = E - v_2 \tag{10-68}$$

But from Eq. (10-61), $v_2 = v_3 + v_4$, so that Eq. (10-68) becomes

$$L_1 \frac{di_1}{dt} = E - v_3 - v_4 \tag{10-69}$$

Next substitute Eq. (10-65) into Eq. (10-62) to obtain

$$i_1 = \frac{v_2}{R_2} + i_3 \tag{10-70}$$

but, according to Eq. (10-63), $i_3 = i_4$; therefore, using Eq. (10-67), Eq. (10-70) becomes

$$i_1 = \frac{v_2}{R_2} + \frac{v_4}{R_4} \tag{10-71}$$

Use Eq. (10-61) to eliminate v_2 with the result

$$i_1 = \frac{v_3 + v_4}{R_2} + \frac{v_4}{R_4} \tag{10-72}$$

Solving this expression for v_4 gives

$$v_4 = i_1 \left\{ \frac{R_2 R_4}{R_2 + R_4} \right\} - v_3 \left\{ \frac{R_4}{R_2 + R_4} \right\} \tag{10-73}$$

Since v_4 is the required output and i_1 and v_3 are the state variables, Eq. (10-73) is the output equation for the state variable formulation; it has the form

$$y = \underline{c} \, \underline{x} + df$$

where $y = v_4$,

$$\underline{x} = \begin{bmatrix} i_1 \\ v_3 \end{bmatrix} \qquad (10\text{-}74)$$

and

$$\underline{c} = \begin{bmatrix} \dfrac{R_2 R_4}{R_2 + R_4} & \dfrac{-R_4}{R_2 + R_4} \end{bmatrix} \qquad (10\text{-}75)$$

To obtain the first state differential equation, substitute Eq. (10-73) into Eq. (10-69) to get

$$L_1 \frac{di_1}{dt} = E - v_3 - i_1 \left\{ \frac{R_2 R_4}{R_2 + R_4} \right\} + v_3 \left\{ \frac{R_4}{R_2 + R_4} \right\} \qquad (10\text{-}76)$$

For the other state differential equation substitute Eq. (10-63) into Eq. (10-66) to obtain

$$C_3 \frac{dv_3}{dt} = i_4 = \frac{v_4}{R_4} \qquad (10\text{-}77)$$

Now substitute Eq. (10-73) for v_4:

$$C_3 \frac{dv_3}{dt} = i_1 \left\{ \frac{R_2}{R_2 + R_4} \right\} - v_3 \left\{ \frac{1}{R_2 + R_4} \right\} \qquad (10\text{-}78)$$

Equations (10-76) and (10-78) can be put in the matrix form

$$\begin{bmatrix} \dfrac{di_1}{dt} \\ \dfrac{dv_3}{dt} \end{bmatrix} = \begin{bmatrix} \dfrac{-R_2 R_4}{L_1(R_2 + R_4)} & \dfrac{-R_2}{L_1(R_1 + R_2)} \\ \dfrac{R_2}{C_3(R_2 + R_4)} & \dfrac{-1}{C_3(R_2 + R_4)} \end{bmatrix} \begin{bmatrix} i_1 \\ v_3 \end{bmatrix} + \begin{bmatrix} \dfrac{1}{L_1} \\ 0 \end{bmatrix} Eu(t) \qquad (10\text{-}79)$$

This expression has the form

$$\underline{\dot{x}} = A\underline{x} + \underline{b}f(t) \qquad (10\text{-}80)$$

where $f(t) = Eu(t)$,

$$\underline{x} = \begin{bmatrix} i_1 \\ v_3 \end{bmatrix} \qquad \underline{b} = \begin{bmatrix} \dfrac{1}{L_1} \\ 0 \end{bmatrix} \qquad (10\text{-}81)$$

and

$$A = \begin{bmatrix} -R_2R_4/L_1 & -R_2/L_1 \\ R_2/C_3 & -1/C_3 \end{bmatrix} \frac{1}{R_2+R_4} \qquad (10\text{-}82)$$

This completes the process of formulating the state equations. How to derive the state equations from the connection equations and the device equations is not obvious for complicated networks. Fortunately, algorithmic methods are available. A number of digital computer programs for doing this have been written and are on file at many university and industrial computer centers.

State variable methods apply equally well to mechanical systems, but the choice of state variables is generally not so critical as it is for electrical networks. Spring elongations and damper velocities are good choices for mechanical state variables.

10.3 EQUIVALENCE TRANSFORMATIONS

In general, once we have a set of state variables, we can obtain an infinite number of other sets of state variables by a procedure called an equivalence transformation.[15,16]

Equivalent Systems

If an nth-order system is described by the state vector $\underline{x}(t)$, then it is also described by the state vector $\underline{q}(t)$ provided $\underline{q}$ and $\underline{x}$ are related by $\underline{x}=T\underline{q}$ where T is any nonsingular square matrix of dimension (n x n). The matrix T is called an equivalence transformation. In terms of the geometric interpretation of the state of a system as a point in an n-dimensional state space, an equivalence transformation corresponds to a rotation of coordinate axes. As there are an infinite number of ways the coordinate axes can be rotated and still provide an adequate framework in which to describe the state of the system, it follows that there are an infinite number of equivalence transformations. To transform the system

$$\dot{\underline{x}} = A\underline{x} + B\underline{f}$$
$$\underline{y} = C\underline{x} + D\underline{f} \qquad (10\text{-}83)$$

we use $\underline{x}=T\underline{q}$ and $\dot{\underline{x}}=T\dot{\underline{q}}$. Substituting these relationships into Eqns. (10-83) yields

$$T\dot{\underline{q}} = AT\underline{q} + B\underline{f} \qquad (10\text{-}84)$$
$$\underline{y} = CT\underline{q} + D\underline{f} \qquad (10\text{-}85)$$

and, after multiplying Eq. (10-84) by T^{-1},

$$\dot{\underline{q}} = T^{-1}AT\underline{q} + T^{-1}B\underline{f} \qquad (10\text{-}86)$$

The final form of the transformed system is

$$\dot{\underline{q}} = \bar{A}\underline{q} + \bar{B}\underline{f}$$
$$\underline{y} = \bar{C}\underline{q} + \bar{D}\underline{f} \qquad (10\text{-}87)$$

where $\bar{A}=T^{-1}AT$, $\bar{B}=T^{-1}B$, $\bar{C}=CT$, and $\bar{D}=D$. Equivalence transforms have many uses, one of which is to decouple coupled systems. The system

$$\begin{bmatrix} \dot{x}_1 \\ \dot{x}_2 \end{bmatrix} = \begin{bmatrix} -1 & 1 \\ 2 & -3 \end{bmatrix} \begin{bmatrix} x_1 \\ x_2 \end{bmatrix} + \begin{bmatrix} 0 \\ 1 \end{bmatrix} f(t) \qquad (10\text{-}88)$$

is coupled because the x_1 equation depends on x_2. Coupled equations are very difficult to solve compared to uncoupled equations such as

$$\begin{bmatrix} \dot{q}_1 \\ \dot{q}_2 \end{bmatrix} = \begin{bmatrix} -5 & 0 \\ 0 & -8 \end{bmatrix} \begin{bmatrix} q_1 \\ q_2 \end{bmatrix} + \begin{bmatrix} 0 \\ 1 \end{bmatrix} f(t) \qquad (10\text{-}89)$$

In this system, the q_1 equation, $\dot{q}_1=-5q_1$, does not contain q_2 and can be solved independently. Simiarly, q_2 can be found without recourse to the q_1 equation. An equivalence transformation which diagonalizes the A matrix of a state-

State Space Analysis

variable system, and thus decouples the state variables, is a powerful analysis tool.

Example 10-6. Verify that the equivalence transformation

$$T = \begin{bmatrix} 4 & 2 \\ 2 & -2 \end{bmatrix} \tag{10-90}$$

decouples the system whose state-variable A matrix is

$$A = \begin{bmatrix} 3 & 4 \\ 2 & 1 \end{bmatrix} \tag{10-91}$$

Solution. The system is decoupled if $\bar{A} = T^{-1}AT$ is a diagonal matrix. First, we need T^{-1} given by

$$T^{-1} = \frac{-1}{12}\begin{bmatrix} -2 & -2 \\ -2 & 4 \end{bmatrix} = \frac{1}{6}\begin{bmatrix} 1 & 1 \\ 1 & -2 \end{bmatrix} \tag{10-92}$$

Using T^{-1} we have

$$\bar{A} = T^{-1}AT = \frac{1}{6}\begin{bmatrix} 1 & 1 \\ 1 & -2 \end{bmatrix}\begin{bmatrix} 3 & 4 \\ 2 & 1 \end{bmatrix}\begin{bmatrix} 4 & 2 \\ 2 & -2 \end{bmatrix} \tag{10-93}$$

$$= \frac{1}{6}\begin{bmatrix} 5 & 5 \\ 1 & 2 \end{bmatrix}\begin{bmatrix} 4 & 2 \\ 2 & -2 \end{bmatrix} \tag{10-94}$$

and

$$\bar{A} = \begin{bmatrix} 5 & 0 \\ 0 & 1 \end{bmatrix} \tag{10-95}$$

Finding The Decoupling Transformation

It is evident that state-variable systems with diagonal A matrices are much more tractable and amenable to analysis by pencil-and-paper methods than the general case. Unfortunately, not all systems can be diagonalized, and even for those which can, the labor involved is considerable. The polynomial $h(\lambda)$ defined by

$$h(\lambda) = \text{Det}(A-\lambda I_n) \tag{10-96}$$

is called the characteristic polynomial of the (n x n) matrix A. The values of λ for which $h(\lambda)=0$, or $\text{Det}(A-\lambda I_n)=0$ are called the <u>eigenvalues</u> of the matrix A. An nth-order system has n such eigenvalues which, as they are the roots of a polynomial, may be real, real and repeated, imaginary, complex, etc. If the eigenvalues are real and nonrepeated, we can always find an equivalence transformation which diagonalizes the system.

Corresponding to the eigenvalues λ_i are a set of <u>eigenvectors</u> $\underline{v}_i$ which satisfy

$$A\underline{v}_i = \lambda_i \underline{v}_i \tag{10-97}$$

for $i=1,2,\ldots n$. Rearranging Eq. (10-97) yields

$$(A-\lambda_i I_n)\underline{v}_i = \underline{0} \tag{10-98}$$

Equations (10-98), one for each eigenvalue, can be used to solve for the elements of the eigenvectors. Any matrix T whose columns are the eigenvectors diagonalizes the A matrix; thus $T=[\underline{v}_1 \underline{v}_2, \ldots \underline{v}_n]$ is a diagonalizing transformation.

Example 10-7. Derive the equivalence transformation given in Example 10-6.

Solution.

$$A = \begin{bmatrix} 3 & 4 \\ 2 & 1 \end{bmatrix} \tag{10-99}$$

$$\text{Det}(A-\lambda I_2) = (\lambda-5)(\lambda+1) \tag{10-100}$$

The eigenvalues are $\lambda_1=5$ and $\lambda_2=-1$. For the first eigenvalue, Eq. (10-98) is $(A-5I_2)\underline{v}_1=\underline{0}$, or

$$\begin{bmatrix} -2 & 4 \\ 2 & -4 \end{bmatrix} \begin{bmatrix} v_{11} \\ v_{12} \end{bmatrix} = \begin{bmatrix} 0 \\ 0 \end{bmatrix} \tag{10-101}$$

from which $v_{11}=2$ and $v_{12}=1$. For the second eigenvalue

$$\begin{bmatrix} 4 & 4 \\ 2 & 2 \end{bmatrix} \begin{bmatrix} v_{11} \\ v_{22} \end{bmatrix} = \begin{bmatrix} 0 \\ 0 \end{bmatrix} \qquad (10\text{-}102)$$

and $v_{21}=1$ with $v_{22}=-1$. The transformation matrix is then

$$T = [\underline{v}_1 \quad \underline{v}_2] = \begin{bmatrix} 2 & 1 \\ 1 & -1 \end{bmatrix} \qquad (10\text{-}103)$$

or any constant multiple thereof. In this instance

$$2T = \begin{bmatrix} 4 & 2 \\ 2 & -2 \end{bmatrix} \qquad (10\text{-}104)$$

which was the transformation given in Example 10-6.

10.4 SOLVING STATE EQUATIONS

If one is interested in obtaining a solution to a set of state-variable equations by the most direct approach then he should turn to the Laplace transform method.[17]

The Laplace Transform Method

When the state differential equation

$$\underline{\dot{x}} = A\underline{x} + B\underline{f} \qquad (10\text{-}105)$$

is Laplace transformed, the result is

$$s\underline{X}(s) - \underline{x}(0) = A\underline{X}(s) + B\underline{F}(s) \qquad (10\text{-}106)$$

where upper-case letters with underbars are the element-by-element Laplace transforms of the corresponding vectors; $\underline{X}(s)=L\{\underline{x}(t)\}$, for example. Observe how the initial conditions vector $\underline{x}(0)$ arises in Eq. (10-106) when we transform $\underline{x}(t)$.

Solving Eq. (10-106) for $\underline{X}(s)$ yields

$$(sI_n - A)\underline{X}(s) = B\underline{F}(s) + \underline{x}(0) \qquad (10\text{-}107)$$

where the (n x n) identity matrix I_n was introduced so that the bracketed quantity is the difference of two square matrices. This procedure is permissible as, for example, it is clear that

$$\begin{bmatrix} 1 & 0 & 0 \\ 0 & 1 & 0 \\ 0 & 0 & 1 \end{bmatrix} \begin{bmatrix} a \\ b \\ c \end{bmatrix} \equiv \begin{bmatrix} a \\ b \\ c \end{bmatrix} \qquad (10\text{-}108)$$

Equation (10-107) is premultiplied by $(sI_n-A)^{-1}$ to obtain

$$\underline{X}(s) = (sI_n-A)^{-1}\{B\underline{F}(s)+\underline{x}(0)\} \qquad (10\text{-}109)$$

The solution for $\underline{x}(t)$ is then

$$\underline{x}(t) = L^{-1}\left\{(sI_n-A)^{-1}B\underline{F}(s)+(sI_n-A)^{-1}\underline{x}(0)\right\} \qquad (10\text{-}110)$$

The matrix

$$\emptyset(t) = L^{-1}\left\{\{sI_n-A\}^{-1}\right\} \qquad (10\text{-}111)$$

is called the <u>state-transition matrix</u> of the system, and $\text{Det}(sI_n-A)$ is the <u>characteristic equation</u> of the system. Once $\underline{x}(t)$ is known, the output $\underline{y}(t)$ is found in the time domain using the output equation $\underline{y}(t)=C\underline{x}(t)+D\underline{f}(t)$.

The Laplace transform method of solution is straightforward and compact; unfortunately, it involves considerable labor for higher-order systems, and one is wise to seek computer assistance for all but the most trivial systems.

<u>Example 10-8</u>. Solve for the state vector $\underline{x}(t)$ for the system $\underline{\dot{x}}=A\underline{x}$ if

$$A = \begin{bmatrix} -1 & 0 \\ 0 & -2 \end{bmatrix} \quad \text{and} \quad \underline{x}(0) = \begin{bmatrix} 3 \\ 5 \end{bmatrix} \qquad (10\text{-}112)$$

<u>Solution</u>. Use the Laplace transform method:

$$(sI_2-A)^{-1} = \begin{bmatrix} \frac{1}{s+1} & 0 \\ 0 & \frac{1}{s+2} \end{bmatrix} \qquad (10\text{-}113)$$

State Space Analysis

Equation (10-109) is now used to get

$$\underline{X}(s) = \begin{bmatrix} \frac{1}{s+1} & 0 \\ 0 & \frac{1}{s+2} \end{bmatrix} \begin{bmatrix} 3 \\ 5 \end{bmatrix} = \begin{bmatrix} \frac{3}{s+1} \\ \frac{5}{s+2} \end{bmatrix} \qquad (10\text{-}114)$$

and, taking inverse transforms,

$$\underline{x}(t) = \begin{bmatrix} 3e^{-t} \\ 5e^{-2t} \end{bmatrix} \qquad (10\text{-}115)$$

Stability. The characteristic equation, $\text{Det}(sI_n - A)$, of a state-variable system is useful in determining whether or not the system is stable.

A state-variable system is stable if the roots of the characteristic equation all lie in the left half of the complex s-plane (or on the $j\omega$-axis for conditional stability).

Example 10-9. Investigate the stability of the state-variable system with A-matrix

$$A = \begin{bmatrix} -4 & 3 \\ 2 & -3 \end{bmatrix} \qquad (10\text{-}116)$$

Solution. First evaluate $sI_2 - A$,

$$sI_2 - A = \begin{bmatrix} s+4 & -3 \\ -2 & s+3 \end{bmatrix} \qquad (10\text{-}117)$$

then find the characteristic equation, $\text{Det}(sI_2 - A) = (s+6)(s+1)$. As both roots are in the left half of the s-plane, the system is stable.

The transfer-function matrix. Just as we were able to find s-domain transfer functions which related input to output for scalar systems (a scalar system is one which is not described by state variables), or $Y(s) = G(s)F(s)$, we can find transfer-function matrices which relate the transformed input vector to the transformed output vector.(18) The relationship is

$$\underline{Y}(s) = G(s)\underline{F}(s) \qquad (10\text{-}118)$$

From Eq. (10-109), if the initial conditions are zero--and transfer functions are always defined for zero initial conditions--we have $\underline{X}(s)=(sI_n-A)^{-1}B\underline{F}(s)$. The transformed output equation for a state-variable system is

$$\underline{Y}(s) = C\underline{X}(s) + D\underline{F}(s) \qquad (10\text{-}119)$$

Substituting $\underline{X}(s)$ into Eq. (10-119) results in

$$\underline{Y}(s) = C(sI_n-A)^{-1}B\underline{F}(s) + D\underline{F}(s) \qquad (10\text{-}120)$$

or

$$\underline{Y}(s) = \{C(sI_n-A)^{-1}B + D\}\underline{F}(s) \qquad (10\text{-}121)$$

Comparing this expression with Eq. (10-118) reveals that the transfer function matrix is

$$G(s) = C(sI_n-A)^{-1}B + D \qquad (10\text{-}122)$$

If the system has r inputs and m outputs then G(s) has dimensions (m x r).

Example 10-10. Find the transfer function matrix for the system which has

$$A = \begin{bmatrix} -1 & 0 \\ 0 & -3 \end{bmatrix} \qquad B = \begin{bmatrix} 0 \\ 1 \end{bmatrix} \qquad (10\text{-}123)$$

$$C = \begin{bmatrix} 1 & 0 \\ 0 & 4 \end{bmatrix} \qquad D = 0$$

Solution. First compute $(sI_2-A)^{-1}$,

$$(sI_2-A)^{-1} = \begin{bmatrix} \frac{1}{s+1} & 0 \\ 0 & \frac{1}{s+3} \end{bmatrix} \qquad (10\text{-}124)$$

State Space Analysis

Now apply Eq. (10-122) to get

$$G(s) = \begin{bmatrix} 1 & 0 \\ 0 & 4 \end{bmatrix} \begin{bmatrix} \frac{1}{s+1} & 0 \\ 0 & \frac{1}{s+3} \end{bmatrix} \begin{bmatrix} 0 \\ 1 \end{bmatrix} = \begin{bmatrix} 0 \\ \frac{4}{s+3} \end{bmatrix} \qquad (10\text{-}125)$$

Solving State Equations In The Time-Domain [19]

We have seen how to solve state-variable equations using Laplace transforms. The approach is straightforward and general, but it does not indicate clearly the time-domain relationships involved. The time-domain is bypassed completely. As we shall see in a later chapter, expressions for the time-domain solutions of state-variable systems are useful when one attempts to use digital computer methods for system analysis.

The force-free solution. Consider the homogeneous state-variable system obtained by setting $\underline{f}(t)=\underline{0}$ in Eq. (10-54). Further, let us ignore the output equation as it is purely algebraic and involves no integration. The resulting system is $\underline{\dot{x}}(t)=A\underline{x}(t)$ with initial condition $\underline{x}(0)$. We now proceed to show that the solution of this state-variable equation is

$$\underline{\dot{x}}(t) = e^{At}\underline{x}(0) \qquad (10\text{-}126)$$

The matrix exponential exp(At) is defined by Eq. (10-38), repeated here in summation form:

$$e^{At} = \sum_{i=0}^{\infty} \frac{(At)^i}{i!} \qquad (10\text{-}127)$$

Substituting the assumed solution, Eq. (10-126), into the differential equation yields

$$\frac{d}{dt}\left\{e^{At}\underline{x}(0)\right\} = Ae^{At}\underline{x}(0) \qquad (10\text{-}128)$$

If Eq. (10-126) is truly a solution, then Eq. (10-128) must be an identity. This is shown by substituting the series expansion for the exponential into the left-hand side of Eq.

(10-128) and showing that it reduces to the right-hand side; thus,

$$\frac{d}{dt}\left\{e^{At}\underline{x}(0)\right\} = \frac{d}{dt}\left\{\sum_{i=0}^{\infty}\frac{A^i t^i}{i!}\underline{x}(0)\right\} \qquad (10\text{-}129)$$

$$= A \sum_{i=1}^{\infty}\frac{A^{i-1} t^{i-1}}{(i-1)!}\underline{x}(0) \qquad (10\text{-}130)$$

Let $i-1=k$. Equation (10-130) then becomes

$$\frac{d}{dt}\left\{e^{At}\underline{x}(0)\right\} = A \sum_{k=0}^{\infty}\frac{(At)^k}{k!}\underline{x}(0) = Ae^{At}\underline{x}(0) \qquad (10\text{-}131)$$

and the result is proved.

The forced response. When $\underline{f}(t)$ is nonzero, we start by assuming a solution of the form

$$\underline{x}(t) = e^{At}\underline{q}(t) \qquad (10\text{-}132)$$

where $\underline{q}(t)$ is an unknown vector function. Differentiating Eq. (10-132) yields

$$\underline{\dot{x}}(t) = Ae^{At}\underline{q}(t) + e^{At}\underline{\dot{q}}(t) \qquad (10\text{-}133)$$

or, using Eq. (10-132),

$$\underline{\dot{x}}(t) = A\underline{x}(t) + e^{At}\underline{\dot{q}}(t) \qquad (10\text{-}134)$$

Comparing Eqns. (10-134) and (10-54) indicates that

$$e^{At}\underline{\dot{q}}(t) = B\underline{f}(t) \qquad (10\text{-}135)$$

To find $\underline{q}(t)$, integrate Eq. (10-135) from $-\infty$ to t, or

$$\underline{q}(t) = \int_{-\infty}^{t} e^{-Ah} B\underline{f}(h)\,dh \qquad (10\text{-}136)$$

State Space Analysis

where h is a dummy variable of integration. Substitute Eq. (10-136) into Eq. (10-132), obtaining

$$\underline{x}(t) = e^{At} \int_{-\infty}^{t} e^{-Ah} \underline{Bf}(h) dh \qquad (10\text{-}137)$$

The integral on the right can be artifically split into two parts; one for $-\infty<h<0$, and one for $0<h<t$. The result is

$$\underline{x}(t) = e^{At} \int_{-\infty}^{0} e^{-Ah} \underline{Bf}(h) dh + e^{At} \int_{0}^{t} e^{-Ah} \underline{Bf}(h) dh \qquad (10\text{-}138)$$

If $\underline{f}(t)$ were $\underline{0}$ for $t>0$, the second integral in Eq. (10-138) would vanish. Comparing the remainder of the equation with Eq. (10-126) reveals that

$$\int_{-\infty}^{0} e^{-Ah} \underline{Bf}(h) dh = \underline{x}(0) \qquad (10\text{-}139)$$

Using this result gives the final solution

$$\underline{x}(t) = e^{At} \underline{x}(0) + e^{At} \int_{0}^{t} e^{-Ah} \underline{Bf}(h) dh \qquad (10\text{-}140)$$

Equation (10-140) is the desired complete solution to the general state-variable system. It agrees with the result obtained using Laplace transforms, Eq. (10-110), although the correspondence is not evident by inspection.

PROBLEMS FOR CHAPTER 10

10-1. Evaluate the following determinants:

(a) $\begin{vmatrix} 1 & 5 \\ 8 & -2 \end{vmatrix}$ (b) $\begin{vmatrix} 2 & 0 & 3 \\ 1 & -5 & 8 \\ 1 & 0 & 2 \end{vmatrix}$ (c) $\begin{vmatrix} 5 & 0 \\ 0 & 10 \end{vmatrix}$

(d) $\begin{vmatrix} 0 & 1 & 0 & 1 \\ 1 & 1 & 0 & 1 \\ 0 & 0 & 1 & 1 \\ 1 & 0 & 1 & 1 \end{vmatrix}$ (e) $\begin{vmatrix} 5 & 0 & -2 & 3 \\ 2 & 8 & 0 & -1 \\ 5 & 1 & 7 & 3 \\ 0 & 0 & 4 & 2 \end{vmatrix}$

10-2. Perform the indicated operations on the following matrices and vectors: (a) Det(A), (b) Det(D), (c) A^T, (d) $\underline{b}^T$, (e) $\underline{c}^T$, (f) A+B, (g) A^2, (h) $\underline{bc}$, (i) $\underline{cb}$, (j) AD, and (k) DA.

$$A = \begin{bmatrix} 1 & 0 \\ 2 & 3 \end{bmatrix} \quad \underline{b} = \begin{bmatrix} 1 \\ 2 \end{bmatrix} \quad \underline{c} = \begin{bmatrix} 5 & -2 \end{bmatrix} \quad D = \begin{bmatrix} 5 & 8 \\ 2 & -6 \end{bmatrix}$$

(10-141)

10-3. Compute the indicated matrix products if the matrices are conformable; if they are not, so state: (a) AB, (b) AC, (c) BA, (d) CA, (e) BC, and (f) CB.

$$A = \begin{bmatrix} 1 & 0 & 3 & -5 \\ 2 & 1 & 4 & 8 \end{bmatrix} \quad B = \begin{bmatrix} 5 & 2 \\ 0 & 1 \\ 1 & 0 \\ 3 & 4 \end{bmatrix} \quad C = \begin{bmatrix} 1 & 5 \\ 2 & 3 \\ 0 & -2 \\ 8 & 4 \\ 2 & 1 \end{bmatrix}$$ (10-142)

10-4. Evaluate A^T, B^T, and C^T for the matrices of Problem 10-3.

10-5. Derive a formula for the inverse of the general (2 x 2) matrix

$$A = \begin{bmatrix} a_{11} & a_{12} \\ a_{21} & a_{22} \end{bmatrix}$$ (10-143)

10-6. Perform the indicated operations for the matrices A and D of Problem 10-2: (a) A^{-1}, (b) D^{-1}, (c) $A^{-1}D^{-1}$, (d) $(AD)^{-1}$, (e) I_2-A, (f) $(I_2-A)^{-1}$, (g) $4I_2+D$, (h) $(4I_2+D)^{-1}$.

10-7. Find the following for the given diagonal matrix A: (a) A^{-1}, (b) $(I_3-A)^{-1}$, (c) $(2I_3+A)^{-1}$, (d) A^{-2}.

$$A = \text{Diag}[5,-2,-4]$$ (10-144)

State Space Analysis

10-8. Invert the following matrices:

$$A = \begin{bmatrix} 1 & 3 & 3 \\ 1 & 4 & 3 \\ 1 & 3 & 4 \end{bmatrix} \quad B = \begin{bmatrix} 1 & 2 & 3 \\ 2 & 4 & 5 \\ 3 & 5 & 6 \end{bmatrix} \quad (10\text{-}145)$$

10-9. Given the matrix

$$A = \begin{bmatrix} 1 & 2 \\ -1 & 0 \end{bmatrix} \quad (10\text{-}146)$$

Compute the following: (a) A^2, (b) A^3, and (c) a three-term approximation to $\exp(A)$.

10-10. Find an exact expression for $\exp(At)$ for the matrix A of Problem 10-7. (Hint: use $\exp(At)=L^{-1}\{(sI_3-A)^{-1}\}$.

10-11. Repeat Problem 10-10 for the matrix A of Problem 10-9.

10-12. Put the following systems of differential equations into state-variable form. Assume zero initial conditions.

(a) $\dot{x}_1 + 3x_1 - 2x_2 = 0$ (10-147)

$\dot{x}_2 + 5x_2 + 3x_1 = 8u(t)$ (10-148)

$y = 4x_1 + 5x_2 - 25u(t)$ (10-149)

(b) $\dot{x}_1 + 2x_1 - x_2 = f(t)$ (10-150)

$\dot{x}_2 = \dot{x}_1 + x_1$ (10-151)

$y = x_1$ (10-152)

(c) $\dot{x}_1 + x_1 + x_2 + x_3 = 0$ (10-153)

$\dot{x}_2 + 2x_2 - x_1 - x_3 = f_1(t)$ (10-154)

$\dot{x}_3 + 3x_3 + x_2 + x_1 = 2f_1(t) - f_2(t)$ (10-155)

$y = x_1 - 2x_2 + 5f_1(t) - f_2(t)$ (10-156)

10-13. Transform the following differential equations into state-variable form:

(a) $\ddot{y} + 5y = f(t)$

$y(0) = 1, \dot{y}(0) = 2$ (10-157)

(b) $\dddot{y} + 2\ddot{y} + 5\dot{y} + 6y = 20u(t)$

$y(0) = 1, \dot{y}(0) = 0, \ddot{y}(0) = 5$ (10-158)

(c) $\ddot{y} + 5y = 10u(t)$

$\ddot{g} + 2y = 0$ (10-159)

$y(0) = \dot{y}(0) = g(0) = \dot{g}(0) = 0$

(d) $\ddot{g} + 3g - h = u(t)$

$\ddot{h} - 6h + g = 0$ (10-160)

(All initial conditions are zero.)

10-14. Derive state variable equations for the following electrical networks. Use inductor currents and capacitor voltages for the state variables. The output variable is $v_o(t)$ in each network.

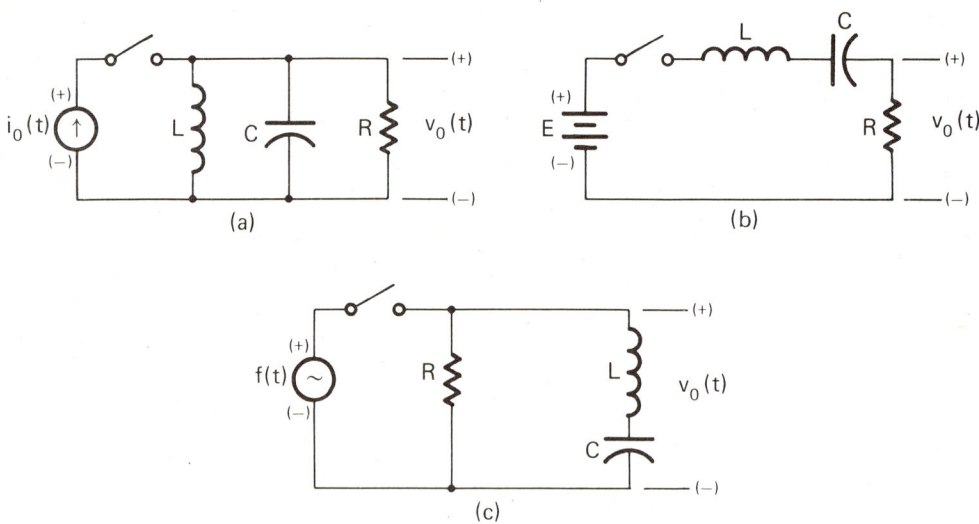

Figure 10-5. Problem 10-14.

State Space Analysis

10-15. Derive a diagonalizing transformation for each of the following matrices:

(a) $A = \begin{bmatrix} 1 & -6 \\ -6 & -4 \end{bmatrix}$ (10-161)

(b) $A = \begin{bmatrix} 2 & 0 & -1 \\ 0 & 2 & 0 \\ -1 & 0 & 2 \end{bmatrix}$ (10-162)

10-16. Diagonalize the systems of Problem 10-14. Use L=1, C=1, and R=1. If the system cannot be diagonalized, show why.

10-17. Solve for the state vector $\underline{x}(t)$ for each of the following systems. Use Laplace transforms.

(a) $A = \begin{bmatrix} -1 & 0 \\ 0 & -2 \end{bmatrix} \quad \underline{b} = \begin{bmatrix} 1 \\ 0 \end{bmatrix}$ (10-163)

Assume $f(t)=u(t)$ and $\underline{x}(0)=\underline{0}$.

(b) $A = \begin{bmatrix} -2 & 0 \\ 0 & -3 \end{bmatrix} \quad \underline{b} = \begin{bmatrix} 0 \\ 0 \end{bmatrix} \quad \underline{x}(0) = \begin{bmatrix} 5 \\ 6 \end{bmatrix}$ (10-164)

(c) $A = \begin{bmatrix} -1 & 0 \\ -2 & 1 \end{bmatrix} \quad \underline{b} = \begin{bmatrix} 0 \\ 1 \end{bmatrix}$ (10-165)

Assume $f(t)=u(t)$ and $\underline{x}(0)=\underline{0}$.

10-18. Solve for the state vector for each of the networks of Problem 10-14. Again use L=1, C=1, and R=1. For part (c) use $f(t)=u(t)$.

10-19. If $C = \begin{bmatrix} 1 & 2 \end{bmatrix}$ and D=0, find the transfer function G(s) for each of the systems given in Problem 10-17. Ignore any nonzero initial conditions.

10-20. Find the transfer function for each of the state-variable systems below:

(a) $A = \begin{bmatrix} -1 & 0 \\ 0 & -2 \end{bmatrix} \quad \underline{b} = \begin{bmatrix} 0 \\ 1 \end{bmatrix} \quad C = \begin{bmatrix} 1 & 0 \end{bmatrix} \quad D = 0$ (10-166)

(b) $A = \begin{bmatrix} -2 & 0 \\ 0 & -3 \end{bmatrix} \quad \underline{b} = \begin{bmatrix} 1 \\ 2 \end{bmatrix} \quad C = \begin{bmatrix} 1 & 0 \\ 0 & 1 \end{bmatrix} \quad D = \begin{bmatrix} 2 \\ 5 \end{bmatrix}$ (10-167)

(c) $A = \begin{bmatrix} -1 & 0 \\ -1 & -1 \end{bmatrix} \quad B = I_2 \quad C = I_2 \quad D = I_2$ (10-168)

REFERENCES FOR CHAPTER 10

(1) Timothy, L. K. and B. E. Bona: State Space Analysis: An Introduction, McGraw-Hill Book Co., Inc., New York, 1968.
(2) Ogata, K.: State Space Analysis of Control Systems, Prentice-Hall, Inc., Englewood Cliffs, New Jersey, 1967.
(3) Hsu, J. C. and A. U. Meyer: Modern Control Principles and Applications, McGraw-Hill Book Co., Inc., New York, 1968.
(4) Tou, J. T.: Modern Control Theory, McGraw-Hill Book, Co., Inc., New York, 1964.
(5) Gantmacher, F. R.: The Theory of Matrices, Chelsea Publishing Co., New York, 1959.
(6) Eisenman, R. L.: Matrix Vector Analysis, McGraw-Hill Book Co., Inc., New York, 1963.
(7) Hohn, F. E.: Elementary Matrix Algebra, The Macmillan Company, New York, 1973.
(8) DeRusso, P. M., Roy, R. J., and C. M. Close: State Variables For Engineers, John Wiley and Sons, Inc., New York, 1965.

(9) Gupta, S. C.: <u>Transform and State Variable Methods In Linear Analysis</u>, John Wiley and Sons, Inc., New York, 1966.

(10) Bellman, R.: <u>Introduction to Matrix Analysis</u>, McGraw-Hill Book Co., Inc., New York, 1970.

(11) Dorf, R. C.: <u>Time-Domain Analysis and Design of Control Systems</u>, Addison-Wesley Publishing Co., Inc., Reading, Massachusetts, 1965.

(12) Calahan, D. A.: <u>Computer-Aided Network Design</u>, McGraw-Hill Book Co., Inc., New York, 1972.

(13) Chen, C. F. and I. J. Haas: <u>Elements of Control Systems Analysis</u>, Prentice-Hall, Inc., Englewood Cliffs, New Jersey, 1968.

(14) Desoer, C. A. and E. S. Kuh: <u>Basic Circuit Theory</u>, McGraw-Hill Book Co., Inc., New York, 1972.

(15) Chen, C. T.: <u>Introduction to Linear System Theory</u>, Holt, Rinehart, and Winston, Inc., New York, 1970.

(16) Pease, M. C.: <u>Methods of Matrix Algebra</u>, Academic Press Inc., New York, 1965.

(17) Gupta, S. C., Bayless, J. W., and B. Reikart: <u>Circuit Analysis With Computer Applications to Problem Solving</u>, Intext Educational Publishers, Scranton, Pennsylvania, 1972.

(18) Chen, C. T.: <u>Analysis and Synthesis of Linear Control Systems</u>, Holt, Rinehart, and Winston, Inc., New York, 1975.

(19) Rohrer, R. A.: <u>Circuit Theory: An Introduction to the State Variable Approach</u>, McGraw-Hill Book Co., Inc., New York, 1970.

CHAPTER **11**

Introduction to Discrete Analysis

As digital computers become ever smaller, faster, and less expensive, the incentives to use them in analyzing continuous systems increase. This is especially true because modern engineering systems have become so complex as to rule out analysis by hand calculation. Virtually all engineering graduates are familiar with some higher-order programming language such as FORTRAN, ALGOL, or BASIC;[1,2] and the university or large corporation without a digital computer is almost nonexistent. As a result, the present day engineer is equipped with both the programming skills and the opportunity to use the digital computer as a tool for systems analysis, provided he can bridge the conceptual gap between the continuous models typical of most dynamic systems and the discrete mathematics required for digital computation. The next three chapters bridge this gap.

11.1 ISSUES IN DISCRETE ANALYSIS

Before plunging into the study of digital computer methods for engineering analysis, we must address a number of peripheral issues which will motivate our selection of specific topics.

The Role Of The Digital Computer

Early uses of digital computers in science and engineering were limited to evaluating complicated functions and performing repetitive calculations.[3] This work was usually done "off-line" in a large computer center. But the coming of small, fast mini- and micro-computers has given the computer new roles in engineering analysis. We can, in fact, envisage three ways a digital computer can be used in the dynamic systems environment.[4,5]

<u>As a tool for data analysis</u>. In this role, the system itself functions while some of the dependent variables are measured and recorded. At some other convenient time or place, discretized versions of these variables are used as input data to a digital computer and are analyzed according to various data-analysis algorithms. Hopefully, the results of the analysis will further an understanding of the behavior of the system.[6]

<u>As a system simulator</u>. In this situation, we discretize the mathematical model of the system--transform the differential equations of motion into difference equations, for example--and then program the discrete model for analysis on the computer.[7] The discrete model produces discrete versions of the dependent variables, which the engineer analyzes at his leisure to learn more about the system. The difference between discrete analysis and digital simultation is

Introduction to Discrete Analysis 473

largely a matter of perspective. Both are derived from classical numerical analysis, the parent discipline of all scientific computer work.

   As part of a real-time system. When operating in a real-time mode, the computer accepts input data from an actual system or process, performs calculations or otherwise modifies the data, and then returns the modified data to the system as a further input.[8,9] The computer is thus an integral part of the system, and as such it must operate fast enough to stay "in step" with the process.[10] Most readers have had some experience with the computer as a tool for data analysis or for simulation, but the real-time role is probably somewhat novel.

   Fortunately, we need not commit ourselves to either of the three computational roles--data analysis, simulation, or real-time computation--as we can present analysis techniques which are applicable to all three, thus better equipping the student for any computational environment he might encounter.

Tools Needed For Discrete Analysis

   As a preliminary step on the road to familiarity with the processes available for analyzing continuous systems on the digital computer, let us preview some of the relevant topics.

   Sampling. Continuous dynamic systems are described mathematically in terms of smoothly varying signals, but digital computers can manipulate only discrete numbers. As a result, before we can study the behavior of a continuous system on the computer, we must find a way to discretize its mathematical model. This process, called sampling, involves more than just listing values of a dependent variable which correspond to discrete values of the independent variable. The discretization process itself can introduce spurious information into the analysis, causing the computer to predict system performance which disagrees with reality. Sections 11.2 and 11.3 treat both the theoretical and practical aspects of the sampling process.

Classical numerical analysis. When we think of using a digital computer for some type of engineering analysis, the first topic which comes to mind is traditional numerical analysis. And although numerical methods are the foundation of modern data analysis, we need a broader perspective if we expect to develop a useful package of systems analysis methods. Basically, we must recast the numerical methods in the framework of system theory. The numerical analyst, for example, is content with theorems and algorithms, but the systems engineer wants discrete transfer functions, block diagrams, etc. The system analyst uses many traditional numerical methods but often in a disguised form. Another difference is that the numerical analyst starts with the assumption that his numbers truly represent the process under study, a luxury the systems engineer often cannot afford.

The z-transform. Just as the Laplace transform is a valuable medium for discussing and representing continuous systems, the z-transform plays a similar role for discrete systems. We will derive useful results in the areas of z-domain transfer functions and block diagrams, pole-zero plots in the z-plane, discrete convolution, and, most important, solving difference equations.

Difference equations. The discrete counterpart of a time-domain differential equation is the sequence-domain difference equation. Indeed, the key to analyzing a continuous system on the digital computer is to transform its differential equations into difference equations which can be handled by the computer. Just as we can choose between time-domain or Laplace methods to solve a differential equation, so we can choose between sequence-domain or z-transform methods to solve a difference equation.

Stability. A dynamic system may be ultra-stable, but the numerical method used to analyze the system may lead to instability and spurious results. An advantage of z-transforms over classical numerical methods is that they often help anticipate such problems.

Data smoothing. When a continuous variable is discretized, a certain amount of noise is introduced, causing the resulting discrete signal to be an inexact replica of the parent continuous signal. The finite length of a digital computer word implies that many numbers cannot be represented exactly and a certain amount of "rounding" occurs. A smooth ramp signal in the time domain becomes a "staircase" when discretized. These and other unwanted perturbations in the discrete signal can be removed with suitable digital filters. A digital filter is analogous to a continuous filter--high-frequency discrete noise can be removed by a low-pass digital filter, for example.

Computer programming. To be sure, if one wants to analyze a system on a digital computer, then he must be able to program that computer. But teaching programming languages would take us too far afield from our main effort of developing techniques for system analysis. However, the discrete analysis methods presented below are algorithmic. That is, they can be transformed into computer programs on a step-by-step basis, although the intermediate stage of flow-charting is often useful. Nor do we introduce the assembly-language programming techniques which, for reasons of speed, are often used in real-time computation.

Further, a new family of so-called "problem oriented" computer languages has appeared in recent years.[11] The term "language" is somewhat misleading because a problem-oriented language is just a computer program designed to solve a specific class of problems. Examples are SCEPTRE[12] and ECAP for analyzing electrical networks and CSMP, DYNAMO, MIMIC, and SL/1 for continuous dynamic systems. The advantages of a problem-oriented language are threefold: (1) the ease of presenting the problem to the computer--the user codes only his input data rather than writing the entire program-- (2) the output is automatically presented in a suitable format, and (3) the pertinent algorithms, such as numerical integration for solving differential equations, are contained in

the program. The disadvantages of problem-oriented languages
arise from the disconcerting phenomenon that certain problems
can confound any numerical method. For this reason one cannot
be a casual user of a problem-oriented language; he must
develop experience and confidence, over a long period of time,
in the ability of the program to handle his specific class of
problems. In summary, problem-oriented languages are not a
panacea, and while they are doubtless useful in a "production"
environment, they cannot substitute for an in-depth understanding of both the problem and the methods applicable to its
solution.

11.2 THE SAMPLING PROCESS

This section introduces the concept of sampling, a
vitally important topic in discrete analysis.[13,14] Many an
attempted analysis has foundered at this early stage.

Discrete-Time Signals

Chapter **2** introduced the idea of a <u>discrete signal</u>, one
which is defined only for discrete values of the independent
variable. Discrete signals are represented mathematically as
sequences of numbers. Recall too from Chapter **2** that if the
amplitude of a discrete signal is quantized, it is called a
<u>digital signal</u> or a <u>digital data function</u>. Because digital
computers have finite word lengths, the discrete signals used
in systems analysis are of necessity quantized and hence are
digital data functions.

If a continuous signal y(t) is sampled at equally spaced
intervals on the time axis, the sequence y(k) results, where

$$y(k) = y(t=kT) = y(kT) \qquad (11-1)$$

and T is the <u>sampling interval</u>. T has dimensions of time and
is usually in milliseconds or microseconds for real-world
systems. We use the notation y(k) in preference to y_k or

Introduction to Discrete Analysis

y(kT) because it facilitates writing complicated subscripts. The numbers y(k) are (1) bounded (finite in magnitude), (2) discretized to a finite precision (rounded off), and (3) defined only for t⩾0. In other words, all problems "start" at t=0.

Example 11-1. Calculate the first 5 values of the sampled sequence f(k) if $f(t)=5t^2$ and T=0.1 s.

Solution. The first five values, obtained from $f(k)=f(kT)=5(0.1k)^2$, are f(0)=0.0, f(1)=0.05, f(2)=0.20, f(3)=0.45, and f(4)=0.80.

The Frequency Content Of A Signal

At the heart of the sampling process is the idea of the frequency content of a time-domain signal.[15,16] Under very general conditions it is possible to represent a signal as a sum (either finite or infinite) of sinusoids of different frequencies. The amplitudes and frequencies of these sinusoids are measures of the frequency content of the parent signal.

The complex Fourier series. If a signal is periodic, that is, if

$$f(t) = f(t + T_o) \qquad (11-2)$$

then f(t) can be represented as a sum of sines and cosines called the Fourier series expansion of f(t). If T_o is the smallest constant for which Eq. (11-2) holds, then T_o is the period of f(t). Equations (2-90) through (2-93) define the Fourier series in its trigonometric form. In the work that follows we use the exponential Fourier series,[17]

$$f(t) = \sum_{n=-\infty}^{\infty} C_n \exp(jn\omega_o t) \qquad (11-3)$$

where $\omega_o = 2\pi/T_o$. The <u>Fourier coefficients</u> C_n are found from the relationship

$$C_n = \frac{1}{T_o} \int_{-T_o/2}^{T_o/2} f(t)\exp(-jn\omega_o t)dt \qquad (11-4)$$

Equations (11-3) and (11-4) can be derived from the trigonometric form by using Euler's identities for the sine and cosine.

To possess a valid Fourier series, the function f(t) must satisfy the so-called <u>Dirichlet conditions</u>.[18] These conditions require that on every finite time interval f(t) has (1) a finite number of maxima and minima, (2) a countable number of finite discontinuities, and (3) a finite number of infinite discontinuities. In addition to satisfying the Dirichlet conditions, f(t) must be such that the integral I given by

$$I = \int_{T_o/2}^{T_o/2} |f(t)|dt \qquad (11-5)$$

has a finite value. These conditions are rather weak and as a consequence periodic signals which arise in most engineering applications have valid Fourier series expansions.

A plot of the magnitudes of the Fourier coefficients, $|C_n|$, versus frequency is called the <u>discrete magnitude spectrum</u> of the signal. (We use the magnitude because C_n may be complex.) A plot of the angles of the C_n is the <u>discrete phase spectrum</u> of the signal. Only periodic functions have discrete spectra.

Example 11-2. Expand the square wave of Fig. 11-1 in a complex Fourier series and sketch the discrete magnitude spectrum.

Introduction to Discrete Analysis

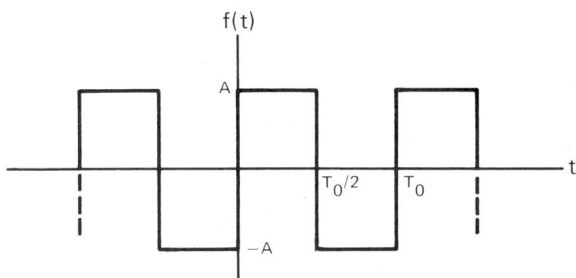

Figure 11-1. As shown in Example 11-2, this periodic square wave can be represented by the complex Fourier series of Eq. (11-9).

Solution. Eq. (11-4) gives the Fourier coefficients

$$C_n = \frac{2}{T_o} \int_0^{T_o/2} (A)\exp(-jn\omega_o t)\,dt \qquad (11\text{-}6)$$

$$= \frac{2Aj}{n\omega_o T_o} \exp(-jn\omega_o t)\Big]_0^{T_o/2} \qquad (11\text{-}7)$$

and finally,

$$C_n = \frac{Aj}{n\pi}\{\exp(-n\pi j)-1\} \qquad (11\text{-}8)$$

When n is odd, $C_n = -2Aj/n\pi$; when n is even $C_n = 0$. The Fourier series is found by substituting the C_n into Eq. (11-3):

$$f(t) = -\frac{2Aj}{\pi} \sum_{n=1,3,5..}^{\infty} \frac{1}{n}\exp(jn\omega_o t) \qquad (11\text{-}9)$$

Figure 11-2 is the magnitude spectrum obtained by plotting $|C_n| = 2A/n\pi$ versus n.

The frequency content of the square wave is infinite, which means that sinusoids of an infinite number of frequencies, all integer multiples of ω_o, must be summed to create the original signal. In this case the magnitudes of the C_n decrease with increasing n, a phenomenon we will make use of in approximating non-bandlimited signals by bandlimited ones.

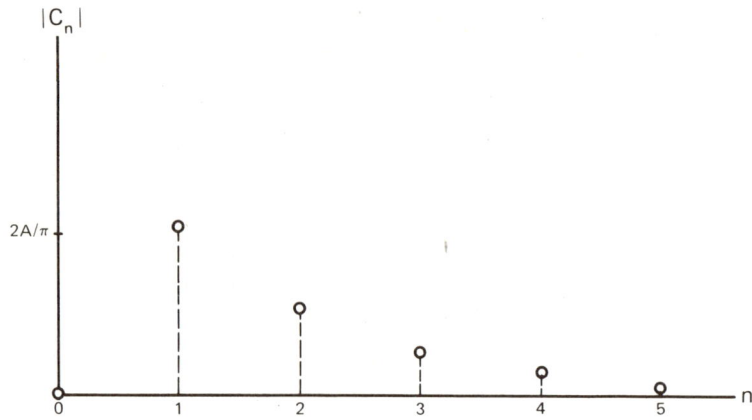

Figure 11-2. A plot of the $|C_n|$ versus n is called a discrete magnitude spectrum. This spectrum is for the square wave of Fig. 11-1.

The Fourier transform. We are now equipped to investigate the frequency content of periodic signals, but what about those which are aperiodic? Here we must turn to the Fourier transform.[18,19] If f(t) is a real or complex time function which satisfies the Dirichlet conditions on $-\infty < t < \infty$, then we can compute its Fourier transform, defined by

$$\Gamma\{f(t)\} = F(j\omega) = \int_{-\infty}^{\infty} f(t) e^{-j\omega t} dt \qquad (11\text{-}10)$$

where the symbol $\Gamma\{\ \}$ denotes the Fourier transform of the bracketed quantity. The transformed function $F(j\omega)$ is generally a complex function of the real variable ω. The inverse transform, $\Gamma^{-1}\{\ \}$, by which f(t) is recovered from $F(j\omega)$, is given by

$$\Gamma^{-1}\{F(j\omega)\} = f(t) = \frac{1}{2\pi} \int_{-\infty}^{\infty} F(j\omega) e^{j\omega t} d\omega \qquad (11\text{-}11)$$

Introduction to Discrete Analysis

The transformed function $F(j\omega)$ can be expressed as a real part and an imaginary part or as a magnitude function (or spectrum) and a phase function. A plot of $|F(j\omega)|$ versus ω is called the <u>magnitude spectrum</u> of $F(j\omega)$. The continuous spectrum associated with an aperiodic signal is analogous to the discrete spectrum of a periodic signal. It shows that the parent signal can be expressed as an infinite sum of sinusoids, but in this case at frequencies which are not necessarily integer multiples of some constant ω_o.

Example 11-3. By taking inverse transforms, show that the Fourier transform of the sinusoid $f(t)=\sin(\omega_o t)$ is given by

$$\Gamma\{\sin(\omega_o t)\} = j\pi\{\delta(\omega+\omega_o) - \delta(\omega-\omega_o)\} \qquad (11-12)$$

Solution. Substitute Eq. (11-12) into Eq. (11-11) to obtain

$$f(t) = \frac{1}{2\pi}\int_{-\infty}^{\infty} j\pi\{\delta(\omega+\omega_o) - \delta(\omega-\omega_o)\}e^{j\omega t}d\omega \qquad (11-13)$$

and, by the sampling property of the impulse function,

$$f(t) = \frac{e^{-j\omega_o t} - e^{j\omega_o t}}{-2j} \qquad (11-14)$$

Using Euler's identity gives the final result, $f(t)=\sin(\omega_o t)$. By a similar procedure we can show that

$$\Gamma\{\exp(j\omega_o t)\} = 2\pi\delta(\omega-\omega_o) \qquad (11-15)$$

$$\Gamma\{\cos(\omega_o t)\} = \pi\{\delta(\omega-\omega_o) + \delta(\omega+\omega_o)\} \qquad (11-16)$$

and

$$\Gamma\{\delta(t-T_o)\} = e^{-j\omega T_o} \qquad (11-17)$$

Example 11-4. Find the time-domain function $g(t)$ which has the "boxcar" function of Fig. 11-3(a) as its Fourier magnitude spectrum.

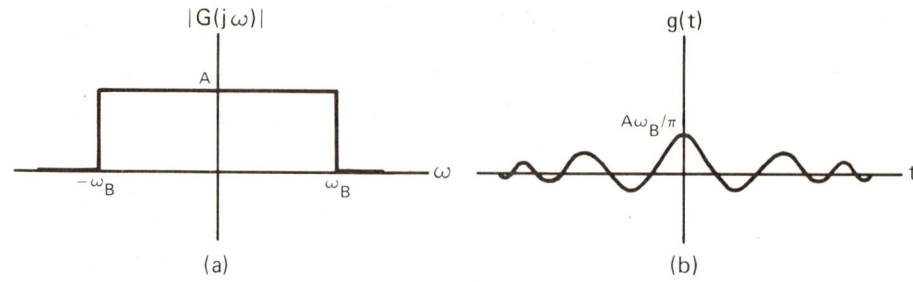

Figure 11-3. The rectangular "boxcar" function in the frequency domain becomes a sinc or sin (x)/x function in the time domain and vice versa. (a) A "boxcar" function in the frequency domain; (b) The inverse Fourier transform of (a).

Solution. The inverse transform is found from Eq. (11-11):

$$g(t) = \frac{1}{2\pi} \int_{-\omega_B}^{\omega_B} Ae^{j\omega t} d\omega = \frac{A}{2\pi jt} e^{j\omega t} \Big]_{-\omega_B}^{\omega_B} \qquad (11\text{-}18)$$

or

$$= \frac{A}{\pi t} \frac{e^{j\omega_B t} - e^{-j\omega_B t}}{2j} \qquad (11\text{-}19)$$

and recognizing the complex exponentials as a sinusoid, $g(t) = A\sin(\omega_B t)/\pi t$. The function $g(t)$ resembles the "sinc" or $\sin(x)/x$ function shown in Fig. 11-3(b).

Bandlimited signals. Having introduced the Fourier series and the Fourier transform, we are now ready to introduce the concept of a signal which is bandlimited. If the magnitude spectrum (either discrete or continuous) of a time-domain signal is everywhere zero outside of some finite region of the frequency axis, then that signal is said to be band-limited. The time-domain signal $g(t)$ of Fig. 11-3(b) is bandlimited because its magnitude spectrum, shown in Fig. 11-3(a), is zero for $\omega > \omega_B$. If a signal is bandlimited it can theoretically be expressed as a sum of sinusoids, the highest frequency of which is less than some upper, or bandlimiting frequency, ω_B. Figure 11-4 shows the magnitude spectra of (a) a bandlimited aperiodic signal, (b) a bandlimited periodic

Introduction to Discrete Analysis

signal, (c) a non-bandlimited aperiodic signal and (d) a non-bandlimited periodic signal.

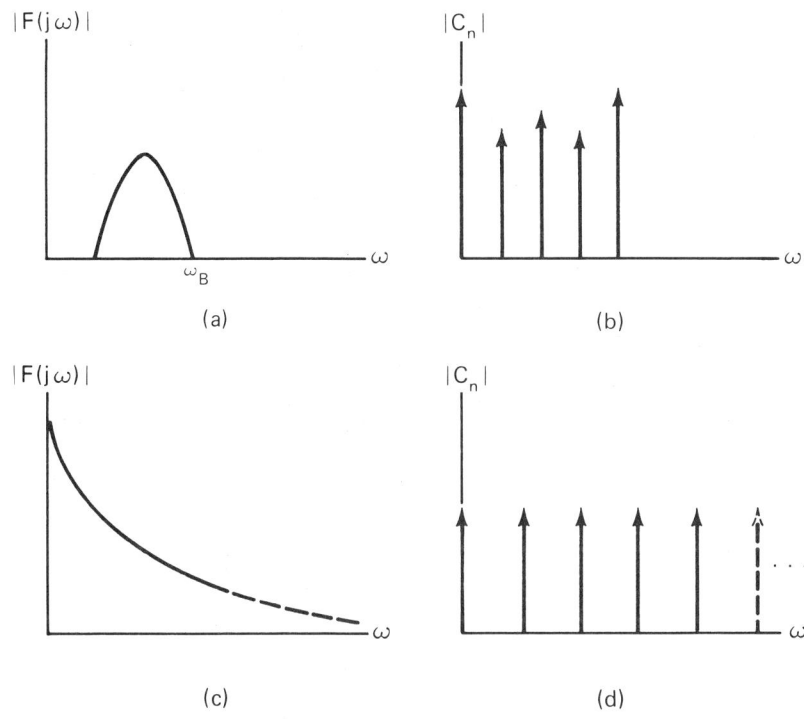

Figure 11-4. Magnitude spectra for bandlimited and non-bandlimited signals. (a) A bandlimited aperiodic signal; (b) A bandlimited periodic signal; (c) A non-bandlimited aperiodic signal; (d) A non-bandlimited periodic signal.

The Sampling Theorem [20]

Let $y(t)$ be a real function defined on $-\infty < t < \infty$. Assume that $y(t)$ is sampled every T time units, generating the sequence $y(k)$. If the Fourier transform of $y(t)$, denoted by $Y(j\omega)$, exists and has the property that $|Y(j\omega)| \equiv 0$ for all $\omega \geqslant \omega_B$, then $y(t)$ can be uniquely recovered from the sampled sequence $y(k)$ if and only if the sampling interval T satisfies the relationship

$$T \leqslant \pi/\omega_B \tag{11-20}$$

If we define the <u>sampling frequency</u> f_s by $f_s = 1/T$ and use $f_B = \omega_B/2\pi$, then Eq. (11-20) assumes the alternate form

$$f_s \geqslant 2f_B \qquad (11-21)$$

Equation (11-21) says that <u>in order to recover the parent continuous signal, we must sample at a frequency at least twice the highest frequency contained in the original signal.</u> This is the sampling theorem. Its proof is outlined in Problem 11-23.

A consequence of the sampling theorem is that a signal can be successfully sampled only if it is bandlimited. Since signals associated with practical systems are seldom bandlimited, this would seem to be a serious restriction. Fortunately, most signals, though not strictly bandlimited, have the property that $|Y(j\omega)| \to 0$ as $\omega \to \infty$ and hence can be approximated by suitably chosen bandlimited signals.

The frequency f_N, defined by $f_N = 2f_B$, is called the <u>Nyquist frequency</u> or the Nyquist rate, giving rise to another statement of the sampling theorem--that to recover the original signal we must sample at a rate greater than the Nyquist frequency.

The alert reader might question the validity of the sampling theorem on the basis of a counter example such as that shown in Fig. 11-5(a).

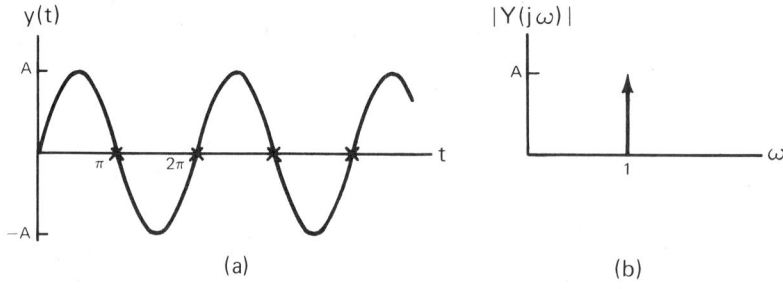

Figure 11-5. Because the magnitude spectrum is not zero at $\omega_B = 1$, the time-domain signal must be sampled at $T < \pi/\omega_B$ if it is to be uniquely specified by its sampled version. (a) A time-domain signal $y(t) = A\sin(t)$ sampled every π time units; (b) The magnitude spectrum for signal (a) is an impulse at $\omega_B = 1$.

Introduction to Discrete Analysis

The sinusoidal signal has frequency $f=1/2\pi$, and, by the sampling theorem, should be sampled every π time units to allow complete recovery. One might argue that the sampling points denoted by "x" in the figure could determine a straight line as well as the required sinusoid. The answer to this seeming paradox is that the sampling theorem states that $|Y(j\omega)|$ must be 0 $\underline{at}$ ω_B (in this case $\omega_B=1$). Since $|Y(j\omega_B)|\neq 0$ in this instance (see Fig. 11-5(b)), we must choose T slightly less than π sec. As a consequence, even if $T=\pi-\varepsilon$, where ε is vanishingly small, we would eventually specify the sinusoid rather than the straight line. In summary if $|Y(j\omega_B)|=0$, $T=\pi/\omega_B$ is satisfactory, but if $|Y(j\omega_B)|\neq 0$, then we must choose $T<\pi/\omega_B$.

The above example shows why for finite record lengths we should sample at a rate somewhat greater than the Nyquist rate for accurate data processing and ten times f_N for aesthetic plots or oscilloscope displays.

Sampling Non-Bandlimited Signals

As mentioned above, most practical signals are not bandlimited. Moreover, if our problem is to solve a differential equation by numerical methods, we do not have $\underline{a}$ $\underline{priori}$ knowledge of the solution and hence cannot determine whether it is a bandlimited function or not. In terms familiar to the numerical analyst, the problem is how to choose a suitable integration step size without solving the differential equation first.

While it is never easy to escape from this dilemma, we can sometimes use energy methods to approximate non-bandlimited signals by bandlimited ones which will produce the same effects in a continuous system. If a bandlimited signal has the same waveform as a non-bandlimited signal, and contains say 99% of the energy of the original signal, then we can be optimistic that it will serve as a suitable substitute. As a consequence, although energy considerations seemingly take us far afield from discrete analysis, they are often the key to selecting a suitable sampling rate.

Power and energy relationships. Two important variations of Parseval's theorem relate power and energy in the time domain to similar quantities in the frequency domain.(22,23) If $y(t)$ is periodic with period T_o and is interpreted as being the voltage across (or the current through) a one-ohm resistor, then the expression

$$P_{ave} = \frac{1}{T_o} \int_{-T_o/2}^{T_o/2} |y(t)|^2 dt \qquad (11-22)$$

gives the average power developed. If $y(t)$ is aperiodic, the total energy W_t is given by

$$W_t = \int_{-\infty}^{\infty} |y(t)|^2 dt \qquad (11-23)$$

We now present three expressions which show how to compute power and energy from frequency-domain quantities.

Case I: Signals with finite energy. If $y(t)$ is Fourier transformable and possesses a finite energy on $-\infty < t < \infty$, then

$$W_t = \frac{1}{2\pi} \int_{-\infty}^{\infty} |Y(j\omega)|^2 d\omega \qquad (11-24)$$

Case II: Periodic signals. If $y(t)$ is a real periodic function with period T_o, then the total energy on $-\infty < t < \infty$ is unbounded but the average power is the same on every period. Parseval's theorem states that

$$P_{ave} = \frac{1}{4\pi^2} \int_{-\infty}^{\infty} |Y(j\omega)|^2 d\omega \qquad (11-25)$$

Case III: Power in terms of Fourier coefficients. A variation of Eq. (11-25) relates the average power over one

Introduction to Discrete Analysis

period to the coefficients C_n of the complex Fourier series:

$$P_{ave} = \sum_{n=-\infty}^{\infty} |C_n|^2 \qquad (11\text{-}26)$$

The following examples show how to use the power relationships as an aid in choosing sampling periods for non-band-limited signals.

Example 11-5. Consider the problem of sampling the output voltage $v(t)$ of the electrical network of Fig. 11-6.[24] Assume switch S is closed at t=0 and that the capacitor is initially charged to the voltage v_C with the polarity shown. Let RC=1 s.

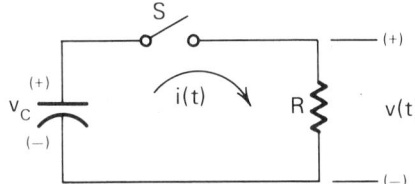

Figure 11-6. Network for Example 11-5. This example shows how energy methods can be used to determine a suitable sampling rate.

Solution. Using the methods of Chapter 4, we can show that $v(t)=v_C \exp(-t/RC)$ or, with RC=1, $v(t)=v_C \exp(-t)$ for $t \geq 0$. Taking the Fourier transform of $v(t)$ yields

$$V(j\omega) = \int_{-\infty}^{\infty} v_C e^{-t} e^{-j\omega t} dt \qquad (11\text{-}27)$$

and, as $v(t)=0$ for $t<0$,

$$V(j\omega) = \frac{-v_C}{1+j\omega} e^{-t(1+j\omega)} \Big]_0^{\infty} = \frac{v_C}{1+j\omega} \qquad (11\text{-}28)$$

The magnitude function is then

$$|V(j\omega)| = \frac{V_C}{\sqrt{1+\omega^2}} \qquad (11\text{-}29)$$

Because $|V(j\omega)|$ is nonzero for all ω, the signal is not band-limited. This means that to sample with no loss of information would require an infinite sampling rate or T=0. We may, however, find a frequency ω_B for which a large fraction, say 90%, of the energy lies between $\omega=0$ and $\omega=\omega_B$. If the total energy is finite, Eq. (11-24) applies and

$$W_t = \frac{1}{2\pi R} \int_{-\infty}^{\infty} |V(j\omega)|^2 d\omega \qquad (11\text{-}30)$$

The 90% criterion leads to

$$0.9W_t = \frac{1}{2\pi R} \int_{-\omega_B}^{\omega_B} |V(j\omega)|^2 d\omega \qquad (11\text{-}31)$$

Using Eq. (11-30) for W_t in Eq. (11-31) yields

$$\int_0^{\omega_B} \frac{V_C^2}{1+\omega^2} d\omega = 0.9 \int_0^{\infty} \frac{V_C^2}{1+\omega^2} d\omega \qquad (11\text{-}32)$$

where symmetry allows us to change the lower limits of integration to 0. Evaluating the integrals yields

$$\tan^{-1}(\omega_B) = 0.9\tan^{-1}(\omega)\Big]_0^{\infty} = 0.9\frac{\pi}{2} \qquad (11\text{-}33)$$

and ω_B=6.31 rad/s. Assuming a bandlimited signal with ω_B the highest frequency, we can choose the sampling period T=π/ω_B=0.498 s. To guard against loss of information in the sampling process (roundoff errors, etc.) we should sample at 2 to 5 times this rate or 0.10<T<0.25 s. If the signal is to be plotted or displayed on an oscilloscope we should consider sampling at 10 times the Nyquist rate or say T=0.05 s.

Introduction to Discrete Analysis

Example 11-6. Find a suitable sampling rate for the square wave of Fig. 11-1.

Solution. We will find a value of T such that the sampled sequence contains 95% of the average power of the original signal. From Example 11-2 the Fourier coefficients for $f(t)$ are $C_n = -2Aj/n\pi$ for n odd and $C_n = 0$ for n even. From Eq. (11-26) the average power is the sum of the $|C_n|^2$, or in this instance

$$P_{ave} = \sum_{n=-\infty}^{\infty} |C_n|^2 = \frac{4A^2}{\pi^2} \sum_{n=-\infty}^{\infty} 1/n^2 \qquad (11\text{-}34)$$

for n odd. Using symmetry to eliminate the negative indices yields

$$P_{ave} = \frac{8A^2}{\pi^2} \sum_{n=1,3,5...}^{\infty} 1/n^2 \qquad (11\text{-}35)$$

It can be shown that the summation has the value $\pi^2/8$, and therefore that $P_{ave} = A^2$. The power contained in the frequency components which correspond to the first few values of n are listed in the following table:

n	$8A^2/\pi^2 n^2$
1	$0.811A^2$
3	$0.090A^2$
5	$0.032A^2$
7	$0.017A^2$
	Total $0.950A^2$

The sum of the contributions for the first 4 odd harmonics is $0.95A^2$ or 95% of the average power. As a result, we can assume that the signal is bandlimited with the highest frequency ω_B given by $\omega_B = 7\omega_o = 7(2\pi/T_o)$. The corresponding sampling interval T is $T = \pi/\omega_B = T_o/14$. Hence we sample 14 times in each period of the square wave. To avoid errors in sampling, we might sample at 2 to 5 times this rate, or 28 to 56 times in each period. If we attempt to reconstruct the original square wave from sampled data which contains only the first four harmonic frequencies, it will resemble a slightly distorted square wave.

Frequency Folding

What happens if we violate the sampling theorem by choosing $f_s < 2f_B$, that is, if we sample at a rate less than twice the highest frequency contained in the spectrum of the signal? It seems plausible that we might simply lose the data at the higher frequencies, but the phenomenon which actually occurs is much more disastrous. We encounter <u>frequency folding</u>, whereby signal components whose frequencies are greater than the so-called folding frequency, defined by $f_f = f_s/2$ or $\omega_f = \pi/T$, appear as lower frequencies in the sampled data.[20,24] This spurious data can never be removed, and it often causes deleterious results in the subsequent analysis. High-frequency noise, for example, which might not have been a problem in the continuous signal, becomes low-frequency noise in the sampled signal.

To see why frequency folding occurs, consider the function $y(t)=A\cos(\omega_0 t)$ which yields the sampled sequence $y(k)=A\cos(\omega_0 kT)$. Decompose ω_0 as

$$\omega_0 = (N + p)\frac{\pi}{T} = (N + p)\omega_f \qquad (11\text{-}36)$$

where N is an integer and p is a real constant in the range $0<p<1$. This means that ω_0 is greater than the folding frequency $\omega_f = \pi/T$. Using Eq. (11-36), $y(k)$ becomes

$$y(k) = A\cos\{\pi k(N + p)\} \qquad (11\text{-}37)$$
$$= A\{\cos(N\pi k)\cos(p\pi k) - \sin(N\pi k)\sin(p\pi k)\} \qquad (11\text{-}38)$$

or, as $\sin(N\pi k)=0$ for all integers N and k,

$$y(k) = A\cos(N\pi k)\cos(p\pi k) \qquad (11\text{-}39)$$

Introduction to Discrete Analysis

If N is even, $\cos(N\pi k)=1$ and Eq. (11-39) becomes

$$y(k) = A\cos(p\pi k) = A\cos\{Tk(p\pi/T)\} \qquad (11-40)$$

The frequency of the sampled sinusoid $y(k)$ is then $\omega_p = p\pi/T$. This in turn means that the original sinusoid, which had frequency ω_o, appears in the sampled data as though it had the lower frequency ω_p. That ω_p is less than ω_o if $N \neq 0$ follows from Eq. (11-36).

If N is odd, then $\cos(N\pi k)=\cos(\pi k)$ and Eq. (11-39) becomes

$$y(k) = A\cos(\pi k)\cos(p\pi k) \qquad (11-41)$$

$$= \frac{A}{2}\{\cos(\pi k - p\pi k) + \cos(\pi k + p\pi k)\} \qquad (11-42)$$

$$= \frac{A}{2}\{\cos(\omega_q kT) + \cos(\omega_r kT)\} \qquad (11-43)$$

where $\omega_q=(1-p)\omega_f$ and $\omega_r=(1+p)\omega_f$. The frequency ω_q is less than the folding frequency for $0<p<1$ and hence contaminates the sampled data; ω_r is greater than ω_f and can be ignored.

In summary, if the time-domain signal contains a component at a frequency greater than the folding frequency, calculate N and p from Eq. (11-36) and use either $\omega_p=p\omega_f$ or $\omega_q=(1-p)\omega_f$ to find the frequency at which the component will appear in the sampled data.

Example 11-7. A function sampled at $T=0.1\pi$ s. contains noise interference at $f_n=60$ Hz. Will the interference contaminate the sampled data and if so at what frequency will it appear?

Solution. The folding frequency is $\omega_f=\pi/T$ or 10 rad/s. The radian frequency of the interference is $\omega_n=2\pi f_n=120\pi$ rad/s. From Eq. (11-36), $N+p=\omega_n/\omega_f=12\pi=37.70$. We identify $N=37$ and $p=0.7$. The relationship $\omega_q=(1-p)\omega_f$ then gives $\omega_q=10(1-0.7)=3.0$ rad/s from which $f_q=\omega_q/2\pi=0.477$ Hz. The interference does contaminate the sampled data and appears in it at a frequency of 0.477 Hz.

The practical consequence of frequency folding (or aliasing as it is often called) is that once the sampling rate has been fixed, the signal must be passed through a continuous low-pass filter to remove components at frequencies greater than the folding frequency. These high frequencies must be removed before the signal is sampled. After the signal has been sampled, the high frequencies will have been folded down to contaminate the sampled data and can be removed only in very special circumstances.

Figure 11-7 shows the magnitude spectrum of a signal and summarizes the relationships among the bandlimiting frequency ω_B, the sampling frequency ω_s, the Nyquist rate ω_N, and the folding frequency ω_f. Recall that $\omega_N = 2\omega_B$ and that $\omega_f = \omega_s/2$. Only frequencies below the folding frequency can be sampled. Frequencies above the folding frequency are folded downward into the interval $0 < \omega < \omega_f$. As a consequence, the low-pass bandlimiting filter should have its cutoff frequency equal to the folding frequency.

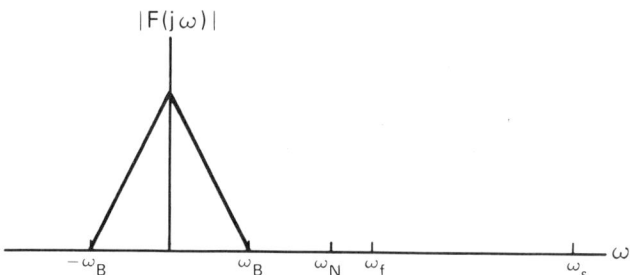

Figure 11-7. Frequency relationships. Recall that the Nyquist frequency is twice the bandlimiting frequency, $\omega_N = 2\omega_B$, and that the folding frequency is half of the sampling frequency, $\omega_f = \omega_s/2$.

11.3 SIGNAL CONVERSION

Section 11.2 introduced the concept of a sampled signal and treated the question of sampling rate. This section addresses another preliminary topic: how the actual sampling is

Introduction to Discrete Analysis

done using digital-to-analog and analog-to-digital converters.[25,26] In the material that follows we use the acronyms DAC and ADC for digital-to-analog and analog-to-digital converters.

Digital-To-Analog Conversion

The reader may well question our interest in this topic when our main goal is the reverse process, analog-to-digital conversion. The reasons are twofold: first, after the discrete data has been processed, we must often reconvert it to continuous form before it is plotted, displayed on an oscilloscope, or returned to the parent system. A second and more important reason for studying DAC's is that many practical analog-to-digital converters use them as component parts.

A DAC is a hardware device whose input is an N-bit digital word (usually presented to the device in parallel) and whose output is a unique voltage level which corresponds to each digital word. The digital words are delivered to the converter at a predetermined clock rate so that the output voltage is a "staircase" waveform. In practical systems DAC's are often followed by low-pass filters which smooth the reconstructed signal. Figure 11-8 shows a symbolic DAC and typical output waveforms. The following paragraphs discuss two commonly used DAC circuits.[27,28]

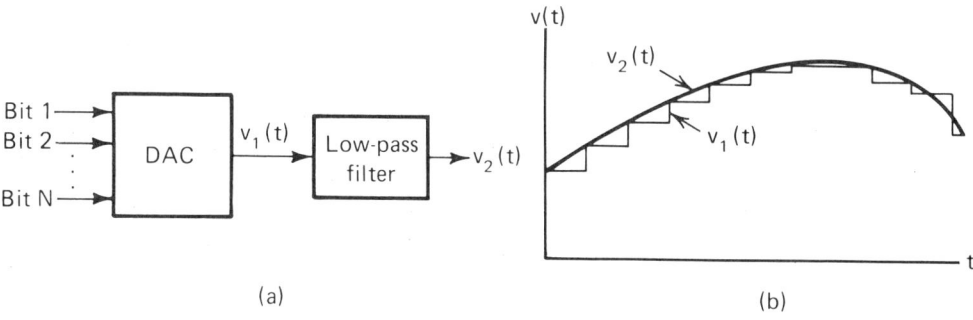

Figure 11-8. A digital-to-analog converter (DAC) and associated waveforms. (a) A DAC is often followed by a low-pass filter to smooth the "staircase" variations in the output; (b) Typical DAC output, $v_1(t)$, and low-pass filter output, $v_2(t)$.

The weighted-resistor DAC. Figure 11-9 is the circuit diagram for a weighted-resistor DAC. A constant reference voltage of E volts is applied at point A in the diagram. The input digital word is stored in the register. In this case the word length is four bits, although in practice longer word lengths are used. If a bit in the register is a logic 1, the associated switch in the resistor network is closed; as a consequence, each bit pattern in the register generates a unique total resistance for the resistor network. This equivalent resistance R_{AB}, developed between points A and B in the diagram, is the input resistance to an operational amplifier with feedback resistance R/2. The output voltage v(t) is thus given by the expression

$$v(t) = -\frac{RE}{2R_{AB}} \qquad (11\text{-}44)$$

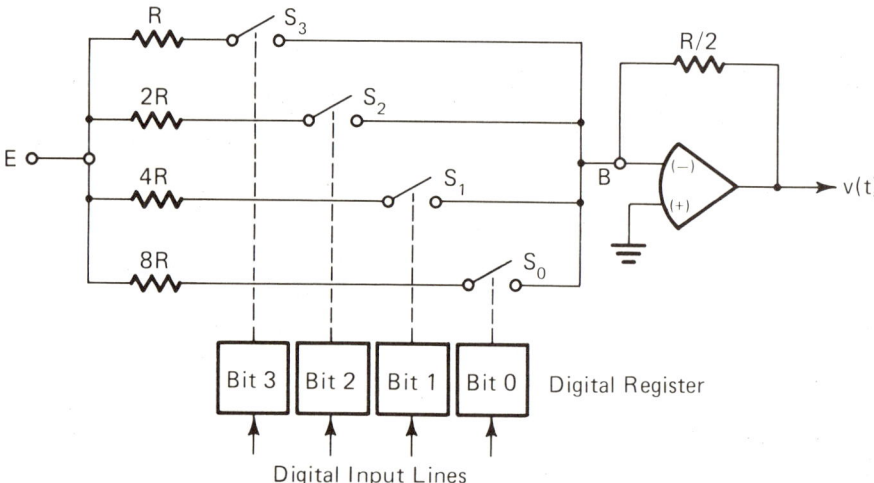

Figure 11-9. A four-bit weighted-resistor DAC. The digital input determines which switches are closed, thus establishing the equivalent input resistance of the operational amplifier.

In general, the resistor values needed to reconstruct the DAC vary from R to $(R)2^{N-1}$, and the output voltage varies in equal-valued steps separated by $1/2^N$ volts. The minimum output is 0 when all bits are logic 0's. The maximum output is

$E(1-1/2^N)$ when all bits are logic 1's. The digital value of the input must be a positive binary number between 0 and 1. For the four-bit converter this implies that the digital input has the form .0000 where bit 0 is on the right and bit 3 is on the left. A sign bit can be included with some extra hardware, and scaling can be used to move the assumed binary point.

The ladder-network DAC. The objection to the weighted-resistor DAC is that the resistor values must vary over a wide range, causing difficulties in the hardware implementation. The ladder-network DAC of Fig. 11-10 avoids this problem at the expense of using double-throw rather than single-throw switches. Only two resistor values are needed for the ladder-network DAC. The **switches** are in the upper position when the corresponding bits are logic 1's and are grounded otherwise. We leave a detailed analysis of this device to the interested reader.[29]

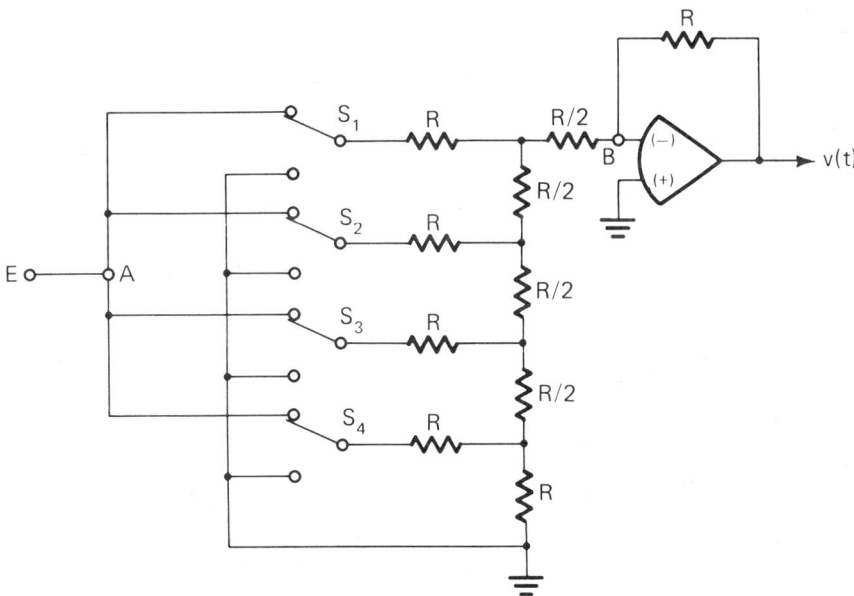

Figure 11-10. A four-bit ladder-network DAC. This DAC uses only two resistor values regardless of the number of bits.

Analog-To-Digital Conversion

An analog-to-digital converter (ADC) converts a continuous or analog signal into a binary word for use by the digital computer.[29,30] This process is more complicated than digital-to-analog conversion, especially for rapidly varying signals. The following paragraphs describe three commonly used ADC's.

<u>The multi-comparator ADC</u>. One approach to analog-to-digital conversion uses an array of comparators and a logic circuit.[30] The comparators respond to the analog input and produce outputs which are coded into a suitable bit pattern by a logic circuit. Figure 11-11 shows a two-bit converter capable of isolating the input value to one of four regions. The comparators C_A, C_B, and C_C compare the continuous input $v_{in}(t)$ with various fractions of the reference voltage $-E$. These fractions are created by applying the reference to a voltage divider. If the sum of the inputs to a comparator is positive, its output is a logic 1, otherwise the output is a logic 0. The remainder of the device consists of logic circuitry to construct a bit pattern whose binary value increases monotonically as $v_{in}(t)$ increases. The ADC of Fig. 11-11 works only for positive inputs in the range $0 < v_{in}(t) < E$. Table 11-1 summarizes the operation.

Table 11.1.

ADC Operation

Input Value	Comparator Output			Bit Values	
$v_{in}(t)/E$	C_A	C_B	C_C	Bit 1	Bit 2
0 to 0.25	0	0	0	0	0
0.25 to 0.50	0	0	1	0	1
0.50 to 0.75	0	1	1	1	0
0.75 to 1.0	1	1	1	1	1

Introduction to Discrete Analysis

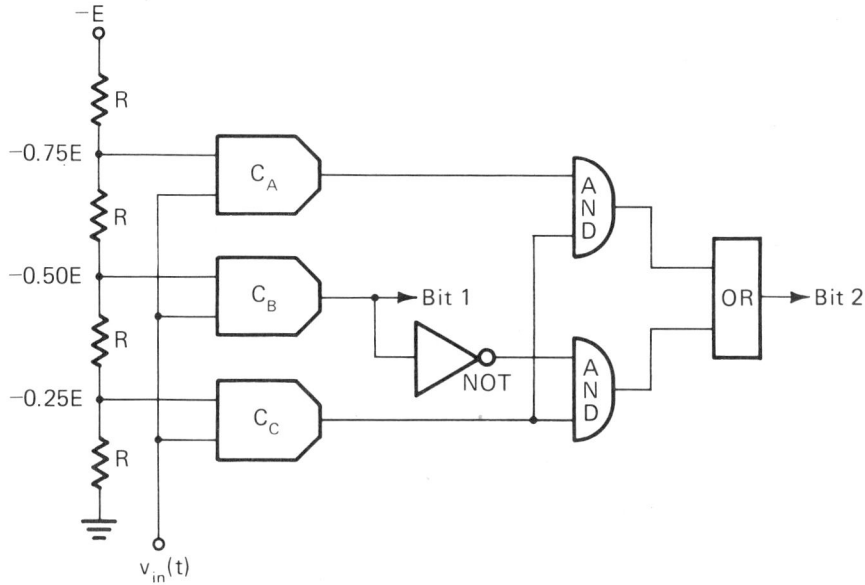

Figure 11-11. This two-bit multi-comparator analog-to-digital converter (ADC) uses three comparators, a voltage divider and a logic network.

While multi-comparator ADC's are fast and easily designed, they are usually impractical because of the large number of comparators needed. An N-bit converter takes 2^N-1 comparators (32,767 for a 15-bit converter)! Clearly, a multi-comparator ADC would be too expensive and too bulky for most real-world applications.

<u>The successive-approximation ADC</u>. The most popular analog-to-digital converter operates by making successive approximations to the unknown analog input voltage.[29,31] Figure 11-12 shows a four-bit successive-approximation converter with an assumed reference of E=1 volt. The device operates as follows:

A "Start" pulse clears the digital register, opens all of the switches, and activates the track-store unit. (A track-store is a device which remembers the value its input had during the "store" period and "tracks", or follows, its input during the track period.) The most significant bit, bit 1, is then changed to a logic 1 and switch S_1 is closed. If this change causes the comparator output to be a logic 1, bit 1 is accepted and S_2 remains closed. If the comparator output is a logic 0, bit 1 is rejected and S_1 is reopened. This process is repeated until all bits have been tested, at which time a "conversion-completed" pulse dispatches the output word from the register.

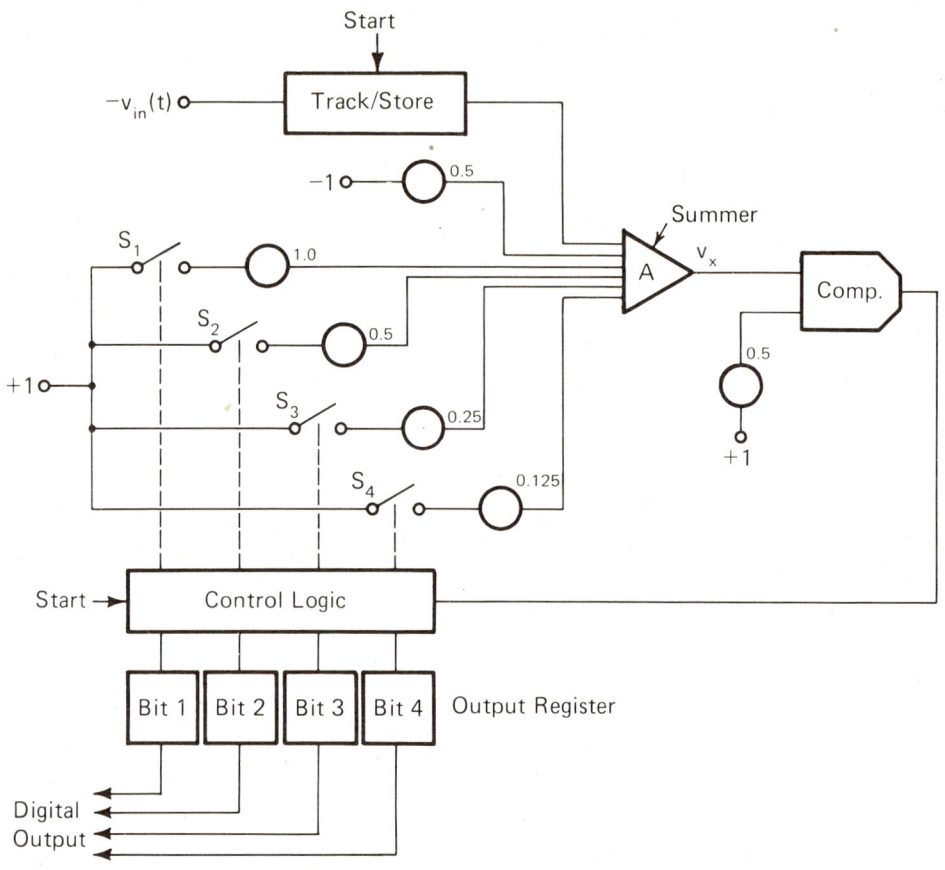

Figure 11-12. A successive-approximation ADC. This, the most popular ADC, is a compromise between complexity and operating speed.

The input $-v_x(t)$ of the summing amplifier A is given by

$$-v_x(t) = -v_{in}(t) - 0.5 + S_1 + 0.5S_2 + 0.25S_3 + 0.125S_4$$

(11-45)

Introduction to Discrete Analysis

where the S's are either 1 or 0 depending on whether the corresponding switch is closed or open. The comparator input $v_c(t)$ is $v_c(t)=v_x(t)+0.5$ or, using Eq. (11-45),

$$v_c(t) = v_{in}(t) + 1.0 - S_1 - 0.5S_2 - 0.25S_3 - 0.125S_4$$

(11-46)

Overall operation of the converter for an input of $v_{in}(t)=0.70$ volts is summarized in Table 11-2. The table indicates the switches which are closed at the beginning of each conversion step and the register value at the end of the step.

Table 11.2.

Successive-Approximation ADC Operation for $v_{in}(t) = 0.70$

Step	Switches Closed	Comparator Input	Comparator Output	Register Value
1	1	+0.700	1	1000
2	1,2	+0.200	1	1100
3	1,2,3	-0.0500	0	1100
4	1,2,4	+0.0750	1	1101

The table shows that when the conversion is completed the bit pattern is 1101. The left-hand bit indicates that the input is positive. The next three bits form a binary 5, or 101. This means that the input is greater than 5 times and less than 6 times the converter resolution R_N which is, in general, $R_N=1/2^{N-1}$. For the four-bit converter, $R_4=1/2^3=0.125$. The output of a binary 5 thus tells us that the input $v_{in}(t)$ lies in the range $5R_N<v_{in}<6R_N$ or $0.625<v_{in}<0.750$. This result is correct because the input was in fact $v_{in}(t)=0.70$.

The sign bit is a logic 1 for positive inputs and a logic 0 for negative inputs. If the sign bit is a 0, the magnitude of the negative input voltage is found by <u>complementing</u> (interchanging 1's and 0's) the other bit positions. This can be verified by a table similar to Table 11-2 but with S_1 open at all times. The bit pattern for an input of 0.70 was 1101 while that for -0.70 would be 0010.

The reader can also verify that the bit pattern forms an ascending binary count, 1000, 1001, 1010, ...1110, 1111, as the continuous input increases from +0.0 to +1.0, again assuming a reference of E=1 volt. (If E has a value other than 1.0 the range of operation is scaled accordingly.) The binary count increments each time the continuous input changes by 0.025E for the four-bit converter. The total time for a conversion is N clock intervals, where N is the number of bits in the binary output word. For this reason the successive-approximation ADC is considerably slower than the multi-comparator ADC (which requires only one clock interval for a conversion), but it has the advantage of requiring only one comparator regardless of the word length.

The time-interval analog-to-digital converter. The time-interval ADC is used in digital voltmeters and other applications where high speed is not essential. It is simple, reliable and inexpensive, but slow compared to the multi-comparator and successive-approximation ADC's. Figure 11-13 is the block diagram for a time-interval ADC. It operates as follows:

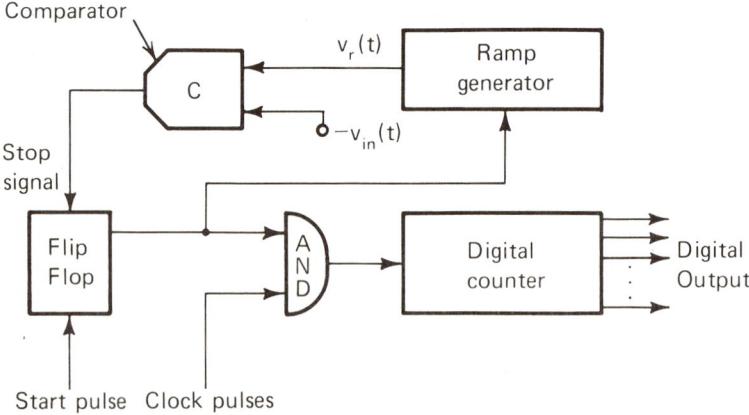

Figure 11-13. A time-interval ADC. This device is used in applications where high operating speed is not essential.

Introduction to Discrete Analysis 501

The start pulse sets the output of the flip-flop to a logic 1. The flip-flop output is directed to the ramp generator which starts generating a ramp-shaped voltage. The flip-flop output is also ANDed with the clock pulses, causing the counter to begin counting at the clock rate, creating a binary output equal to the number of clock pulses which have elapsed since the start pulse. The comparator C continually compares the output $v_r(t)$ of the ramp generator with the analog input $v_{in}(t)$. The instant the ramp generator output first exceeds the input, the comparator supplies a logic 1 which deactivates the flip-flop, thus changing its output from a logic 1 to a logic 0. This in turn halts the counter whose output is now proportional to the magnitude of $v_{in}(t)$. The next start pulse resets the counter and the ramp generator and initiates another conversion.

For an accurate conversion the integration rate of the ramp generator must be large compared to the maximum slope of the continuous input. Figure 11-14 shows typical waveforms associated with the device. The start-conversion pulses occur at the times labeled S_i in the figure.

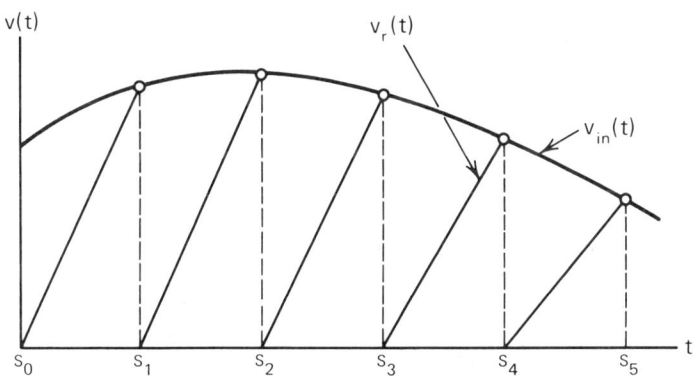

Figure 11-14. Waveforms for the time-interval ADC. When $v_r(t)$ first equals $v_{in}(t)$ the counter halts.

11.4 SUMMARY

This chapter introduced discrete analysis and discussed sampling and signal conversion. If our objective is to analyze the data from an actual process or from a working dynamic system, then both of these topics are important. If, on the other hand, we seek to analyze a proposed system design by simulating its mathematical model on a digital computer, then data conversion is not needed. We still must select a sufficiently high sampling rate to insure accurate, stable, and convergent operation of our numerical algorithms. With these preliminary topics out of the way, we turn to the problem of developing a language of analysis for discrete system models--the z-transform.

PROBLEMS FOR CHAPTER 11

11-1. List the first five values of the sampled versions of the following functions, assuming a sampling period T: (a) $y(t)=u(t)$, (b) $f(t)=\exp(-t/5)$, (c) $g(t)=3\cos(2\pi t/5)$, and (d) $f(t)=3+5t^2$.

11-2. Expand the functions of Fig. 11-15 in exponential Fourier series:

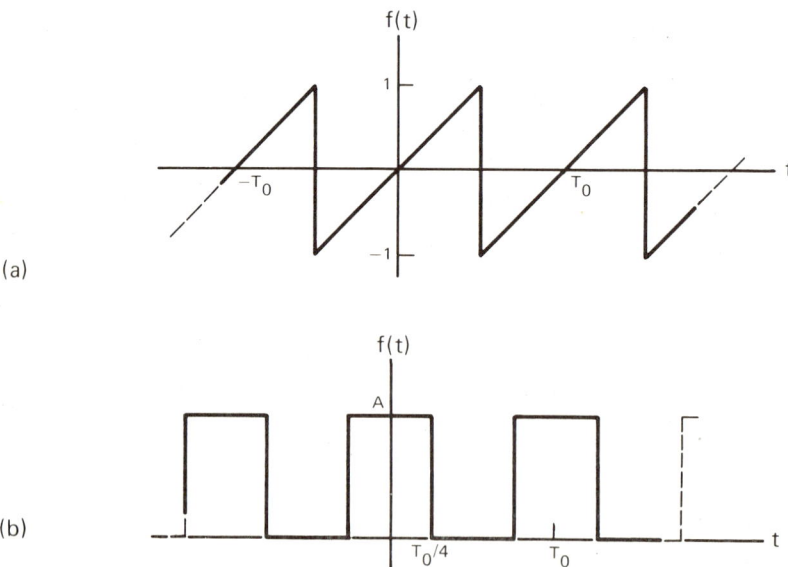

Figure 11-15. Problem 11-2.

Introduction to Discrete Analysis

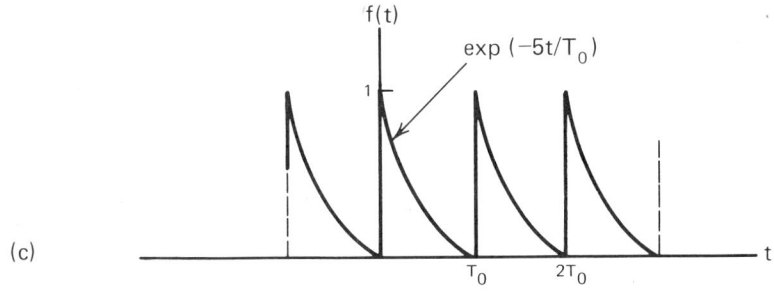

(c)

Figure 11-15. Problem 11-2, continued.

11-3. Expand the function of Problem 2-37(b) in a complex Fourier series.

11-4. Expand each of the functions of Problem 8-33 in a complex Fourier series.

11-5. Find the Fourier transforms of the functions shown in Fig. 11-16.

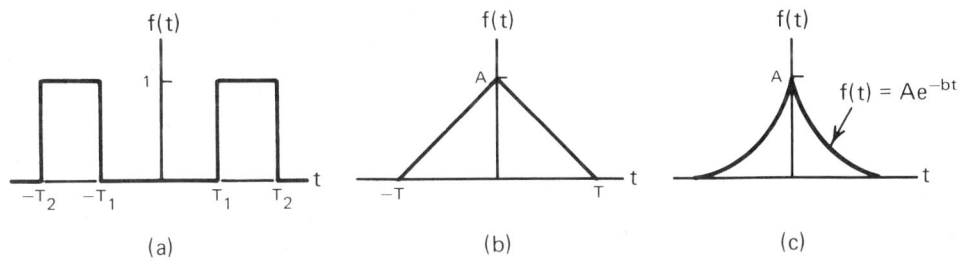

Figure 11-16. Problem 11-5.

11-6. Find the Fourier transforms of the following time-domain functions: (a) $f(t)=5\cos(2t)+12\sin(2t)$, (b) $f(t)=A\exp(-bt)u(t)$, (c) $g(t)=u(t)-u(t-T)$, and (d) $y(t)=8t\{u(t)-u(t-2)\}$.

11-7. The inverse Fourier transform of the magnitude spectrum of a system transfer function is the impulse response function of the system. Find the impulse response $g(t)$ for: (a) the ideal high-pass filter, (b) the ideal low-pass filter, and (c) the ideal bandpass filter. These responses are shown in Fig. 11-17.

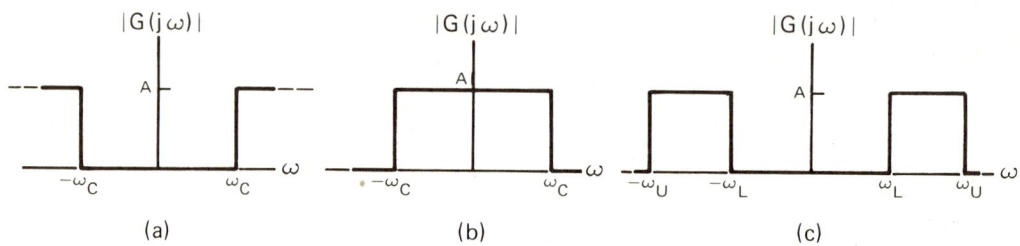

(a) (b) (c)

Figure 11-17. Problem 11-7. (a) The ideal high-pass magnitude response; (b) The ideal low-pass magnitude response; (c) The ideal band-pass magnitude response.

11-8. Calculate the Nyquist frequency in rad/s for each of the following signals: (a) $f(t)=\sin(3t)+\sin(t/3)$, (b)

$$f(t) = \sum_{i=0}^{20} \cos(2\pi i t) \qquad (11\text{-}47)$$

(c) the periodic function $f(t)$ whose complex Fourier coefficients are shown in Fig. 11-18. Assume $T_o=0.01$ s.

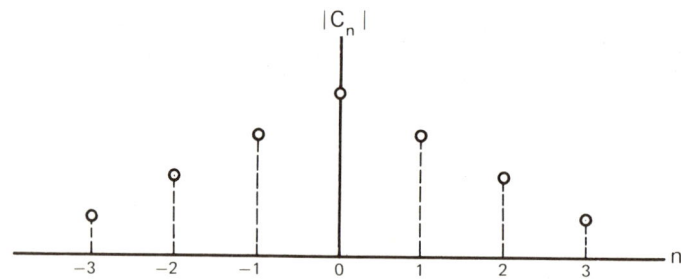

Figure 11-18. Problem 11-8(c).

(d) the function whose Fourier transform is shown in Fig. 11-19.

Introduction to Discrete Analysis

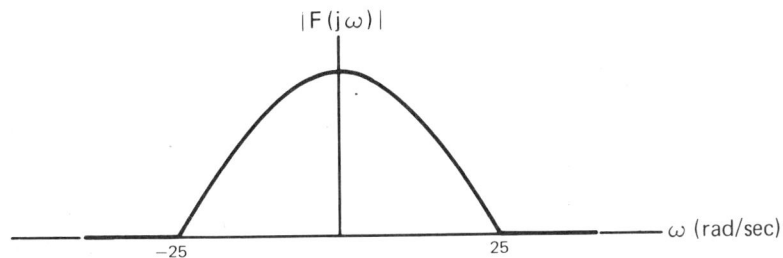

Figure 11-19. Problem 11-8(d).

11-9. Compute the total energy W_t dissipated if the voltage $v(t) = \sqrt{t\exp(-t)}$ volts is applied to a one-ohm resistor at t=0.

11-10. Compute the fraction of the average power contained in the first five harmonics of the function sketched in Problem 11-2(b).

11-11. Choose a sampling interval T which will insure that the sampled version of the solution of the differential equation $\dot{y}+5y=0$ with $y(0)=1$ will contain 99% of the energy of the continuous solution. (Hint: take Laplace transforms; then let $s=j\omega$ and follow Example 11-5.)

11-12. A function is sampled with a sampling frequency of $f_s=100$ Hz. The signal is known to contain interference at 60 Hz. (a) Will aliasing occur? (b) If so, at what frequency will the interference appear in the sampled signal?

11-13. Repeat Problem 11-12 if $f_s=10$ kHz and the interference frequency is 9.8 kHz.

11-14. A continuous signal is to be sampled with T= 0.0025 s. What must be the cutoff frequency in rad/s of an ideal low-pass bandlimiting filter which will preclude any possibility of aliasing in the sampled signal?

11-15. Design a three-bit weighted-resistor DAC.

11-16. Design a three-bit multi-comparator ADC. Make a table similar to Table 11-1 for the operation of the device.

11-17. Calculate the binary output of the successive approximation ADC of Fig. 11-12 if (a) $v_{in}(t)=0.865$ V, (b) $v_{in}(t)=0.25$ V, and (c) $v_{in}=0.48$ V. Assume E=1.0 V.

11-18. Calculate the resolution of an eight-bit successive-approximation ADC.

11-19. If the sampled function $\underline{y}(t)$ is defined by

$$\underline{y}(t) = \sum_{n=-\infty}^{\infty} y(t)\delta(t-nT) \qquad (11\text{-}48)$$

prove that

$$\underline{Y}(j\omega) = \frac{1}{T} \sum_{n=-\infty}^{\infty} Y\{j\omega-n\omega_s\} \qquad (11\text{-}49)$$

where $\omega_s = 2\pi/T$.

11-20. Prove Eq. (11-24).
11-21. Prove Eq. (11-26).
11-22. Prove Eq. (11-25).
11-23. The results of Problem 11-19 imply that if $Y(j\omega)$ has the magnitude spectrum shown in Fig. 11-20(a), then $\underline{Y}(j\omega)$ has the magnitude spectrum of Fig. 11-20(b). Show that if $\omega_s > 2\omega_B$, an ideal low-pass filter can be used to recover $y(t)$ from its sampled version $\underline{y}(t)$. Also show that if $\omega_s < 2\omega_B$, $y(t)$ cannot be recovered by a low-pass filter. This problem is a heuristic proof of the sampling theorem.

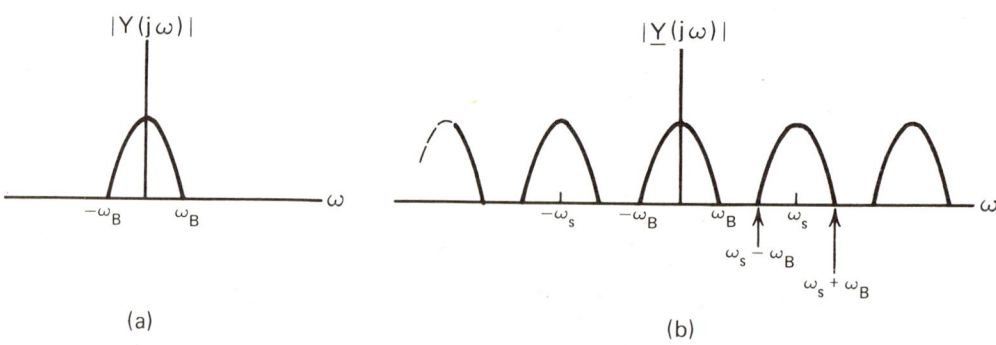

Figure 11-20. Problem 11-23.

Introduction to Discrete Analysis

11-24. The periodic function f(t) expressed in terms of its complex Fourier series expansion is

$$f(t) = \sum_{n=-\infty}^{\infty} c_n e^{jn\omega_0 t} \qquad (11\text{-}50)$$

Find the Fourier transform, $F(j\omega)$, of this expression.

REFERENCES FOR CHAPTER 11

(1) Sanderson, P.C.: *Computer Languages: A Practical Guide to the Chief Programming Languages*, Philosopical Library Inc., New York, 1970.

(2) Heaps, H.S.: *An Introduction to Computer Lanauages*, Prentice-Hall, Inc., Englewood Cliffs, New Jersey, 1972.

(3) Freiberger, W. F. and W. Prager: *Applications of Digital Computers*, Ginn and Company, Inc., Boston, Massachusetts, 1963.

(4) Korn, G. A.: *Minicomputers For Scientists and Engineers*, McGraw-Hill Book Co., Inc., New York, 1973.

(5) Perone, S. P. and D. O. Jones: *Digital Computers In Scientific Instrumentation*, McGraw-Hill Book Co., Inc., New York, 1973.

(6) Green, B. F., Jr.: *Digital Computers In Research*, McGraw-Hill Book Co., Inc., New York, 1963.

(7) Naylor, T. H., Balintfy, J. L., Burdick, D. S., and K. Chu: *Computer Simulation Techniques*, John Wiley and Sons, Inc., New York, 1966.

(8) Gruenberger, F., ed.: *The Transition To On-Line Computing*, The Thompson Book Co., Inc., Washington, D.C., 1967.

(9) Coury, F. F., ed: *A Practical Guide to Minicomputer Applications*, IEEE Press, New York, 1972.

(10) Gruenberger, F. and D. Babcock: *Computing With Mini Computers*, Melville Publishing Company, Los Angeles, California, 1973.

(11) Kuo, F. F. and J. F. Kaiser: *System Analysis By Digital Computer*, John Wiley and Sons, Inc., New York, 1966.

(12) Bowers, J. C. and S. R. Sedore: *SCEPTRE: A Computer Program For Circuit and Systems Analysis*, Prentice-Hall Inc., Englewood Cliffs, New Jersey, 1971.

(13) Chirlian, P. M.: *Signals, Systems and the Computer*, Intext Educational Publishers, Inc., New York, 1973.

(14) Taub, H. and D. L. Schilling: *Principles of Communications Systems*, McGraw-Hill Book Co., Inc., New York, 1971.

(15) Stein, S. and J. J. Jones: *Modern Communication Principles*, McGraw-Hill Book Co., Inc., New York, 1967.

(16) Lucky, R. W., Salz, J. and E. J. Weldon, Jr.: *Principles of Data Communication*, McGraw-Hill Book Co., Inc., New York, 1968.

(17) Kaplan, W.: *Operational Methods For Linear Systems*, Addison-Wesley Publishing Co., Inc., Reading, Massachusetts, 1962.

(18) Sneddon, I. N.: *Fourier Transforms*, McGraw-Hill Book Co., Inc., New York, 1951.

(19) Papoulis, A.: *The Fourier Integral And Its Applications*, McGraw-Hill Book Co., Inc., New York, 1962.

(20) Otnes, R. K. and L. Enochson: *Digital Time Series Analysis*, John Wiley and Sons, Inc., New York, 1972.

(21) Cooper, G. R. and C. D. McGillem: *Methods of Signals and System Analysis*, Holt, Rinehart and Winston, Inc., New York, 1967.

(22) Mathews, J. and R. L. Walker: *Mathematical Methods of Physics*, W. A. Benjamin, Inc., New York, 1964.

(23) Sokolnikoff, I. S. and R. M. Redheffer: *Mathematics of Physics and Modern Engineering*, McGraw-Hill Book Co., Inc., New York, 1958.

(24) Cadzow, J. A.: *Discrete-Time Systems*, Prentice-Hall, Inc., Englewood Cliffs, New Jersey, 1973.

(25) Schmid, H.: *Electronic Analog/Digital Conversions*, Van Nostrand Reinhold Company, New York, 1970.

(26) Hoeschele, D. F., Jr.: Analog-to-Digital/Digital-to-Analog Conversion Techniques, John Wiley and Sons, Inc., New York, 1968.

(27) Diefenderfer, J. A.: Principles of Electronic Instrumentation, W. B. Saunders Company, Philadephia, Pennsylvania, 1972.

(28) Huskey, H. D. and G. A. Korn, eds.: Computer Handbook, McGraw-Hill Book Co., Inc., New York, 1962.

(29) Durling, A.: Computational Techniques: Analog Digital and Hybrid Systems, Intext Educational Publishers, Inc., New York, 1974.

(30) Graeme, J. G., Tobey, G. E., and L. P. Huelsman, eds.: Operational Amplifiers: Design And Applications, McGraw-Hill Book Co., Inc., New York, 1971.

(31) Bekey, G. A. and W. J. Karplus: Hybrid Computation, John Wiley and Sons, Inc., New York, 1968.

CHAPTER 12

System Analysis by z-Transform

The most powerful tool for analyzing discrete-time systems is the z-transform, which plays a role in discrete analysis analogous to that of the Laplace transform in continuous analysis. Just as the Laplace transform changes differential equations into algebraic equations in the complex s-domain, the z-transform changes <u>difference equations</u> into algebraic equations in the complex z-domain. And where the Laplace transform leads to block diagrams, transfer functions, and convolution by multiplication for continuous systems, the z-transform does the same for discrete systems. Consequently, as a language with which to discuss the concepts of discrete analysis, the z-transform is unsurpassed.[1,2,3]

12.1 THE z-TRANSFORM DEFINED

This section defines the z-transform, shows how it relates to the Laplace transform, derives the transforms of several useful functions, and in general lays the foundation for using z-transforms for systems analysis.[4,5]

The Sampled Function

The Laplace transform F(s) of a time-domain function f(t) is defined by Eq. (5-1), repeated here as

$$L\{f(t)\} = F(s) = \int_0^\infty f(t)e^{-st}dt \qquad (12\text{-}1)$$

where s is the complex variable $s=\sigma+j\omega$. Using the sampling property of the impulse function, we showed in Chapter 5 that $L\{\delta(t)\}=1$. We also proved the time-delay theorem, Eq. (5-46), by which

$$L\{f(t-T)\} = e^{-sT}F(s) \qquad (12\text{-}2)$$

The significance of Eq. (12-2) is that a delay of T units in the time domain is equivalent to multiplication by exp(-sT) in the transform domain. Combining Eq. (12-2) and $L\{\delta(t)\}=1$ shows that the Laplace transform of an impulse function delayed by kT time units (k is an integer) is

$$L\{\delta(t-kT)\} = e^{-kTs} \qquad (12\text{-}3)$$

Consider next the sampled sequence y(k) for k=0,1,2,... . This sequence gives rise to a <u>sampled time function</u> y(t) according to the relationship

$$\underline{y}(t) = y(0)\delta(t) + y(1)\delta(t-T) + \ldots\ldots y(k)\delta(t-kT) + \ldots$$
$$(12\text{-}4)$$

System Analysis by z-Transform

or, using summation notation,

$$\underline{y}(t) = \sum_{k=0}^{\infty} y(k)\delta(t-kT) \qquad (12\text{-}5)$$

Figure 12-1(b) shows the sampled function $\underline{y}(t)$ associated with the parent continuous function $y(t)$ of Fig. 12-1(a). Note that $\underline{y}(t)$ is nonzero only at the sample points $t=kT$ and has the same value as the sequence $y(k)$ at those points.

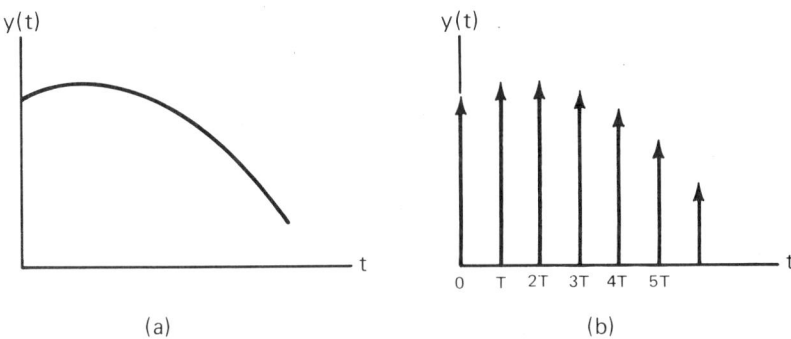

Figure 12-1. Continuous and sampled time functions. (a) A continuous time function; (b) A sampled function represented by a sequence of impulses.

We have introduced $\underline{y}(t)$ as an artifice to insure having a well-behaved time function which corresponds to the sampled sequence $y(k)$. Clearly, evaluating $\underline{y}(t)$ is a satisfactory way of discretizing a continuous function $y(t)$. Now let us compute $\underline{Y}(s)$, the Laplace transform of $\underline{y}(t)$. Substituting Eq. (12-5) into Eq. (12-1) yields

$$L\{\underline{y}(t)\} = \underline{Y}(s) = \int_0^{\infty} \sum_{k=0}^{\infty} y(k)\delta(t-kT) e^{-st} dt \qquad (12\text{-}6)$$

Under very loose conditions it is permissible to interchange the integral and the summation in Eq. (12-6) to obtain

$$\underline{Y}(s) = \sum_{k=0}^{\infty} \underline{y}(k) \int_{0}^{\infty} \delta(t-kT) e^{-st} dt \qquad (12-7)$$

But, according to the sampling property of the impulse function, the integral has the value exp(-skT), so

$$\underline{Y}(s) = \sum_{k=0}^{\infty} \underline{y}(k) e^{-skT} \qquad (12-8)$$

Equation (12-8) is our stepping stone from the Laplace transform to the z-transform.

The z-Transform

Define the complex variable z by the relationship

$$z = e^{-sT} \qquad (12-9)$$

from which s=ln(z)/T. Substituting Eq. (12-9) into Eq. (12-8) results in

$$Y(z) = \underline{Y}(s) \Big]_{s=\ln(z)/T} = \sum_{k=0}^{\infty} \underline{y}(k) z^{-k} \qquad (12-10)$$

Equation (12-10) is fundamentally important: it tells us that to compute the z-transform of a sampled function we multiply each sample y(k) by z^{-k} and sum the terms to ∞. As a notational convenience, we introduce Z{ } as the z-transform of the bracketed quantity. In addition, if the parent signal y(t) is sampled at interval T, then the following expressions all have identical meanings:

$$Z\{y(t)\} = Z\{\underline{y}(t)\} = Z\{y(k)\} = Y(z) \qquad (12-11)$$

System Analysis by z-Transform

In other words, we assume that the z-transforms of time functions, sampled functions and sequences are the same. To avoid notational confusion we will refer only to the z-transforms of sequences throughout the remainder of the book so that $Y(z)=Z\{y(k)\}$. The sampled function $\underline{y}(t)$, introduced to help bridge the gap between the Laplace and z-transforms, is not mentioned again.

The sampled sequence associated with a continuous time function is found by making the substitution $t=kT$, where k is an integer and T is the sampling period. This procedure discretizes the continuous time axis into a sequence of points spaced T units apart. If, for example, $y(t)=A\exp(-bt)$, then the associated sampled function is $y(k)=A\exp(-bkT)$. Whereas $y(t)$ was a function in the time domain, we say that $y(k)$ is a function in the <u>sequence domain</u>. Figure 12-2 shows both functions.

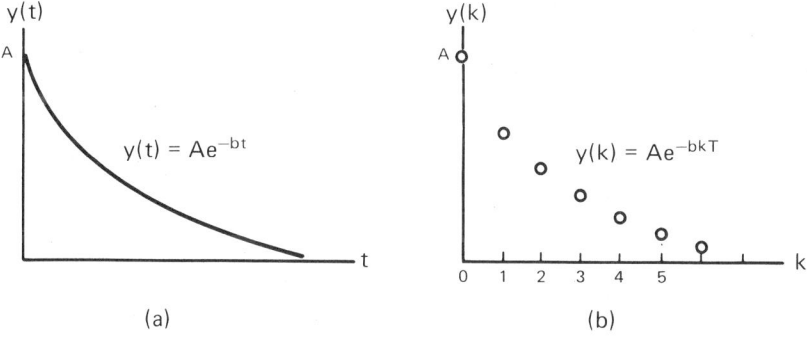

Figure 12-2. Time- and sequence-domain exponential functions. (a) The time-domain exponential function $y(t) = A\exp(-bt)$; (b) The sequence-domain exponential function $y(k) = A\exp(-bkT)$.

A final point: Recall that multiplication by $\exp(-sT)$ in the s-domain corresponds to a T-unit delay in the time domain, and that from the definition of z we have $z^{-1}=\exp(-sT)$. As a result, multiplication by z^{-k} in the z-domain corresponds to a delay of k sample intervals in the sequence domain.

The following examples show how to use Eq. (12-10) to evaluate z-transforms

Example 12-1. Define the discrete impulse function $\delta(k-i)$ by the relationship $\delta(k-i)=1$ for $k=i$ and $\delta(k-i)=0$ for $k \neq i$. Now compute $Z\{\delta(k)\}$. (The discrete impulse function is identical to the Kronecker delta function familiar to mathematicians. It plays the same role in discrete analysis as does the conventional impulse function in continuous analysis.)

Solution. Use Eq. (12-10):

$$Z\{\delta(k)\} = \sum_{k=0}^{\infty} \delta(k) z^{-k} \qquad (12\text{-}12)$$

As $\delta(k)$ is nonzero only for $k=0$,

$$Z\{\delta(k)\} = (1)z^0 + (0)z^{-1} + (0)z^{-2} + \ldots \qquad (12\text{-}13)$$

and $Z\{\delta(k)\}=1$.

Example 12-2. Calculate the z-transform of the unit impulse sequence

$$p(k) = \sum_{i=0}^{\infty} \delta(k-i) \qquad (12\text{-}14)$$

shown in Fig. 12-3(a).

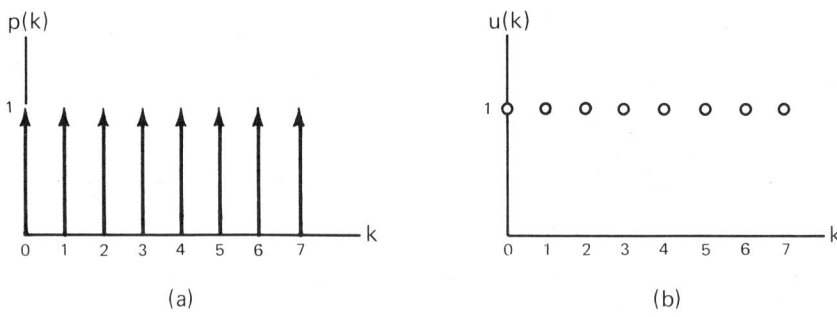

Figure 12-3. The discrete impulse sequence and the sampled unit step function have the same values at the sampling instants. (a) The unit impulse sequence; (b) The discrete step sequence.

System Analysis by z-Transform

Solution. Because $p(k)$ has the value 1 at every sample point, Eq. (12-10) yields

$$P(z) = \sum_{k=0}^{\infty} (1) z^{-k} \qquad (12\text{-}15)$$

Example 12-3. Calculate the z-transform of the sampled unit step function $u(k)$ of Fig. 12-3(b). An expression for $u(k)$ is $u(k)=(1)^k$ for $k \geqslant 0$.

Solution. Because $u(k)$ has the value 1 at the sample points, the z-transform is

$$Z\{u(k)\} = \sum_{k=0}^{\infty} (1) z^{-k} \qquad (12\text{-}16)$$

The z-transforms of the sampled step function and the impulse sequence are identical. This is because the functions have the same values at the sampling instants. The z-transform is not unique in the sense that it tells us nothing about the value of a function between the sampling instants. The only way to avoid ambiguity is to choose T to satisfy the sampling theorem.

Example 12-4. Find the z-transform of the exponential function $f(t)=\exp(-bt)$, assuming it is sampled at interval T.

Solution. The sampled sequence associated with $f(t)$ is $f(k)=\exp(-bkT)$ and Eq. (12-10) yields

$$F(z) = \sum_{k=0}^{\infty} e^{-bkT} z^{-k} \qquad (12\text{-}17)$$

These examples show that the z-transform of a function is easily computed as an infinite series. In many instances, however, it is possible to express the series in closed form. When this can be done, the resulting expressions are easier to work with than are the series expansions.

z-Transforms In Closed Form

Many of the sampled functions encountered in linear systems work can be written in the form $f(k)=A^k$, where A is a constant which may be either real or complex. Equation (12-10) gives the z-transform of $f(k)$ as

$$F(z) = \sum_{k=0}^{\infty} A^k z^{-k} \qquad (12\text{-}18)$$

To express $F(z)$ in closed form, define the partial series

$$S_n = 1 + Az^{-1} + A^2 z^{-2} + \ldots A^{n-1} z^{-(n-1)} \qquad (12\text{-}19)$$

Next compute $Az^{-1}S_n$:

$$Az^{-1}S_n = Az^{-1} + A^2 z^{-2} + \ldots A^n z^{-n} \qquad (12\text{-}20)$$

Subtracting Eq. (12-20) from Eq. (12-19) gives

$$S_n - Az^{-1}S_n = 1 - A^n z^{-n} \qquad (12\text{-}21)$$

Solving Eq. (12-21) for S_n results in

$$S_n = \frac{1-(Az^{-1})^n}{1-Az^{-1}} \qquad (12\text{-}22)$$

Since $F(z)=S_\infty$, we have

$$F(z) = \lim_{n \to \infty} S_n = \lim_{n \to \infty} \frac{1-(Az^{-1})^n}{1-Az^{-1}} \qquad (12\text{-}23)$$

If $|Az^{-1}|<1$, or, equivalently, $|z|>|A|$, then Eq. (12-23) converges as $n \to \infty$ to the value

$$F(z) = \frac{1}{1-Az^{-1}} = \frac{z}{z-A} \qquad (12\text{-}24)$$

If, on the other hand, $|z|<|A|$, the series diverges and $F(z)$ is unbounded. The borderline case occurs when $Az^{-1}=1$ or

System Analysis by z-Transform

$|z|=|A|$. In this case we go back to Eq. (12-18) and write

$$F(a) = \sum_{k=0}^{\infty} A^k A^{-k} = \sum_{k=0}^{\infty} (1) \qquad (12\text{-}25)$$

This expression diverges as $k \to \infty$ and has no closed form.

In summary, $Z\{A^k\}=z/(z-A)$ provided $|z|>|A|$. In general, values for z for which the z-transform of a function converges define the <u>region of convergence</u>; values which cause the series expansion to diverge define the <u>region of divergence</u>.

The Complex z-Plane

Just as the s-plane is useful in conjunction with the Laplace transform, the complex z-plane leads to additional insight into discrete analysis. Our first use of the z-plane is to provide a geometrical interpretation of convergence and divergence.[6]

The z-plane is a complex plane with axes that correspond to the real and imaginary parts of the complex variable z, Re{z} and Im{z}. According to this interpretation, the function $f(k)=A^k$ is seen to converge for values of z outside of a circle of radius $|A|$ and to diverge for values of z within or on the circle. (The equation $|z|=|A|$ is that of a circle when z is complex.) Figure 12-1 illustrates these ideas.

z-Transforms always have regions of convergence and divergence, although we generally ignore them in the text because we seldom need to evaluate z-domain expressions for specific values of z. The variable z is usually left in symbolic form as is s when we work with Laplace transforms.

<u>Example 12-5</u>. Express $Z\{u(k)\}$ in closed form.

<u>Solution</u>. The function u(k) can be written as $u(k)=1^k$, which is of the form $f(k)=A^k$ with A=1. As a result, Eq.

(12-24) applies and the closed-form expression is

$$Z\{u(k)\} = \frac{z}{z-1} \qquad (12\text{-}26)$$

Example 12-6. Find a closed-form expression for the z-transform of the sampled exponential function $f(k) = \exp(-bkT)$.

Solution. The given function can be written as

$$f(k) = \{e^{-bT}\}^k \qquad (12\text{-}27)$$

which has the form $f(k) = A^k$ with $A = \exp(-bT)$; hence Eq. (12-24) applies and

$$F(z) = \frac{z}{z - e^{-bT}} \qquad (12\text{-}28)$$

The transform converges for values of z outside of a circle of radius $\exp(-bT)$ in the z-plane.

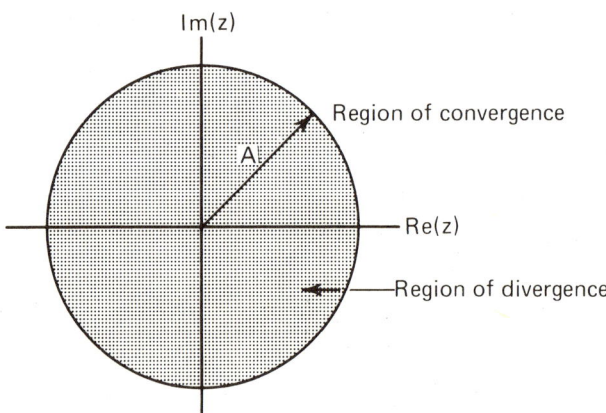

Figure 12-4. The series expansion for $Z\{A^k\}$ converges for values of z which lie outside of a circle of radius A in the complex z-plane.

Example 12-7. Find a closed-form expression for the z-transform of the sampled sinusoid $y(k) = \sin(\omega_o kT)$.

System Analysis by z-Transform

Solution. Write the sinusoid in terms of complex exponentials:

$$y(k) = \frac{\exp(j\omega_o kT) - \exp(-j\omega_o kT)}{2j} \quad (12\text{-}29)$$

$$= \frac{y_1(k) - y_2(k)}{2j} \quad (12\text{-}30)$$

The term

$$y_1(k) = \exp(j\omega_o T)^k \quad (12\text{-}31)$$

has the form $y_1(k) = A^k$ with $A = \exp(j\omega_o T)$. We have, therefore,

$$Y_1(z) = \frac{z}{z - \exp(j\omega_o T)} \quad (12\text{-}32)$$

Similarly,

$$Y_2\{(z)\} = \frac{z}{z - \exp(-j\omega_o T)} \quad (12\text{-}33)$$

and as $Y(z) = \{Y_1(z) - Y_2(z)\}/2j$,

$$Y(z) = \left\{ \frac{z}{z - \exp(j\omega_o T)} - \frac{z}{z - \exp(-j\omega_o T)} \right\} \frac{1}{2j} \quad (12\text{-}34)$$

Since $|\exp(j\omega_o T)| = 1$, $Y(z)$ converges for $|z| > 1$. Finally, Euler's identities can be used to rearrange Eq. (12-34) as

$$Z\{\sin(\omega_o kT)\} = \frac{z\sin(\omega_o T)}{z^2 - 2z\cos(\omega_o T) + 1} \quad (12\text{-}35)$$

Similar methods can be used to find closed-form expressions for the z-transforms of a number of other useful functions. Appendix F lists several z-transform pairs.

s-Plane z-Plane Relationships

The defining relationship $z = \exp(sT)$ can be used to map the s-plane into the z-plane as follows: let $s = \sigma + j\omega$ and express the complex variable z by $z = \alpha + j\beta$ where α and β are real variables. Substitute for s and z in the defining relationship to obtain

$$\alpha + j\beta = e^{(\sigma+j\omega)T} = e^{\sigma T}\{\cos(\omega T) + j\sin(\omega T)\} \qquad (12\text{-}36)$$

from which $\alpha = \exp(\sigma T)\cos(\omega T)$, $\beta = \exp(\sigma T)\sin(\omega T)$, and

$$\alpha^2 + \beta^2 = (e^{\sigma T})^2 \qquad (12\text{-}37)$$

Equation (12-37) is the equation of a circle of radius $\exp(\sigma T)$ in the z-plane. Points for which $\sigma=0$, which correspond to the $j\omega$-axis in the s-plane, map onto the <u>unit circle</u> in the z-plane since for $\sigma=0$ Eq. (12-37) becomes $\alpha^2 + \beta^2 = 1$. When σ is negative, we have points in the left half of the s-plane and inside the unit circle in the z-plane. Figure 12-5 illustrates these ideas.

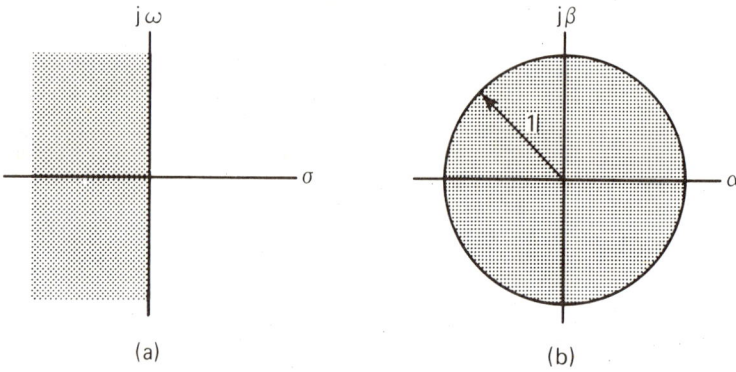

(a)   (b)

Figure 12-5. The interior of the unit circle in the z-plane maps onto the left half of the s-plane.

The implications of this mapping for stability should be obvious. Stable continuous systems have s-domain transfer functions whose poles lie in the left half of the s-plane; hence discrete systems have z-domain transfer functions whose poles lie inside the unit circle in the z-plane. We pursue the topic of stability in a later section.

System Analysis by z-Transform

12.2 THE INVERSE z-TRANSFORM

When using Laplace transforms we must often recover y(t) from Y(s). Similarly, we often need to recover the sampled sequence y(k) from its z-transformed version Y(z). Three methods for doing this are: (1) the inversion formula, (2) partial fractions, and (3) the power-series method. We will consider only the latter two techniques.

The Partial-Fraction Method

Recall that an s-domain function can be decomposed by partial fractions to the form

$$y(s) = \frac{A}{s+a} + \frac{B}{s+b} + \frac{C}{s+c} + \ldots \qquad (12\text{-}38)$$

with the inverse transform

$$y(t) = Ae^{-at} + Be^{-bt} + Ce^{-ct} + \ldots \qquad (21\text{-}39)$$

A similar method applies to the z-transform, but because $Z\{a^k\} = z/(z-a)$ we prefer to expand the quotient $Y(z)/z$ rather than $Y(z)$ alone. If, for example,

$$\frac{Y(z)}{z} = \frac{A}{z-p_1} + \frac{B}{z-p_2} + \frac{C}{z-p_3} \qquad (12\text{-}40)$$

then

$$Y(z) = \frac{Az}{z-p_1} + \frac{Bz}{z-p_2} + \frac{Cz}{z-p_3} \qquad (12\text{-}41)$$

and the sampled sequence y(k) is

$$y(k) = A(p_1)^k + B(p_2)^k + C(p_3)^k \qquad (12\text{-}42)$$

The following examples demonstrate the partial-fraction method for inverse z-transforms. All of the standard partial-fraction techniques apply--the methods for handling repeated roots, complex roots, etc.,--and Appendix F is used to find the inverse transforms of the individual terms.

Example 12-8. Find the inverse z-transform of the function

$$Y(z) = \frac{z}{(z-1)(z-0.25)} \qquad (12\text{-}43)$$

Solution. Recalling that we prefer to expand $Y(z)/z$ we write

$$\frac{Y(z)}{z} = \frac{1}{(z-1)(z-0.25)} = \frac{A}{z-1} + \frac{B}{(z-0.25)} \qquad (12\text{-}44)$$

$$= \frac{A(z-0.25)+B(z-1)}{(z-1)(z-0.25)} \qquad (12\text{-}45)$$

and

$$\frac{1}{(z-1)(z-0.25)} = \frac{z(A+B)-(B+0.25A)}{(z-1)(z-0.25)} \qquad (12\text{-}46)$$

Equating like powers of z in Eq. (12-46) yields the simultaneous equations $A+B=0$ and $-(0.25A+B)=1$. These are solved to obtain $A=4/3$ and $B=-4/3$. The partial-fraction expansion is then

$$Y(z) = \frac{4}{3}\left\{\frac{z}{z-1} - \frac{z}{z-0.25}\right\} \qquad (12\text{-}47)$$

Equation (12-24) can be used to evaluate both of these terms with the result

$$y(k) = \frac{4}{3}\{(1)^k-(0.25)^k\} \qquad (12\text{-}48)$$

where we could have written the first term on the right as 1 or $u(k)$ instead of $(1)^k$. The alert reader may wonder what happens if the numerator does not contain a convenient power of z so we can expand $Y(z)/z$. The following example shows one way to handle this situation.

Example 12-9. Find the inverse z-transform of $F(z)=1/\{z(z-0.5)\}$.

Solution. We force the form $F(z)/z$ by writing

$$\frac{F(z)}{z} = \frac{1}{z^2(z-0.5)} = \frac{A}{z} + \frac{B}{z^2} + \frac{C}{z-0.5} \qquad (12\text{-}49)$$

System Analysis by z-Transform

Finding a common denominator and equating numerators gives

$$1 = z^2(A + C) + z(B - 0.5A) - 0.5B \qquad (12\text{-}50)$$

The simultaneous equations, $A+C=0$, $B-0.5A=0$, and $-0.5B=1$, are solved for $A=-4$, $B=-2$, and $C=4$. The partial-fraction expansion is

$$F(z) = -4 - \frac{2}{z} + \frac{4z}{z-0.5} \qquad (12\text{-}51)$$

and the sampled sequence is

$$f(k) = -4\delta(k) - 2\delta(k-1) + 4(0.5)^k \qquad (12\text{-}52)$$

where entries 1 and 2 of Appendix F were used for the first two terms, respectively. In a later section we show a simpler way to handle problems of this type.

The Power-Series Method

This method gives the sampled sequence point-by-point rather than in closed form. Recall that

$$Y(z) = \sum_{k=0}^{\infty} y(k) z^{-k} \qquad (12\text{-}53)$$

If we expand $Y(z)$ as a power series in z^{-k}, then the coefficients of the terms in the power series are the sequence points. The expansion is done by long division after $Y(z)$ has been arranged in the polynomial form

$$Y(z) = \frac{a_m z^m + a_{m-1} z^{m-1} + \ldots\ldots a_1 z + a_0}{b_n z^n + b_{n-1} z^{n-1} + \ldots\ldots b_1 z + b_0} \qquad (12\text{-}54)$$

For most real-world systems $n > m$. Dividing numerator by denominator gives the series

$$Y(z) = C_1 z^{m-n} + C_2 z^{m-n-1} + C_3 z^{m-n-2} + \ldots \qquad (12\text{-}55)$$

or, using summation notation,

$$Y(z) = \sum_{k=n-m}^{\infty} C_k z^{-k} \qquad (12\text{-}56)$$

Comparing Eqns. (12-53) and (12-56) shows $y(k)=C_k$ for $k=n-m$, $n-m+1$, $n-m+2$, ... If $n-m \neq 0$, say $n-m=j$, then $y(k)=0$ for $k=0,1,\ldots j-1$. The power-series method is easier to implement on a digital computer than the partial-fraction technique, a fact which somewhat offsets the inconvenience of having the inverse transform as a list of points rather than in closed form.

Example 12-10. Find the inverse transform of $F(z)= z/(z-1)$ by the power-series method.
Solution. By long division,

$$\begin{array}{r} 1+z^{-1}+z^{-2}+\ldots \\ z-1 \overline{\smash{\big)}\ z\phantom{-1-z^{-1}-z^{-1}-z^{-2}}} \\ \underline{z-1\phantom{-1-z^{-1}-z^{-1}-z^{-2}}} \\ 1\phantom{-z^{-1}-z^{-1}-z^{-2}} \\ \underline{1-z^{-1}\phantom{-z^{-1}-z^{-2}}} \\ z^{-1}\phantom{-z^{-1}-z^{-2}} \\ \underline{z^{-1}-z^{-2}} \\ z^{-2} \end{array}$$

The coefficients of the quotient are all 1, hence

$$F(z) = \sum_{k=0}^{\infty} (1) z^{-k} \qquad (12\text{-}57)$$

and the sampled sequence is $f(k)=(1)^k$.

12.3 THEOREMS OF THE z-TRANSFORM

Listed below are a number of useful theorems of the z-transform. Some of the theorems have proofs which are trivial and hence are omitted. The proofs of others are too difficult for our introductory treatment.[7]

System Analysis by z-Transform

Linearity

The z-transform is linear and hence satisfies the principle of superposition:

$$Z\{ay_1(k) + by_2(k)\} = aZ\{y_1(k)\} + bZ\{y_2(k)\} \quad (12\text{-}58)$$

The proof is trivial and is left as an exercise.

The Shifting Theorems

Two theorems show how sequences can be shifted to the right or left on the sequence axis by multiplying by appropriate powers of z.

The right-shifting theorem. If $y(k)=0$ for $k<0$, then

$$Z\{y(k-n)\} = z^{-n} Z\{y(k)\} \quad (12\text{-}59)$$

To prove this theorem, we have, by definition,

$$Z\{y(k-n)\} = \sum_{k=0}^{\infty} y(k-n) z^{-k} \quad (12\text{-}60)$$

which can be rewritten as

$$Z\{y(k-n)\} = z^{-n} \sum_{k=0}^{\infty} y(k-n) z^{-(k-n)} \quad (12\text{-}61)$$

If $y(k)=0$ for $k<0$, then $y(k-n)=0$ for $k<n$ and the lower limit on the summation of Eq. (12-61) can be changed from 0 to n:

$$Z\{y(k-n)\} = z^{-n} \sum_{k=n}^{\infty} y(k-n) z^{-(k-n)} \quad (12\text{-}62)$$

Introduce the new index $j=k-n$ and Eq. (12-62) becomes

$$Z\{y(k-n)\} = z^{-n} \sum_{j=0}^{\infty} y(j) z^{-j} \quad (12\text{-}63)$$

If we recognize the summation on the right as $Z\{y(k)\}$, then

Eq. (12-63) is identiical to Eq. (12-59) and the theorem is proved.

<u>The left-shifting theorem</u>. This theorem states that

$$Z\{y(k+n)\} = z^n\left[Y(z) - \sum_{k=0}^{n-1} y(k)z^{-k}\right] \qquad (12\text{-}64)$$

The proof is similar to that for the right-shifting theorem and is left as an exercise.

Complex Translation

The relationship

$$Z\{a^k y(k)\} = Y(za^{-k}) \qquad (12\text{-}65)$$

which follows at once from Eq. (12-10), is useful in developing the transforms of composite functions.

The Limit Theorems

Two limit theorems enable us to evaluate the initial and final values of the sequences associated with z-domain functions without finding their inverse transforms.

<u>The initial-value theorem</u>. The initial sequence value $y(0)$ is given by

$$y(0) = \lim_{z \to \infty}\{Y(z)\} \qquad (12\text{-}66)$$

provided the limit exists. The proof is obvious after writing $Y(z)$ as a power series.

<u>The final-value theorem</u>. If the function $(1-z^{-1})Y(z)$ has no poles outside of the unit circle in the z-plane, then

$$y(\infty) = \lim_{k \to \infty}\{y(k)\} = \lim_{z \to 1}\{(1-z^{-1})Y(z)\} \qquad (12\text{-}67)$$

We omit the proof because it involves some tedious algebra which would add little to our understanding of the z-transform.

System Analysis by z-Transform

The z-transform has many other useful and interesting properties, but we have introduced only those theorems which apply directly to our goal of learning how to analyze continuous systems by discrete methods. The interested reader can pursue z-transforms in depth in any of a number of excellent sources.[1,3,6]

Example 12-11. Find the z-transform of the delayed step sequence $u(k-j)$ defined by

$$u(k-j) = \underbrace{0,0,0,\ldots.0}_{j\text{ zeros}},1,1,\ldots. \qquad (12\text{-}68)$$

Solution. Use the right-shifting theorem, Eq. (12-59), to obtain

$$Z\{u(k-j)\} = z^{-j} Z\{u(k)\} = \frac{z^{-j} z}{z-1} \qquad (12\text{-}69)$$

and $Z\{u(k-j)\} = z^{1-j}/(z-1)$.

Example 12-12. Find the z-transform of the delayed impulse sequence $\delta(k-j)$ defined by

$$\delta(k-j) = \underbrace{0,0,0\ldots.0}_{j\text{ zeros}},1,0,0,0,\ldots \qquad (12\text{-}70)$$

Solution. Again apply the right-shifting theorem to get $Z\{\delta(k-j)\} = z^{-j} Z\{\delta(k)\}$ and $Z\{\delta(k-j)\} = z^{-j}$.

Delayed steps and impulses often appear in the solution sequences of the difference equations which describe continuous systems.

Example 12-13. Use Eq. (12-65) and the results of Example 12-7 to find the z-transform of

$$y(k) = e^{-akT} \sin(\omega_o kT) \qquad (12\text{-}71)$$

Solution. Use Eqns. (12-35) and (12-65), to obtain

$$Y(z) = \frac{z e^{aT} \sin(\omega_o T)}{z^2 e^{2aT} - 2z e^{aT} \cos(\omega_o T) + 1} \qquad (12\text{-}72)$$

Example 12-14. Given the function

$$Y(z) = \frac{z^2}{(z-1)(z^2 - 0.5z + 0.8)} \qquad (12\text{-}73)$$

determine $y(0)$ and $y(\infty)$ without finding the inverse z-transform.

Solution. Equation (12-66) gives the initial value as $y(0)=Y(\infty)$. In this example $y(0)=0$. Equation (12-67), the final-value theorem, yields

$$y(\infty) = \lim_{z \to 1}\{(1 - z^{-1})Y(z)\} \qquad (12\text{-}74)$$

$$= \lim_{z \to 1}\left\{\frac{z}{z^2 - 0.5z + 0.8}\right\} \qquad (12\text{-}75)$$

which results in $y(\infty)=0.7692$.

This completes our introduction to z-transform theory. In the remainder of the chapter we turn to the more pragmatic problem of using z-transforms to analyze discrete systems.

12.4 DISCRETE SYSTEMS

A discrete system is a system whose dependent variables are digital data functions.[2,8] Like the definition of a system, the definition of a discrete system is not meant to be rigorous but is intended as a crude categorization. The following are examples of discrete systems:

1. A physical device whose variables are discrete.
2. A digital filter realized either as hardware or as a computer program.
3. A numerical algorithm implemented on a digital computer.
4. A discrete transform.
5. The stock market.
6. A continuous system whose variables have been discretized.

From these examples we see that discrete systems are not restricted to any particular engineering or scientific discipline, and more important, they can be either physical devices or mathematical abstractions.

Many of the familiar properties of continuous systems apply to discrete systems--linearity and time invariance for example. In addition, a number of the techniques used in continuous systems analysis can be extended or reinterpreted for use in working with discrete systems. Among these are convolution, transforms, state variables, frequency-domain analysis, transfer functions, and block diagrams.

Because our purpose is to learn how to analyze continuous dynamic systems with the aid of a digital computer, it should be clear that pursuing discrete systems does not lead us astray. In fact, the <u>only</u> way to study a continuous system by digital methods is to discretize the mathematical model of the system. That a knowledge of z-transforms and discrete analysis enables the student to solve other kinds of problems is, as the advertising people say, "an extra, added bonus."

Difference Equations

A discrete system can be modelled mathematically by one or more difference equations of the form

$$y(k-n) + a_{n-1} y(k-n-1) + \ldots a_1 y(k-1) + a_0 y(k) =$$
$$b_0 f(k) + b_1 f(k-1) + \ldots b_m f(k-m) \qquad (12\text{-}76)$$

with the initial conditions $y(-1), y(-2), \ldots y(-n)$. The following comments apply to the system represented by Eq. (12-76):[9,10]

1. The system is linear if all of the a_i are constants, linear and <u>sequence-varying</u> if at least one of the a_i is a function of the index k, nonlinear if one of the a_i depends on $y(k)$, and nonlinear and sequence-varying if one of the a_i depends on both k and $y(k)$.

2. The system is of the nth order because to evaluate y(k) we need n values of y(k) from the past. We also need m+1 values of f(k).

3. The function f(k) is the input or forcing function of the system.

4. The difference equation describes a single-input single-output system because there is a single dependent variable y(k) and a single difference equation.

The parallelism between differential and difference equations is fortunate in that many of our hard-won continuous analysis techniques apply with minor modifications to the discrete world. With little extra effort, the student proficient in analyzing continuous systems can become equally proficient in analyzing their discrete counterparts.

Solving Difference Equations

A difference equation is said to be "solved" by the sequence y(k) if, when substituted into the difference equation, the sequence points y(k) reduce the equation to an identity.[11] (The identity must hold for all points in the sequence y(k).)

Example 12-15. Show that the sequence $y(k)=(0.5)^k$ is a solution of the difference equation

$$y(k) - 0.5y(k-1) = \delta(k) \qquad (12\text{-}77)$$

with y(-1)=0.

Solution. When k=0, $\delta(k)$=1 and Eq. (12-77) becomes y(0)-0.5y(-1)=1. Substituting $y(0)=(0.5)^0=1$ along with y(-1)=0 reduces this expression to an identity as required. For k≥1, $\delta(k)$=0 and Eq. (12-77) becomes y(k)-0.5y(k-1)=0. Substituting $y(k)=(0.5)^k$, leads to $(0.5)^k - 0.5(0.5)^{k-1}=0$ which is also an identity.

If one is content with a solution expressed as a list of points in the sequence domain, then any difference equation can be solved iteratively. Just evaluate each succeeding value of

System Analysis by z-Transform

y(k) in terms of the previous ones using the given difference equation and its initial conditions. The process is easily implemented on a digital computer.

Example 12-16. Derive the first five values of y(k) for the difference equation $y(k)-0.5y(k-1)=2^k$ with $y(-1)=2$.

Solution. Starting with k=0, the solution sequence is computed from the given difference equation:

$$y(0) = 2^0 + 0.5(2) = 2 \qquad (12\text{-}78)$$

$$y(1) = 2^1 + 0.5(2) = 3 \qquad (12\text{-}79)$$

$$y(2) = 2^2 + 0.5(3) = 5.5 \qquad (12\text{-}80)$$

$$y(3) = 2^3 + 0.5(5.5) = 10.75 \qquad (12\text{-}81)$$

$$y(4) = 2^4 + 0.5(10.75) = 21.375 \qquad (12\text{-}82)$$

.
.
.

This straightforward but tedious solution technique leaves us with two intriguing questions: (1) Under what conditions does a <u>closed-form</u> expression exist for y(k), a formula which will generate any value of y(k) without our first having to generate all of the preceeding ones, and (2) given that such a closed form is known to exist, how does one proceed to find it? How, for example, can we derive $y(k)=(0.5)^k$ as the solution to Eq. (12-77)? Both of these questions are answered by a judicious use of the z-transform.

z-Transform Solutions Of Difference Equations

The procedure for solving a difference equation using z-transforms parallels that for solving a differential equation by Laplace transforms. We take the z-transform of the difference equation, solve for the transform of the dependent variable, say Y(z), and then compute the inverse z-transform to obtain the solution sequence y(k).[12,13] Let us apply this procedure to the difference equation

$$y(k) - 1.5y(k-1) + 0.5y(k-2) = \delta(k) \qquad (12\text{-}83)$$

with $y(-1)=y(-2)=0$. Because the z-transform is linear, superposition applies and we can transform the difference equation term by term. This gives

$$Z\{y(k)\} = Z\{\delta(k)\} + 1.5Z\{y(k-1)\} - 0.5Z\{y(k-2)\} \quad (12\text{-}84)$$

Examine each term of Eq. (12-84). The transform of the discrete impulse is $Z\{\delta(k)\}=1$. The term $Z\{y(k)\}$ is, by definition, $Y(z)$, and the transforms $Z\{y(k-1)\}$ and $Z\{y(k-2)\}$ are found using the right-shifting theorem, Eq. (12-59), to be $Z\{y(k-1)\}=z^{-1}Y(z)$ and $Z\{y(k-2)\}=z^{-2}Y(z)$. Using these results Eq. (12-84) becomes

$$Y(z) = 1 + 1.5z^{-1}Y(z) - 0.5z^{-2}Y(z) \qquad (12\text{-}85)$$

Solving for $Y(z)$ yields

$$Y(z) = \frac{1}{1 - 1.5z^{-1} + 0.5z^{-2}} = \frac{z^2}{(z-1)(z-0.5)} \qquad (12\text{-}86)$$

Expand $Y(z)/z$ by partial fractions with the result

$$\frac{Y(z)}{z} = \frac{2}{z-1} - \frac{1}{z-0.5} \qquad (12\text{-}87)$$

Restore the z to the right-hand side and

$$y(k) = 2(1)^k - (0.5)^k \qquad (12\text{-}88)$$

For most purposes the closed-form solution is superior to a list of points in the sequence domain. For one thing, the initial and steady-state values are immediately available as $y(0)=1$ and $y(\infty)=2$. The steady-state value may not be evident from the point-by-point solution. Finally, notice that the correct values for $y(-1)$ and $y(-2)$ cannot be found from Eq. (12-88) by substituting $k=-1$ and $k=-2$. This is because z-transform solutions are good only for $k \geqslant 0$, a consequence of the definition of the z-transform by an infinite summation starting at $k=0$.

System Analysis by z-Transform

Example 12-17. Use z-transforms to solve the difference equation $y(k)=u(k)+y(k-1)$ with $y(-1)=0$.

Solution. Take z-transforms to obtain

$$Y(z) = \frac{z}{z-1} + z^{-1}Y(z) \qquad (12\text{-}89)$$

from which

$$Y(z) = \frac{z^2}{(z-1)^2} \qquad (12\text{-}90)$$

Entry 12 of Appendix F gives the output sequence $y(k) = (k+1)(1)^k$.

Initial conditions. When z-transforms are used to solve difference equations, the initial conditions are by default $y(-1)=y(-2)=\ldots y(-n)=0$. The output sequence starts at $k=0$, and $y(0)$ is a computed value, not an initial condition. If the initial conditions are not all 0, we can incorporate them by using an extension of the right-shifting theorem to the situation where $y(k) \neq 0$ for $k<0$. Consider, for example, the term $y(k-1)$ and assume that $y(-1) \neq 0$. By definition,

$$Z\{y(k-1)\} = \sum_{k=0}^{\infty} y(k-1)z^{-k} \qquad (12\text{-}91)$$

Expand the sum by writing the first term separately:

$$Z\{y(k-1)\} = y(-1) + \sum_{k=1}^{\infty} y(k-1)z^{-k} \qquad (12\text{-}92)$$

Let $j=k-1$ and Eq. (12-92) becomes

$$Z\{y(k-1)\} = y(-1) + \sum_{j=0}^{\infty} y(j)z^{-(j+1)} \qquad (12\text{-}93)$$

$$= y(-1) + z^{-1}\sum_{j=0}^{\infty} y(j)z^{-j} \qquad (12\text{-}94)$$

Recognizing the summation as the definition of Y(z) yields the result

$$Z\{y(k-1)\} = y(-1) + z^{-1}Y(z) \qquad (12-95)$$

Following a similar procedure we can show that

$$Z\{y(k-2)\} = y(-2) + z^{-1}y(-1) + z^{-2}Y(z) \qquad (12-96)$$

and so on.

Example 12-18. Solve the difference equation

$$y(k) = 1.5y(k-1) - 0.5y(k-2) \qquad (12-97)$$

with $y(-1)=1$ and $y(-2)=5$.

Solution. z-Transforming Eq. (12-97) using the extended right-shifting theorem yields

$$Y(z) = 1.5\{y(-1)+z^{-1}Y(z)\} - 0.5\{y(-2)+z^{-1}y(-1)+z^{-2}Y(z)\}$$

$$(12-98)$$

Inserting the given values for y(-1) and y(-2) and solving for Y(z) yields

$$y(z) = \frac{-z(z + 0.5)}{z^2 - 1.5z + 0.5} \qquad (12-99)$$

$$= \frac{-z(z + 0.5)}{(z - 1)(z - 0.5)} = \frac{-3z}{z - 1} + \frac{2z}{z - 0.5}$$

$$(12-100)$$

The output sequence is then $y(k)=-3(1)^k+2(0.5)^k$. Again, the solution sequence is good only for $k \geq 0$; it does not give the correct values for y(-1) and y(-2). However, y(0) is correct, as can be verified from Eq. (12-97).

When a set of nonzero values $y(0), y(1), \ldots y(n-1)$ are associated with an nth-order difference equation, the values for $k=0, 1, \ldots n-1$ will not be computed correctly by the z-transform. The solution must be forced to fit the initial conditions by some other means. The best approach is to use the initial values and the difference equation to calculate a suitable set of values, $y(-1), y(-2), \ldots y(-n)$, and then use the extended right-shifting theorem. The procedure is best illustrated by an example.

System Analysis by z-Transform

Example 12-19. Solve the difference equation

$$y(k) - 0.25y(k-1) = u(k) \qquad (12\text{-}101)$$

if $y(0)=56$.

Solution. Substitute $k=0$ into Eq. (12-101) and solve for the correct value of $y(-1)$; thus, $y(0)-0.25y(-1)=1$ and $y(-1)=220$. Transform Eq. (12-101) using the extended shifting theorem to obtain

$$Y(z) - 0.25\{y(-1) + z^{-1}Y(z)\} = \frac{z}{z-1} \qquad (12\text{-}102)$$

from which

$$Y(z) = \frac{55z}{z-0.25} + \frac{4z/3}{z-1} - \frac{z/3}{z-0.25} \qquad (12\text{-}103)$$

The solution sequence is then

$$y(k) = 55(0.25)^k + \frac{4(1)^k}{3} - \frac{(0.25)^k}{3} \qquad (12\text{-}104)$$

As a check, when $k=0$ Eq. (12-104) gives $y(0)=56$ as required.

12.5 CONVOLUTION

Convolution was introduced briefly in Chapter 5 where we pointed out that s-domain multiplication corresponds to convolution in the time domain. We did not dwell on time-domain convolution because it is used more for theoretical work than as a practical analysis method. Multiplying in the s-domain is generally easier. The situation is different with discrete functions.[14,15] Integration is replaced by summation with the result that convolution of discrete functions is emminently practical, particularly if a digital computer is available. Moreover, just as in the continuous world, discrete convolution provides additional theoretical insight into the behavior of linear systems.

The Convolution Summation

Recall from Chapter 5, Eq. (5-285), that if a linear system has input f(t), the output y(t) can be found from the relationship

$$y(t) = \int_0^t f(q)h(t-q)dq \qquad (12-105)$$

where q is a dummy variable of integration and h(t - q) is the response of the system at time t due to an impulse function input applied at time q. The roles of f(t) and h(t) in Eq. (12-105) have been reversed with respect to f(t) and g(t) in Eq. (5-285). The two forms are equivalent as can be shown by making the change of variables r=t-q in Eq. (12-105). If we allow for the possibility that the system is <u>anticipatory</u>, or that it could have an output before the input is applied, Eq. (12-105) takes the more general form

$$y(t) = \int_0^\infty f(q)h(t-q)dq \qquad (12-106)$$

Although anticipatory systems do not exist in nature, the generalization makes the subsequent derivation easier without jeopardizing the applicability of the results to real-world systems.

For discrete systems work we must use sampled functions in Eq. (12-106) and replace the integral by a summation. The result is

$$y(i) = \sum_{k=0}^\infty f(k)h(i-k) \qquad (12-107)$$

Equation (12-107) is the discrete convolution summation. To relate it to the z-transform we proceed as follows: Expand Eq. (12-107) as

$$y(i) = f(0)h(i) + f(1)h(i-1) + f(2)h(i-2) + \ldots \qquad (12-108)$$

System Analysis by z-Transform

By definition, Y(z) is given by

$$Y(z) = \sum_{i=0}^{\infty} y(i)z^{-i} \qquad (12\text{-}109)$$

Substitute Eq. (12-108) into Eq. (12-109) to get

$$Y(z) = \sum_{i=0}^{\infty} \{f(0)h(i) + f(1)h(i-1) + \ldots\}z^{-i} \qquad (12\text{-}110)$$

$$= \sum_{i=0}^{\infty} f(0)h(i)z^{-i} + \sum_{i=0}^{\infty} f(1)h(i-1)z^{-i} + \ldots \qquad (12\text{-}111)$$

$$= f(0)\sum_{i=0}^{\infty} h(i)z^{-i} + f(1)\sum_{i=0}^{\infty} h(i-1)z^{-i} + \ldots \qquad (12\text{-}112)$$

Use the shifting property of the z-transform to obtain

$$Y(z) = f(0)\sum_{i=0}^{\infty} h(i)z^{-i} + f(1)z^{-1}\sum_{i=0}^{\infty} h(i)z^{-i} + \ldots \quad (12\text{-}113)$$

Each summation is recognized as being H(z), so that Eq. (12-113) becomes

$$Y(z) = \{f(0) + f(1)z^{-1} + f(2)z^{-2} + \ldots\}H(z) \quad (12\text{-}114)$$

Finally, writing the bracketed quantity as a summation gives

$$Y(z) = \sum_{k=0}^{\infty} f(k)z^{-k}H(z) \qquad (12\text{-}115)$$

We recognize the summation as the z-transform of f(k). The final expression is then

$$Y(z) = F(z)H(z) \qquad (12\text{-}116)$$

Equation (12-116) says that multiplication of z-transforms is equivalent to convolution of sequences, thus giving us a powerful tool for analyzing linear discrete systems. To exploit this tool, we introduce the concept of a discrete transfer function in the next section.

A final word: The impulse response sequence h(k) used in the convolution summation is the response of the system to an impulse sequence $\delta(k)=1,0,0\ldots$ Remember too that the convolution summation applies only when the initial conditions on the system are zero. This restriction is equivalent to the requirement that s-domain transfer functions are defined only for zero initial conditions.

Example 12-20. Find the impulse response of the system described by the difference equation $y(k)=f(k)+0.5y(k-1)$ with $y(-1)=0$.

Solution. Let $f(k)=\delta(k)$; then take z-transforms to obtain

$$Y(z) = 1 + 0.5z^{-1}Y(z) \qquad (12\text{-}117)$$

Solve for $Y(z)$ which, because of the choice of $f(k)=\delta(k)$, is the impulse response $H(z)$; thus,

$$Y(z) = H(z) = \frac{z}{z - 0.5} \qquad (12\text{-}118)$$

and $h(k)=(0.5)^k$. Equipped with h(k) and the convolution summation, we could find the response of the system to any other input sequence, and the process could be programmed on a digital computer.

Example 12-21. A discrete linear system has the impulse response sequence found in Example 12-20. Find the response of the system to the input sequence $f(k)=2u(k)$.

Solution. As the z-transform of f(k) takes a simple form in this instance, we prefer z-domain multiplication to the convolution summation. Thus, $Y(z)=H(z)F(z)$ where, from Example 12-20,

$$H(z) = \frac{z}{z - 0.5} \qquad (12\text{-}119)$$

System Analysis by z-Transform

From Appendix F, $F(z)=2z/(z-1)$. Multiplying $H(z)$ and $F(z)$ yields

$$Y(z) = \frac{2z^2}{(z-1)(z-0.5)} \qquad (12\text{-}120)$$

Expanding $Y(z)/z$ by partial fractions results in

$$Y(z) = \frac{4z}{(z-1)} - \frac{2z}{(z-0.5)} \qquad (12\text{-}121)$$

from which $y(k)=4(1)^k-2(0.5)^k$.

12.6 DISCRETE TRANSFER FUNCTIONS

In Chapter 7 we showed that a continuous linear system can be described by an s-domain transfer function $G(s)$ which relates the output function $Y(s)$ to the input $F(s)$ by the expression $Y(s)=G(s)F(s)$, provided the system contains no stored energy. A parallel situation exists for discrete systems.[16,17]

Consider the nth-order difference equation

$$y(k)+a_1 y(k-1)+\ldots a_n y(k-n)=b_0 f(k)+b_1 f(k-1)+\ldots b_m f(k-m) \qquad (12\text{-}122)$$

with initial conditions $y(-1)=y(-2)=\ldots y(-n)=0$ and with $m<n$. We have taken $a_0=1$ with no loss of generality. Take z-transforms of Eq. (12-122) term by term, applying the shifting theorem as needed, to obtain

$$\{1+a_1 z^{-1}+a_2 z^{-2}+\ldots a_n z^{-n}\}Y(z) = \{b_0+b_1 z^{-1}+\ldots b_m z^{-m}\}F(z) \qquad (12\text{-}123)$$

Solve for $Y(z)$:

$$Y(z) = \frac{b_0 + b_1 z^{-1} + b_2 z^{-2} + \ldots b_m z^{-m}}{1 + a_1 z^{-1} + a_2 z^{-2} \ldots \ldots a_n z^{-n}} F(z) \qquad (12\text{-}124)$$

Equation (12-124) has the form

$$Y(z) = G(z)F(z) \qquad (12\text{-}125)$$

where

$$G(z) = \frac{b_0 + b_1 z^{-1} + \ldots + b_m z^{-m}}{1 + a_1 z^{-1} + a_2 z^{-2} + a_n z^{-n}} \qquad (12\text{-}126)$$

This expression can be converted to positive powers of z by multiplying numerator and denominator by z^n. $G(z)$ is called the <u>discrete transfer function</u> or the <u>pulse transfer function</u> of the system described by Eq. (12-122). Figure 12-6 shows the block diagram that goes with Eq. (12-125).

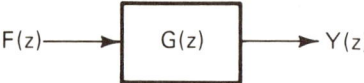

Figure 12-6. The z-domain transfer function, which satisfies the relationship $Y(z) = F(z)G(z)$.

It is evident from Eq. (12-125) that if $F(z)=1$ (which implies that the system input is the discrete impulse sequence) then $G(z)$ is the z-transform of the impulse response sequence of the system.

Example 12-22. Find the transfer function of the discrete system described by the difference equation

$$y(k) - 0.5y(k-1) + 0.125y(k-2) = f(k) - 5f(k-1) \qquad (12\text{-}127)$$

Solution. z-Transform Eq. (12-127) to obtain

$$(1 - 0.5z^{-1} + 0.125z^{-2})Y(z) = (1 - 5z^{-1})F(z) \qquad (12\text{-}128)$$

and solve for

$$G(z) = \frac{Y(z)}{F(z)} = \frac{1 - 5z^{-1}}{1 - 0.5z^{-1} + 0.125z^{-2}} \qquad (12\text{-}129)$$

or, in terms of positive powers of z,

$$G(z) = \frac{z^2 - 5z}{z^2 - 0.5z + 0.125} \qquad (12\text{-}130)$$

System Analysis by z-Transform

Nonzero Initial Conditions

Even if the initial conditions are nonzero we can still derive a system transfer function by introducing an artificial change to the input function F(z). This is done by applying the method of Section 12.4 for handling nonzero initial conditions. The procedure is best illustrated by an example.

Example 12-23. Derive a transfer function for the system described by $y(k)-0.45y(k-1)=f(k)$ with $y(-1) \neq 0$.

Solution. Transform the difference equation using the extended right-shifting theorem. The result is

$$Y(z) - 0.45\{y(-1) + z^{-1}Y(z)\} = F(z) \qquad (12\text{-}131)$$

from which

$$Y(z)(1 - 0.45z^{-1}) = F(z) + 0.45y(-1) \qquad (12\text{-}132)$$

If we now define a pseudo-input function $\underline{F}(z)$ by $\underline{F}(z)=F(z)+0.45y(-1)$ then the transfer function between $\underline{F}(z)$ and $Y(z)$ is

$$G(z) = \frac{Y(z)}{\underline{F}(z)} = \frac{z}{z - 0.45} \qquad (12\text{-}133)$$

The Algebra of Discrete Transfer Functions

The main use of transfer functions is in representing discrete systems as interconnections of elementary transfer function blocks.[18] It is easy to show that z-domain transfer functions satisfy the same "algebra" as do s-domain transfer functions. Transfer functions in series multiply as shown in Fig. 12-7, those in parallel add (Fig. 12-8), and the feedback configuration satisfies the familiar G/(1+GH) relationship of Fig. 12-9.

If a complicated system consists of transfer function blocks, we can use these algebraic relationships to reduce the system to a single overall transfer function. The procedure parallels that discussed in Chapter 7 for s-domain transfer functions. We will not repeat it here.

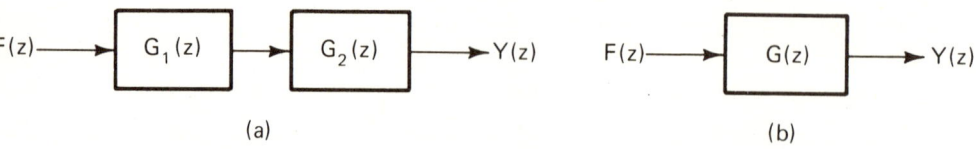

Figure 12-7. System (b) is equivalent to system (a). (a) Discrete transfer functions in series (or cascade); (b) An equivalent system.

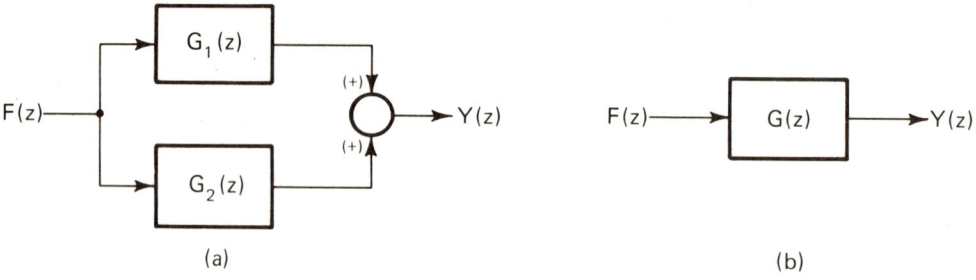

Figure 12-8. Transfer functions in parallel combine by addition. System (b) is equivalent to system (a) if $G(z) = G_1(z) + G_2(z)$. (a) Discrete transfer functions in parallel; (b) An equivalent system.

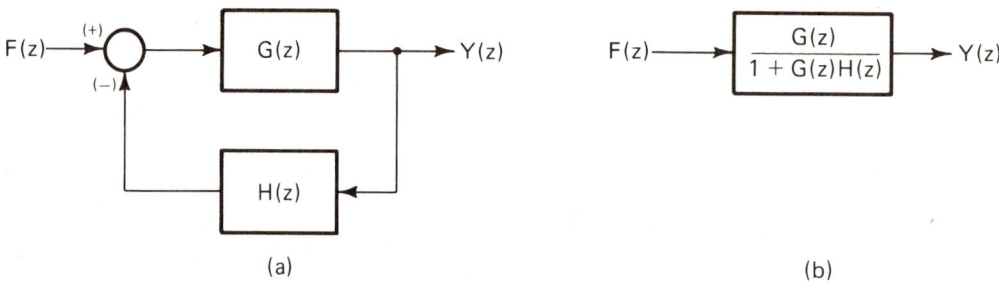

Figure 12-9. Transfer functions in the feedback configuration. (a) Discrete transfer functions in the feedback configuration; (b) An equivalent system.

12.7 POLES, ZEROS AND TRANSIENTS

We showed in Section 12.1 that the $j\omega$-axis in the s-plane maps into the unit circle in the z-plane, and that the left half of the s-plane lies within the unit circle. As a result,

System Analysis by z-Transform

a stable system--one for which a bounded input sequence produces a bounded output sequence--must have all of the poles of its z-domain transfer function within the unit circle. Moreover, the relative locations of the poles of the transfer function within the unit circle have a profound effect on the type of transient response the system will exhibit. This section classifies system response according to pole-zero location.

Poles and Zeros of G(z)

The general z-domain transfer function, Eq. (12-126), is a rational function (a ratio of polynomials) in z which can be arranged in the factored form

$$G(z) = K \frac{(z-z_1)(z-z_2) \ldots (z-z_m)}{(z-p_1)(z-p_2) \ldots (z-p_n)} \qquad (12\text{-}134)$$

where the z_i are the zeros of the transfer function, the p_i are its poles, and K is a constant gain factor. The poles and zeros are either real or, if complex, occur in complex-conjugate pairs. In either case it is the poles p_i which must lie within the unit circle for stability. This means that if a pole p_i is real then $|p_i|<1$. If p_i is complex, $p_i = c+jd$, for example, then the requirement for stability is $\sqrt{c^2+d^2}<1$. The location of the zeros does not effect stability.

Transient And Steady-State Response

Let Y(z) be the output sequence of a system whose transfer function G(z) is given by Eq. (12-134) and whose input F(z) has the form

$$F(z) = \frac{N(z)}{(z-q_1)(z-q_2) \ldots (z-q_r)} \qquad (12\text{-}135)$$

where N(z) is a numerator polynomial. The output Y(z) is then

$$Y(z) = \frac{N(z)}{(z-q_1)(z-q_2)\ldots(z-q_r)} \left\{ \frac{K(z-z_1)(z-z_2)\ldots(z-z_m)}{(z-p_1)(z-p_2)\ldots(z-p_n)} \right\}$$
(12-136)

Equation (12-136) can be expanded by partial fractions as $Y(z)=K_o+Y_q(z)+Y_p(z)$ where $Y_q(z)$ is due to the input sequence and is given by

$$Y_q(z) = \frac{Q_1 z}{z-q_1} + \frac{Q_2 z}{z-q_2} + \ldots + \frac{Q_r z}{z-q_r} \qquad (12\text{-}137)$$

The function $Y_p(z)$ is due to the poles of G(z) and is given by

$$Y_p(z) = \frac{P_1 z}{z-p_1} + \frac{P_2 z}{z-p_2} + \ldots + \frac{P_n z}{z-p_n} \qquad (12\text{-}138)$$

The output sequence y(k) is then $y(k)=K_o \delta(k)+y_q(k)+y_p(k)$ where

$$y_q(k) = Q_1(q_1)^k + Q_2(q_2)^k + \ldots + Q_r(q_r)^k \qquad (12\text{-}139)$$

and

$$y_p(k) = P_1(p_1)^k + P_2(p_2)^k + \ldots + P_n(p_n)^k \qquad (12\text{-}140)$$

If the input sequence is bounded as $k \to \infty$, then $|q_i|<1$ for $i=0,1,2,\ldots r$. If y(k) is a bounded sequence (if the system is stable) then the p_i must also be less than 1 in magnitude, which means that the poles of G(z) must lie within the unit circle.

Many practical systems have input sequences which never decay to zero--step, ramp, or sinusoidal sequences, for example. Hence we define the steady-state response as that part of the output sequence generated by the input sequence $y_q(k)$. And because for stability the terms due to the transfer function poles must tend to zero, we define the transient response as that part of the output sequence $y_p(k)$ associated with the poles of the transfer function.

System Analysis by z-Transform

Pole Location And Transient Response

The type of transient response is determined by the locations of the poles of G(z) relative to the unit circle in the z-plane. The following paragraphs investigate the transient response associated with various types of poles.(2,14)

Real poles. If G(z) contains a non-repeated real pole at z=p, the partial-fraction expansion for Y(z) contains the term

$$Y_1(z) = \frac{Az}{z - p} \qquad (12\text{-}141)$$

where A is a real constant. Equation (12-141) gives rise to the sequence

$$y_1(k) = A(p)^k \qquad (12\text{-}142)$$

The response $y_1(k)$ takes one of the following forms:

* A sequence of increasing numbers if p>1 (unstable).
* A constant sequence of magnitude A if p=1 (conditionally stable).
* A monotonically decreasing sequence for 0<p<1 (stable).
* A sequence of decreasing numbers which alternate in sign if -1<p<0 (stable).
* A sequence of magnitude A with terms which alternate in sign for p=-1 (conditionally stable).
* An increasing sequence which alternates in sign if p<-1 (unstable).

If p=0 the response is the discrete impulse Aδ(k). Refer to Figs. 12-10 and 12-11. For each case a plot of the pole location is shown along with a typical output sequence.

Complex conjugate poles. If a transfer function contains a pair of complex-conjugate poles, Y(z) contains a term $Y_2(z)$ of the form

$$Y_2(z) = \frac{A_1 z}{z - p_1} + \frac{A_2 z}{z - p_2} \qquad (12\text{-}143)$$

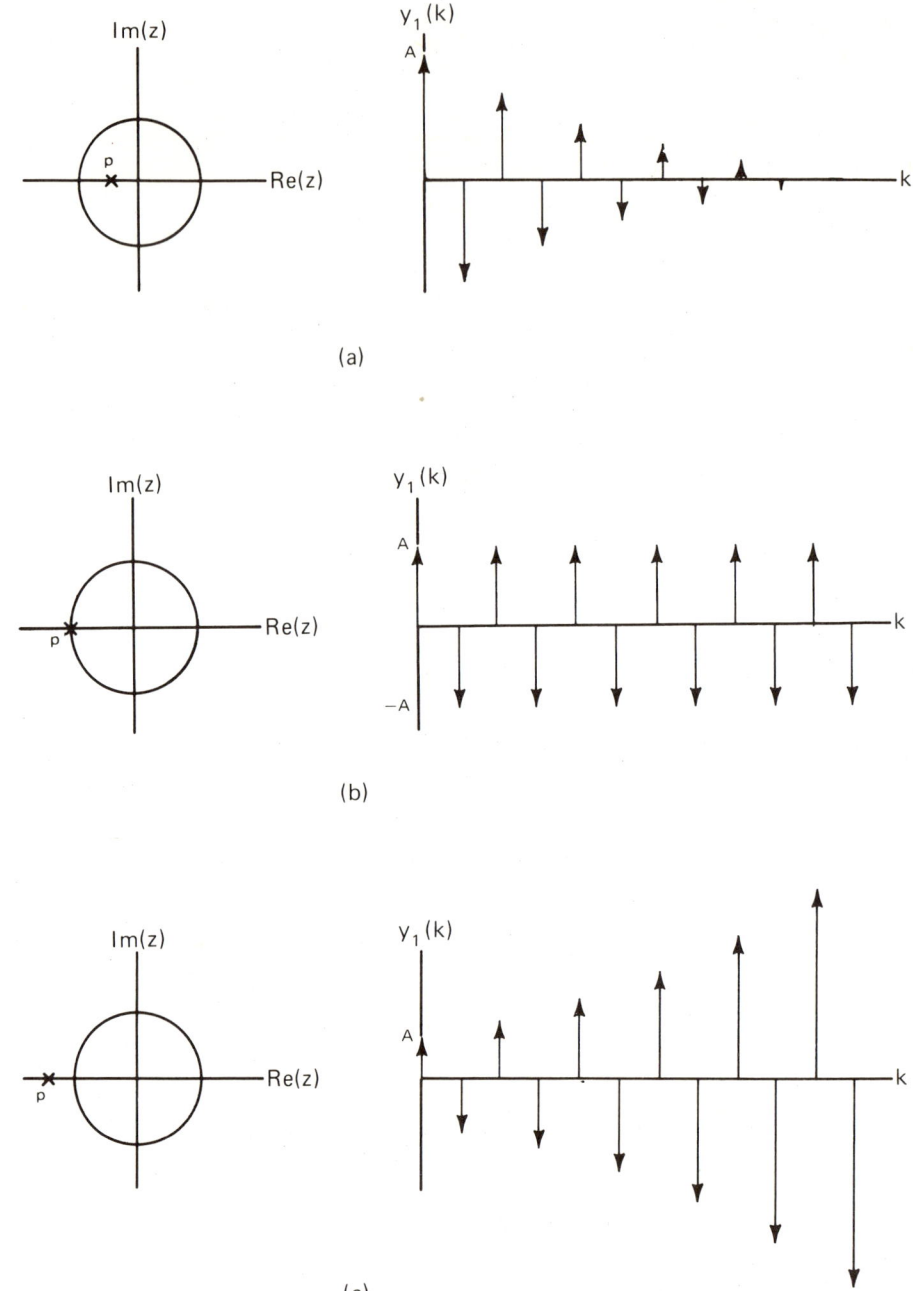

Figure 12-10. Pole-zero plots and output sequences for $Y_1(z) = Az/(z - p)$. (a) Output sequence for $-1 < p < 0$; (b) Output sequence for $p = -1$; (c) Output sequence for $p \leq -1$.

System Analysis by z-Transform

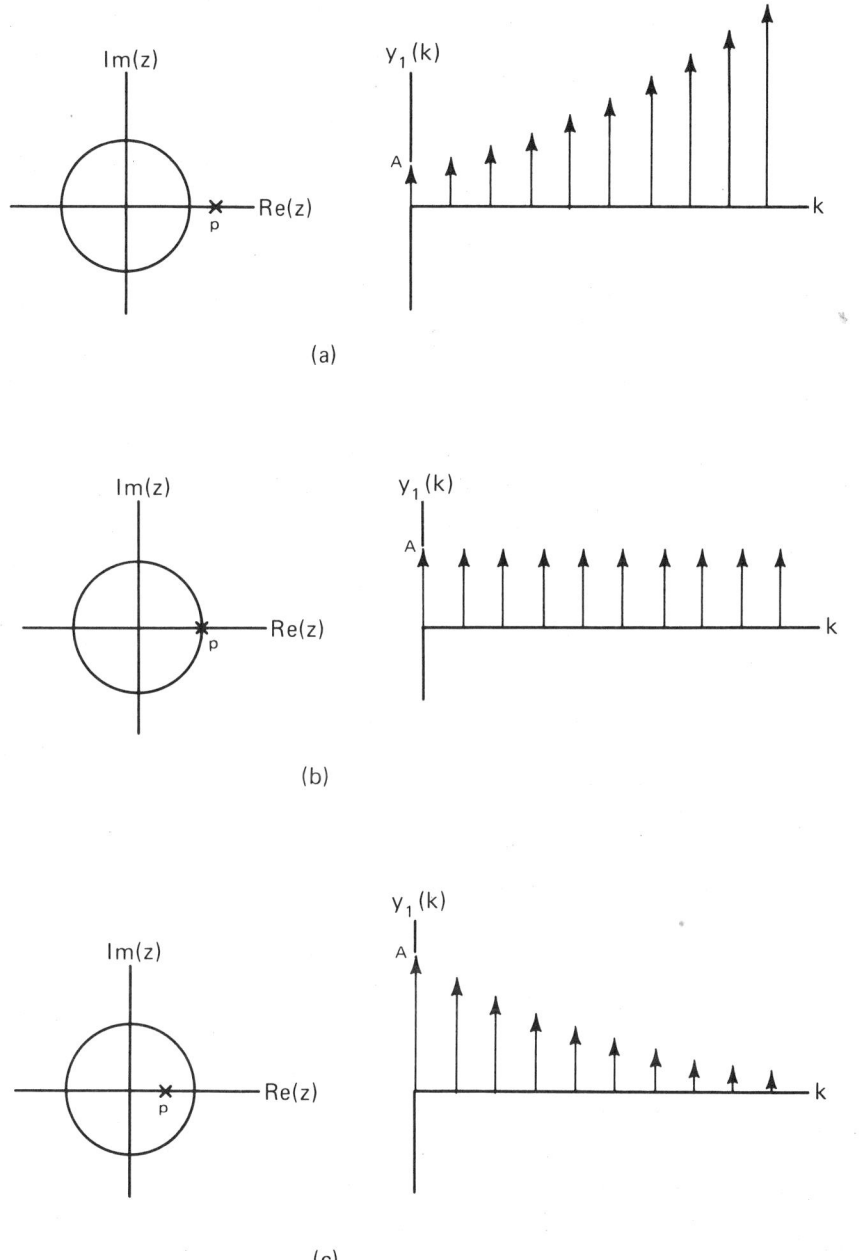

Figure 12-11. Pole-zero plots and output sequences for $Y_1(z) = Az/(z - p)$. (a) Output sequence for $p > 1$; (b) Output sequence for $p = 1$; (c) Output sequence for $0 < p < 1$.

where $p_2 = p_1^*$ (* denotes complex conjugation). The associated sequence $y_2(k)$ is $y_2(k) = A_1(p_1)^k + A_2(p_2)^k$, but, as the poles are complex, this expression is not too revealing. Let $p_1 = C\exp(j\theta)$ and $p_2 = C\exp(-j\theta)$. Using these expressions, $y_2(k)$ can be written

$$y_2(k) = A_1\{Ce^{j\theta}\}^k + A_2\{Ce^{-j\theta}\}^k \tag{12-144}$$

$$= C^k\{A_1 e^{j\theta k} + A_2 e^{-j\theta k}\} \tag{12-145}$$

or, using Euler's relationships for the exponentials,

$$y_2(k) = C^k\{A_1\{\cos(k\theta)+j\sin(k\theta)\} + A_2\{\cos(k\theta)-j\sin(k\theta)\}\} \tag{12-146}$$

or

$$y_2(k) = C^k\{\cos(k\theta)(A_1+A_2) + j\sin(k\theta)(A_1-A_2)\} \tag{12-147}$$

The requirement that $y_2(k)$ be real--and this is true because $y_2(k)$ is the solution of a difference equation with real coefficients--requires that (A_1+A_2) and $j(A_1-A_2)$ both be real. This is possible only if $A_2 = A_1^*$. Let $A_1 = a+jb$; then $A_1+A_2 = 2a$, $j(A_1-A_2) = -2b$, and Eq. (12-147) becomes

$$y_2(k) = 2C^k\{(a)\cos(k\theta) - (b)\sin(k\theta)\} \tag{12-148}$$

Let $A = \sqrt{a^2+b^2}$ and write Eq. (12-148) as

$$y_2(k) = 2C^k A\{(a/A)\cos(\theta) - (b/A)\sin(k\theta)\} \tag{12-149}$$

If we now define $\emptyset$ by the triangle of Fig. 12-12, we have $a/A = \cos(\emptyset)$, $b/A = \sin(\emptyset)$, and

$$y_2(k) = 2C^k A\{\cos(\emptyset)\cos(k\theta) - \sin(\emptyset)\sin(k\theta)\} \tag{12-150}$$

The final expression for $y_2(k)$ is then

$$y_2(k) = 2AC^k \cos(k\theta + \emptyset) \tag{12-151}$$

System Analysis by z-Transform

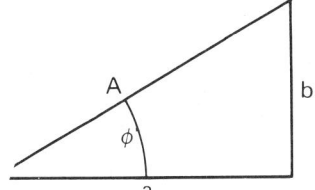

Figure 12-12. A right triangle to define ϕ in Eq. (12-151).

where $A=|A_1|$ and $\phi=\tan^{-1}(b/a)=\text{Arg}\{A_1\}$. Finally, recalling that $C=|p_1|$, we can write Eq. (12-151) in terms of the given quantities p_1 and A_1:

$$y_2(k) = 2|A_1||p_1|^k \cos(k\theta + \phi) \quad (12\text{-}152)$$

The response is a discrete version of the damped sinusoid. The envelope, which is $|p_1|^k$ rather than an exponential, is the key to the nature of the response:

*If $|p_1|>1$, the pole p_1 is outside of the unit circle, the response is unstable, and $y_2(k)$ grows without bound as k increases.

*If $|p_1|=1$, the pole p_1 is on the unit circle, the response is conditionally stable, and $y_2(k)$ is a pure discrete sinusoidal oscillation.

*If $|p_1|<1$, the pole p_1 lies within the unit circle, the response is stable, and $y_2(k)$ decays with increasing k. Figure 12-13 illustrates some of these cases.

Dominant poles. For a stable system all transfer function poles are located inside the unit circle. Those nearest the boundary of the unit circle are called <u>dominant poles</u> because the associated transient terms take longer to die out. Consider the poles in Fig. 12-14.

The pole at z=0.9 is associated with a term in the output sequence having the value $A(0.9)^k$, while the pole at z=0.1 is associated with $B(0.1)^k$. Clearly as k→∞ the latter term subsides more quickly. Hence the pole at z=0.9 is said to be dominant over the one at z=0.1.

The concept of dominant poles often lets the analyst neglect poles near the origin of the z-plane and get good

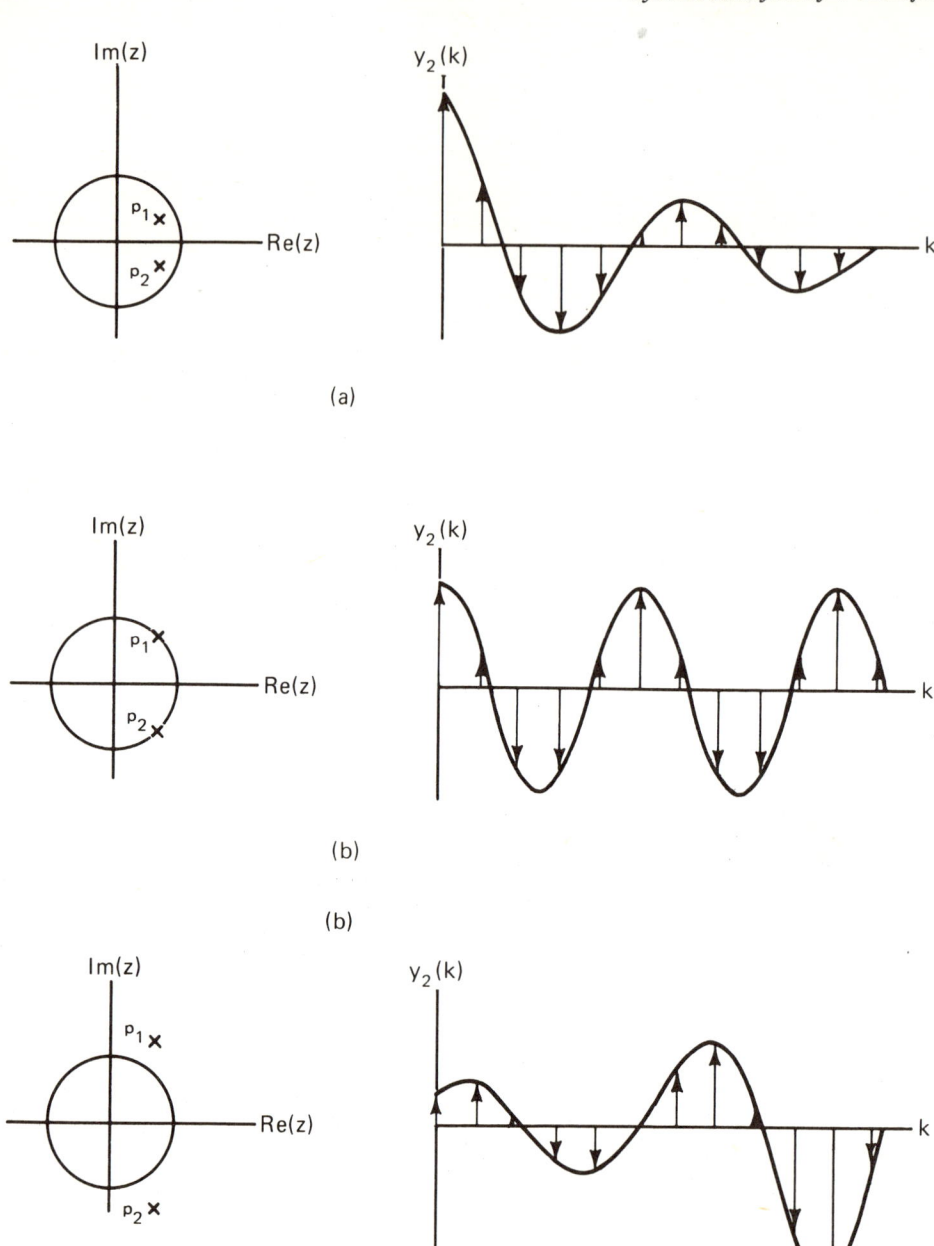

Figure 12-13. Pole-zero plots and output sequences for complex poles using Eq. (12-152).
(a) A stable response: $|p_1| < 1$; (b) A conditionally stable response: $|p_1| = 1$;
(c) An unstable response: $|p_1| > 1$.

System Analysis by z-Transform

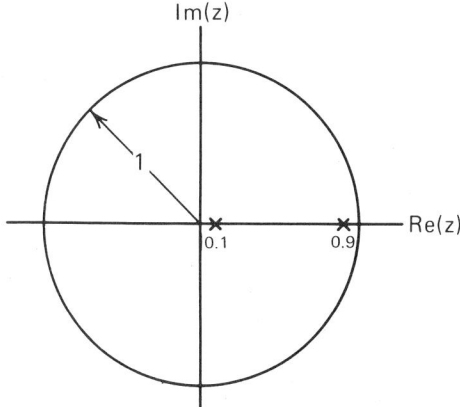

Figure 12-14. Poles near the unit circle are said to be dominant over those near the origin because the output sequences associated with the former take longer to subside. In this illustration the pole at z = 0.9 dominates the pole at z = 0.1.

results by considering only the dominant poles. This procedure can simplify the labor involved in analyzing a discrete system.

Example 12-24. Find the unit step response with and without neglecting the remote pole at z=0.01 of the system whose transfer function is

$$G(z) = \frac{z^2}{(z - 0.01)(z - 0.9)} \qquad (12\text{-}153)$$

Solution. For a step input, $F(z)=z/(z-1)$ and $Y(z)/z$ is

$$\frac{Y(z)}{z} = \frac{z^2}{(z - 1)(z - 0.01)(z - 0.9)} \qquad (12\text{-}154)$$

$$= \frac{10.1}{z - 1} + \frac{0.0001135}{z - 0.01} - \frac{9.101}{z - 0.9} \qquad (12\text{-}155)$$

The exact output sequence is

$$y(k) = 10.10(1)^k + 0.0001135(0.01)^k - 9.101(0.9)^k$$

$$(12\text{-}156)$$

If we neglect the pole at z=0.01 (and the corresponding z in the numerator) Y(z)/z is

$$\frac{Y(z)}{z} = \frac{z}{(z-1)(z-0.9)} = \frac{10}{z-1} - \frac{9}{z-0.9} \qquad (12\text{-}157)$$

Hence, $y(k)=10(1)^k - 9(0.9)^k$, which, for large k, is a good approximation to the exact solution.

12.8 SYNTHESIS OF TRANSFER FUNCTIONS

Transfer functions in the z-domain can be synthesized as difference equations or as interconnections of gains and delays. Such realizations are an important preliminary step in systems analysis by digital computer, in the design of digital controllers, and in hardware realizations of discrete systems.[1,6,8]

Synthesis, however, is somewhat tangential to our main objective of learning how to analyze continuous systems by discrete methods, and we will not plunge deeply into the subject. Rather than a general development of transfer function synthesis, we will be content with specific examples.

Difference-Equation Realizations

If one is interested in analyzing a discrete system which is specified as an interconnection of transfer functions, the most direct approach is to transform the transfer functions to difference equations and then use a digital computer to find a point-by-point solution. The methods parallel those for synthesizing s-domain transfer functions.[19,20]

<u>The direct realization</u>. To synthesize a z-domain transfer function as a difference equation, one merely cross-multiplies and takes the inverse transform. If, for example, we have

$$\frac{Y(z)}{F(z)} = G(z) = \frac{z+1}{z^2 - 0.8z - 1.2} \qquad (12\text{-}158)$$

System Analysis by z-Transform

the first step is to express G(z) in negative powers of z, or

$$\frac{Y(z)}{F(z)} = \frac{z^{-1} + z^{-2}}{1 - 0.8z^{-1} - 1.2z^{-2}} \quad (12\text{-}159)$$

Cross-multiply to obtain

$$(1 - 0.8z^{-1} - 1.2z^{-2})Y(z) = (z^{-1} + z^{-2})F(z) \quad (12\text{-}160)$$

Finally, take inverse z-transforms to arrive at the required difference equation

$$y(k) - 0.8y(k-1) - 1.2y(k-2) = f(k-1) + f(k-2) \quad (12\text{-}161)$$

The initial conditions are $y(-1)=y(-2)=0$ and the iteration starts with $k=0$. A point-by-point solution of Eq. (12-161) can be obtained from a digital computer program.

Cascade programming. If the poles of a transfer function are known; that is, the denominator is given in factored form, then it is often advantageous to synthesize the transfer function as a cascade of simpler transfer function blocks. Suppose we have the transfer function

$$G(z) = \frac{Y(z)}{F(z)} = \frac{4z^2}{(z - 0.8)(z - 0.6)} \quad (12\text{-}162)$$

Introduce a new variable W(z) and write

$$\frac{Y(z)}{F(z)} = \frac{Y(z)}{W(z)} \frac{W(z)}{F(z)} = \frac{2z}{(z - 0.8)} \frac{2z}{(z - 0.6)} \quad (12\text{-}163)$$

where

$$\frac{W(z)}{F(z)} = \frac{2z}{z - 0.8} = \frac{2}{1 - 0.8z^{-1}} \quad (12\text{-}164)$$

and

$$\frac{Y(z)}{W(z)} = \frac{2z}{z - 0.6} = \frac{2}{1 - 0.6z^{-1}} \quad (12\text{-}165)$$

Realize each transfer function as a difference equation. The result is the pair of coupled equations

$$w(k) - 0.8w(k-1) = 2f(k) \quad (12\text{-}166)$$

and

$$y(k) - 0.6y(k-1) = 2w(k) \qquad (12\text{-}167)$$

with $y(-1)=w(-1)=0$. Again, the iteration starts with $k=0$. Note that we cannot find $y(0)$ until after we have found $w(0)$. Equations (12-166) and (12-167) are easily implemented on a digital computer. This method of programming transfer functions, which is called <u>cascade programming</u>, can be generalized to transfer functions of any order. The only difficulty is that we must factor the numerator and denominator of $G(z)$.

<u>Parallel programming</u>. Transfer functions can also be synthesized as a sum rather than a product of simpler functions. This is done by first expanding $G(z)$ as a partial fraction. For the system above,

$$\frac{G(z)}{z} = \frac{4z}{(z-0.8)(z-0.6)} = \frac{16}{z-0.8} - \frac{12}{z-0.6} \qquad (12\text{-}168)$$

from which

$$G(z) = \frac{16}{1-0.8z^{-1}} - \frac{12}{1-0.6z^{-1}} \qquad (12\text{-}169)$$

Let

$$Y_1(z) = \frac{16F(z)}{1-0.8z^{-1}} \qquad (12\text{-}170)$$

and

$$Y_2(z) = \frac{12F(z)}{1-0.6z^{-1}} \qquad (12\text{-}171)$$

Equations (12-170) and (12-171) are converted to the difference equations $y_1(k)-0.8y_1(k-1)=16f(k)$, $y_2(k)-0.6y_2(k-1)=12f(k)$, and the results summed to obtain $y(k)$, or $y(k)=y_1(k)+y_2(k)$. If a digital computer is used, either $y_1(k)$ or $y_2(k)$ can be computed first. When using this method, we must not only factor the denominator of $G(z)$ but also do the partial-fraction expansion. For this and other computational reasons the partial-fraction method, also called <u>parallel programming</u>, is used less often than cascade programming.

System Analysis by z-Transform

Feed-forward programming.[21,22] The difference-equation realization of

$$G(z) = \frac{z + 1}{z^2 - 0.8z - 1.2} \qquad (12\text{-}172)$$

as given in Eq. (12-161), requires us to store delayed values of the input function, $f(k-1)$ and $f(k-2)$. In many implementations it is impractical to store the input, hence we seek a synthesis technique which avoids this difficulty. Starting with Eq. (12-172), define a new sequence $w(k)$ such that

$$G(z) = \frac{Y(z)}{F(z)} = \frac{Y(z)}{W(z)} \frac{W(z)}{F(z)} = \frac{z^{-1} + z^{-2}}{1 - 0.8z^{-1} - 1.2z^{-2}} \qquad (12\text{-}173)$$

where

$$\frac{W(z)}{F(z)} = \frac{1}{1 - 0.8z^{-1} - 1.2z^{-2}} \qquad (12\text{-}174)$$

and

$$\frac{Y(z)}{F(z)} = z^{-1} + z^{-2} \qquad (12\text{-}175)$$

Inverse transform Eq. (12-174), obtaining the difference equation

$$w(k) - 0.8w(k-1) - 1.2w(k-2) = f(k) \qquad (12\text{-}176)$$

Equation (12-175) gives the output sequence $y(k)$ as

$$y(k) = w(k-1) + w(k-2) \qquad (12\text{-}177)$$

As Eq. (12-176) is solved iteratively on the computer, the delayed values of $w(k)$ are available for computing the sequence $y(k)$--and all of this can be done without storing the delayed values of $f(k)$. The feed-forward method will always work as long as the order of the numerator of $G(z)$ does not exceed that of the denominator, as is the case for physically realizable systems. The principal advantage of feed-forward programming is that it requires less total delays, and hence less storage than the direct difference-equation method. In the example above the direct method used four storage locations, those for $y(k-1)$, $y(k-2)$, $f(k-1)$, and f(k-2), while the feed-forward method required only two, those for w(k-1) and w(k-2). Minimizing the number of storage locations can be extremely important in the design of special-purpose computers to analyze or control large systems.

Block Diagram Realizations

Transfer functions in the z-domain can also be synthesized as interconnections of gains and delays, where a delay is a block containing z^{-1}. Such realizations are an important preliminary step in the design of hardware digital filters, and they also lead to a better understanding of discrete system theory. Figures 12-15, 12-16, 12-17, and 12-18 show the block diagram realizations which correspond to direct, cascade, parallel and feed-forward programming. The realizations can be verified by elementary block diagram reduction similar to that used in Chapter 7 for continuous transfer functions.

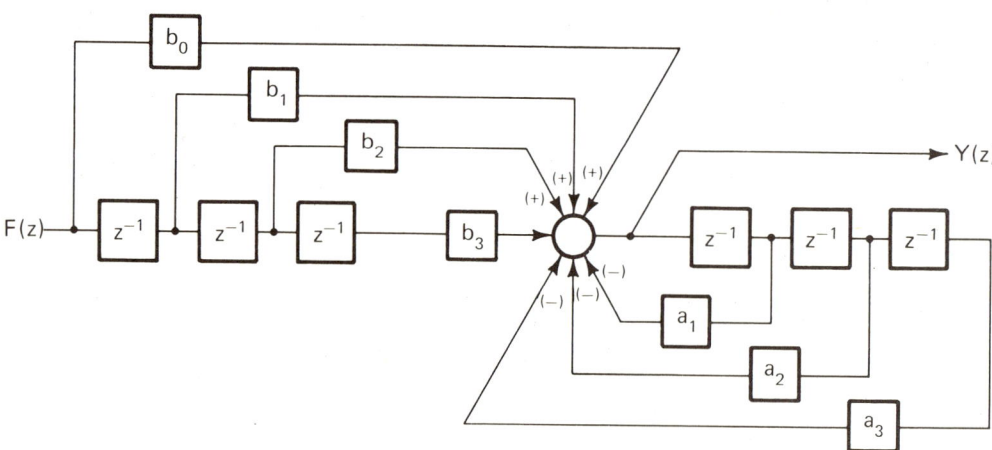

Figure 12-15. Direct realization of $G(z) = \dfrac{b_0 + b_1 z^{-1} + b_2 z^{-2} + b_3 z^{-3}}{1 + a_1 z^{-1} + a_2 z^{-2} + a_3 z^{-3}}$

System Analysis by z-Transform

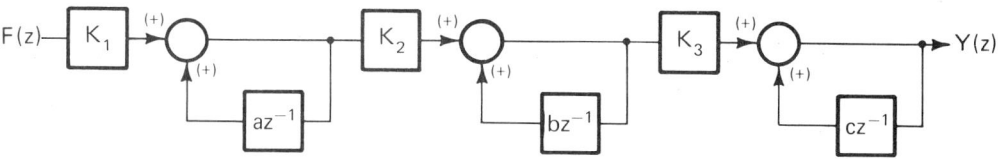

Figure 12-16. Cascade realization of $G(z) = \dfrac{K_1 K_2 K_3}{(1 - az^{-1})(1 - bz^{-1})(1 - cz^{-1})}$

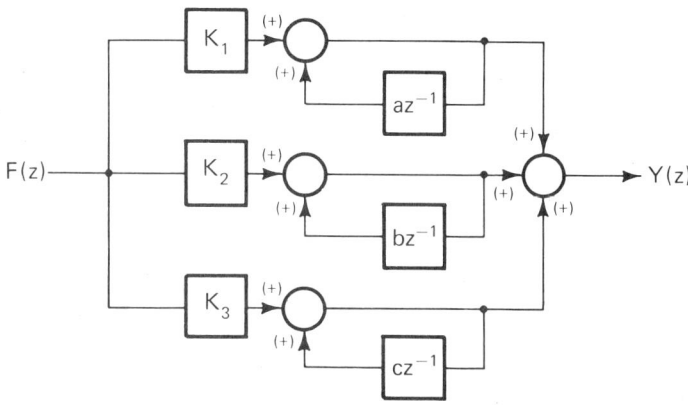

Figure 12-17. Parallel programming realization of $G(z) = \dfrac{K_1}{1 - az^{-1}} + \dfrac{K_2}{1 - bz^{-1}} + \dfrac{K_3}{1 - cz^{-1}}$

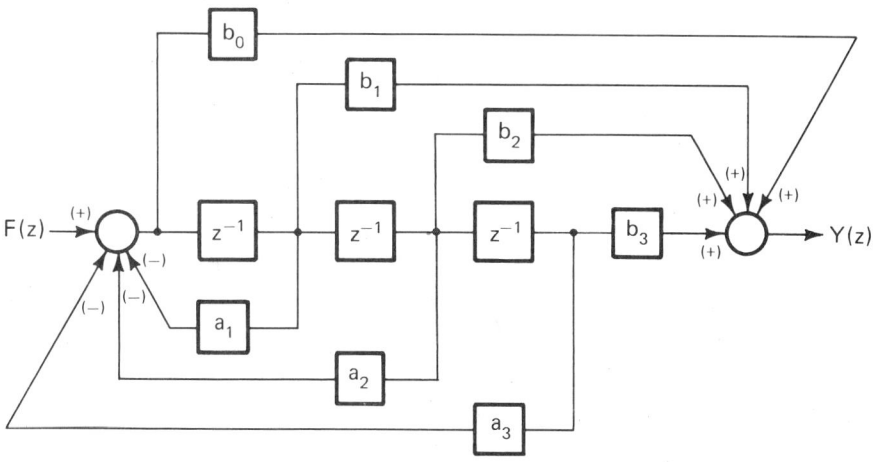

Figure 12-18. Feed-forward programming realization of $G(z) = \dfrac{b_0 + b_1 z^{-1} + b_2 z^{-2} + b_3 z^{-3}}{1 + a_1 z^{-1} + a_2 z^{-2} + a_3 z^{-3}}$

The feed forward program used only 3 storage locations for the third-order transfer function, whereas the direct program used 6.

PROBLEMS FOR CHAPTER 12

12-1. Write the infinite series expressions for the z-transforms of the sampled versions of the following functions: (a) $f(t)=t$, (b) $f(t)=t^2$, (c) $f(t)=t\exp(-at)$, and (d) $f(t)=A\cos(\omega_o t)$.

12-2. Write <u>finite</u> series expressions for the z-transforms of the following functions: (a) $f(t)=\delta(t)$, (b) $f(t)=u(t)-u(t-5T)$, (c) $g(t)=t\{u(t)-u(t-2T)\}$, and (d) $h(t)=u(t-T)-u(t-4T)$.

12-3. Use the partial-series method of Eqns. (12-19) through (12-24) to derive a closed-form expression for $Z\{kT\}$.

12-4. Repeat Problem 12-3 for $Z\{(kT)^2\}$.

12-5. Derive closed-form expressions for the z-transforms of the following sequences. In each case specify the radius of convergence. (a) $y(k)=k$, (b) $y(k)=A(b)^k$, (c) $y(k)=k(b)^k$, (d) $y(k)=\cos(\omega_o kT)$, (e) $y(k)=\exp(-kT)$, and (f) $y(k)=\sin(\omega_o kT+\emptyset)$.

12-6. Derive a closed-form expression for the z-transform of the delayed sequence $u\{(k-n)T\}$. Do not use the shifting theorems.

12-7. Derive a closed-form expression for the z-transform of the function $f(k)=(0.5)^k$ for $k=0,1,2,\ldots 25$ and $f(k)=(0.25)^k$ for $k=26,27,\ldots$. Also find the radius of convergence.

12-8. Derive a closed-form expression for the z-transform of the sequence $y(k)=1/(-3)^k$.

12-9. Find the locus of points in the z-plane which corresponds to the vertical line $\sigma=-2.5$ in the s-plane if $T=0.5$.

12-10. Identify the poles and zeros of the following

System Analysis by z-Transform

z-domain functions:

(a) $F(z) = \dfrac{1 + z^{-1}}{3 + 4z^{-1} + z^{-2}}$ (12-178)

(b) $F(z) = \dfrac{z^{-2}}{1 - (5/8)z^{-1} + (1/16)z^{-2}}$ (12-179)

(c) $F(z) = \dfrac{z^{-1}(1 - 2z^{-1})}{1 - 0.5z^{-1} + 2z^{-2}}$ (12-180)

12-11. Derive Entry 12 of Appendix **F**.

12-12. Use partial fractions to find the inverse z-transforms of the following functions:

(a) $F(z) = \dfrac{z}{(z - 1)(z - 0.125)}$ (12-181)

(b) $F(z) = \dfrac{z^2}{(z - 0.5)(z + 0.25)}$ (12-182)

(c) $F(z) = \dfrac{z - 2}{z(z - 0.4)(z + 0.25)}$ (12-183)

(d) $F(z) = \dfrac{z}{(z - 1)(z - 0.5)^2}$ (12-184)

12-13. Find the inverse z-transforms of the following functions:

(a) $Y(z) = \dfrac{z}{\{z - \exp(j2)\}\{z - \exp(-j2)\}}$ (12-185)

(b) $Y(z) = \dfrac{z(z - 1)}{(z+1)\{z-\exp(j0.25)\}\{z-\exp(-j0.25)\}}$ (12-186)

(c) $Y(z) = \dfrac{z}{(z -1)(z^2 - z + 0.5)}$ (12-187)

12-14. Use the power-series method to find the inverse z-transform of each of the following functions: (a) $F(z) = z/(z-5)$, (b) $H(z) = (z+1)/(z-2)$, and (c) $G(z) = (z+1)/(z^2-2)$.

12-15. Show that the inverse z-transform of $y(z) = A/z$ is the sequence $y(k) = 0, A, 0, 0, \ldots$.

12-16. Generalize the result of Problem 12-15 to $Y(z) = A/z^n$.

12-17. Prove the left-shifting theorem, Eq. (12-64).

12-18. Prove the initial-value theorem, Eq. (12-66).

12-19. Prove the final-value theorem, Eq. (12-67).

12-20. Prove that taking the z-transform is a linear operation.

12-21. Find the inverse z-transforms of the following functions. Use the right-shifting theorem to avoid expanding terms of the form $1/z^n$.

(a) $\quad F(z) = \dfrac{z + 1}{z^2(z - 0.125)}$ \hfill (12-188)

(b) $\quad Y(z) = \dfrac{1}{z^3(z - a)}$ \hfill (12-189)

(c) $\quad H(z) = \dfrac{1}{z^{25}(z + 0.5)}$ \hfill (12-190)

12-22. Solve for the first 4 points of the solution sequence for each of the following difference equations. Start each solution with k=0.

(a) $\quad y(k) - 0.25y(k-1) = 0 \qquad y(-1)=2$ \hfill (12-191)

(b) $\quad h(k) - h(k-1) + 2h(k-2) = u(k) \qquad h(-1)=h(-2)=0$ \hfill (12-192)

(c) $\quad g(k) = g(k-2)/5 \qquad g(-1)=0 \qquad g(-2)=8$ \hfill (12-193)

(d) $\quad y(k) = \sin\{\pi y(k-1)/4\} \qquad y(-1)=0$ \hfill (12-194)

(e) $\quad y(k) + \{y(k-1)\}^2 = u(k) \qquad y(-1)=2$ \hfill (12-195)

(f) $\quad (k+1)h(k) + 0.5h(k-1) = 0 \qquad h(-1)=1$ \hfill (12-196)

12-23. Find closed-form solutions for the following difference equations. Use z-transforms, assume the solutions start with k=0, and take all initial values, y(-1) etc., as 0.

(a) $\quad y(k) = u(k) + 0.4y(k-1)$ \hfill (12-197)

(b) $\quad y(k) - 0.45y(k-2) = 5\delta(k)$ \hfill (12-198)

(c) $\quad q(k) - 0.5q(k-1) + 0.06q(k-2) = u(k)$ \hfill (12-199)

(d) $\quad h(k) - 0.5h(k-1) + 0.04h(k-2) = \delta(k)$ \hfill (12-200)

12-24. Find closed-form solutions for the following difference equations using z-transform methods:

(a) $y(k) - 0.9y(k-1) = 2\delta(k)$ $y(-1)=2$ (12-201)

(b) $y(k) = 0.25y(k-1) + u(k)$ $y(-1)=5$ (12-202)

(c) $y(k) = 1.3y(k-1) - 0.4y(k-2)$ $y(-1)=1$ $y(-2)=0$ (12-203)

(d) $y(k) = 0.8y(k-2)$ $y(-1)=2$ $y(-2)=5$ (12-204)

12-25. Derive the correct value for $y(-1)$ to impose the condition $y(0)=0$ on the difference equations of Problem 12-24(a) and (b).

12-26. Derive the correct values for $y(-1)$ and $y(-2)$ to impose the initial conditions $y(0)=A$ and $y(1)=B$ on the difference equations of Problem 12-24(c) and (d).

12-27. The difference equation

$$y(k) - 0.5y(k-1) + 0.25y(k-2) = u(k) - u(k-1)$$

(12-205)

has initial conditions $y(0)=5$ and $y(-1)=2$. Find the correct value for $y(-2)$.

12-28. Find the impulse response sequence $g(k)$ for the difference equations:

(a) $4y(k) - y(k-1) = f(k)$ (12-206)

(b) $y(k) = f(k) - \omega_n^2 y(k-2) + 2\zeta\omega_n y(k-1)$ (12-207)

where $\zeta<1$.

12-29. A discrete linear system has the impulse response $g(k)=1-(0.5)^k$. Find the response of the system to the input $f(k)=u(k)$.

12-30. Find the discrete transfer functions for the systems having the following impulse response sequences. Express each transfer function as a rational function: (a) $g(k)=(0.5)^k$, (b) $g(k)=(0.375)^k+(-0.24)^k$, (c) $g(k)=(1)^k+(0.125)^k$, (d) $g(k)=ku(k)$, and (e) $g(k)=\cos(\omega_o kT)$.

12-31. Find the transfer functions of the systems described by the difference equations of Problem 12-23.

13-32. Reduce the following systems to a single transfer function block:

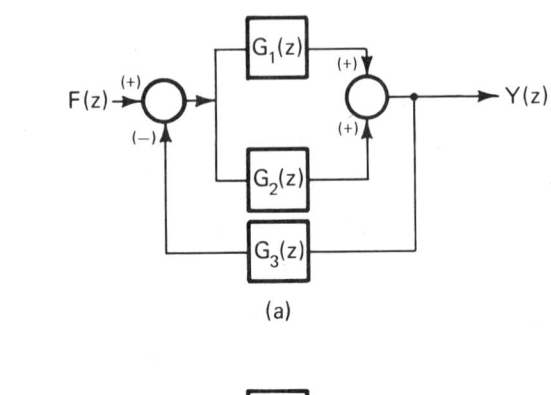

(a)

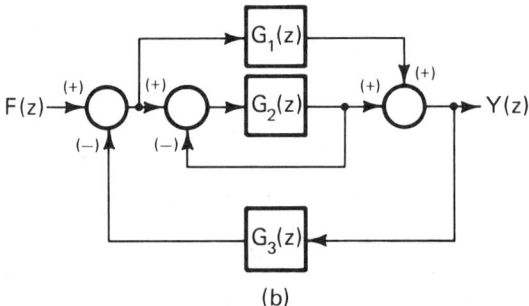

(b)

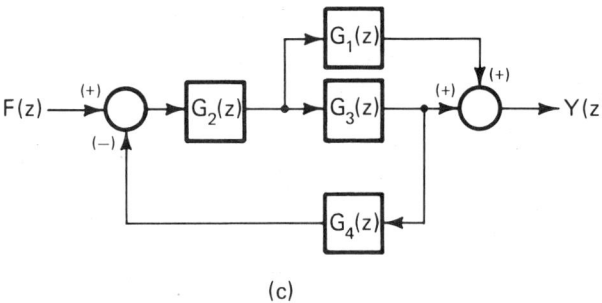

(c)

Figure 12-19. Problem 12-32.

12-33. Find the transfer function and the pseudo input function for the system described by

$$y(k) - 0.25y(k-1) + 0.1y(k-2) = f(k) \qquad (12\text{-}208)$$

with y(-1)=2 and y(-2)=5.

12-34. Find the difference equations which represent the systems whose transfer functions are given below. Assume $G(z)=Y(z)/F(z)$.

(a) $G(z) = \dfrac{z^3}{(z-1)(z^2+0.12z)}$ (12-209)

(b) $G(z) = \dfrac{1}{z^2 - 0.35z + 0.008}$ (12-210)

(c) $G(z) = \dfrac{z(z-1)}{z^2 - 0.55z + 0.975}$ (12-211)

12-35. Solve the difference equation $y(k)=(1-a)f(k)+ay(k-1)$ for (a) $f(k)=u(k)$ and (b) $f(k)=(-1)^k$. Assume $y(-1)=0$.

12-36. A low-pass filter has the difference equation $y(k)=0.2f(k)+0.8y(k-1)$. If two of these filters are cascaded, find (a) the overall transfer function and (b) the unit step response. Assume $y(-1)=0$.

12-37. Derive an expression for $y(k)$ in the form

$$y(k) = A^k\{\cos(Bk+\emptyset)\} \qquad (12\text{-}212)$$

if

$$Y(z) = \dfrac{Rz}{z-p} + \dfrac{R^*z}{z-p^*} \qquad (12\text{-}213)$$

where R is complex and p has the form $p=c+jd$.

12-38. Find the impulse response sequence of the system whose transfer function is

$$G(z) = \dfrac{z^2}{(z-0.02)(z-0.8)} \qquad (12\text{-}214)$$

(a) with (b) without the remote pole. Compare the exact and approximate responses.

12-39. Find, to within a gain constant K, the transfer functions of the systems having the pole-zero diagrams shown:

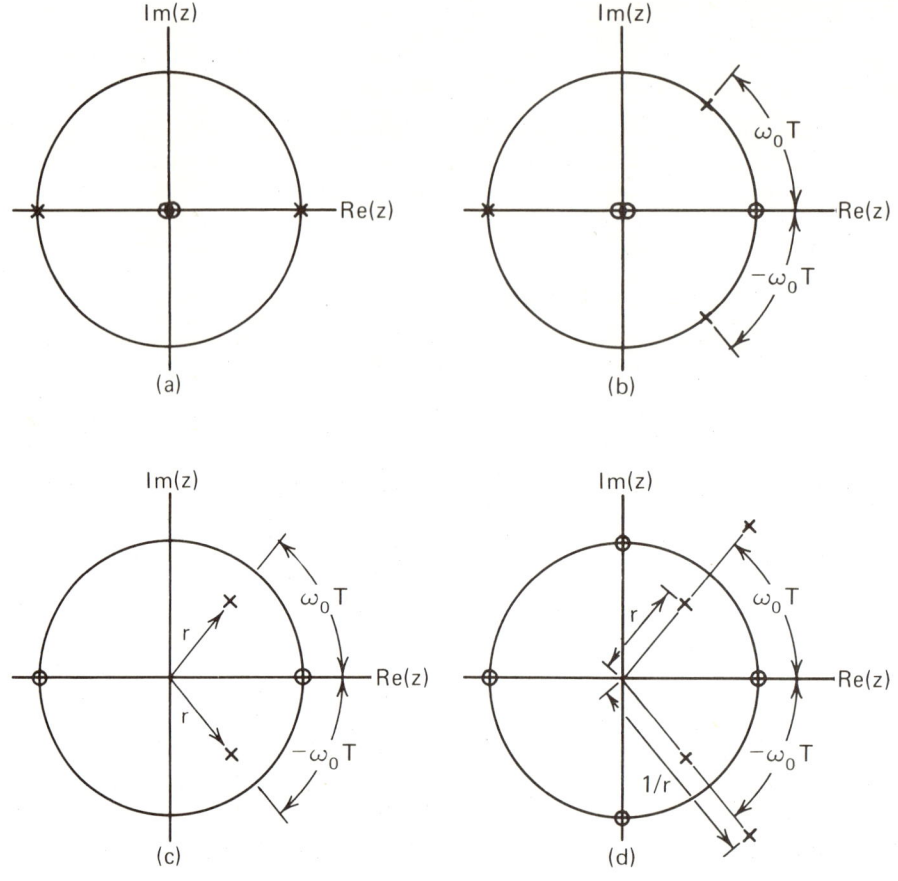

Figure 12-20. Problem 12-39.

12-40. Develop block diagram realizations for the following transfer functions using the indicated programming method:

(a) $G(z) = \dfrac{5z^2}{(z-1)(z-0.5)}$ (Cascade Programming) (12-215)

(b) $G(z) = \dfrac{z}{z-1} + \dfrac{0.5z}{z-0.5}$ (Parallel Programming) (12-216)

(c) $G(z) = \dfrac{z^2}{(z-0.5)(z-0.2)}$ (Parallel Programming) (12-217)

(d) $G(z) = \dfrac{z^2 + 2z + 1}{z^2 + 0.6z + 0.05}$ (Feed-forward Programming) (12-218)

REFERENCES FOR CHAPTER 12

(1) Kuo, B. E.: <u>Analysis and Synthesis of Sampled-Data Control Systems</u>, Prentice-Hall, Inc., Englewood Cliffs, New Jersey, 1963.

(2) Cadzow, J.A.: <u>Discrete-Time Systems</u>, Prentice-Hall, Inc., Englewood Cliffs, New Jersey, 1973.

(3) Gold, B. and C. M. Rader: <u>Digital Processing of Signals</u>, McGraw-Hill Book Co., Inc., New York, 1969.

(4) Director, S. W. and R. A. Rohrer: <u>Introduction to Systems Theory</u>, McGraw-Hill Book Co., Inc., New York, 1972.

(5) Gabel, R. A. and R. A. Roberts: <u>Signals and Linear Systems</u>, John Wiley and Sons, Inc., New York, 1973.

(6) Jury, E. I.: <u>Sampled-Data Control Systems</u>, John Wiley and Sons, Inc., New York, 1958.

(7) Tou, J.T.: <u>Digital and Sampled-Data Control Systems</u>, McGraw-Hill Book Co., Inc., New York, 1959.

(8) Otnes, R. K. and L. Enochson: <u>Digital Time Series Analysis</u>, John Wiley and Sons, New York, 1972.

(9) Henrici, P.: <u>Discrete Variable Methods in Ordinary Differential Equations</u>, John Wiley and Sons, New York, 1962.

(10) Miller, K. S.: <u>An Introduction to the Calculus of Finite Differences and Difference Equations</u>, Henry Holt and Company, Inc., New York, 1960.

(11) Salvadori, M. G.: <u>Numerical Methods in Engineering</u>, Prentice-Hall, Inc., Englewood Cliffs, New Jersey, 1961.

(12) DeRusso, P. M., Roy, R. J., and C. M. Close: <u>State Variables for Engineers</u>, John Wiley and Sons, Inc., New York, 1965.

(13) Brown, W. M.: <u>Analysis of Linear Time Invariant Systems</u>, McGraw-Hill Book Co., Inc., New York, 1963.

(14) Opperheim, A. V. and R. W. Schafer: <u>Digital Signal Processing</u>, Prentice-Hall, Inc., Englewood Cliffs, New Jersey, 1975.

(15) Rabiner, L. R. and B. Gold: <u>Theory and Application of Digital Signal Processing</u>, Prentice-Hall, Inc., Englewood Cliffs, New Jersey, 1975.

(16) Monroe, A. J.: *Digital Processes for Sampled-Data Systems*, John Wiley and Sons, Inc., New York, 1962.

(17) Ragazzini, J. R. and G. F. Franklin: *Sampled-Data Control Systems*, McGraw-Hill Book Co., Inc., New York, 1958.

(18) Rosko, J. S.: *Digital Simulation of Physical Systems*, Addison-Wesley Publishing Co., Inc., Reading, Massachusetts, 1972.

(19) Eveleigh, V. W.: *Introduction to Control System Design*, McGraw-Hill Book Co., Inc., New York, 1972.

(20) Elgerd, O. I.: *Control Systems Theory*, McGraw-Hill Book Co., Inc., New York, 1967.

(21) Beck, C.: "Treating Transfer Functions On Analog Computers," *Electrical Manufacturing*, Vol. 62, Oct. 1958.

(22) Hausner, A.: *Analog and Analog/Hybrid Computer Programming*, Prentice-Hall, Inc., Englewood Cliffs, New Jersey, 1971.

CHAPTER 13

Numerical Solution of Differential Equations

To this point in our study of computational methods for systems analysis we have introduced the sampling process, by which one can discretize continuous variables without losing information. And we have introduced a language of discrete analysis, the z-transform. Our task will be completed when we learn how to discretize not just variables, but the continuous mathematical models which describe continuous dynamic systems. More specifically, we must know how to transform differential equations into difference equations which can be solved in closed form by z-transforms or iteratively on a digital computer. Our treatment is not exhaustive: the interested student can pursue the study of numerical solution of differential equations in greater depth in any of a number of excellent works.[1,2,3]

Remember that when a continuous variable is discretized the sampling period T must be chosen small enough to satisfy the sampling theorem. The alert reader may point out that if one had had enough a priori knowledge about a system to select a suitable sampling frequency, he would not need to analyze it in the first place. While this is in general a valid objection, sometimes the nature of the system or of the differential equations will provide a clue to a satisfactory sampling period. If that fails there is always trial and error. Solve the problem with sampling interval T and then with T/2. If the solutions differ drastically, try T/4 and so on until successive solutions are approximately the same. Then stop.

13.1 DISCRETIZATION IN THE TIME DOMAIN

The most direct way to derive a difference equation whose solution is a sampled version of the solution of the parent continuous differential equation is to apply discrete approximations to the derivative.[4]

Discrete Approximations To The Derivative

Recall that the derivative $dy(t)/dt$ of a function $y(t)$ is defined by

$$\frac{dy(t)}{dt} = \lim_{t \to 0} \frac{y(t) - y(t-\Delta t)}{\Delta t} \qquad (13\text{-}1)$$

where Δt is a time increment. Geometrically, as shown in Fig. 13-1, $dy(t)/dt$ is the slope of the tangent to the curve at time t, and the quantity $\{y(t)-y(t-\Delta t)\}/\Delta t$ is an approximation

Numerical Solution of Differential Equations

to the slope. If Δt is small, we can make the approximation

$$\frac{dy(t)}{dt} \approx \frac{y(t) - y(t-\Delta t)}{\Delta t} \quad (13\text{-}2)$$

Equation (13-2) implies that we can approximate the slope at point P in Fig. 13-1 by the slope of the line P-Q, and that the approximation improves as Δt gets smaller.

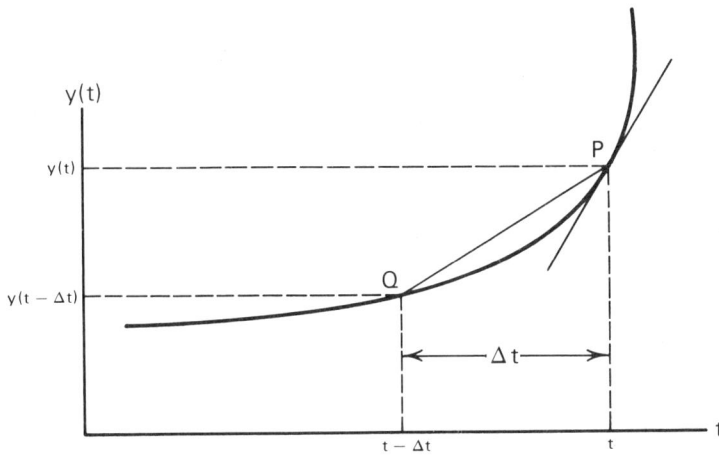

Figure 13-1. The derivative at point P can be approximated by the slope of the line PQ provided Δt is small.

If y(t) is sampled, so that in place of the continuous function we have the sampled sequence y(i), then we identify Δt with the sampling period T; that is, we let T=Δt and take t=iT. Using this notation and dropping the approximately equal sign with the understanding that dy(i)/dt is a discrete approximation to dy(t)/dt, Eq. (13-2) becomes

$$\frac{dy(i)}{dt} = \frac{y(i) - y(i-1)}{T} \quad (13\text{-}3)$$

Following a similar procedure, we can derive an expression for the discrete approximation to the second derivative; thus

$$\frac{d^2 y(t)}{dt^2} \approx \frac{dy(t)/dt - dy(t-\Delta t)/dt}{\Delta t} \quad (13\text{-}4)$$

or, using sampled-sequence notation,

$$\frac{d^2 y(i)}{dt} = \frac{dy(i)/dt - dy(i-1)/dt}{T} \tag{13-5}$$

The first term on the right is the first derivative, Eq. (13-3), and the second term is the first derivative evaluated at (i-1)T, or

$$\frac{dy(i-1)}{dt} = \frac{y(i-1) - y(i-2)}{T} \tag{13-6}$$

Substituting Eqns. (13-3) and (13-6) into Eq. (13-5) and rearranging yields

$$\frac{d^2 y(i)}{dt^2} = \frac{y(i) - 2y(i-1) + y(i-2)}{T^2} \tag{13-7}$$

A similar procedure can be used to find discrete approximations for derivatives of all orders. A generalization of Eq. (13-7) for the nth-order derivative is

$$\frac{d^n y(i)}{dt^n} = \frac{1}{T^n} \sum_{k=0}^{n} a_k y(i-k)(-1)^k \tag{13-8}$$

where the a_k are the binomial coefficients (the coefficients of $(p+q)^n$).

From Differential To Difference Equations

To discretize a differential equation of the form

$$\frac{d^n y(t)}{dt^n} + a_{n-1}\frac{d^{n-1} y(t)}{dt^{n-1}} + \ldots + a_1 \frac{dy(t)}{dt} + a_0 y(t) = f(t) \tag{13-9}$$

replace each derivative by Eq. (13-8) evaluated for the correct order. The result is an nth-order difference equation. The process is best illustrated by an example.

Example 13-1. Approximate the differential equation

$$\dot{y}(t) + 4y(t) = \cos(10\pi t) \tag{13-10}$$

with y(0)=0 by a suitable difference equation.

Numerical Solution of Differential Equations

Solution. Because the given equation is linear, we know that the sinusoidal forcing function will produce a steady-state sinusoidal response at the same frequency. Further, the transient response will contain an exponential term of the form Aexp(-4t) which subsides to 0 in approximately five time constants, or 5(0.25)=1.25 s. Therefore, we should be safe with a sampling period of T=0.01 s, which implies a sampling frequency 20 times the frequency of the forcing function. This value of T also produces 125 solution points during the transient exponential decay, more than enough for a plot or display. (These are the kinds of considerations one uses in selecting a sampling period.)

Substitute the discrete approximation to the first derivative, Eq. (13-3), for $\dot{y}(t)$ and replace t by iT to obtain

$$\frac{y(i) - y(i-1)}{T} + 4y(i) = \cos(10\pi iT) \qquad (13\text{-}11)$$

Equation (13-11) could be solved point-by-point starting with i=1 and the given initial condition y(0)=0. If z-transforms are to be used, however, we must derive the initial value y(-1).

Initial values. After transforming an nth-order differential equation into a difference equation, we must transform the continuous initial conditions, $y(0)$, $\dot{y}(0)$, $\ddot{y}(0)$,... $y(0)^{(n-1)}$, into a suitable set of initial values, y(-1),y(-2), ...y(-n), for the difference equation. This is done by evaluating Eq. (13-8) iteratively.[5] Consider the approximation to the first derivative

$$\dot{y}(i) = \frac{y(i) - y(i-1)}{T} \qquad (13\text{-}13)$$

If we evaluate this expression for i=0 and solve for y(-1), the resulting expression

$$y(-1) = y(0) - T\dot{y}(0) \qquad (13\text{-}14)$$

gives y(-1) in terms of the continuous initial conditions y(0) and $\dot{y}(0)$. For the second derivative Eq. (13-8) with n=2 and i=0 gives

$$\ddot{y}(0) = \frac{y(0) - 2y(-1) + y(-2)}{T^2} \qquad (13\text{-}15)$$

from which

$$y(-2) = T^2\ddot{y}(0) - y(0) + 2y(-1) \qquad (13\text{-}16)$$

Equation (13-16) gives y(-2) in terms of the continuous initial conditions y(0) and $\ddot{y}(0)$ and the initial value y(-1) found from Eq. (13-14). For the third derivative Eq. (13-8) with n=3 and i=0 yields

$$\dddot{y}(i) = \frac{y(0) - 3y(-1) + 3y(-2) - y(-3)}{T^3} \qquad (13\text{-}17)$$

Solving for y(-3) yields

$$y(-3) = T^3\dddot{y}(0) - y(0) + 3y(-1) - 3y(-2) \qquad (13\text{-}18)$$

Following this procedure we can find all of the initial values needed. One difficulty with this procedure is that to find y(-2) for a second-order system we need $\ddot{y}(0)$ from the continuous system. But the initial conditions for a second-order continuous system are y(0) and $\dot{y}(0)$; $\ddot{y}(0)$ is normally not given. We can circumvent this difficulty by substituting y(0) and $\dot{y}(0)$ into the original differential equation to find $\ddot{y}(0)$. Example 13-2 illustrates this idea.

Example 13-2. Discretize the second-order differential equation $\ddot{y}+3\dot{y}+2y=u(t)$ with y(0)=1 and $\dot{y}(0)$=2. Use T=1 and find the initial values y(-1) and y(-2).

Solution. Applying the approximations for the first and second derivatives yields

$$\frac{y(k)-2y(k-1)+y(k-2)}{T^2} + 3\frac{y(k)-y(k-1)}{T} + 2y(k) = u(k) \qquad (13\text{-}19)$$

or, with T=1,

$$6y(k) - 5y(k-1) + y(k-2) = u(k) \qquad (13\text{-}20)$$

Numerical Solution of Differential Equations

To obtain y(-1), use Eq. (13-14):

$$y(-1) = y(0) - T\dot{y}(0) = 1 - 2 = -1 \quad (13-21)$$

We can use Eq. (13-16) to find y(-2) but first we must determine $\ddot{y}(0)$ from the given differential equation; thus,

$$\ddot{y}(0) + 3\dot{y}(0) + 2y(0) = u(0) = 1 \quad (13-22)$$

from which, using the given initial conditions, $\ddot{y}(0) = -7$. Equation (13-16) now gives

$$y(-2) = T^2\ddot{y}(0) - y(0) + 2y(-1) \quad (13-23)$$

from which y(-2)=-10. Equation (13-20) could now be solved using the initial values y(-1)=-1 and y(-2)=-10 with the assurance that its solution would be a sampled version of the solution of the parent differential equation.

Example 13-3. Solve the differential equation of Example 13-2 by Laplace transform methods and the difference equation by z-transforms. Then compare the solutions.

Solution. The given differential equation is Laplace transformed to obtain

$$s^2 Y(s) - sy(0) - \dot{y}(0) + 3\{sY(s) - y(0)\} + 2Y(s) = \frac{1}{s} \quad (13-24)$$

Inserting y(0)=1 and $\dot{y}(0)$=2 and solving for Y(s) results in

$$Y(s) = \frac{s(s+5) + 1}{s(s+1)(s+2)} \quad (13-25)$$

and, after a partial-fraction expansion,

$$Y(s) = \frac{1}{2s} + \frac{3}{s+1} - \frac{5/2}{s+2} \quad (13-26)$$

The time-domain solution is then

$$y(t) = 0.5u(t) + 3e^{-t} - 2.5e^{-2t} \quad (13-27)$$

The difference equation from Example 13-2,

$$6y(k) - 5y(k-1) + y(k-2) = u(k) \quad (13-28)$$

with y(-1)=-1 and y(-2)=-10, is z-transformed using the extended right-shifting theorem to insert the initial values. The result is

$$6Y(z) - 5\{y(-1)+z^{-1}Y(z)\} + y(-2) + y(-1)z^{-1} + z^{-2}Y(z) = \frac{z}{z-1}$$

(13-29)

or

$$(6-5z^{-1}+z^{-2})Y(z) = 5y(-1) - y(-1)z^{-1} - y(-2) + \frac{z}{z-1} \quad (13\text{-}30)$$

Inserting numerical values and solving for Y(z)/z yields

$$\frac{Y(z)}{z} = \frac{6z^2-4z-1}{6(z-1)(z^2-5z/6+1/6)} \quad (13\text{-}31)$$

$$= \frac{6z^2-4z-1}{6(z-1)(z-1/2)(z-1/3)} \quad (13\text{-}32)$$

$$= \frac{1/2}{z-1} + \frac{3}{z-1/2} - \frac{5/2}{z-1/3} \quad (13\text{-}33)$$

The output sequence is then

$$y(k) = \frac{1}{2(1)^k} + \frac{3}{(2)^k} - \frac{5}{2(3)^k} \quad (13\text{-}34)$$

The table below compares the first few values of the exact solution evaluated for t=kT with the corresponding points from the difference equation solution. The third column lists the per cent error between the exact and approximate solutions.

k	Exact Solution (t=kT)	Approximate Solution	Per Cent Error
0	1	1	0.0
1	1.265	1.167	7.7
2	0.8602	0.972	-13.0
3	0.643	0.782	-21.6
4	0.554	0.657	-18.6

The large error between the exact and approximate solutions is caused by the large sampling period T. If a digital computer were used we could afford the luxury of a small sampling period, say T=0.0001. Here we chose T=1 to avoid obscuring the procedure with numbers.

The Taylor's Series Method

The approximation to the first derivative proposed in the preceeding section was based on connecting the points y(t) and y(t-T) by a straight line whose slope was used to approximate the derivative. This procedure is equivalent to expanding y(t) in a Taylor's series in the neighborhood of the point y{(k-1)T} and discarding all but the first two terms.[6,7] The Taylor's series expansion of y(t) in the vicinity of the point $t=t_o$ is given by

$$y(t) = y(t_o) + \frac{(t-t_o)\dot{y}(t_o)}{1!} + \frac{(t-t_o)^2 \ddot{y}(t_o)}{2!} + \ldots \quad (13\text{-}35)$$

If t is near t_o, then magnitudes of the terms decrease rapidly due to the factors $(t-t_o)^k$. Let $t=(k-1)T$ and $t_o=kT$. Then Eq. (13-35) becomes

$$y(k-1) = y(k) + \frac{(-T)\dot{y}(k)}{1!} + \frac{(-T)^2 \ddot{y}(k)}{2!} + \ldots \quad (13\text{-}36)$$

Saving only the first two terms (which is reasonable for small T) and solving for the derivative gives

$$\dot{y}(k) \approx \frac{y(k) - y(k-1)}{T} \quad (13\text{-}37)$$

which is identical to the geometrical approximation of Eq. (13-3).

In summary, discretization based on a straight-line approximation to the slope is equivalent to expanding the function in a Taylor's series and preserving the first two terms. We have introduced this comparison because one of the standard ways of measuring the quality a numerical integration method is to compare it to a Taylor's series.[8]

13.2 RECTANGULAR INTEGRATION

Figure 13-2 illustrates a numerical integration scheme based on approximating the area under a curve as a sum of rectangles.

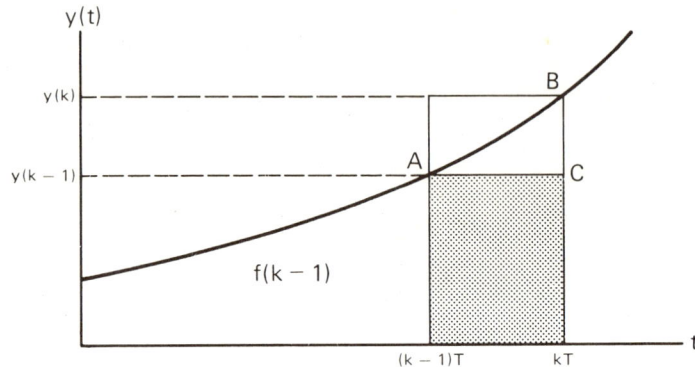

Figure 13-2. Rectangular integration approximates the area from $(k-1)T$ to kT by the shaded rectangle. The area of the "triangle" ABC is lost.

Define the area under the curve $y(t)$ from $t=0$ to $t=(k-1)T$ by $f(k-1)$, or

$$f(k-1) = \int_0^{(k-1)T} y(t)dt \qquad (13\text{-}38)$$

The area $f(k)$ to the left of $t=kT$ is approximately equal to the sum of Eq. (13-38) and the area of the shaded rectangle; thus,

$$f(k) = f(k-1) + Ty(k-1) \qquad (13\text{-}39)$$

The lost area is the area of "triangle" ABC. (We could also use the "outer" rectangle with the result $f(k)=f(k-1)+Ty(k)$. For small T both relationships give comparable results.)

The Derivative Operator

As Eq. (13-38) implies that $f(t)$ is the integral of $y(t)$, then $y(t)$ can be interpreted as the derivative of $f(t)$.

Numerical Solution of Differential Equations

Solving Eq. (13-39) for y(k-1) gives

$$y(k-1) = \frac{f(k) - f(k-1)}{T} \tag{13-40}$$

which is identical to Eq. (13-3) if we interpret y(t) as the derivative of f(t). Equation (13-40) can be further interpreted as a differentiating system whose input is f(t) and whose output is y(t)=df(t)/dt. (See Fig. 13-3(a).)

Figure 13-3. The discrete approximation to the derivative, $G_1(z) = (z - 1)/T$, can be interpreted as a differentiating system in the z-domain.

Taking the z-transform of Eq. (13-40) yields

$$z^{-1}Y(z) = \frac{1 - z^{-1}}{T}F(z) \tag{13-41}$$

or, treating the operation as a discrete differentiating system with transfer function $G_1(s)$, we have

$$G_1(z) = \frac{Y(z)}{F(z)} = \frac{1 - z^{-1}}{Tz^{-1}} = \frac{z - 1}{T} \tag{13-42}$$

Figure 13-3(b) shows Eq. (13-42) interpreted as a discrete differentiator.

The second derivative has the transfer function $G_2(z)$ obtained by squaring $G_1(z)$, or

$$G_2(z) = \left\{\frac{z - 1}{T}\right\}^2 \tag{13-43}$$

The nth derivative has the transfer function

$$G_n(z) = \left\{\frac{z - 1}{T}\right\}^n \tag{13-44}$$

We can use Eq. (13-44) to transform continuous differential equations by replacing the derivatives with appropriate z-domain transfer functions and then solving the resulting transformed equations, either by z-transforms or by difference equation methods. The results are identical to those of the previous section except that we have bypassed the intermediate step of writing the difference equation. This method of numerical integration is equivalent to Euler's method from classical numerical analysis.[9]

The First-Order System

Let us apply rectangular integration to the differential equation

$$\dot{y}(t) + ay(t) = u(t) \tag{13-45}$$

with $y(0)=0$. Replace $\dot{y}(t)$ by Eq. (13-42) and z-transform the other variables. This yields

$$\left\{\frac{z-1}{T} + a\right\}Y(z) = U(z) \tag{13-46}$$

or

$$\{z + (aT-1)\}Y(z) = TU(z) \tag{13-47}$$

where $U(z)=Z\{u(k)\}$. It appears that we can now solve Eq. (13-47) for $Y(z)$ and then compute the output sequence $y(k)$, but the situation is not so simple. We must first convert Eq. (13-47) back to difference-equation form to insure that $y(-1)=0$, the automatic assumption of the z-transform, will produce the correct value of $y(0)=0$. Converting Eq. (13-47) to negative powers of z gives

$$\{1 + z^{-1}(aT-1)\}Y(z) = z^{-1}U(z)T \tag{13-48}$$

and the corresponding difference equation is

$$y(k) + (aT-1)y(k-1) = Tu(k-1) \tag{13-49}$$

Evaluating Eq. (13-49) for $k=0$ with $y(0)=0$ yields $y(0)+(aT-1)y(-1)=0$, from which $y(-1)=0$. Hence we can solve Eq.

Numerical Solution of Differential Equations

(13-47) by z-transforms with the assurance that the solution will have the correct initial value. (Of course, if a point-by-point solution is to be generated on a digital computer, Eq. (13-49) is the equation to be programmed.) Using $U(z) = z/(z-1)$, Eq. (13-47) is solved for

$$\frac{Y(z)}{z} = \frac{T}{(z-1)\{z-(1-aT)\}} \tag{13-50}$$

$$= \frac{1}{a(z-1)} - \frac{1}{a\{z-(1-aT)\}} \tag{13-51}$$

and the output sequence is

$$y(k) = \frac{1}{a}\left\{1-(1-aT)^k\right\} \tag{13-52}$$

If T is chosen such that $|1-aT|<1$, the output sequence converges. We should also choose $1-aT>0$, else the solution sequence, though convergent, will alternate in sign and not look much like the desired exponential (see Fig. 13-4). The net restriction then is $0<1-aT<1$ or $T<1/a$.

As a check on the result, we know that the exact solution of the continuous equation is

$$y(t) = \frac{1}{a}\left\{1 - e^{-at}\right\} \tag{13-53}$$

The initial and steady-state values, $y(0)=0$ and $y(\infty)=1/a$, agree with Eq. (13-52). The question that remains is how good an approximation to $\exp(-at)$ is $(1-aT)^k$? Expanding $\exp(-at)$ as a power series gives

$$e^{-at} = 1 - at + \frac{(at)^2}{2!} - \frac{(at)^3}{3!} + \ldots \tag{13-54}$$

Taking the first two terms gives the approximation

$$e^{-at} \simeq 1 - at \tag{13-55}$$

Evaluating at $t=T$ and raising both sides to the k-th power results in

$$e^{-akT} \simeq (1 - aT)^k \tag{13-56}$$

The effect of rectangular integration has thus been to give a solution in which $\exp(-at)$ has been approximated by the first two terms of its series expansion. Figure 13-4 shows the exact solution, the output sequence of the difference equation for aT=1.5 (stable but alternating), and for aT=0.5 (a good solution). The value a=1 was used for the plots.

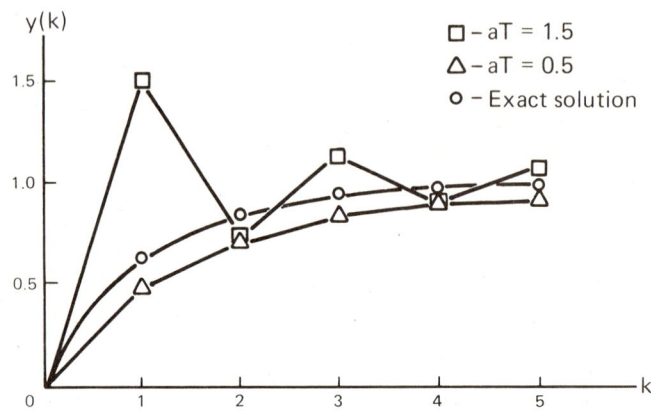

Figure 13-4. Comparative solutions for the differential equation $\dot{y} + ay = u(t)$. Equation (13-53) is the exact solution and Eq. (13-52) is the approximation obtained by rectangular integration.

The following example shows how to handle a nonzero initial condition.

Example 13-4. Solve $\dot{y}(t)+y(t)=u(t)$ with $y(0)=2$ using rectangular integration. Leave T in symbolic form.

Solution. Assuming the variables are sampled and substituting the discrete derivative operator $G(z)=(z-1)/T$ yields

$$\left\{\frac{z-1}{T} + 1\right\}Y(z) = U(z) \tag{13-57}$$

with the difference equation

$$y(k) + (T-1)y(k-1) = Tu(k-1) \tag{13-58}$$

Evaluating at k=0 gives $y(0)+(T-1)y(-1)=0$ and, with $y(0)=2$, $y(-1)=2/(1-T)$. Transform Eq. (13-58), remembering to insert $y(-1)$ by using the extended right-shifting theorem:

$$Y(z) + (T-1)\left\{\frac{2}{1-T} + z^{-1}Y(z)\right\} = \frac{Tz^{-1}z}{z-1} \tag{13-59}$$

Solving for Y(z) and performing the indicated partial-fraction expansion yields

$$Y(z) = \frac{z}{z-1} + \frac{z}{z-(1-T)} \qquad (13\text{-}60)$$

with the output sequence

$$y(k) = (1)^k + (1-T)^k \qquad (13\text{-}61)$$

The sampling period T must lie in the range 0<T<1 for a stable nonoscillatory solution. The exact solution of the original differential equation is y(t)=1+exp(-t) which implies that we have used the approximation exp(-T)≈1-T. This approximation is good for small T.

13.3 TRAPEZOIDAL INTEGRATION

Trapezoidal integration as the name implies is based on approximating the area under a curve by a series of trapezoids.[10] As the amount of "lost" area is less for a trapezoidal than for a rectangular approximation, we would expect improved accuracy for the same sampling period. Figure 13-5 shows the geometry for trapezoidal integration.[11]

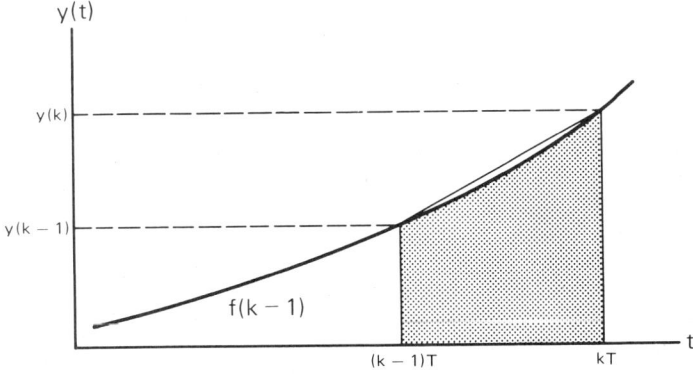

Figure 13-5. The trapezoidal integration formula approximates the area between (k − 1)T and kT by the area of the shaded trapezoid.

Again, the area from t=0 to t=(k-1)T is given by f(k-1) according to Eq. (13-38). The area f(k) from t=0 to t=kT is given approximately by f(k-1) plus the area of the shaded trapezoid in Fig. 13-5. Recalling that the area of a trapezoid is 1/2 of the sum of the bases times the altitude, we have

$$f(k) = f(k-1) + \frac{T}{2}\{y(k-1) + y(k)\} \qquad (13-62)$$

The Trapezoidal Derivative Operator

If f(t) is the integral of y(t), then y(t) must be the derivative of f(t). Solve Eq. (13-62) for y(k) with the result

$$y(k) = \frac{2}{T}\{f(k) - f(k-1)\} - y(k-1) \qquad (13-63)$$

where y(k) is an approximation to the derivative of f(k). Taking the z-transform of Eq. (13-63) yields

$$Y(z)(1+z^{-1}) = \frac{2F(z)}{T}(1-z^{-1}) \qquad (13-64)$$

Defining a discrete differentiator with input F(z), output Y(z), and transfer function $G_1(z)$, we have

$$G_1(z) = \frac{2(1-z^{-1})}{T(1+z^{-1})} = \frac{2(z-1)}{T(z+1)} \qquad (13-65)$$

Equation (13-65) is equivalent to Tustin's approximation to the derivative.[12] The nth-order derivative has the transfer function

$$G_n(z) = \frac{2}{T}\left\{\frac{(1-z^{-1})}{(1+z^{-1})}\right\}^n = \frac{2}{T}\left\{\frac{(z-1)}{(z+1)}\right\}^n \qquad (13-66)$$

A continuous differential equation can be solved by replacing the derivative terms with Eq. (13-66) evaluated for appropriate values of n and assuming the dependent variable and forcing function are sampled. The initial conditions are found iteratively using Eq. (13-8).

Numerical Solution of Differential Equations 585

The First-Order System

As an example, consider again the first-order differential equation

$$\dot{y}(t) + ay(t) = u(t) \qquad (13\text{-}67)$$

with y(0)=0. Substitute Eq. (13-65) for the derivative and assume y(t) and u(t) are sampled:

$$\left\{ \frac{2}{T} \frac{(1-z^{-1})}{(1+z^{-1})} + a \right\} Y(z) = U(z) \qquad (13\text{-}68)$$

We cannot solve Eq. (13-68) using z-transforms without first making sure the initial condition "fits". The difference equation, obtained by taking inverse z-transforms in Eq. (13-68), is

$$(2+aT)y(k) - (2-aT)y(k-1) = T\{u(k)+u(k-1)\} \qquad (13\text{-}69)$$

Evaluating this expression for k=0 and substituting y(0)=0 reveals that y(-1)=T/(aT-2). (If we had solved Eq. (13-68) by z-transforms the sequence value y(0) would have been incorrect because the z-transform method assumes y(-1)=0 by default.) Taking the z-transform of Eq. (13-69), but this time inserting the correct value for y(-1) by using the extended right-shifting theorem, yields

$$(2+aT)Y(z) - (2-aT)\left\{ \frac{T}{aT-2} + z^{-1}Y(z) \right\} = T\frac{(1+z)}{(z-1)} \qquad (13\text{-}70)$$

Define p=(2-aT)/(2+aT) and solve for Y(z)/z:

$$\frac{Y(z)}{z} = \frac{T}{2+aT}\left\{ \frac{z+1}{(z-1)(z-p)} - \frac{1}{z-p} \right\} \qquad (13\text{-}71)$$

After a partial-fraction expansion and some algebraic simplification we arrive at

$$Y(z) = \frac{1}{a}\left\{ \frac{z}{z-1} - \frac{z}{z-p} \right\} \qquad (13\text{-}72)$$

from which

$$y(k) = \frac{1}{a}\left\{ (1)^k - (p)^k \right\} \qquad (13\text{-}73)$$

The solution converges without oscillation if 0<p<1, or

$$0 < \frac{1 - aT/2}{1 + aT/2} < 1 \qquad (13\text{-}74)$$

from which T<2/a. Compare this with T<1/a for rectangular integration. The larger permissible sampling period for trapezoidal integration means that we can solve the problem twice as fast on a digital computer because we need take only half as many solution points. Finally, how does Eq. (13-73) compare with the exact solution y(t)={1-exp(-at)}/a, which in sampled form is

$$y(kT) = \frac{1}{a}(1 - e^{-akT}) \qquad (13\text{-}75)$$

The series expansion for exp(-aT) is

$$e^{-aT} = 1 - aT + \frac{(aT)^2}{2!} - \frac{(aT)^3}{3!} + \ldots \qquad (13\text{-}76)$$

while, after long division, we can write

$$p = \frac{1 - aT/2}{1 + aT/2} = 1 - aT + \frac{(aT)^2}{2} - \frac{(aT)^3}{4} + \ldots \qquad (13\text{-}77)$$

Comparing Eqns. (13-76) and (13-77) indicates that approximating exp(-aT) by (1-aT/2)/(1+aT/2) is exact for the first three terms of the series (as compared with rectangular integration which was good only for the first two terms). As a result, we can expect better accuracy for the same sampling period if we use trapezoidal integration.

Example 13-5. Repeat Example 13-4 using trapezoidal integration, but this time compare the exact, rectangular, and trapezoidal solutions for T=0.2.

Solution. Taking z-transforms and substituting Eq. (13-65) for $\dot{y}(t)$ yields

$$\left\{\frac{2}{T}\frac{1-z^{-1}}{1+z^{-1}} + 1\right\}Y(z) = U(z) \qquad (13\text{-}78)$$

The associated difference equation is

$$(T+2)y(k) + (T-2)y(k-1) = T\{u(k) + u(k-1)\} \qquad (13\text{-}79)$$

Numerical Solution of Differential Equations

Substituting y(0)=2 when k=0 yields y(-1)=(4+T)/(2-T). Transforming the difference equation, but this time inserting the correct value for y(-1), results in

$$\{(T+2) + (T-2)z^{-1}\}Y(z) = T + 4 + T\frac{z+1}{z-1} \qquad (13-80)$$

and

$$\frac{Y(z)}{z} = \frac{1}{2+T}\left\{\frac{4+T}{z-b} + \frac{T(z+1)}{(z-1)(z-b)}\right\} \qquad (13-81)$$

where b=(2-T)/(2+T). After a partial-fraction expansion,

$$Y(z) = \frac{z}{z-1} + \frac{z}{z-b} \qquad (13-82)$$

from which we obtain the output sequence

$$y(k) = (1)^k + \left\{\frac{2-T}{2+T}\right\}^k \qquad (13-83)$$

The exact solution, sampled at t=kT, is y(kT)=1+exp(-kT) and the result using rectangular integration was $y(k)=1+(1-T)^k$. Inserting T=0.2 yields the three expressions:

Exact: $\qquad y_E(k) = 1 + e^{-0.2k} \qquad$ (13-84)

Rectangular: $\qquad y_R(k) = 1 + (0.8)^k \qquad$ (13-85)

Trapezoidal: $\qquad y_T(k) = 1 + (0.8182)^k \qquad$ (13-86)

The table below compares these three solution for the first few values of k.

k	$y_E(k)$	$y_R(k)$	$y_T(k)$
0	2.0000	2.0000	2.0000
1	1.8187	1.8000	1.8182
2	1.6703	1.6400	1.6695
3	1.5488	1.512	1.5477
4	1.4493	1.4096	1.4482
5	1.3679	1.3277	1.3667

The improved accuracy of trapezoidal over rectangular integration is evident even from this elementary example.

Finally, one need not stop at trapezoids in developing approximate integration formulas. Simpson's rule approximates the area under a curve by connecting a parabola between three points.[13] A z-domain integration formula based on Simpson's rule gives a more accurate approximate solution than the other two methods for the same step size, but at the price of more complicated algebra.

13.4 STATE-VARIABLE SYSTEMS

As introduced in Chapter 10, state variables are a generalized approach to formulating the mathematical models of systems which can be described by linear constant-coefficient differential equations. From the standpoint of numerical analysis, state-variable systems have the advantage of requiring only a first-order approximation to the derivative regardless of the order of the system.[14,15] The price we pay for this simplicity of order is the complexity of matrix notation.

The standard state-variable system is described by the equations

$$\dot{\underline{x}}(t) = A\underline{x}(t) + B\underline{f}(t) \qquad (13-87)$$

$$\underline{y}(t) = C\underline{x}(t) + D\underline{f}(t) \qquad (13-88)$$

where A, B, C and D are (n x n), (n x m), (r x n), and (r x m) matrices with constant coefficients. The quantities $\underline{f}(t)$, $\underline{y}(t)$, and $\underline{x}(t)$ are the input, output, and state vectors, respectively, with dimensions (m x 1), (r x 1), and (n x 1). The (n x 1) column vector $\underline{x}(0)$ specifies the initial conditions.

Numerical Solution Of State Equations

In Chapter 10 we showed that the time-domain solution of Eqns. (13-87) and (13-88) is

$$\underline{x}(t) = e^{A(t-t_o)}\underline{x}(t_o) + e^{At}\int_{t_o}^{t} e^{-Aq}B\underline{f}(q)dq \qquad (13-89)$$

Numerical Solution of Differential Equations 589

where we have changed the starting time from t=0 to $t=t_o$. The state vector $\underline{x}(t)$ is the solution at time t due to an input $\underline{f}(t)$ applied at time t_o. Once $\underline{x}(t)$ is known, the output vector $\underline{y}(t)$ is obtained algebraically from Eq. (13-88).

Equation (13-89) is good for any starting time t_o. As a result, we can choose $t_o=(k-1)T$ and $t=kT$ and express the discrete state vector $\underline{x}(k)$ in terms of $\underline{x}(k-1)$ and an integral, or

$$\underline{x}(k) = e^{AT}\underline{x}(k-1) + e^{ATk}\int_{(k-1)T}^{kT} e^{-Aq}B\underline{f}(q)dq \qquad (13-90)$$

If we regard $\underline{f}(t)$ as a constant over the time interval $(k-1)T<t<kT$ (which is permissible if f(t) is a well-behaved function and T is small), and assign it the value f(k-1), then Eq. (13-90) becomes

$$\underline{x}(k) = e^{AT}\underline{x}(k-1) + e^{ATk}\left\{\int_{(k-1)T}^{kT} e^{-Aq}dq\right\}B\underline{f}(k-1) \qquad (13-91)$$

After performing the indicated integration in Eq. (13-91), we have

$$\underline{x}(k) = e^{AT}\underline{x}(k-1) + e^{AT}\left[-e^{-Aq}A^{-1}\right]\bigg|_{(k-1)T}^{kT} B\underline{f}(k-1) \qquad (13-92)$$

which simplifies to

$$\underline{x}(k) = e^{AT}\underline{x}(k-1) + \left[e^{AT} - I_n\right]A^{-1}B\underline{f}(k-1) \qquad (13-93)$$

Equation (13-93) has the form

$$\underline{x}(k) = P\underline{x}(k-1) + Q\underline{f}(k-1) \qquad (13-94)$$

where $P=\exp(AT)$ and $Q=\{\exp(AT)-I_n\}A^{-1}B$. The matrices P and Q are constant and need be computed only once during the solution.[16] Both matrices, however, contain the matrix exponential exp(AT) which can be computed by Laplace methods from the relationship

$$e^{AT} = L^{-1}\{(sI_n - A)^{-1}\}\Big|_{t=T} \quad (13\text{-}95)$$

Equation (13-95) is a difficult computation for higher-order systems with an arbitrary A matrix. In fact, evaluating Eq. (13-95) is equivalent to finding an exact solution of the continuous system. To avoid this difficulty, and arrive at an approximation which is easily implemented on a digital computer, we observe that exp(AT) has the series expansion

$$e^{AT} = I_n + \frac{(AT)^1}{1!} + \frac{(AT)^2}{2!} + \frac{(AT)^3}{3!} + \ldots \quad (13\text{-}96)$$

If AT is small, the first few terms of Eq. (13-96) are a suitable approximation to the matrix exponential and can be used in calculating the matrices P and Q. These terms can be evaluated numerically on a digital computer without the need for Laplace transforms. Many of the classical numerical analysis techniques, when applied to state-variable equations, are equivalent to approximating the matrix exponential by the first few terms of its series expansion.

Finally, Eq. (13-88) is discretized to

$$\underline{y}(k) = C\underline{x}(k) + D\underline{f}(k) \quad (13\text{-}97)$$

Once $\underline{x}(k)$ is known, Eq. (13-97) can be evaluated on a digital computer using arithmetic. The following three examples illustrate the methods of this section for a system with a diagonal A matrix, thus giving a clear insight into the nature of the approximations used.

Example 13-6. Discretize the state vector $\underline{x}(t)$ for the system

$$\begin{bmatrix} \dot{x}_1 \\ \dot{x}_2 \end{bmatrix} = \begin{bmatrix} -1 & 0 \\ 0 & -5 \end{bmatrix} \begin{bmatrix} x_1 \\ x_2 \end{bmatrix} \quad \begin{bmatrix} x_1(0) \\ x_2(0) \end{bmatrix} = \begin{bmatrix} 2 \\ 3 \end{bmatrix} \quad (13\text{-}98)$$

Use the exact expression for exp(AT) and leave T in symbolic form.

Numerical Solution of Differential Equations

Solution. As $\underline{f}(t)=0$, we see from Eq. (13-94) that the matrix Q is not needed, only P=exp(AT). Use

$$e^{At} = L^{-1}\{(sI_2 - A)^{-1}\} \tag{13-99}$$

$$= L^{-1}\begin{bmatrix} \frac{1}{s+1} & 0 \\ 0 & \frac{1}{s+5} \end{bmatrix} = \begin{bmatrix} e^{-t} & 0 \\ 0 & e^{-5t} \end{bmatrix} \tag{13-100}$$

Evaluating this expression for t=T yields the matrix P. The sampled solution is then $\underline{x}(t)=P\underline{x}(k-1)$ and from Eq. (13-94)

$$\begin{bmatrix} x_1(k) \\ x_2(k) \end{bmatrix} = \begin{bmatrix} e^{-T} & 0 \\ 0 & e^{-5T} \end{bmatrix} \begin{bmatrix} x_1(k-1) \\ x_2(k-1) \end{bmatrix} \tag{13-101}$$

This expression can be evaluated iteratively starting with k=1 and the initial-condition vector $\underline{x}(0)$. We were able to find the exact expression for exp(AT) in this example because of the diagonal form of the matrix A. In the general case, this might be too complicated or too time consuming, so we must turn to an approximate method.

Rectangular Integration Of State Equations

We can avoid the necessity of evaluating exp(AT) by using a matrix equivalent of rectangular integration.[16] Expand the state vector $\underline{x}(t)$ in a Taylor's series in the neighborhood of t_o with the result

$$\underline{x}(t) = \underline{x}(t_o) + \frac{1}{1!}\underline{\dot{x}}(t_o)(t-t_o) + \frac{1}{2!}\underline{\ddot{x}}(t_o)(t-t_o)^2 + \ldots \tag{13-102}$$

Let t=kT, $t_o=(k-1)T$, and save the first two terms of the series; the result is the approximation

$$\underline{x}(k) = \underline{x}(k-1) + T\underline{\dot{x}}(k-1) \tag{13-103}$$

Substituting Eq. (13-87) evaluated for t=(k-1)T for $\underline{\dot{x}}(k-1)$ in Eq. (13-103) yields

$$\underline{x}(k) = \underline{x}(k-1) + T\{A\underline{x}(k-1) + B\underline{f}(k-1)\} \tag{13-104}$$

or

$$\underline{x}(k) = (I_n + AT)\underline{x}(k-1) + TB\underline{f}(k-1) \quad (13-105)$$

Equation (13-105), also known as Euler's method, can be used to solve iteratively for the state vector, starting with k=1 and $\underline{x}(0)$. Rectangular integration gives the same accuracy as Eq. (13-94) if exp(AT) in the latter is approximated by the first two terms of its series expansion. To achieve an accurate solution with rectangular integration, we must choose T very small, but this increases the number of solution steps needed to cover a specified time interval. The result is slow running time on the computer. We can improve the situation by choosing the more accurate trapezoidal integration formula.

Example 13-7. Repeat Example 13-6 using rectangular integration.

Solution. We need $(I_2 + AT)$ or

$$I_2 + AT = \begin{bmatrix} 1 & 0 \\ 0 & 1 \end{bmatrix} + \begin{bmatrix} -1 & 0 \\ 0 & -5 \end{bmatrix} T = \begin{bmatrix} 1-T & 0 \\ 0 & 1-5T \end{bmatrix} \quad (13-106)$$

Use Eq. (13-105) with $\underline{f}(k-1)=\underline{0}$ to obtain

$$\begin{bmatrix} x_1(k) \\ x_2(k) \end{bmatrix} = \begin{bmatrix} 1-T & 0 \\ 0 & 1-5T \end{bmatrix} \begin{bmatrix} x_1(k-1) \\ x_2(k-1) \end{bmatrix} \quad (13-107)$$

Comparing Eqns. (13-101) and (13-107) reveals that we have indeed used the approximation $\exp(AT)=I_2+AT$.

Trapezoidal Integration Of State Equations

Consider the problem of finding the area A between t=(k-1)T and t=kT under a curve which is a plot of $\dot{x}(t)$ versus t. This area is given exactly by

$$A = \int_{(k-1)T}^{kT} \dot{x}(t)dt = \int_{(k-1)T}^{kT} dx(t) = x(k) - x(k-1) \quad (13-108)$$

Numerical Solution of Differential Equations

According to the trapezoidal rule from calculus, the area defined by Eq. (13-108) can be approximated by

$$A = \frac{T}{2}\{\dot{x}(k) + \dot{x}(k-1)\} \qquad (13\text{-}109)$$

Equating Eqns. (13-108) and (13-109) gives

$$x(k) - x(k-1) = \frac{T}{2}\{\dot{x}(k) + \dot{x}(k-1)\} \qquad (13\text{-}110)$$

Equation (13-110) is valid if $x(k)$ is replaced by the $(n \times 1)$ column vector $\underline{x}(k)$. Interpreting Eq. (13-110) as a vector equation and substituting the sampled version of Eq. (13-87) leads to

$$\underline{x}(k) = \underline{x}(k-1) + \frac{T}{2}\{A\underline{x}(k) + B\underline{f}(k) + A\underline{x}(k-1) + B\underline{f}(k-1)\} \qquad (13\text{-}111)$$

The solution for $\underline{x}(k)$ is

$$\underline{x}(k) = \left\{I_n - \frac{AT}{2}\right\}^{-1}\left(\left\{I_n + \frac{AT}{2}\right\}\underline{x}(k-1) + \frac{TB}{2}\{\underline{f}(k-1) + \underline{f}(k)\}\right) \qquad (13\text{-}112)$$

The trapezoidal integration method defined by Eq. (13-112) is accurate and stable. It requires computation of the constant matrices $(I_n - AT/2)^{-1}$ and $(I_n + AT/2)$, both of which need be evaluated only once before the solution starts. The entire process is readily automated on a digital computer and is accurate to the first three terms of the series expansion for $\exp(AT)$.

Example 13-8. Repeat Example 13-7 using trapezoidal integration.

Solution. We need

$$I_2 + \frac{AT}{2} = \begin{bmatrix} 1 & 0 \\ 0 & 1 \end{bmatrix} + \frac{T}{2}\begin{bmatrix} -1 & 0 \\ 0 & -5 \end{bmatrix} \qquad (13\text{-}113)$$

$$= \begin{bmatrix} 1 - \frac{T}{2} & 0 \\ 0 & 1 - \frac{5T}{2} \end{bmatrix} \qquad (13\text{-}114)$$

and

$$(I_2 - \frac{AT}{2})^{-1} = \begin{bmatrix} \frac{1}{1 + T/2} & 0 \\ 0 & \frac{1}{1 - 5T/2} \end{bmatrix} \qquad (13\text{-}115)$$

Equation (13-112) gives the solution

$$\underline{x}(k) = \begin{bmatrix} \frac{1 - T/2}{1 + T/2} & 0 \\ 0 & \frac{1 - 5T/2}{1 + 5T/2} \end{bmatrix} \underline{x}(k-1) \qquad (13\text{-}116)$$

This result shows that the approximation

$$e^{-AT} \simeq (I_2 - \frac{AT}{2})(I_2 + \frac{AT}{2})^{-1} \qquad (13\text{-}117)$$

has been used. As before, the diagonal form of the A matrix shows the type of approximation to exp(-AT) that was used. In the general case things are more obscure.

Stability [17]

A state-variable system is stable (a bounded input produces a bounded output) if the roots of the characteristic equation

$$\text{Det}(sI_n - A) = 0 \qquad (13\text{-}118)$$

all have negative real parts. If A is an (n x n) matrix, the characteristic equation has n roots. As noted in Chapter 10, these roots, $s_1, s_2, \ldots, s_n$, are called the <u>eigenvalues</u> of the system. We now consider the problem of determining the stability of the discrete approximation of a state-variable system. We shall see that the numerical approximation can be unstable even though the parent system is stable.

Consider the homogeneous (unforced) system $\underline{\dot{x}}(t) = A\underline{x}(t)$ with solution

$$\underline{x}(t) = e^{At}\underline{x}(0) \qquad (13\text{-}119)$$

The sampled version of Eq. (13-119) is obtained from Eq. (13-93) with $\underline{f}(k-1)=0$:

$$\underline{x}(k) = e^{AT}\underline{x}(k-1) \qquad (13\text{-}120)$$

Using a series expansion for exp(AT), the sampled solution becomes

$$\underline{x}(k) = \left\{I_n + (AT) + \frac{(AT)^2}{2!} + \ldots + \frac{(AT)^j}{j!}\right\}\underline{x}(k-1) \qquad (13\text{-}121)$$

Equation (13-121) can be z-transformed to yield

$$\underline{X}(z) = \left\{I_n + (AT) + \frac{(AT)^2}{2!} + \ldots \frac{(AT)^j}{j!}\right\}z^{-1}\underline{X}(z) \qquad (13\text{-}122)$$

where $\underline{X}(z)=Z\{\underline{x}(z)\}$. After writing $\underline{X}(z)=I_n\underline{X}(z)$, Eq. (13-122) becomes

$$\{zI_n - R\}\underline{X}(z) = \underline{0} \qquad (13\text{-}123)$$

where $\underline{0}$ is the (n x 1) zero column vector, and the matrix R is given by

$$R = \sum_{i=0}^{j} \frac{(AT)^j}{j!} I_n \qquad (13\text{-}124)$$

We state without proof (although the result is plausible from our previous z-transform work) that the sampled homogeneous system is stable if and only if the roots of the z-domain characteristic equation

$$\text{Det}(zI_n - R) = 0 \qquad (13\text{-}125)$$

lie inside of the unit circle in the z-plane or on the unit circle for conditional stability. It can be shown by methods beyond the scope of our treatment that the roots of the z-domain characteristic equation are

$$z_i = 1 + (Ts_i) + \frac{1}{2!}(Ts_i)^2 + \frac{1}{3!}(Ts_i)^3 + \ldots \frac{1}{j!}(Ts_i)^j \qquad (13\text{-}126)$$

where the s_i are the eigenvalues of the continuous system as given by Eq. (13-118). For a stable numerical solution, the sampling period T must be chosen such that the z_i in Eq. (13-126) lie inside of the unit circle in the z-plane, or $|z_i|<1$ for $i=1,2,\ldots n$. We might also want to impose the restriction that the real z-domain eigenvalues are positive so the solution will be nonoscillatory as well as stable.

Suppose, for example, we use a first-order approximation to the exponential so that Eq. (13-126) becomes $z_i=1+Ts_i$. In the most general case, an eigenvalue s_i might be complex and of the form $s_i=a+jb$ so that $z_i=1+T(a+jb)$. The stability requirement, $|z_i|<1$, is then

$$\sqrt{(1 + aT)^2 + (Tb)^2} = |z_i| < 1 \qquad (13-127)$$

A stable numerical solution results if Eq. (13-127) is satisfied for every eigenvalue s_i.

Higher-order approximations for exp(AT) give larger regions of stability, implying that we can choose a larger value of T. This in turn lets us generate the solution sequence point by point on a digital computer with less solution steps needed to cover a given time interval. If an exact expression is used for exp(AT), the numerical solution is stable for any value of T as long as the system itself is stable.

Example 13-9. A homogeneous state-variable system has

$$A = \begin{bmatrix} -2 & 0 \\ 0 & -3 \end{bmatrix} \qquad (13-128)$$

Determine the range of T for a stable nonoscillatory solution if a two-term approximation is used in Eq. (13-126).

Solution. The characteristic equation of the continuous system

$$\text{Det}(sI_2 - A) = (s + 2)(s + 3) = 0 \qquad (13-129)$$

leads to the eigenvalues $s_1=-2$ and $s_2=-3$. We must choose T such that both of the z-domain eigenvalues $z_1=1+Ts_1$ and $z_2=$

Numerical Solution of Differential Equations

$1+Ts_2$ lie inside of the unit circle in the z-plane and are positive. These conditions require $0<(1-2T)<1$ and $0<(1-3T)<1$. The first of these inequalities leads to $0<T<1/2$. The other yields $0<T<1/3$. Taking the more restrictive case, we see that the permissible range of T for a stable nonoscillatory numerical solution is $0<T<1/3$.

Example 13-10. Repeat Example 13-9 if a three-term approximation is used for exp(AT). Compare the permissible range for T with that for Example 13-9.

Solution. For a three-term approximation to exp(AT), the z-domain eigenvalues are

$$z_1 = 1 + Ts_1 + \frac{1}{2}(Ts_1)^2 \qquad (13-130)$$

and

$$z_2 = 1 + Ts_2 + \frac{1}{2}(Ts_2)^2 \qquad (13-131)$$

where $s_1=-2$ and $s_2=-3$. In order for z_1 to be positive and lie within the unit circle, T must be chosen to satisfy $0<1-2T+2T^2<1$, and for $0<|z_2|<1$ we must have $0<1-3T+9T^2/2<1$. If $T=2/3$, $z_2=1$ and $z_1=5/9$. Hence, any value of T in the range $0<T<2/3$ yields a stable nonoscillatory numerical solution. The result for Example 13-9 was $0<T<1/3$. Using a three-term rather than a two-term approximation for exp(AT) lets us choose T twice as large.

The stability results apply only if the system is unforced. Just as was the case with scalar systems, if the input is unbounded we could have an unbounded output no matter how inherently stable the system.

13.5 RUNGE–KUTTA INTEGRATION

Our introductory treatment of numerical techniques for solving differential equations would be incomplete without some mention of the widely used Runge-Kutta method.[18] This method, though strictly applicable only to first-order differential equations, can be applied to higher-order equations if they are represented in state-variable form.[19] The Runge-Kutta method assumes a differential equation of the form

$$\dot{y}(t) = f(y,t) \tag{13-132}$$

with a nonzero initial condition $y(0)$. The form of Eq. (13-132) allows for the possibility that the differential equation is nonlinear, time-varying, or both. Assuming sampled variables, Eq. (13-132) takes the form

$$\dot{y}(k) = f(y(k),kT) \tag{13-133}$$

The Runge-Kutta procedure is to evaluate successively the four quantities

$$g_1(k) = f\{y(k),kT\} \tag{13-134}$$

$$g_2(k) = f\{y(k) + \frac{T}{2}g_1(k), (k + \frac{1}{2})T\} \tag{13-135}$$

$$g_3(k) = f\{y(k) + \frac{T}{2}g_2(k), (k + \frac{1}{2})T\} \tag{13-136}$$

and

$$g_4(k) = f\{y(k) + Tg_3(k), (k + 1)T\} \tag{13-137}$$

after which the estimate for the next point $y(k+1)$ is

$$y(k+1) = \frac{T}{6}\{g_1(k) + 2g_2(k) + 2g_3(k) + g_4(k)\} + y(k) \tag{13-138}$$

These expressions are used iteratively starting with k=1 and $y(0)$. The method is readily implemented by a digital computer program. For state-variable systems, $y(k)$, $f(y,kT)$ and the $g_i(k)$ are vectors.

An important difference between Runge-Kutta and the other methods we have studied is the requirement to evaluate the derivative $\dot{y}(t)$ <u>between</u> the sample points (note the $(k+1/2)T$ terms in the expressions for $g_2(k)$ and $g_3(k)$). This presents no difficulty if $y(t)$ is known in continuous form, but if it is sampled with a fixed T, as is the case with many real-time systems, the half-sample points are not available. (We might consider sampling at twice the required frequency and solving

Numerical Solution of Differential Equations 599

the differential equation at every other point to avoid this problem.)

Accuracy Of Runge-Kutta Integration

It is impossible to obtain a z-domain transfer function for the Runge-Kutta formula when it is applied to an arbitrary differential equation. We can, however, gain some insight by applying the method to the specific equation $\dot{y}(t)+ay(t)=0$ with initial condition $y(0)$.[20] In this instance

$$f(y,t) = -ay(t) \qquad (13\text{-}139)$$

Hence, using sampled variables, $g_1(k)=-ay(k)$, and

$$g_2(k) = -a\{y(k) - \tfrac{aT}{2}y(k)\} \qquad (13\text{-}140)$$

$$= -ay(k)(1-q) \qquad (13\text{-}141)$$

where q is defined by $q=aT/2$. For $g_3(k)$, Eq. (13-136) gives

$$g_3(k) = -a\{y(k) - \tfrac{aT}{2}y(k)(1-q)\} \qquad (13\text{-}142)$$

$$= -ay(k)\{1 - q(1-q)\} \qquad (13\text{-}143)$$

Finally,

$$g_4(k) = -a\Big\{y(k) - aTy(k)\{1 - q(1-q)\}\Big\} \qquad (13\text{-}144)$$

$$= -a\{1 - 2q + 2q^2(1-q)\}y(k) \qquad (13\text{-}145)$$

Using these expressions, Eq. (13-138) becomes

$$y(k+1) = y(k)+\tfrac{T}{6}y(k)\Big[-a-2a(1-q)-2a\{1-q(1-q)\}-a\{1-2q+2q^2(1-q)\}\Big] \qquad (13\text{-}146)$$

After replacing q with $aT/2$ and combining terms, Eq. (13-146) becomes

$$y(k+1) = y(k)\Big\{1 - (aT) + \tfrac{(aT)^2}{2} - \tfrac{(aT)^3}{6} + \tfrac{(aT)^4}{24}\Big\} \qquad (13\text{-}147)$$

The response has the form $y(k+1)=y(k)h(T)$ where

$$h(T) = 1 - (aT) + \tfrac{(aT)^2}{2!} - \tfrac{(aT)^3}{3!} + \tfrac{(aT)^4}{4!} \qquad (13\text{-}148)$$

Taking the z-transform of y(k+1)=y(k)h(T) using the left-shifting theorem to handle y(k+1) results in

$$zY(z) - zy(0) = Y(z)h(T) \qquad (13-149)$$

and

$$Y(z) = \frac{zy(0)}{z - h(T)} \qquad (13-150)$$

The output sequence is therefore $y(k)=y(0)\{h(T)\}^k$. The sampled version of the exact solution is

$$y(k) = y(0)\{e^{-aT}\}^k \qquad (13-151)$$

so that the Runge-Kutta method has used the approximation exp(-aT)=h(T). It is evident from Eq. (13-148) that h(T) is identical to the first <u>five</u> terms of the series expansion for the exponential function. We would therefore expect Runge-Kutta integration to give better accuracy than either rectangular or trapezoidal integration for this class of problems. And as we must use a digital computer anyhow, we might as well program the more accurate method.

Finally, Eq. (13-150) shows that the sampled solution is stable if h(T)<1. It can be shown that this inequality is satisfied for values of T such that aT<2.8. (Compare this with the maximum permissible values of aT<1 and aT<2 for two-term and three-term approximations to exp(-aT) for the same problem.) The larger permissible value for T implies faster solutions along with the greater accuracy. Finally, as noted above, Runge-Kutta integration can be extended to higher-order systems if they are expressed in state-variable form.

PROBLEMS FOR CHAPTER 13

13-1. Use Eq. (13-8) to find approximations for the third and fourth derivatives.

13-2. Discretize the following differential equations using the methods of Section 13.1. Leave T in symbolic form.

Evaluate the initial conditions which correspond to k=-1,-2, ...-n.

(a) $\ddot{y} + 5\dot{y} + 25y = u(t)$ $y(0)=\dot{y}(0)=0$ (13-152)

(b) $\dot{q} + 10q = \cos(5t)$ $q(0)=2.5$ (13-153)

(c) $\dddot{y} + \ddot{y} + \dot{y} + y = u(t)$ $y(0)=1,$ (13-154)

$\dot{y}(0)=2$, $\ddot{y}(0)=3$.

13-3. A third-order differential equation has initial conditions $\ddot{y}(0)=1$, $\dot{y}(0)=5$, and $y(0)=10$. Find the correct values for $y(-1)$, $y(-2)$, and $y(-3)$ using T=0.1 s. When substituted into the differential equation the initial conditions yield $\dddot{y}(0)=2$.

13-4. Find closed-form sampled solutions for the following differential equations using z-transforms and rectangular integration.

(a) $\dot{y} + 4y = u(t); \quad T=0.375; \quad y(0)=0$ (13-155)

(b) $\dot{q} + 16q = 0; \quad T=0.0046875; \quad q(0)=5$ (13-156)

(c) $\ddot{r} + 5\dot{r} + 6r = u(t); \quad T=0.5; \quad r(0)=\dot{r}(0)=0$ (13-157)

13-5. Derive a parabolic integration formula based on Simpson's rule for approximating the definite integral. (Simpson's rule can be found in any standard calculus textbook.)[13]

13-6. Derive a difference equation of the form

$$y(k) + Ay(k-1) + By(k-2) = Ef(k-j) \quad (13\text{-}158)$$

with initial conditions $y(-1)=C$ and $y(-2)=D$, starting with the second-order equation

$$\ddot{y}(t) + 2\zeta\omega_n \dot{y}(t) + \omega_n^2 y(t) = f(t) \quad (13\text{-}159)$$

The continuous equation has nonzero initial conditions $y(0)$ and $\dot{y}(0)$. Use rectangular integration.

13-7. Solve the differential equations of Problem 13-4(a) and (b) using trapezoidal integration. Use T=1/6 for the (a) problem and T=1/12 for (b).

13-8. Solve the differential equation $\dot{y}+y=0$ with $y(0)=1$ using trapezoidal integration. Use T=1/5.

13-9. Derive an expression for the error between the exact and approximate solutions for the differential equation $\dot{y}(t)+ay(t)=0$. Use $y(0)=b$ and rectangular integration.

13-10. Repeat Problem 13-9 using trapezoidal integration and compare the error expressions for the two methods.

13-11. Use Eq. (13-94) to discretize the unit-step response of the state-variable system for which

$$A = \begin{bmatrix} -1 & 0 \\ 0 & -2 \end{bmatrix} \quad \underline{b} = \begin{bmatrix} 2 \\ 1 \end{bmatrix} \quad \underline{x}(0) = \begin{bmatrix} 0 \\ 1 \end{bmatrix} \qquad (13\text{-}160)$$

Solve only for the state vector $\underline{x}(k)$.

13-12. Use Eq. (13-94) to discretize the unit-step response of the state-variable system for which

$$A = \text{Diag}\begin{bmatrix} -1 & -2 & -4 \end{bmatrix}; \quad \underline{b}^T = \begin{bmatrix} 0 & 0 & 1 \end{bmatrix}; \quad \underline{x}(0)^T = \begin{bmatrix} 0 & 0 & 0 \end{bmatrix} \qquad (13\text{-}161)$$

13-13. Find a three-term approximation to $\exp(AT)$ for:

(a) $A = \begin{bmatrix} -1 & 0 \\ 0 & -2 \end{bmatrix}$ \qquad (b) $A = \begin{bmatrix} 1 & 0 \\ 2 & -3 \end{bmatrix}$ \qquad (13-162)

(c) $A = \begin{bmatrix} -1 & 0 & 0 \\ 0 & -5 & 0 \\ 0 & 0 & 2 \end{bmatrix}$ \qquad (d) $A = \begin{bmatrix} 2 & -1 \\ -3 & 4 \end{bmatrix}$ \qquad (13-163)

13-14. Find an open-form (iterative) solution for the state-variable system of Problem 13-11 using rectangular integration.

13-15. Repeat Problem 13-14 using trapezoidal integration.

13-16. Solve Eq. (13-94) for a closed-form expression for $\underline{x}(k)$. Use z-transforms and assume $\underline{x}(0)=\underline{0}$.

13-17. Use z-transforms to solve Eq. (13-105) for a closed-form expression for $\underline{x}(k)$. Assume $\underline{x}(0)=\underline{0}$.

13-18. Repeat Problem 13-16 if $\underline{x}(0) \neq \underline{0}$.

13-19. Repeat Problem 13-17 if $\underline{x}(0) \neq \underline{0}$.

Numerical Solution of Differential Equations

13-20. Solve the state-variable equations of Problem 13-11 by the z-transform method developed in Problem 13-18.

13-21. A homogeneous state-variable system has A= $\text{Diag}\begin{bmatrix} -5, & -10 \end{bmatrix}$. Determine the range of T for a stable numerical solution if we use (a) a two-term approximation to exp(AT), and (b) a three-term approximation to exp(AT).

13-22. Find the first four terms of the solution of the differential equation $\dot{y}+2y=0$ using Runge-Kutta integration. Use y(0)=1 and T=0.1.

13-23. Find a closed-form solution for the differential equation $\dot{y}(t)+5y(t)=0$ with y(0)=2 using Runge-Kutta integration with T=0.1. (Hint: use Eq. (13-148) and $y(k)= y(0)\{h(T)\}^k$.)

13-24. Determine the range of T for a stable numerical solution of the state-variable system for which

$$A = \begin{bmatrix} -1 & 0 \\ -2 & -4 \end{bmatrix} \quad \underline{b} = \begin{bmatrix} 0 \\ 0 \end{bmatrix} \quad (13\text{-}164)$$

if the approximation for exp(AT) has (a) two terms, and (b) three terms.

13-25. Find the first four terms of the solution of the state-variable system with

$$A = \begin{bmatrix} -1 & 0 \\ 0 & -2 \end{bmatrix} \quad \underline{b} = \begin{bmatrix} 0 \\ 0 \end{bmatrix} \quad \underline{x}(0) = \begin{bmatrix} 5 \\ 4 \end{bmatrix} \quad (13\text{-}165)$$

Use a matrix interpretation of the Runge-Kutta method. Let T=0.2.

REFERENCES FOR CHAPTER 13

(1) Milne, W. E.: Numerical Solution of Differential Equations, John Wiley and Sons, Inc., New York, 1953.
(2) Fox, L.: Numerical Solution of Ordinary and Partial Differential Equations, Addison-Wesley Publishing Co., Inc., Reading, Massachusetts, 1962.

(3) Lapidus, L. and J. H. Seinfeld: *Numerical Solution of Ordinary Differential Equations*, Academic Press, Inc., New York, 1971.

(4) James, M. L., Smith, G. M. and J. C. Wolford: *Applied Numerical Methods for Digital Computation With FORTRAN*, International Textbook Company, Scranton, Pennsylvania, 1967.

(5) Rosko, J. S.: *Digital Simulation of Physical Systems*, Addison-Wesley Publishing Co., Inc., Reading, Massachusetts, 1972.

(6) Fox, A. N.: *Fundamentals of Numerical Analysis*, The Ronald Press Company, New York, 1963.

(7) Gerald, C. F.: *Applied Numerical Analysis*, Addison-Wesley Publishing Co., Inc., Reading, Massachusetts, 1970.

(8) Conte, S. D.: *Elementary Numerical Analysis*, McGraw-Hill Book Co., Inc., New York, 1965.

(9) Fox, L. and D. F. Mayers: *Computing Methods for Scientists and Engineers*, Oxford University Press, New York, 1968.

(10) Grove, W. E.: *Brief Numerical Methods*, Prentice-Hall, Inc., Englewood Cliffs, New Jersey, 1966.

(11) Durling, A: *Computational Techniques*, INTEXT Educational Publishers, Inc., New York, 1974.

(12) Cadzow, J. A.: *Discrete-Time Systems*, Prentice-Hall, Inc., Englewood Cliffs, New Jersey, 1973.

(13) Adams, L. J. and P. A. White: *Analytic Geometry and Calculus*, Oxford University Press, New York, 1961.

(14) Huelsman, L. P.: *Basic Circuit Theory With Digital Computations*, Prentice-Hall, Inc., Englewood Cliffs, New Jersey, 1972.

(15) Koenig, H. E., Tokad, Y., and H. K. Kesavan: *Analysis of Discrete Physical Systems*, McGraw-Hill Book Co., Inc., New York, 1967.

(16) Calahan, D. A.: *Computer-Aided Network Design*, McGraw-Hill Book Co., Inc., New York, 1972.

(17) DeRusso, P. M., Roy, R. J., and C. M. Close: <u>State Variables for Engineers</u>, John Wiley and Sons, Inc., New York, 1965.

(18) Hamming, R. W.: <u>Numerical Methods for Scientists and Engineers</u>, McGraw-Hill Book Co., Inc., New York, 1962.

(19) Pennington, R. A.: <u>Introductory Computer Methods and Numerical Analysis</u>, The Macmillan Company, New York, 1965.

(20) Bekey, G. A. and W. J. Karplus: <u>Hybrid Computation</u>, John Wiley and Sons, Inc., New York, 1968.

Appendices

APPENDIX A. COMPLEX ALGEBRA

Complex quantities often appear as roots of the characteristic equation associated with a linear differential equation. As a result, virtually all of the mathematical skills needed by the continuous systems analyst involve complex quantities, either variables or constants, in some form or another. This appendix introduces the algebra of complex variables. Our treatment is limited to those topics needed for an understanding of the material in the book. The form of the presentation is that of a review rather than a rigorous development because the reader is presumed to have some prior familiarity with the subject.

A.1 DEFINITIONS AND REPRESENTATIONS

Rectangular Form

The <u>rectangular form</u> of a complex variable z is

$$z = x + jy \qquad (A-1)$$

where x and y are continuous real variables and $j = \sqrt{-1}$. (We do not use the common notation $i = \sqrt{-1}$ to avoid conflict with the use of i for electrical current.) The quantities x and y are the <u>real</u> and <u>imaginary</u> parts respectively of the complex variable z. The notations

$$x = \text{Re}(z) \qquad (A-2)$$

and

$$y = \text{Im}(z) \qquad (A-3)$$

are used to indicate taking the real and imaginary parts of a complex quantity.

A complex variable can be represented as a point in a two-dimensional complex plane whose axes are labeled x and jy. Figure A-1 is a plot of the complex constant $A = a + jb$.

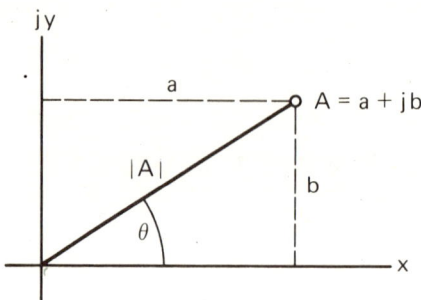

Figure A-1. The complex number A = a + jb corresponds to the point (a,b) in the complex plane.

Polar Form

The geometrical interpretation of Fig. A-1 leads to another way of representing a complex quantity. As can be seen from the figure, the point A can be identified by the length of the ray from the origin to the point and the angle θ measured clockwise from the positive axis to the ray. The magnitude or modulus of A, denoted by |A|, is found using the Pythagorean theorem:

$$|A| = \sqrt{a^2 + b^2} \qquad (A-4)$$

The angle or argument of A is given by

$$\theta = \text{Arg}(A) = \underline{/A} = \tan^{-1}(b/a) \qquad (A-5)$$

The polar form of the complex constant A is then

$$A = |A|\underline{/\theta} \qquad (A-6)$$

The angle θ in the polar form is usually specified in degrees rather than radians. In general, the magnitude of a complex quantity z is given by

$$z = \sqrt{\text{Re}(z)^2 + \text{Im}(z)^2} \qquad (A-7)$$

Appendices

and the angle by

$$\angle z = \tan^{-1}\left\{\frac{\text{Im}(z)}{\text{Re}(z)}\right\} \qquad (A\text{-}8)$$

Equations (A-4) and (A-5) imply a method to convert from rectangular to polar form; to convert from polar to rectangular we use

$$a = |A|\cos(\theta) \qquad (A\text{-}9)$$

and

$$b = |A|\sin(\theta) \qquad (A\text{-}10)$$

These relationships are also evident from Fig. A-1. In fact, combining Eqns. (A-9) and (A-10) leads to still another form for a complex quantity, the <u>trigonometric form</u>

$$A = |A|\{\cos(\theta) + j\sin(\theta)\} \qquad (A\text{-}11)$$

The Exponential Form

A final and extremely useful way to represent a complex quantity is the <u>exponential form</u>. This form is based on the identity

$$e^{j\theta} = \cos(\theta) + j\sin(\theta) \qquad (A\text{-}12)$$

which can be proved by equating the series expansions for the quantities on the left- and right-hand sides of Eq. (A-12). Comparing Eqns. (A-11) and (A-12) leads to

$$A = |A|e^{j\theta} \qquad (A\text{-}13)$$

When the exponential form is used, θ is given in radians rather than in degrees.

A.2 THE ALGEBRA OF COMPLEX QUANTITIES

Addition

Let A and B be the complex numbers A=a+jb and B=c+jd. The quantities are added (or subtracted) by the rule

$$C = A + B = (a+jb) + (c+jd) \qquad (A\text{-}14)$$

and, finally, $C=(a+c)+j(b+d)$. Complex quantities expressed in polar or exponential form cannot be added or subtracted directly: they must first be converted to rectangular form. Figure A-2 shows that under the vector interpretation, complex addition follows the familiar "head-to-tail" rule for vector addition.

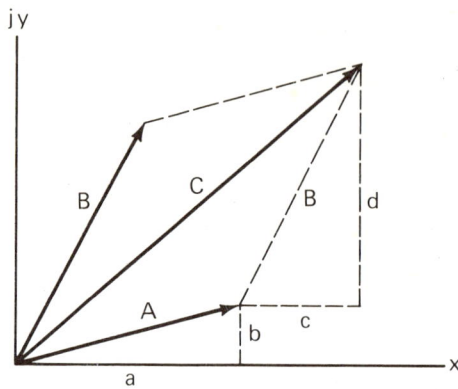

Figure A-2. Addition of complex quantities is analogous to the addition of vectors. The "head-to-tail" rule for vector addition applies.

Multiplication By A Constant

If $A=a+jb$ and c is a real constant, then

$$cA = c(a+jb) = ca+jcb \qquad (A-15)$$

If A is given in polar form,

$$cA = c|A|e^{j\theta} \qquad (A-16)$$

Geometrically, the effect of multiplying by a constant is to change the length of the vector in the complex plane.

Multiplication And Division

If A and B are the complex quantities $A=a+jb$ and $B=c+jd$, then the product $C=AB$ is found from

$$C = AB = (a+jb)(c+jd) \qquad (A-17)$$

Appendices

and, finally,

$$C = (ac-bd) + j(ad+bc) \quad (A-18)$$

where we have used the result that $(j)^2 = -1$. The quotient D=A/B can be found only after the denominator is <u>rationalized</u>; thus,

$$D = \frac{A}{B} = \frac{a+jb}{c+jd} = \frac{(a+jb)}{(c+jd)} \frac{(c-jd)}{(c-jd)} \quad (A-19)$$

and

$$D = \frac{(ac+bd)+j(bc-ad)}{c^2+d^2} \quad (A-20)$$

Multiplication and division are easier if the quantities are given in polar form. For example,

$$C = AB = \{|A|e^{j\theta}\}\{|B|e^{j\phi}\} \quad (A-21)$$

and

$$C = |A||B|e^{j(\theta+\phi)} \quad (A-22)$$

The product has a magnitude equal to the product of the magnitudes of the multipliers and an angle equal to the sum of their angles. Similarly, the quotient

$$D = \frac{A}{B} = \frac{|A|e^{j\theta}}{|B|e^{j\phi}} \quad (A-23)$$

becomes

$$D = \{|A|/|B|\}e^{j(\theta-\phi)} \quad (A-24)$$

where the magnitude of D is the quotient of the magnitudes and the angle of D is the difference of the angles.

The Complex Conjugate

The complex conjugate z* of the complex variable z is defined by

$$z^* = x - jy \quad (A-25)$$

if z=x+jy. That is, we merely replace j by -j to form the complex conjugate. If z is given in polar form, say

$z=|z|\exp(j\theta)$, then the complex conjugate is again found by replacing j by -j, or

$$z^* = |z|e^{-j\theta} \qquad (A-26)$$

Finally, if z is known in the trigonometric form,

$$z = |z|\{\cos(\theta) + j\sin(\theta)\} \qquad (A-27)$$

the conjugate is

$$z^* = |z|\{\cos(\theta) - j\sin(\theta)\} \qquad (A-28)$$

It is also easy to show that $zz^* = |z|$.

APPENDIX B. ELECTRICAL UNITS

Variable or Parameter	Symbol	Dimensions	Preferred Units	Units Symbol
Charge	q	Q	Coulomb	C
Voltage	v, e, E	FL/Q	Volt	V
Current	i, I	Q/T	Ampere	A
Resistance	R	FLT/Q^2	Ohm	Ω
Capacitance	C	Q^2/FL	Farad	F
Inductance	L	FLT^2/Q^2	Henry	H
Power	p, P	FL/T	Watt	W
Frequency	f	1/T	Hertz	Hz
Radian Frequency	ω	1/T	Radian per second	rad/s
Time	t	T	second	s
Energy	W	FL	Joule	J

APPENDIX C. UNIT PREFIXES

Factor by which unit is multiplied	Prefix	Prefix Symbol
10^{12}	tera	T
10^{9}	giga	G
10^{6}	mega	M
10^{3}	kilo	k
10^{2}	hecto	h
10	deka	da
10^{-1}	deci	d
10^{-2}	centi	c
10^{-3}	milli	m
10^{-6}	micro	µ
10^{-9}	nano	n
10^{-12}	pico	p
10^{-15}	femto	f
10^{-18}	atto	a

APPENDIX D. LAPLACE TRANSFORM PAIRS

	$f(t)$	$F(s)$
1	$\delta(t)$	1
2	$u(t)$	$\dfrac{1}{s}$
3	$tu(t)$	$\dfrac{1}{s^2}$
4	$\dfrac{t^{n-1}}{(n-1)!}$	$\dfrac{1}{s^n}$
5	e^{-at}	$\dfrac{1}{s+a}$
6	$\dfrac{1}{(n-1)!}t^{n-1}e^{-at}$	$\dfrac{1}{(s+a)^n}$
7	$\dfrac{1}{a}(1-e^{-at})$	$\dfrac{1}{s(s+a)}$
8	$\sin(\omega t)$	$\dfrac{\omega}{s^2+\omega^2}$
9	$\cos(\omega t)$	$\dfrac{s}{s^2+\omega^2}$
10	$1-\cos(\omega t)$	$\dfrac{\omega^2}{s(s^2+\omega^2)}$
11	$\alpha - \sqrt{\alpha^2+\omega^2}\cos\{\omega t+\tan^{-1}(\omega/\alpha)\}$	$\dfrac{\omega^2(s+\alpha)}{s(s^2+\omega^2)}$
12	$\dfrac{\omega_n}{\sqrt{1-\zeta^2}}e^{-\zeta\omega_n t}\sin\left\{\omega_n\sqrt{1-\zeta^2}\,t\right\}$	$\dfrac{\omega_n^2}{s^2+2\zeta\omega_n s+\omega_n^2}$
13	$1 - \dfrac{1}{\sqrt{1-\zeta^2}}e^{-\zeta\omega_n t}\sin\left\{\omega_n\sqrt{1-\zeta^2}\,t+\phi\right\}$ $\phi = \cos^{-1}(\zeta)$	$\dfrac{\omega_n^2}{s(s^2+2\zeta\omega_n s+\omega_n^2)}$
14	$e^{-at}\sin(\omega t)$	$\dfrac{\omega}{(s+a)^2+\omega^2}$
15	$e^{-at}\cos(\omega t)$	$\dfrac{s+a}{(s+a)^2+\omega^2}$

APPENDIX E. PROPERTIES OF THE LAPLACE TRANSFORM

	$f(t)$	$F(s)$
1	$f(t)$	$\int_0^\infty e^{-st} f(t) dt$
2	$\dot{f}(t)$	$sF(s) - f(0^+)$
3	$f^{(n)}(t)$	$s^n F(s) - s^{n-1} f(0^+) - s^{n-2} \dot{f}(0^+) - \ldots - f^{(n-1)}(0^+)$
4	$\int_0^t f(\tau) d\tau + f(0^+)$	$\frac{1}{s} F(s) + \frac{1}{s} f(0^+)$
5	$\int_0^t f_1(t-\tau) f_2(\tau) d\tau$	$F_1(s) F_2(s)$
6	$t f(t)$	$\frac{-dF(s)}{ds}$
7	$t^n f(t)$	$(-1)^n F^{(n)}(s)$
8	$\frac{1}{t} f(t)$	$\int_0^\infty F(s) ds$
9	$e^{at} f(t)$	$F(s-a)$
10	$f(t-b) u(t-b)$	$e^{-bs} F(s)$
11	$\frac{1}{c} f\left\{\frac{t}{c}\right\}$	$F(cs)$

APPENDIX F. z-TRANSFORM PAIRS

	$f(k)$	$F(z)$
1	$\delta(k)$	1
2	$\delta(k-i)$	z^{-i}
3	$u(k)$	$\dfrac{z}{z-1}$
4	$u(k-i)$	$z^{-i}\left\{\dfrac{z}{z-1}\right\}$
5	$ku(k)$	$\dfrac{z}{(z-1)^2}$
6	$k^2 u(k)$	$\dfrac{z(z+1)}{(z-1)^3}$
7	e^{-akT}	$\dfrac{z}{z-\exp(-aT)}$
8	ke^{-akT}	$\dfrac{z\exp(-aT)}{\{z-\exp(-aT)\}^2}$
9	$1 - e^{-akT}$	$\dfrac{z\{1-\exp(-aT)\}}{(z-1)\{z-\exp(-aT)\}}$
10	a^k	$\dfrac{z}{z-a}$
11	ka^k	$\dfrac{az}{(z-a)^2}$
12	$(k+1)a^k$	$\dfrac{z^2}{(z-a)^2}$
13	$(k+1)(k+2)a^k/2$	$\dfrac{z^3}{(z-a)^3}$
14	$\sin(\omega kT)$	$\dfrac{z\sin(\omega T)}{z^2-2z\cos(\omega T)+1}$
15	$\cos(\omega kT)$	$\dfrac{z\{z-\cos(\omega T)\}}{z^2-2z\cos(\omega T)+1}$
16	$e^{-akT}\sin(\omega kT)$	$\dfrac{z\exp(-aT)\sin(\omega T)}{z^2-2z\exp(-aT)\cos(\omega T)+\exp(-2aT)}$
17	$e^{-akT}\cos(\omega kT)$	$\dfrac{z^2-z\exp(-aT)\cos(\omega T)}{z^2-2z\exp(-aT)\cos(\omega T)+\exp(-2aT)}$

APPENDIX G. BODE PLOTS FOR COMMON FACTORS

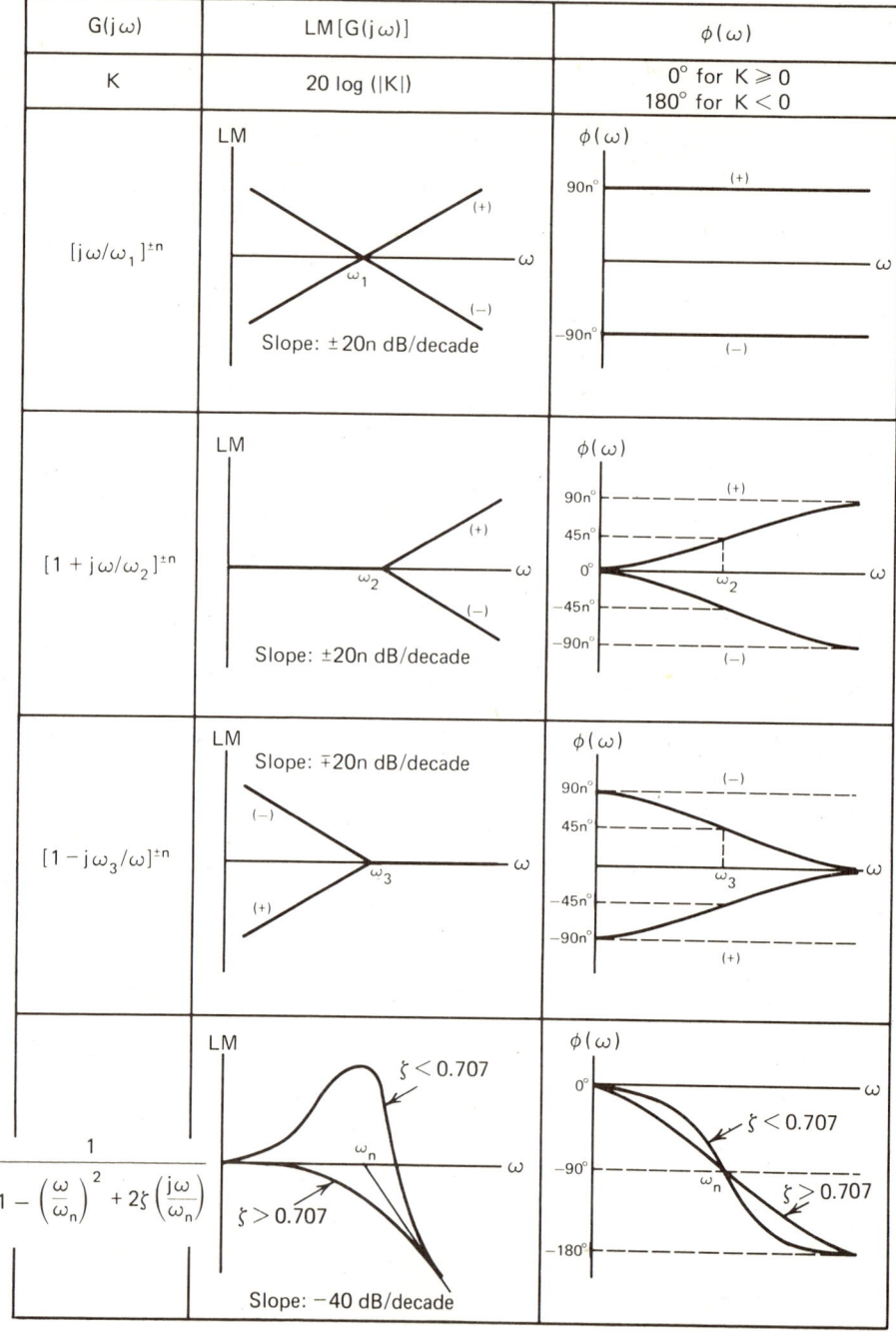

APPENDIX H.   MECHANICAL SYSTEMS

Our emphasis on electrical examples in the preceeding chapters has not meant to imply that linear systems analysis works only for electrical circuits. The analysis techniques apply to a variety of other engineering disciplines. In particular, many kinds of mechanical systems are readily analyzed by the methods presented above.[1,2] This appendix introduces mechanical analysis with emphasis on component models and writing differential equations.

H.1 FUNDAMENTAL LAWS AND QUANTITIES

The first step in analyzing a mechanical system is to sketch its block diagram and assign names and directions to the spatial quantities of interest. We use symbols such as $y(t)$, $\dot{y}(t)$, and $\ddot{y}(t)$ for translational displacement, velocity, and acceleration and $\theta(t)$, $\dot{\theta}(t)$, and $\ddot{\theta}(t)$ for angular displacement, velocity, and acceleration. Positive reference directions must be chosen for both translational and rotational quantities. The dimensions and units of the mechanical quantities are summarized in Appendices I and J.

Newton's Second Law Of Motion

Mechanical systems satisfy Newton's second law of motion, which states that the sum of the applied forces equals the time rate of change of linear momentum.[3] For translational systems, this law is expressed as

$$\sum \underline{f} = \frac{d(m\underline{v})}{dt} \qquad (H-1)$$

where $\underline{f}$, m, and $\underline{v}$ represent force, mass, and velocity respectively, and the product $m\underline{v}$ is the <u>linear momentum</u> of the system. The underbar denotes <u>vector</u> quantities--those which have direction as well as magnitude. Performing the indicated differentiation gives

$$\sum \underline{f} = m\frac{d\underline{v}}{dt} + \underline{v}\frac{dm}{dt} \qquad (H-2)$$

For constant-mass systems $dm/dt=0$ and Newton's second law becomes

$$\sum \underline{f} = m\underline{a} \qquad (H-3)$$

where $\underline{a}=d\underline{v}/dt$ is the acceleration of the mass caused by the sum of the applied forces. In this form Newton's second law defines mass as the constant of proportionality between force and acceleration. The summation is taken over all forces which act on mass m. If a system contains more than one mass, then an equation like Eq. (H-3) must be written for each mass. Equation (H-3) can be decomposed into three <u>scalar</u> (magnitude only) equations, one for each coordinate axis, or

$$\sum f_x = m\ddot{x} \qquad (H-4)$$

$$\sum f_y = m\ddot{y} \qquad (H-5)$$

and

$$\sum f_z = m\ddot{z} \qquad (H-6)$$

where $\ddot{x}$, $\ddot{y}$ and $\ddot{z}$ are the components of the vector acceleration $\underline{a}$. A useful way to read the first equation is "The sum of the forces in the plus x direction equals the mass times the acceleration in the plus x direction".

For a rigid body rotating about a fixed axis, the scalar form of Newton's second law is

$$\sum M = J\ddot{\theta} \qquad (H-7)$$

where M is a <u>moment</u> or torque, J is the <u>moment of inertia</u> about the axis of rotation, and $\ddot{\theta}$ is the <u>angular acceleration</u>.

Appendices

A moment is the product of a force perpendicular to the axis of rotation and the distance from the point of application to the axis of rotation.

Moment of inertia. The moment of inertia J needed to describe a rotating mass is defined by the relationship

$$J = \int_{\text{mass}} r^2 dm \qquad (H-8)$$

where dm is an element of mass, r is the radius from the axis of rotation to the element of mass, and the integral is taken over the entire mass of the body. The moment of inertia is the constant of proportionality between an applied moment and the resulting angular acceleration. As such, it is the rotational equivalent of mass.

H.2 MECHANICAL COMPONENTS

Translational Mechanical Components

In addition to masses, translational mechanical systems are assumed to contain linear springs and viscous dampers.[4] These components are useful mathematical abstractions; practical springs and dampers exhibit varying degrees of nonlinearity. Many mechanical systems which contain no actual springs or dampers can be accurately modeled by assuming the presence of these devices. An example is the automobile tire, whose compression can be modeled by a parallel combination of a spring and a damper.

The linear spring. The linear spring of Fig. H-1(a) satisfies the relationship

$$f_k = -k(y_1 - y_2) \qquad (H-9)$$

where f_k is the force exerted by the spring on mass m_1 and y_1 and y_2 are the coordinates of the ends of the spring. The difference $(y_1 - y_2)$ is the amount the spring is stretched or elongated. If $y_1 = y_2$ the spring is unstretched (at rest) and exerts no force. The spring constant k, which is the constant proportionality between the force and the elongation, is a positive quantity. The minus sign before k in Eq. (H-9) emphasizes that the force always opposes any attempt to elongate the spring from its unstretched configuration.

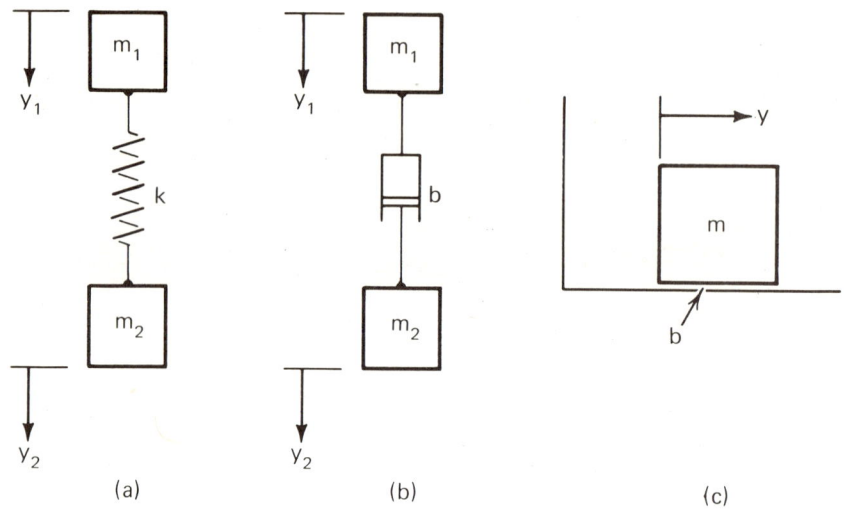

Figure H-1. Translational mechanical systems are often comprised of masses, springs, and dampers. (a) A linear spring connected between masses m_1 and m_2; (b) A viscous damper connected between masses m_1 and m_2; (c) An alternate notation for viscous damping.

The viscous damper. A practical damper consists of a cylinder filled with some kind of viscous fluid. A piston exerts pressure on the fluid, which is allowed to escape through an orifice in the piston. Figure H-2 shows this construction.

The force exerted by the damper depends on the velocity of the piston relative to the cylinder. The viscous damper, Fig. H-1(b), is modeled mathematically by

$$f_b = -b(\dot{y}_1 - \dot{y}_2) \tag{H-10}$$

where f_b is the force the damper exerts on mass m_1, $\dot{y}_1$ and $\dot{y}_2$ are velocities of the ends of the damper, and the difference $\dot{y}_1 - \dot{y}_2$ is the velocity of the piston relative to the

Appendices

cylinder. The damper coefficient b is the constant of proportionality (always positive) between the velocity differential and the resulting force. The origins and positive directions for y_1 and y_2 are arbitrary, but once chosen they must be maintained. Actual displacements and velocities may be either positive or negative with respect to the directions chosen. The minus sign before b in Eq. (H-10) emphasizes that the damper always opposes any attempt to change the velocity of the piston.

In later work we will at times use another notation for viscous damping. The mass of Fig. H-1(c), for example, slides with viscous damping along the surface upon which it rests, as indicated by the damper coefficient written beside the region of contact. The viscous damper relationship of Eq. (H-10) still applies, although in this case the $\dot{y}_2$ term is not needed.

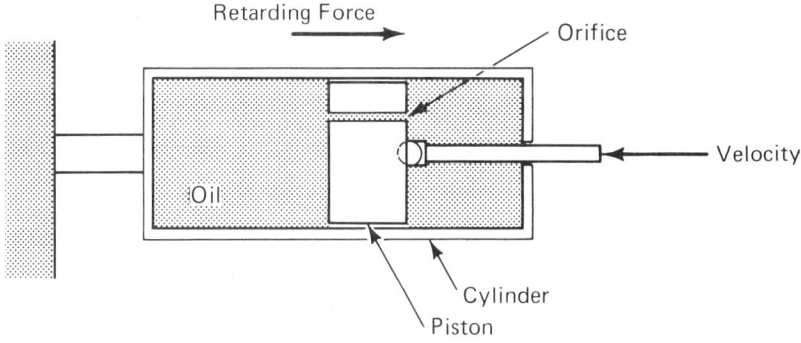

Figure H-2. A practical viscous damper consists of a piston which moves in an oil-filled cylinder.

Rotational Mechanical Components

Analagous to the mass, linear spring, and viscous damper of the translational mechanical system are the moment of inertia, torsion spring, and rotary damper of the rotational system. Again, we emphasize that the rotary spring and rotary damper are useful abstractions. An actual rotational system may contain nothing which looks like a spring or damper. The twist of a solid steel shaft, for example, can be modeled in terms of a rotary spring coefficient.

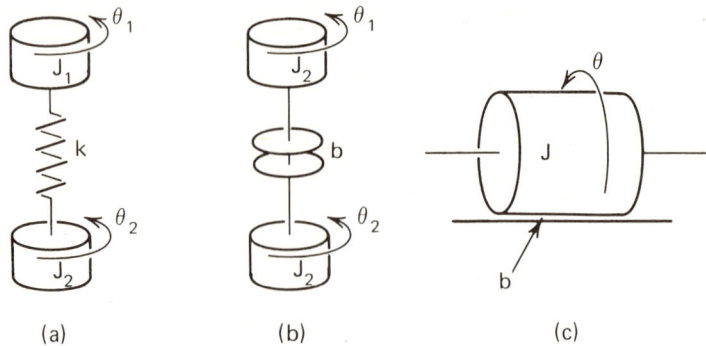

Figure H-3. Rotational mechanical systems are often comprised of torsion springs, rotary dampers and rotating masses. (a) A torsion spring connected between two rotating masses; (b) A rotary damper connected between two rotating masses; (c) An alternate notation for rotary damping.

The rotary spring. The rotary spring of Fig. H-3(a) satisfies the equation

$$M_k = -k(\theta_1 - \theta_2) \qquad (H-11)$$

where M_k is the moment produced by the spring on rotating mass J_1, k is the spring constant, and $\theta_1 - \theta_2$ is the amount of twist of one end of the spring relative to the other. The positive constant k is the constant of proportionality between the amount the spring is twisted and the resulting moment or torque. The minus sign before k in Eq. (H-11) emphasizes that the spring always opposes any attempt to twist it.

The rotary damper. The rotary damper, Fig. H-3(b), can be visualized as a fluid-filled cylinder which contains two flywheels connected to shafts protruding from the ends of the cylinder. When one flywheel is rotated, the fluid transfers torque to the other flywheel. The damper is modeled mathematically by the relationship

$$M_b = -b(\dot{\theta}_1 - \dot{\theta}_2) \qquad (H-12)$$

Appendices 625

where M_b is the moment produced by the device, b is the damper coefficient, and $\dot{\theta}_1 - \dot{\theta}_2$ is the relative velocity between the two flywheels. If both flywheels rotate at the same velocity the relative torque is zero as we would expect. The symbolic representation of the rotary damper, Fig. H-3(b), shows the two flywheels connected to the shafts. The damper coefficient b is always positive but is preceeded by a minus sign in Eq. (H-12) to emphasize that the damper always opposes any attempt to change the relative velocities of the flywheels.

In addition to the symbol of Fig. H-3(b), we will make some use of an alternate notation for rotary damping. The rotating mass J in Fig. H-3(c) rotates with viscous damping as indicated by the damper coefficient shown. This model is useful for representing such phenomena as air friction with which no actual damping device is associated.

H.3 TRANSLATIONAL SYSTEMS

This section considers the problem of writing differential equations which describe translational mechanical systems. The intuitive method presented below lets one write the differential equations for simple problems by inspecting the system diagram. The method breaks down for complex systems where the powerful techniques of Lagrangian dynamics are more suitable.[5]

Equations Of Motion For Translational Systems [6, 7]

The basic tool for writing the equations of motion for an N-mass translational system is Newton's second law in the form

$$m_i \ddot{y}_i = f_{i1} + f_{i2} + \ldots f_{in_i} \tag{H-13}$$

for $i=1,2,\ldots N$. In Eq. (H-13), n_i is the number of forces acting on the i-th mass, and the f_{ij} are the individual forces acting on the ith mass. To find the forces for the right-hand side of Eq. (H-13) for mass m_i proceed as follows:

(1) Assume all other masses are held fixed and visualize mass m_i as being displaced slightly in the positive $\dot{y}_i$ direction.

(2) Spring forces, which always oppose the positive motion of mass y_i, are written on the right-hand side of Eq. (H-13) as $-k(y_i - y_j)$ where the y_j are the coordinates of the opposite ends of the springs connected to mass m_i.

(3) Viscous damper forces, which always oppose the assumed positive motion of mass m_i, are written on the right-hand side of Eq. (H-13) as $-k(\dot{y}_i - \dot{y}_j)$ where the $\dot{y}_j$ are the velocities of the opposite ends of the dampers connected to mass m_i.

The procedure appears complex due to the indices and the summation. As we shall see from the following examples, however, the method is easier to use than to explain.

Example H-1. Write the equations of motion for the system of Fig. H-4. The mass is pulled to the right y_0 units and released at t=0. As indicated, the mass moves along the ground plane with a retarding force which can be represented as viscous damping.

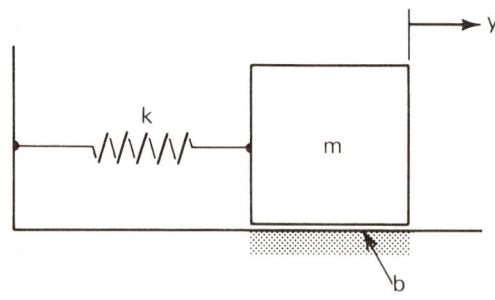

Figure H-4. System for Example H-1.

Solution. If we visualize the mass as being moved to the right, we see that both the spring and damper forces

Appendices

oppose the motion. Using the procedure above and Eq. (H-13) we write

$$m\ddot{y} = -ky - b\dot{y} \qquad (H-14)$$

Because the mass is assumed to be moved in the positive y direction, both y and $\dot{y}$ are positive, and the spring and damper forces oppose the motion as required. Once the signs for the spring and damper forces have been correctly assigned, they will continue to be correct for all subsequent motion. Rewriting the differential equation in standard form (all dependent variable terms on the left and a coefficient of 1 for the highest-order derivative) yields

$$\ddot{y} + \frac{b}{m}\dot{y} + \frac{k}{m}y = 0 \qquad (H-15)$$

The initial condition for $y(t)$ is $y(0)=y_o$, and, as the mass is released from rest, its initial velocity is 0, or $\dot{y}(0)=0$.

Example H-2. Write the equations of motion for the system of Fig. H-5. The blocks slide with viscous friction. Assume zero initial conditions.

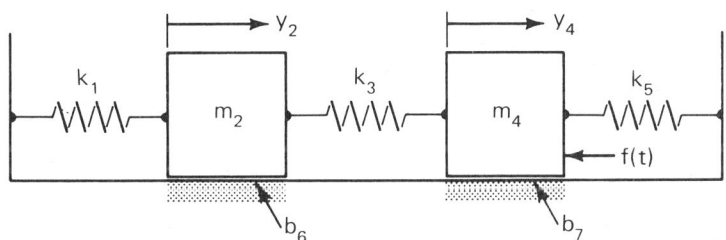

Figure H-5. System for Example H-2.

<u>Solution</u>. Following the inspection procedure, we assume that m_2 moves to the right with m_4 fixed. Then springs k_1 and k_3 and damping b_6 all oppose the motion; hence we have

$$m_2\ddot{y}_2 = -k_1y_2 - k_3(y_2-y_4) - b_6\dot{y}_2 \qquad (H-16)$$

The term $-k_1y_2$ is the force due to spring k_1; the negative sign indicates that this force is in the minus y_2 direction when m_2 is moved in the positive y_2 direction. Similarly, the force $-b_6\dot{y}_2$ accounts for the opposing viscous damper force due to b_6. The term $-k_3(y_2-y_4)$ is the force due to the spring k_3. By assuming that m_4 remains fixed, $y_4=0$ and a positive y_2 yields an opposing spring force as required. When the system undergoes its actual motion, if y_4 exceeds y_2 at any instant, the spring force will aid the motion, as indicated by the fact that the term $-k_3(y_2-y_4)$ would be positive. Once the force terms have been written correctly, the signs will be correct for any subsequent combination of positions and velocities.

Now assume m_4 moves to the right with m_2 fixed. In this case springs k_3 and k_5 and damping b_7 all oppose the motion; the required equation is therefore

$$m_4\ddot{y}_4 = -k_5 y_4 - k_3(y_4-y_2) - b_7\dot{y}_4 - f(t) \qquad (H-17)$$

Here the force due to spring k_3 has been written as $-k_3(y_4-y_2)$ rather than $-k_3(y_2-y_4)$ as was the case for mass m_2. This is not inconsistent because the spring force acts in opposite directions on masses m_2 and m_4.

To show the advantage of using the inspection method even for complex problems, consider the system in the following example.

Example H-3. Write the differential equations of motion for the system of Fig. H-6. Assume zero initial conditions.

Solution. Let mass m_2 move in the positive y_2 direction with the other masses held fixed. Springs k_1 and k_2 and damper b_1 all oppose the motion. Newton's second law for mass m_2 is

$$m_2\ddot{y}_2 = -k_1(y_2-y_1) - k_2(y_2-y_3) - b_1(\dot{y}_2-\dot{y}_3) \qquad (H-18)$$

Again, the "self" terms (y_2 and $\dot{y}_2$ in the m_2 equation) always appear first when writing spring and damper terms. Next hold m_2 and m_3 fixed and assume m_1 moves downward. Springs k_1 and k_3 and damper b_2 oppose the motion. The m_1 equation is

$$m_1\ddot{y}_1 = -k_3 y_1 - k_1(y_1-y_2) - b_2(\dot{y}_1-\dot{y}_3) + f(t) \qquad (H-19)$$

Appendices

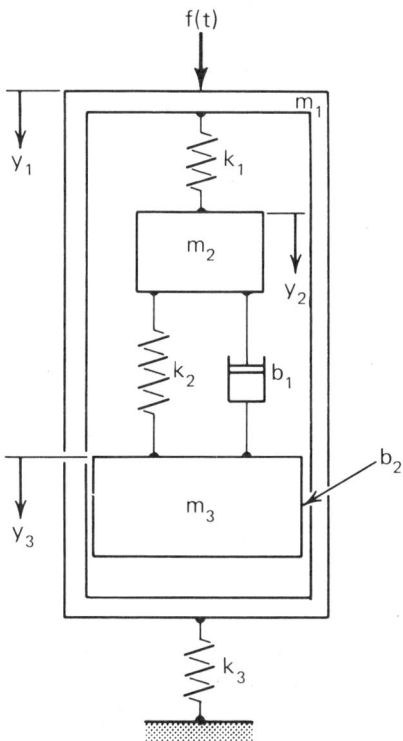

Figure H-6. System for Example H-3.

The forcing function f(t) is written with a positive sign because it acts in the plus y_1 direction. Again, because all forces oppose the motion, they are written with negative signs and the self terms, y_1 and $\dot{y}_1$ in this case, come first when writing expressions for forces caused by components connected to other elements.

Finally, hold m_1 and m_2 constant and assume m_3 moves downward. Spring k_2 and dampers b_1 and b_2 oppose the motion; the differential equation is

$$m_3\ddot{y}_3 = -k_2(y_3-y_2) - b_1(\dot{y}_3-\dot{y}_2) - b_2(\dot{y}_3-\dot{y}_1) \tag{H-20}$$

The forcing function f(t) appears only in the m_1 equation as it is applied directly to m_1, but this does not mean that m_2 and m_3 remain stationary. The effect of the forcing function is coupled to the other masses by the springs and dampers.

Equations (H-18), (H-19), and (H-20) must be solved simultaneously to analyze the motion of the system.

What About Gravity?

The reader might question why we failed to include the weights of the three masses as downward forces in the equations of motion for Example H-3. The reason is that if the origins of the y_1, y_2, and y_3 coordinate systems are at the equilibrium position, gravitational forces always cancel, as seen from the following argument: Fig. H-7(a) shows an unstretched spring of length L_1.

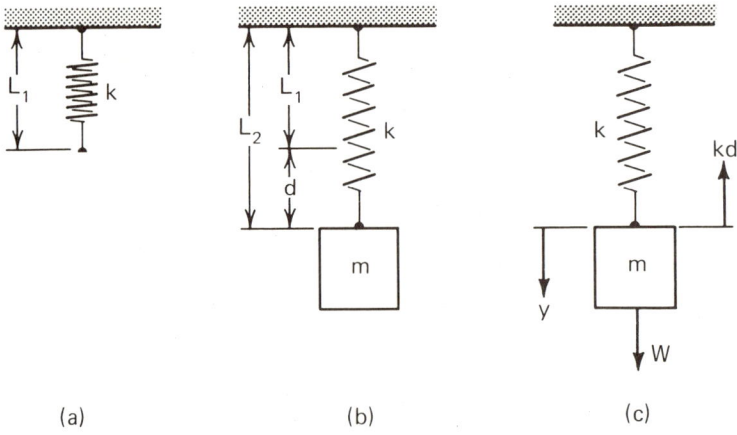

(a)   (b)   (c)

Figure H-7. This system illustrates that if the origin of the coordinate axes is at the equilibrium position, the weight of a hanging mass can be neglected. (a) An unstretched spring with length L_1; (b) When mass m is attached, the spring of (a) has length L_2; (c) The weight of mass m is exactly balanced by the spring tension kd.

When the mass m with weight W is attached, the stretched length is L_2 (Fig. H-7(b)). Because the system is in equilibrium, the weight W is exactly balanced by the spring force $k(L_2-L_1)=kd$, or $W=kd$. If the origin for y is chosen from the equilibrium position (Fig. H-7(c)), the equation of motion is

$$m\ddot{y} = -ky + W - kd \tag{H-21}$$

Appendices

However, as W is exactly equal to kd, we can omit both terms and write

$$m\ddot{y} = -ky \qquad (H-22)$$

This approach is completely general: whenever a set of coordinate axes has its origin at the equilibrium position, we can neglect the weights of hanging masses.

Massless Points

Occasionally a system has moving points which are not associated with masses. These occur when a forcing function is applied to the end of a spring or a damper rather than to a mass, or when a spring and a damper are connected in series rather than in parallel. In such cases we assume the presence of a fictitious mass, write the equations of motion, then let the fictitious mass become zero. Example H-4 illustrates the procedure.

Example H-4. Write equations of motion for the system of Fig. H-8.

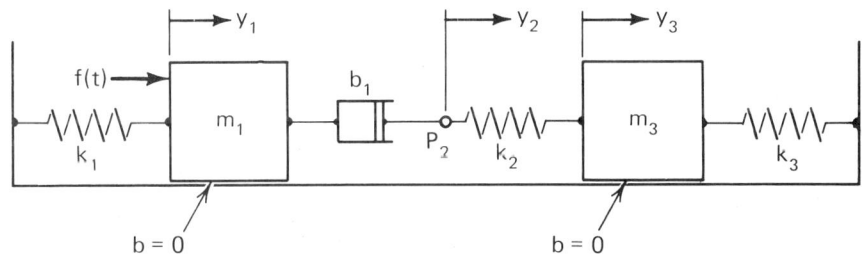

Figure H-8. System for Example H-4. This system has a massless point at P_2 because P_2 moves independently of m_1 and m_3.

Solution. Point P_2 is a massless point because the spring and the damper are connected in series. Assume a fictitious mass m_2 at P_2 and write the equations of motion by the inspection method:

$$m_1\ddot{y}_1 = -k_1 y_1 - b_1(\dot{y}_1 - \dot{y}_2) + f(t) \qquad (H-23)$$

$$m_2\ddot{y}_2 = -b_1(\dot{y}_2-\dot{y}_1) - k_2(y_2-y_3) \quad \text{(H-24)}$$

$$m_3\ddot{y}_3 = -k_2(y_3-y_2) - k_3 y_3 \quad \text{(H-25)}$$

Set $m_2=0$ in Eq. (H-24) to obtain the actual equation

$$0 = -b_1(\dot{y}_2-\dot{y}_1) - k_2(y_2-y_3) \quad \text{(H-26)}$$

Equations (H-23), (H-25) and (H-26) can be solved simultaneously to analyze the motion of the system.

H.4 ROTATIONAL SYSTEMS

The procedure for rotational mechanical systems parallels that for translational systems except that angular quantities and moments are used. The following examples illustrate the procedure.[8,9]

Example H-5. Write the equations of motion for the system of Fig. H-9. Assume zero initial conditions. The externally applied moment M(t) attempts to rotate the mass.

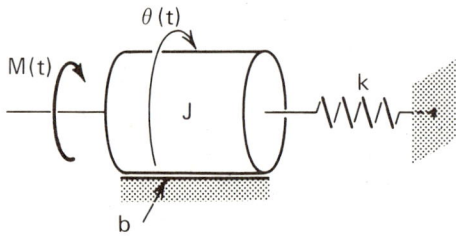

Figure H-9. System for Example H-5.

Solution. Assume the cylinder rotates in the positive θ direction. The torsion spring and the viscous damping force resist the motion, while the externally applied moment M(t) aids it. Newton's second law of motion, "The sum of the moments in the plus θ direction equals the moment of inertia times $\ddot{\theta}$," yields

$$J\ddot{\theta} = -b\dot{\theta} - k\theta + M(t) \quad \text{(H-27)}$$

Rearranging Eq. (H-27) in standard form gives

$$\ddot{\theta} + \frac{b}{J}\dot{\theta} + \frac{k}{J}\theta = \frac{M(t)}{J} \qquad (H-28)$$

with $\theta(0)=0$ and $\dot{\theta}(0)=0$.

Example H-6. Write the differential equations of motion for the system of Fig. H-10. Assume the J_1 cylinder is rotated A units in the negative θ_1 direction and released to move under the influence of the external moment $M(t)$. All other initial conditions are zero.

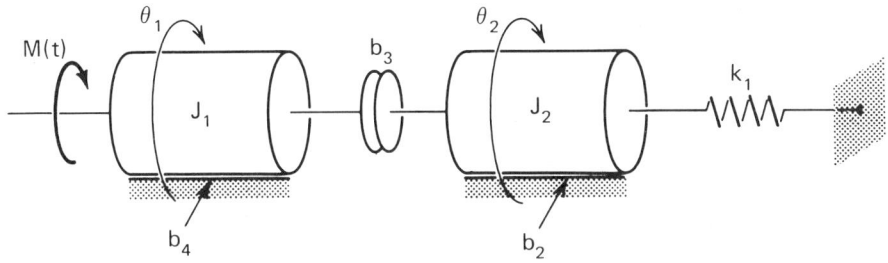

Figure H-10. System for Example H-6.

Solution. Hold J_2 fixed and assume that J_1 moves in the positive θ_1 direction. The damping caused by b_3 and b_4 opposes the motion while $M(t)$ aids it. The differential equation of motion is therefore

$$J_1\ddot{\theta}_1 = -b_4\dot{\theta}_1 - b_3(\dot{\theta}_1-\dot{\theta}_2) + M(t) \qquad (H-29)$$

In the J_1 equation, the "self" or θ_1 terms must appear first when the difference terms are needed, $\dot{\theta}_1-\dot{\theta}_2$ in this instance. Next assume that J_1 remains fixed while J_2 rotates slightly in the plus θ_2 direction. The moments due to k_1, b_2, and b_3 oppose the motion. The resulting equation is

$$J_2\ddot{\theta}_2 = -k_1\theta_2 - b_2\dot{\theta}_2 - b_3(\dot{\theta}_2-\dot{\theta}_1) \qquad (H-30)$$

In the J_2 equation every difference term must be written with the "self' or θ_2 terms first. As with translational systems, this procedure is not inconsistent because the moments act oppositely on J_1 and J_2, and one would expect a sign difference such as is produced by writing $-b_3(\dot{\theta}_1-\dot{\theta}_2)$ in one equation and $-b_3(\dot{\theta}_2-\dot{\theta}_1)$ in the other. Equations (H-29) and (H-30), along with the initial condition $\theta_1(0)=-A$, describe motion of the system.

Massless Points

A massless point in a rotational system is handled by assuming a fictitious moment of inertia at the massless point, writing the equation of motion by the inspection procedure, and then letting the fictitious moment of inertia become zero. The procedure is illustrated by the following example.

Example H-7. Write the differential equations of motion for the system of Fig. H-11. Assume zero initial conditions.

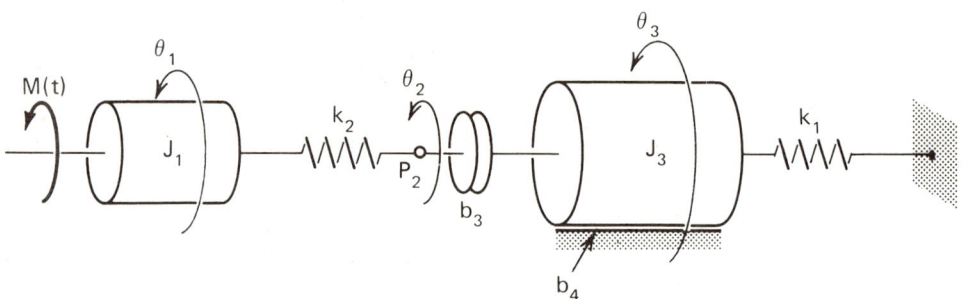

Figure H-11. Figure for Example H-7. This system has a massless point at P_2 because the angular position at P_2 differs from t either θ_1 or θ_3.

Solution. A massless point exists at P_2 because the spring k_2 is connected in series with the rotary damper b_3. Assume that a fictitious mass with moment of inertia J_2 exists at point P_2. The equations of motion are written by the inspection method:

$$J_1\ddot{\theta}_1 = -k_2(\theta_1-\theta_2) + M(t) \tag{H-31}$$

$$J_3\ddot{\theta}_3 = -k_1\theta_3 - b_4\dot{\theta}_3 - b_3(\dot{\theta}_3-\dot{\theta}_2) \tag{H-32}$$

Appendices

$$J_2\ddot{\theta}_2 = -k_2(\theta_2-\theta_1) - b_3(\dot{\theta}_2-\dot{\theta}_3) \qquad (H-33)$$

Set $J_2=0$ in Eq. (H-33) to give the actual equation

$$0 = -k_2(\theta_2-\theta_1) - b_3(\dot{\theta}_2-\dot{\theta}_3) \qquad (H-34)$$

Equations (H-31), (H-32), and (H-34) are solved simultaneously to analyze the motion of the system.

Gears

Rotational mechanical systems often contain gear trains. These are used to change the speeds of rotating shafts and to amplify torques. The analysis of gears proceeds as follows: Refer to Fig. H-12.(10)

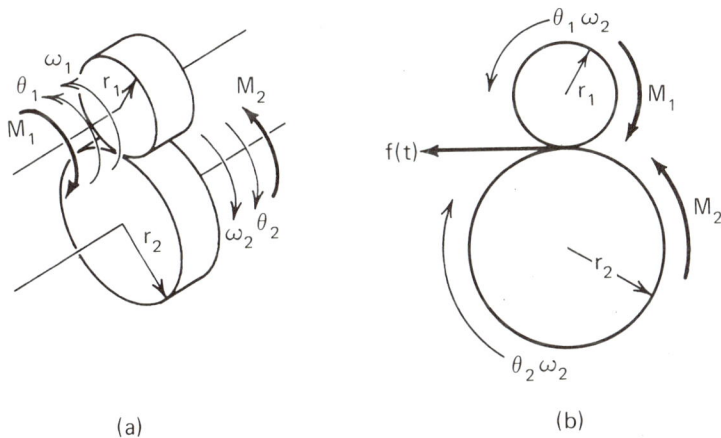

Figure H-12. Equations (H-36), (H-38), and (H-40) are based on these reference conventions for angular positions, velocities and moments.

If the gears roll without slipping, then the tangential speeds must be the same. Let ω_1 and ω_2 be the angular velocities of the gears and v be the common tangential velocity. Then,

$$\omega_1 r_1 = v \quad \text{and} \quad \omega_2 r_2 = v \qquad (H-35)$$

or, after rearranging,

$$\frac{\omega_2}{\omega_1} = \frac{r_1}{r_2} \tag{H-36}$$

Thus the angular velocities of the gears are in inverse proportion to their respective radii. Because $\omega_1 = d\theta_1/dt$ and $\omega_2 = d\theta_2/dt$, Eq. (H-37) can be written

$$d\theta_2 = \frac{r_1}{r_2} d\theta_1 \tag{H-37}$$

Integrating Eq. (H-37) between 0 and t and assuming zero initial angles yields

$$\frac{\theta_2}{\theta_1} = \frac{r_1}{r_2} \tag{H-38}$$

Finally, Newton's third law (for every action there is an equal and opposite reaction) tells us that each gear exerts a force f(t) on the other at the point of tangency (Fig. H-12(b)). Because a moment is, by definition, a force times a moment arm, and if we assume gear 1 is trying to turn gear 2, we have

$$M_1 = f(t)r_1 \quad \text{and} \quad M_2 = -f(t)r_2 \tag{H-39}$$

with reference directions shown, or

$$\frac{M_1}{M_2} = -\frac{r_1}{r_2} \tag{H-40}$$

The minus sign in Eq. (H-40), which accounts for the opposing directions of M_1 and M_2, is valid only if the moment reference directions oppose the θ reference directions for each gear.

Equations (H-36), (H-38) and (H-40), as summarized in Table H-1, and the references of Fig. H-12 describe the gear train.

Example H-8. Write equations of motion for the system of Fig. H-13. The gears have radii r_2 and r_3.

Appendices

Table H-1.

Gear Relationships*

Angular Displacement	$\dfrac{\theta_2}{\theta_1}$	$= \dfrac{r_1}{r_2}$
Angular Velocity	$\dfrac{\omega_2}{\omega_1}$	$= \dfrac{r_1}{r_2}$
Moments	$\dfrac{M_1}{M_2}$	$= -\dfrac{r_1}{r_2}$

*Refer to Fig. H-12 for reference directions.

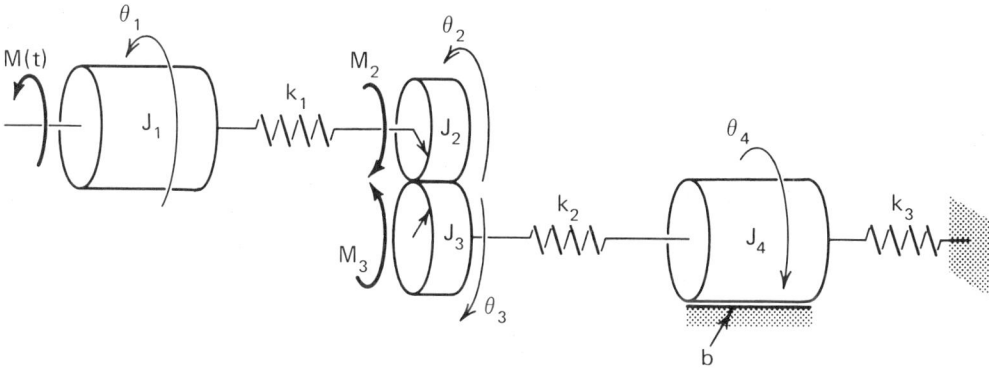

Figure H-13. A rotational system with gears for Example H-8.

Solution. Assume the reference directions shown. M_2 and M_3 are the reactive moments of the gears on each other. The equations of motion are written by the inspection method:

$$J_1\ddot{\theta}_1 = M(t) - k_1(\theta_1-\theta_2) \tag{H-41}$$

$$J_2\ddot{\theta}_2 = -k_1(\theta_2-\theta_1) - M_2 \tag{H-42}$$

$$J_3\ddot{\theta}_3 = -k_2(\theta_3-\theta_4) - M_3 \tag{H-43}$$

$$J_4\ddot{\theta}_4 = -k(\theta_4-\theta_3) - b\dot{\theta}_4 - k_3\theta_4 \tag{H-44}$$

We can eliminate one equation because θ_2 and θ_3 are related by the gear ratio. From Eq. (H-38)

$$\theta_3 = \frac{\theta_2 r_2}{r_3} = N\theta_2 \tag{H-45}$$

where $N=r_2/r_3$ is the gear ratio. Similarly, from Eq. (H-40),

$$\frac{M_2}{M_3} = -\frac{r_2}{r_3} = -N \tag{H-46}$$

or $M_2 = -M_3 N$. Solve Eq. (H-42) for M_2:

$$M_2 = -J_2 \ddot{\theta}_2 - k_1(\theta_2 - \theta_1) \tag{H-47}$$

Eliminate M_2 and θ_2 using Eqns. (H-45) and (H-46) to obtain

$$-M_3 N = -\frac{J_2}{N}\ddot{\theta}_3 - k_1\left\{\frac{\theta_3}{N} - \theta_1\right\} \tag{H-48}$$

Substitute this expression into Eq. (H-43) to get

$$J_3 \ddot{\theta}_3 = -k_2(\theta_3 - \theta_4) - \frac{J_2}{N^2}\ddot{\theta}_3 - \frac{k_1}{N}\left(\frac{\theta_3}{N} - \theta_1\right) \tag{H-49}$$

and, after collecting terms,

$$\left(\frac{J_2}{N^2} + J_3\right)\ddot{\theta}_3 + k_2(\theta_3 - \theta_4) + k_1\left(\frac{\theta_3}{N^2} - \frac{\theta_1}{N}\right) = 0 \tag{H-50}$$

Equation (H-50) shows that the moment of inertia of one gear is coupled into that of the other by the turns ratio squared ($\ddot{\theta}_3$ is multiplied by $J_3 + J_2/N^2$ rather than just J_3). Equations (H-41), (H-44), and (H-50) can be solved simultaneously to analyze the motion.

H.5 NONLINEAR SYSTEMS

To this point we have treated the mechanical world as if it consisted solely of masses, springs, and dampers engaged in translational or rotational motion. The resulting mathematical models consist of linear constant-coefficient differential equations. Although such models are often faithful approximations to reality, we must at times broaden our

Appendices

horizons to consider systems which give rise to nonlinear differential equations.(11)

Nonlinear terms can arise in mechanical systems from two major sources. First, some spring or damper phenomena cannot be approximated by linear models. Aerodynamic drag, for example, is often taken as being proportional to velocity squared, giving rise to differential equations of motion which contain the nonlinear term $(\dot{y})^2$. The force due to sliding friction depends on the direction rather than the velocity of the moving body, causing the nonlinear term $\dot{y}/|\dot{y}|$ to appear in the equation of motion. Second, if the motion is neither purely rotational nor purely translational, nonlinear terms often arise due to the need to resolve forces and moments into components along the coordinate axes. Consider the pendulum in Example H-9.

Example H-9. Write the differential equation which describes the motion of the simple pendulum of Fig. H-14. The pendulum is given an initial angular displacement of $\theta(0)$ radians to the right and released at t=0.

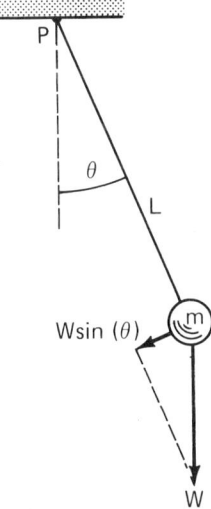

Figure H-14. This simple pendulum for Example H-9 illustrates the occurrence of nonlinearities even in elementary mechanical systems.

Solution. We can view the pendulum as a problem of rotation about the pivot point P. The moment of inertia J of a point mass rotating with radius L about a point is $J=mL^2$. The component of the weight perpendicular to the pendulum is $W\sin(\theta)$. If the mass is moved in the positive θ direction, the moment $WL\sin(\theta)$ due to the weight opposes the motion. The rotational form of Newton's second law for the pendulum is then

$$mL^2\ddot{\theta} = -WL\sin(\theta) \qquad (H-51)$$

or

$$\ddot{\theta} + \frac{g}{L}\sin(\theta) = 0 \qquad (H-52)$$

The quantity $g=W/m$ is the acceleration of gravity in appropriate units. If $\theta(t)$ is restricted to a few degrees, we can use the approximation $\sin(\theta) \approx \theta$, where θ is in radians, and obtain the linear model

$$\ddot{\theta} + \frac{g}{L}\theta = 0 \qquad (H-53)$$

with the initial condition $\theta(0)$.

Linearization

The technique of replacing $\sin(\theta)$ by θ for small angles is an example of a more general process known as <u>linearization</u>.[12] To take advantage of the power of linear system theory the analyst must understand the process of linearization in some depth. He must be familiar with the conditions under which a nonlinear model can be transformed into an approximately equivalent linear one, and he must of course be able to carry out the mathematical processes involved in the transformation.

To be sure, a nonlinear system can never be replaced by a linear one which is equivalent for all values of the dependent variables; but we can restrict the ranges of the dependent variables and find a linear system which behaves like the nonlinear one for the restricted range of operation. The

Appendices

first step in this process is to identify an <u>operating point</u>, a set of quiescent values of the dependent variables from which minor excursions will be permitted. In the pendulum example the operating point is the rest position wherein the pendulum hangs vertically downward. In other systems the operating point might correspond to the initial or t=0 position.

<u>The Taylor's series method</u>. If a function possesses derivatives of all orders at a point, it can be expanded in a Taylor's series in the neighborhood of that point. The general form for the expansion of f(x) in the neighborhood of x_o is

$$f(x) = f(x_o) + \frac{(x-x_o)f'(x_o)}{1!} + \frac{(x-x_o)^2 f''(x_o)}{2!} + \ldots \quad \text{(H-54)}$$

where the prime denotes differentiation with respect to x and

$$f'(x_o) = f'(x)\Big]_{x=x_o} \quad \text{(H-55)}$$

Using summation notation, Eq. (H-54) can be written

$$f(x) = \sum_{k=0}^{\infty} \frac{(x-x_o)^k}{k!} f^{(k)}(x_o) \quad \text{(H-56)}$$

where $f^{(k)}$ is the kth derivative of f(x) with respect to x. Clearly, when x is close to x_o the terms containing higher powers of $(x-x_o)$ decrease rapidly in magnitude. This makes it reasonable to truncate the series after the first two terms. The first two terms of the Taylor's series express f(x) as a linear function of x, or

$$f(x) \simeq f(x_o) + f'(x_o)(x-x_o) \quad \text{(H-57)}$$

Provided x remains near x_o, the right-hand side of Eq. (H-57) is a good approximation to the nonlinear function f(x) on the left. How close must x remain to x_o? The answer depends on the radius of convergence of the series and the permissible error in the approximation. The maximum error, $E(x_o)$, in truncating the Taylor's series after two terms is

$$E(x_o) = \frac{(x-x_o)^2}{2} f'(\varepsilon) \tag{H-58}$$

where $x<\varepsilon<x+x_o$. The following examples show how the Taylor's series method can be used to advantage in linearizing simple systems.

Example H-10. Linearize the equation $y=x^2$ in the neighborhood of the point $x=2$ using the Taylor's series method.

Solution. Expanding $y=x^2$ in a Taylor's series around the point $x=2$, $y=4$ yields

$$y = 4 + 4(x-2) + 2\frac{(x-2)^2}{2} \tag{H-59}$$

Saving only the first two terms gives the approximate equation

$$y = 4x - 4 \tag{H-60}$$

Figure H-15 shows that the effect of this linearization has been to replace the function $y=x^2$ by a straight line tangent to the curve at $x=2$.

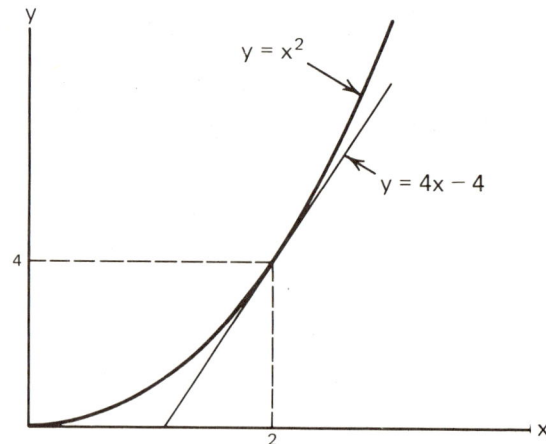

Figure H-15. When $y = x^2$ is linearized in the neighborhood of the point (2, 4) the result is the straight line $y = 4x - 4$.

Appendices 643

For values of x near x=2, points on the line y=4x-4 will approximate those on the curve $y=x^2$.

Example H-11. Define a "hard" spring as one for which the restoring force is $-ky^3$ rather than $-ky$. The equation of motion for the mass and hard-spring system of Fig. H-16 is

$$m\ddot{y} + ky^3 = f(t) \qquad (H-61)$$

Figure H-16. In this system for example H-11 the spring is assumed to be "hard". It satisfies the relationship $f_k = ky^3$.

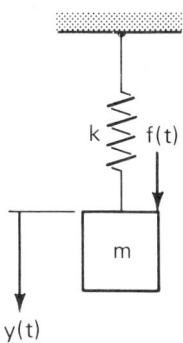

Assume the mass is moved y_o units downward and released to move under the influence of the forcing function $f(t)$. Linearize the system in the neighborhood of y_o.

Solution. The given differential equation is nonlinear because the coefficient of the y term is a function of y. Expanding $g(y) = y^3$ in a Taylor's series around y_o yields

$$g(y) = g(y_o) + \frac{(y-y_o)g'(y_o)}{1!} + \frac{(y-y_o)^2 g''(y_o)}{2!} + \ldots \qquad (H-62)$$

$$= y_o^3 + 3(y-y_o)y_o^2 + \ldots \qquad (H-63)$$

Preserving only the first two terms and substituting the result into the differential equation gives

$$m\ddot{y} + k\{y_o^3 + 3(y-y_o)y_o^2\} = f(t) \qquad (H-64)$$

or

$$m\ddot{y} + (3ky_o^2)y = f(t) + 2ky_o^3 \qquad (H-65)$$

with initial conditions $y(0)=y_o$ and $\dot{y}(0)=0$. Equation (H-65) is linear because the coefficient of the y term is now a constant. The constant term $2ky_o^2$ on the right-hand side has become a part of the forcing function. It acts as a constant force applied to the mass starting at t=0. The linearized solution will approximate that of the original nonlinear equation as long as y(t) is near y_o.

We were able to make the approximation $\sin(\theta) \approx \theta$ in the pendulum example because the Taylor's series expansion for $\sin(\theta)$ near $\theta=\theta_o$ is

$$\sin(\theta) = \sin(\theta_o) + (\theta-\theta_o)\cos(\theta_o) + \ldots \qquad (H-66)$$

and for $\theta_o=0$ we have $\sin(\theta) \approx \theta$, provided, of course, that θ is given in radians. This is a better approximation than is generally supposed. The error in using θ for $\sin(\theta)$ is 4.7% when $\theta=\pi/6$ and approximately 0.5% when $\theta=\pi/18$.

PROBLEMS FOR APPENDIX H

H-1. The forces $\underline{f}_1=2\underline{i}+3\underline{j}+4\underline{k}$, $\underline{f}_2=5\underline{k}$, $\underline{f}_3=\underline{i}-\underline{j}$, all given in newtons, act on a 10-kilogram point mass in an xyz-coordinate system. Find the vector acceleration of the mass in meters per second squared.

H-2. Use Eq. (H-1) to show that in the mks system a kilogram is equivalent to one newton-second-squared per meter.

H-3. Derive expressions for the moments of inertia of the following bodies:

(a) A point mass m at a distance R from the center of rotation.

(b) A circular disc of radius R, thickness L, and mass m about a transverse axis.

(c) A slender rod of length L and mass m rotating about its midpoint.

H-4. Use Eq. (H-8) to show that another set of units for J is kilogram-meters squared.

H-5. If energy has the dimensions of force times distance, use Newton's second law to show that the change in energy of an accelerating mass m during the time interval t_1-t_2 is given by

$$W = \frac{m(v_2^2 - v_1^2)}{2} \qquad (H-67)$$

where v_1 and v_2 are the velocities at times t_1 and t_2.

H-6. Show that the energy W_s, stored in a spring which is elongated y units from its unstretched position is $W_s = ky^2/2$.

H-7. Verify the dimensional compatibility of the terms in the following mechanical differential equations:

(a) $\ddot{y} + \frac{b}{m}\dot{y} + \frac{k}{m}y = \frac{f(t)}{m}$ \qquad (H-68)

(b) $\ddot{\theta} + \frac{b}{J}\dot{\theta} + \frac{k}{J}\theta = \frac{M(t)}{J}$ \qquad (H-69)

H-8. Write differential equations of motion for the translational mechanical systems of Fig. H-17. Assume that all initial conditions are zero and that the coordinate system for each hanging mass has its origin at the equilibrum point.

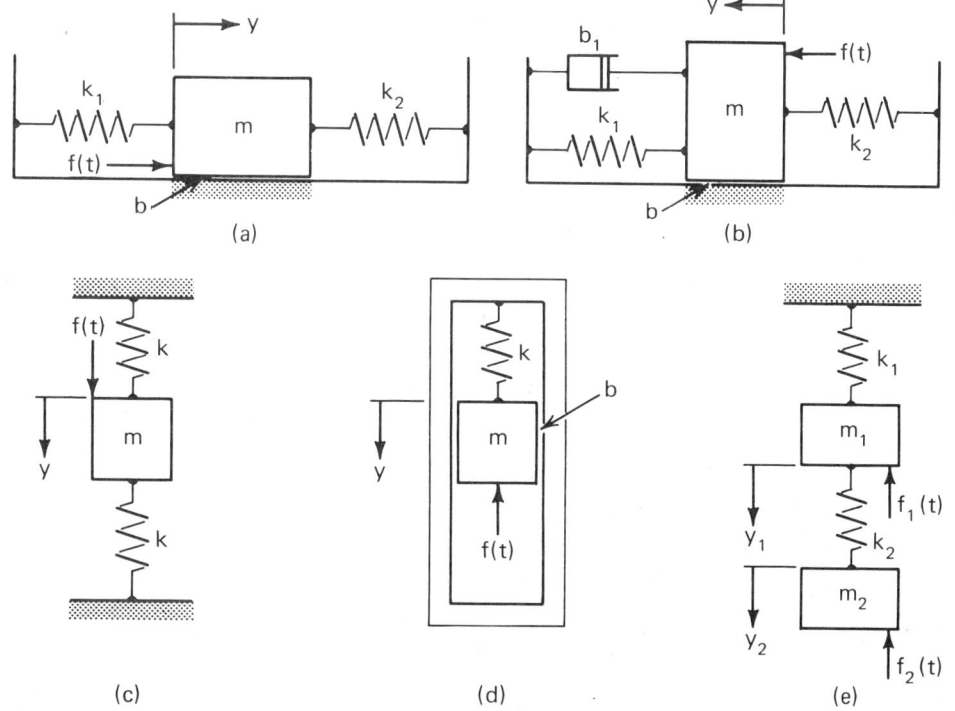

Figure H-17. Problem H-8.

H-9. Mass m_2 in Fig. H-18 is pulled down D units from equilibrium and released. Write the differential equations which describe the subsequent motion. Assume $y_1(0)=0$.

Figure H-18. Problem H-9.

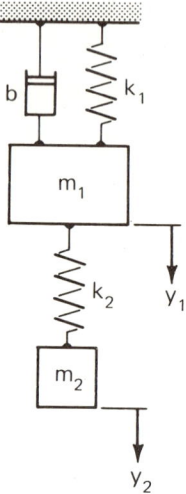

H-10. The car in Fig. H-19 moves to the right with constant velocity V. If it is stopped suddenly at t=0, write the differential equation for the subsequent motion of the restrained block.

Figure H-19. Problem H-10.

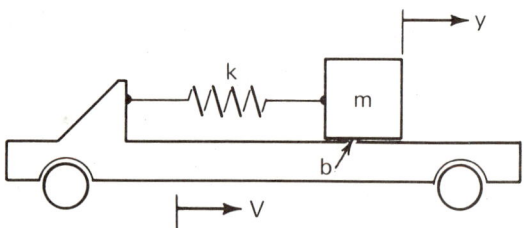

H-11. Write differential equations of motion for the system of Fig. H-20. Assume zero initial conditions.

H-12. Compare the differential equations of motion for the systems of Fig. H-21. Assume $y(0)=A$ and $\dot{y}(0)=0$ for each system.

H-13. An aircraft nosewheel can be modeled by the spring-damper system shown in Fig. H-22. If the effect of landing is

Appendices

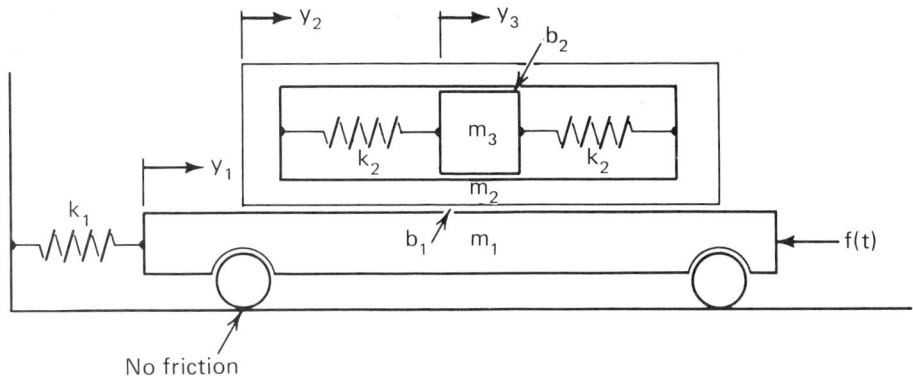

Figure H-20. Problem H-11.

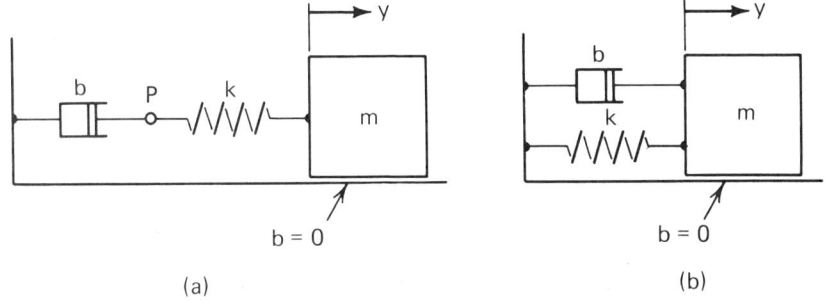

Figure H-21. Problem H-12.

Figure H-22. Problem H-13.

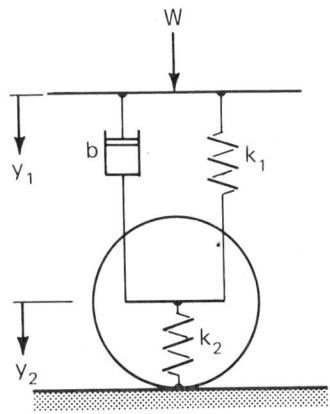

suddenly to apply the weight W of the aircraft, write the differential equations which describe the subsequent motion.

Let the origins for y_1 and y_2 be at the final rest position of the system. Neglect the mass of the wheel but use $m_1 = W/g$ for the mass of the aircraft.

H-14. Write differential equations of motion for the rotational systems of Fig. H-23. Assume zero initial conditions:

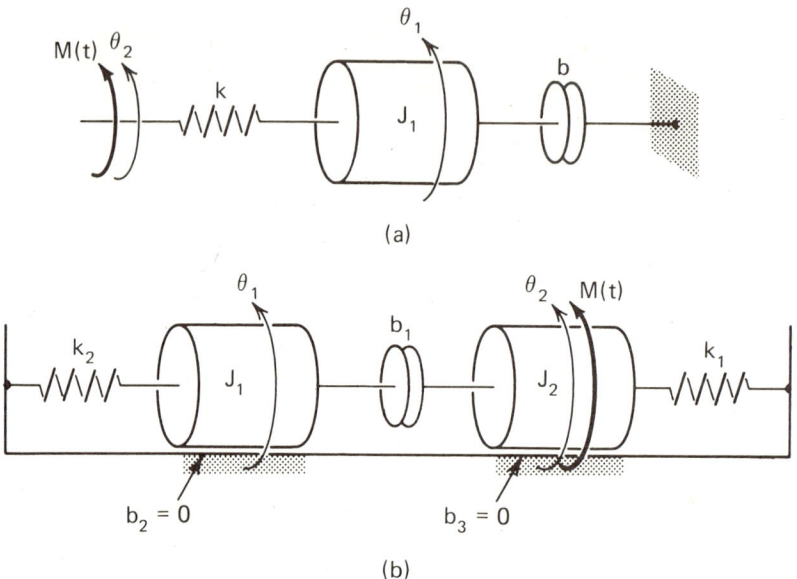

Figure H-23. Problem H-14.

H-15. Write the differential equations for the systems shown in Fig. H-24. Assume zero initial conditions.

H-16. Compare the equations of motion for the two systems of Fig. H-25. Assume $\theta(0) = A$ and $\dot{\theta}(0) = 0$.

H-17. The spool in Fig. H-26 is free to rotate on a shaft connected to a turntable. There is viscous friction between the spool and the turntable.
If the system rotates with velocity Ω rad/s and the turntable is suddenly stopped, write the differential equation for the subsequent velocity $\omega(t)$ of the spool.

H-18. Write the differential equations of motion for the system shown in Fig. H-27. Assume zero initial conditions. The gears have radii r_1, r_2, and r_3 respectively.

Figure H-24. Problem H-15.

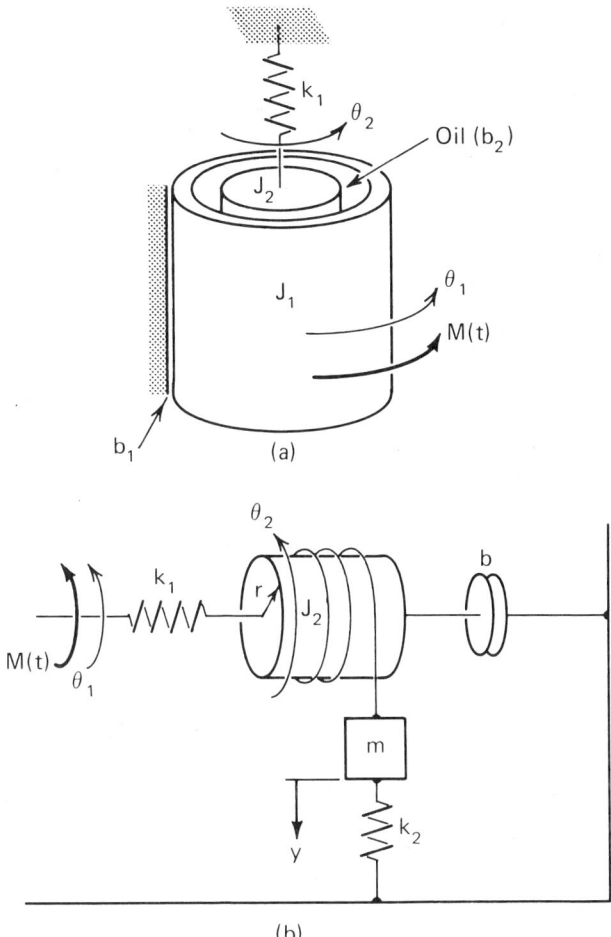

Figure H-25. Problem H-16.

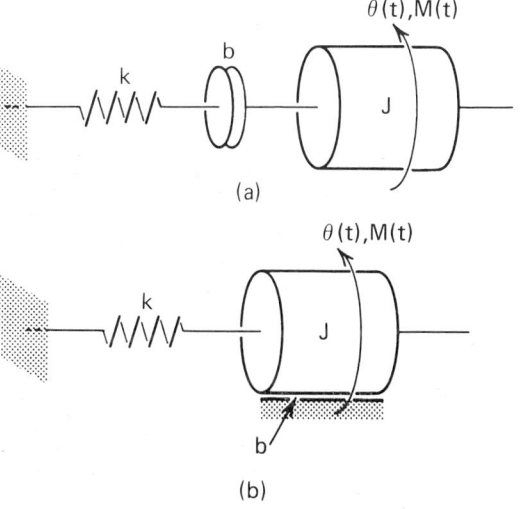

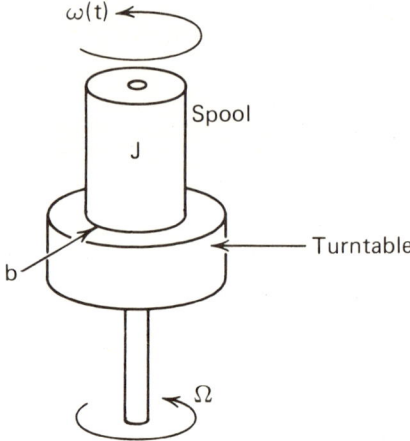

Figure H-26. Problem H-17.

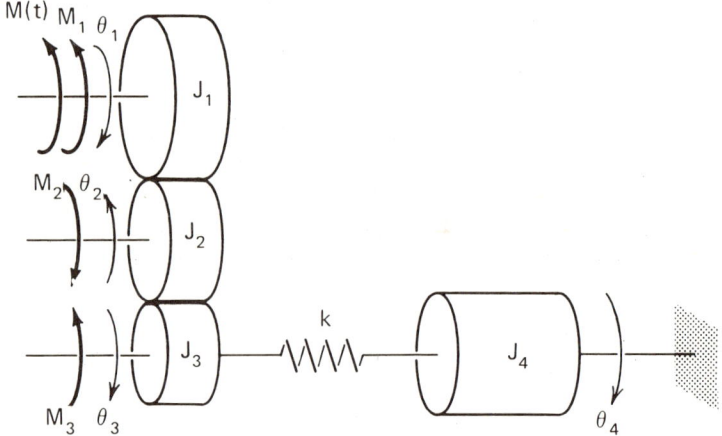

Figure H-27. Problem H-18.

H-19. A propeller shaft can be modeled as shown in Fig. H-28. Write the equations of motion if a step function moment $M(t)=Ru(t)$ is applied. The wind resistance is represented by viscous damping b. Assume zero initial conditions.

H-20. Linearize the following functions near y_o using the Taylor's series method:

(a) $f(y) = my + b$ (H-70)

(b) $f(y) = e^y$ (H-71)

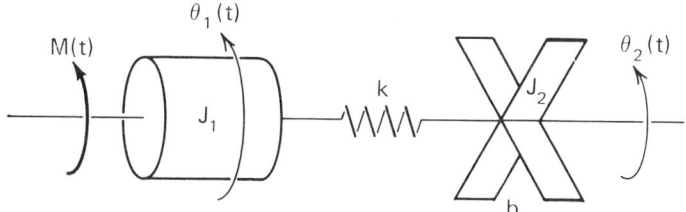

Figure H-28. Problem H-19.

(c) $f(y) = y^n$ \hfill (H-72)

(d) $f(y) = A\cos(y) + B\sin(y)$ \hfill (H-73)

(e) $f(y) = \dfrac{1}{y}$ \hfill (H-74)

(f) $f(y) = Ay^2 + By + C$ \hfill (H-75)

H-21. Linearize the following differential equations in the neighborhood of y_o using the Taylor's series method:

(a) $\dot{y} + \dfrac{1}{y} = f(t)$ \hfill (H-76)

(b) $\ddot{y} + y^3 = f(t)$ \hfill (H-77)

(c) $\ddot{y} + \cos(y) = g(t)$ \hfill (H-78)

(d) $\ddot{y} + ye^y = f(t)$ \hfill (H-79)

REFERENCES FOR APPENDIX H

(1) Freberg, C. R. and E. N. Kemler: <u>Elements of Mechanical Vibrations</u>, John Wiley and Sons, Inc., New York, 1949.

(2) Hansen, H. M. and P. F. Chenea: <u>Mechanics of Vibrations</u>, John Wiley and Sons, Inc., New York, 1952.

(3) Best, C. L. and W. G. McLean: <u>Analytical Mechanics for Engineers--Dynamics</u>, International Textbook Co., Inc., Scranton, Pennsylvania, 1970.

(4) Gibson, J. E. and F. B. Tuteur: <u>Control System Components</u>, McGraw-Hill Book Co., Inc., New York, 1958.

(5) Guillemin, E. A.: <u>Theory of Linear Physcial Systems</u>, John Wiley and Sons, Inc., New York, 1963.

(6) Meriam, J. L.: <u>Dynamics</u>, John Wiley and Sons, Inc., New York, 1966.

(7) Higdon, A. and W. B. Stiles: <u>Engineering Mechanics Vol. 2: Dynamics</u>, Prentice-Hall, Englewood Cliffs, New Jersey, 1962.

(8) Timoshenko, S. and D. H. Young: <u>Advanced Dynamics</u>, McGraw-Hill Book Co., Inc., New York, 1948.

(9) Bishop, R. E. D. and D. C. Johnson: <u>The Mechanics of Vibration</u>, Cambridge University Press, 1960.

(10) Thaler, G. J. and R. G. Brown: <u>Analysis and Design of Feedback Control Systems</u>, McGraw-Hill Book Co., Inc., New York, 1960.

(11) Shinners, S. M.: <u>Modern Control System Theory and Application</u>, Addison-Wesley Publishing Co., Inc., 1973.

(12) Seifert, W. W. and C. W. Steeg, Jr., eds.: <u>Control Systems Engineering</u>, McGraw-Hill Book Co., New York, 1960.

APPENDIX I. TRANSLATIONAL MECHANICAL UNITS

Variable or Parameter	Symbol	Dimensions	Preferred Units	Units Symbol
Time	t	T	Second	s
Force	f	F	Newton	N
Displacement	y	L	Meter	m
Velocity	$\dot{y}, v$	L/T	Meter per second	m/s
Acceleration	$\ddot{y}$	L/T^2	Meter per second squared	m/s^2
Mass	m	M	Kilogram	kg
Damper Coefficient	b	FT/L	Newton second per meter	N·s/m
Spring Coefficient	k	F/L	Newton per meter	N/m
Energy	W	FL	Joule	J

APPENDIX J. ROTATIONAL MECHANICAL UNITS

Variable or Parameter	Symbol	Dimensions	Preferred Units	Units Symbol
Moment (torque)	M	FL	Newton meter	$N \cdot m$
Angular Displacement	θ	None	Radian	rad
Angular Velocity	$\dot{\theta}, \omega$	$1/T$	Radian per second	rad/s
Angular Acceleration	$\ddot{\theta}$	$1/T^2$	Radian per second squared	rad/s^2
Moment of Inertia	J	FLT^2	Kilogram meter squared	$kg \cdot m^2$
Rotary Damper Coefficient	b	FLT	Newton meter second per radian	$N \cdot m \cdot s/rad$
Torsional Spring Coefficient	k	FL	Newton meter per radian	$N \cdot m/rad$

Answers to Selected Problems

CHAPTER 2

(2-9) $A=Ta-1$, $B=T$ (2-11) (a) $f(t)=u(t-3)+u(t-5)-2u(t-7)$,
(b) $f(t)=5u(t)-6u(t-2)+11u(t-4)$,
(c) $f(t)=10\{u(t-\pi)-u(t-5\pi)\}$,

(d) $f(t)=A\sum_{k=0}^{\infty}\{1-2u(t-kT_o-T_o/4)+2u(t-kT_o-3T_o/4)\}$

(2-13) (a) $y(t)=5\cos(50\pi t-126.9°)$, (b) $y(t)=5\sin(50\pi t-36.9°)$
(2-15) (a) $T_o=1$ s, $A=7.07$, $\emptyset=45°$, (b) $T_o=2$ s, $A=36.1$,
$\emptyset=-33.7°$, (c) $T_o=50$ s, $A=7.07$, $\emptyset=-45°$, (d) $T_o=0.05$ s,
$A=300$, $\emptyset=90°$ (2-17) $A=2500\omega_o/(625+\omega_o^2)$ (2-19) $T_c=4.33$ s

(2-21) $A=C$, $p=B$ (2-23) $K_1=2.5\exp(-j\pi/4)$, $K_2=2.5\exp(j\pi/4)$, $a=-(3-20\pi j)$, $b=-(3+20\pi j)$ (2-25) $b=1.3$, $A=j$, $B=-1$, $\omega=120\pi$ rad/s
(2-27) (a) $\dot{f}(t)=\delta(t-3)+\delta(t-5)-2\delta(t-7)$,
(b) $\dot{f}(t)=5\delta(t)-6\delta(t-2)+11\delta(t-4)$ (2-29) $b=5$, $A=4$
(2-31) $t_m=\{\pi/2+\emptyset+\tan^{-1}(\omega_o T_c)\}/\omega_o$,

$$|y_m| = \frac{A\omega_o T_c \exp(-t_m/T_c)}{\sqrt{1+(\omega_o T_c)^2}}$$

(2-33) (a) $f(t)=(A/T_o)\sum_{k=0}^{\infty}r(t-kT_o)\{u(t-kT_o)-u(t-kT_o-T_o)\}$,

(b) $f(t)=(10/3)\{r(t)-r(t-3)\}-15u(t)+r(t-3)-r(t-8)$,
(c) $f(t)=2u(t-3)+r(t-3)-2r(t-5)+r(t-7)-2u(t-7)$,
(d) $f(t)=3t\{u(t-4)-u(t-6)\}$ (2-35) (a) $y(t)=-10\sin(2t)+5\cos(2t)$,

(b) $y(t)=1+4\exp(-5t)$ (2-36) $A_n=\sqrt{a_n^2+b_n^2}$, $\emptyset_n=\tan^{-1}(b_n/a_n)$
(2-38) (a) Microfarad, (b) Picofarad, (c) Millihenry,
(d) Milliampere, (e) Kilowatt, (f) Microhenry, (g) Kilohm,
(h) Megohm, (i) Kilohertz, (j) Millisecond, (k) Millivolt,
(l) Megawatt, (m) Microampere, (n) Microsecond,
(o) Nanosecond

CHAPTER 3

(3-1) (a) $p=4$ kW, (b) $W=14.4$ MJ
(3-3) (a) $P_{ave}=0.0$ W, (b) $W=5.0$ J (3-5) (a) $v_x=4$ V,
(b) $v_x=0$ V (3-7) (a) $i_1=i_2$, $i_2+i_5=i_3+i_4$, $0=i_5+i_6$,
(b) $i_1=i_2$, $i_2+i_4=i_3$, $i_3+i_5+i_7=0$, $0=i_4+i_5+i_6$ (3-9) (a) $R=200$ Ω
(b) $R=50$ Ω (3-11) (a) $R=1/3$ Ω, (b) $R=1.25$ Ω, (c) $R_{eq}=5/6$ Ω
(3-13) $v(t)=-18.8\exp(-3t)$ mV (3-15) (a) $v_L(t_1)=0$, (b) $v_L(t_1)=$
$-\pi/t_1$ V, (c) $i_L(t_1)=5.0$ A, (d) $i_L(t_1)=4.0$ A, (e) $i_L(t_1)=$
$2t_1/\pi$ A (3-27) (a) $i_1R_1+R_2(i_1-i_2)=E$, $R_2(i_2-i_1)+L_3di_2/dt+$
$i_2R_4=0$, (b) $dv_1/dt=R_1di_1/dt+(i_1-i_2)/C$, $-dv_2/dt=R_4di_2/dt+$
$L_3d^2i_2/dt^2+(i_2-i_1)/C$, (c) $dv(t)/dt=(R_1+R_2)di_1/dt+(i_1-i_2)/C_6$,

Answers to Selected Problems 657

$0=(i_2-i_1)/C_6+L_5(d^2i_2/dt^2-d^2i_3/dt^2)+R_3di_2/dt$, $0=R_4i_3+L_5(di_3/dt-di_2/dt)$, (d) $E=R_1i_1+L_4di_1/dt-di_2/dt)+R_3(i_1-i_3)$, $L_4(di_2/dt-di_1/dt)+R_2i_2+R_6(i_2-i_3)+v(t)=0$, $R_6(di_3/dt-di_2/dt)+i_3/C_5+R_3(di_3/dt-di_2/dt)=dv(t)/dt$ (3-29) (a) $di(t)/dt=\dot{v}_a/R+v_a/L+C\ddot{v}_a$, (b) $(v_a-E)/R_1+v_a/R_2+C_3\dot{v}_a=i(t)$, (c) $i_1(t)=v_a/R_2+(v_a-v_b)/R_3$, $C_5(\ddot{v}_b-\ddot{v}_c)+(\dot{v}_b-\dot{v}_a)/R_3+v_b/L_4=0$, $C_5(\dot{v}_c-\dot{v}_b)+v_c/R_6=0$, (d) $\dot{v}_a(1/R_1+1/R_2+1/R_3)+di_6/dt+C_5\ddot{v}_a+v_a/L_4=0$ (3-31) $E_1/R_1=v_2(t)/R_2+v_a(1/R_1+1/R_2)+C_2\dot{v}_a$ (3-33) $v_0=45.9$ V (3-35) (a) $I_N=0.833$ A, $R_N=60$ Ω, (b) $I_N=1.125$ A, $R_N=57.14$ Ω (3-37) $v_T=1.43$ V, $R_T=1.57$ Ω, (3-39) $v_0(0^+)=100$ V, $i_1(0^+)=2.0$ A, $i_0(0^+)=0$ (3-41) $v_C(0^+)=100$ V, $i_L(0^+)=20$ A, $di_L(0^+)/dt=-25$ A/s, $dv_C(0^+)/dt=0$ V/s, $v_C(\infty)=50$ V, $i_L(\infty)=10$ A (3-43) $i_L(0^-)=i_L(0^+)=-50$ A, $v_0(0^+)=-750$ V (3-45) $v_{gap}=2\times10^{10}$ V (3-47) (a) $v_0=2v_a$, (b) $v_0=v_b-v_a$, (c) $v_0=2v_b-3v_a$

CHAPTER 4

(4-3) (a) $y(t)=4\exp(-10t)$, (b) $v(t)=-5\exp(-3t)$, (c) $q(t)=5\exp(-2t)$, (d) $y(t)=5+4t$ (4-5) $e_0(t)=50\exp(-10^4 t)$ (4-7) $e_0(t)=-(ER_2/R_1)\exp(-R_2 t/L)$ (4-9) $i(t)=(E/R)\{1-\exp(-Rt/(L_1+L_2))\}$ (4-11) $L=0.02$ H, $R_2=1.5$ Ω (4-13) (a) $y(t)=(2/3)(1-\exp(-3t))$, (b) $\theta(t)=3\exp(-t)-\exp(-2t)$, (c) $q(t)=\{2t-1-19\exp(-3t)\}/4$ (4-17) $\ddot{y}+3\dot{y}=0$, $y(0)=0$, $\dot{y}(0)=-3$ (4-19) $y(t)=\{1+(\omega_n t-1)\exp(-\omega_n t)\}b/\omega_n^2$

CHAPTER 5

(5-5) $F(s)=(2s+a+b)/(s^2+(a+b)s+ab)$ (5-7) $F(s)=1/s^2-\exp(-5s)(2+1/s)/s$ (5-15) (a) $f(\infty)=0.5$, (b) $f(\infty)=0$, (c) $f(\infty)\to\infty$,

(d) $f(\infty)=2.5$ (5-17) (a) $F(s)=2\omega s/(s^2+\omega^2)^2$, (b) $F(s)=2/s^3$, (c) $F(s)=2/(s+a)^3$

(5-19) (a) $Y(s) = \dfrac{1}{s(s^2+2\zeta\omega_n s+\omega_n^2)}$, (b) $1/(s^2(s+4))+1/(s+1)(s+4)$,

(c) $Y(s)=s/(s^2+16)^2$, (d) $Y(s)=\dfrac{s-\exp(-2s)}{s(s^3+4s^2+5s)}$ (5-21) (a) $f(t)=\exp(-at)\sin(bt)/b$, (b) $f(t)=\exp(-at)\{\cos(bt)-(a/b)\sin(bt)\}$

(5-23) $e_o(t)=50\exp(-10^4 t)$ (5-25) $e_o(t)=-(ER_2/R_1)\exp(-R_2 t/L)$

(5-29) (a) $y(t)=2(1-\exp(-3t))/3$, (b) $\theta(t)=3\exp(-t)-\exp(-2t)$,

(c) $q(t)=(2t-1-19\exp(-2t))/4$, (d) $\theta(t)=4\{1-\exp(-2.5t)(\cos(10.9t))$ $=0.23\sin(10.9t))\}$ (5-33) (a) $v_o(t)=18$ V,

(b) $v_o(t)=60$ V, (c) $v_o(t)=10\sin(377t)$ (5-35) (a) $v_T=33.3$ V, $R_T=16.67$ Ω, (b) $v_T(t)=55\sin(377t)$, $R_T=100$ Ω (5-37)

$V_1(s) = \dfrac{E(s+8)}{s(s^2+10s+11)}$,

$V_2(s)=\dfrac{5E}{s(s^2+10s+11)}$ (5-39) $r(t)=\exp(-3t)-\exp(-5t)$ (5-43) $f(t)=t^{n-1}\exp(-at)/(n-1)!$

CHAPTER 6

(6-1) (a) Poles: $s=j4$, $s=-j4$; Zeros: $s=0$, $s=-1$, (b) Poles: $s=-2$, $s=3$; Zeros: $s=0$, $s=j2$, $s=-j2$, (c) Poles: $s=-2+j3.46$, $s=-2-j3.46$; Zeros: $s=0$, $s=0$, $s=-2$, (d) Poles: $s=0$, $s=0$, $s=j3$, $s=-j3$, $s=-1+j2.65$, $s=-1-j2.65$; No finite zeros (6-3) (a) $|F(-1)|=1/6$, $\mathrm{Arg}\{F(-1)\}=-180°$, (b) $|F(j4)|-1.626$, $\mathrm{Arg}\{F(j4)\}=-26.57°$, (c) $|F(1+j2)|=1.8$, $\mathrm{Arg}\{F(1+j2)\}=33.69°$

(6-5) $\omega_{max}=\omega_n\sqrt{1-2\zeta^2}$ (6-7) (a) Poles: $s=-K$; No finite zeros,

(b) Poles: $s=0$, $s=0$, $s=-K$; No finite zeros, (c) Poles: $s=-a+\sqrt{a^2-K}$, $s=-a-\sqrt{a^2-K}$; No finite zeros, (d) Poles: $s=0$; zero; $s=-K$, (e) Poles: $s=-K$, $s=-K$; No finite zeros, (f) Poles: $s=-K+\sqrt{K^2-4}$, $s=-K-\sqrt{K^2-4}$; No finite zeros (6-9) (a) $g(t)=\{4-5\exp(-2t)+\exp(-10t)\}/16$ (b) $g(t)=\{1-\cos(4t)+2\sin(4t)\}/8$, (c) $g(t)=0.0625+0.26\exp(-2t)\cos(3.46t-103.9°)$ (6-10) Error=$\{\exp(-at)-10\exp(-10at)\}/90a$ (6-12) (a) Stable, (b) Conditionally Stable, (c) Unstable, (d) Unstable, (e) Stable, (f) Conditionally Stable, (g) Conditionally Stable, (h) Unstable, (i) Unstable (6-14) $a>0$, $b>0$, $c>0$, $d>0$, $ab>c$, $c(ab-c)-a^2>0$.

CHAPTER 7

(7-1) (a) $G(s)=2.5/s$, (b) $G(s)=(s^2+s+10)/(s^2+5s+6)$, (c) $G(s)=\dfrac{(a/d)(s^2+bs/a+c/a)}{s^3+es^2/d+fs/d+g/d}$, (d) $G(s)=\dfrac{6(s^2+11s/3+3)}{s^3+6s^2+11s+6}$ (7-3) (a) $G(s)=1/(s^2+4s+18)$, (b) $G(s)=1/s^2(s+4)$, (c) $G(s)=(s+2)/(s^2+6s+32)$ (7-8) $G(s)=G_1(s)G_2(s)\ldots G_N(s)$ (7-15) $K=199.5$ (7-16) $K=25.0 \times 10^4$, $\zeta=0.05$

CHAPTER 8

(8-1) (a) $A+B-C=-1+j16$, (b) $AB=-13+j13$, (c) $|A|=3.61$, (d) $\text{Arg}\{A\}=56.31°$, (e) $A/C=-0.2+j0.35$, (f) $B^2=-24+j10$, (g) $\sqrt{A}=1.9\exp(j0.491)$ (8-3) (a) $V_1/V_2=2\exp(j1.476)$, (b) $V_1+V_2=0.17+j11.59$, (c) $V_1V_2=50\exp(j2.52)$, (d) $\sqrt{V_1}/V_2^2=0.126\exp(-j0.047)$ (8-5) $y(t)=0.05-0.0069\exp(-5t)-0.043\cos(2t)-0.017\sin(2t)$ (8-7) (a) $y_{ss}(t)=28.3\cos(2t-45°)$, (b) $y_{ss}(t)=1.109\cos(2t-56.3°)$, (c) $y_{ss}(t)=11.31\cos(2t-45°)$, (d) $y_{ss}(t)=$

10.24cos(2t+91.8°) (8-9) (a) $G(s)=1/(RCs+1)$, (b) $|G(j\omega)|= 1/\sqrt{(\omega RC)^2+1}$, $\text{Arg}\{G(j\omega)\}=-\tan^{-1}(\omega RC)$, (c) $v_{oss}(t)= K\sin(\omega_o t-\tan^{-1}(\omega_o RC))/\sqrt{(\omega_o RC)^2+1}$, (d) $|v_{oss}(\infty)|=0$ (8-11) (a) $\bar{Y}(j2)=28.3\exp(2jt-j\pi/4)$, (b) $\bar{Y}(j2)=1.109\exp(2jt-j0.983)$, (c) $\bar{Y}(j2)=11.31\exp(2jt-j\pi/4)$, (d) $\bar{Y}(j2)=10.24\exp(2jt+j1.6)$ (8-19) $\bar{I}_1=4.38\exp(j0.266)$ A, $\bar{I}_2=1.96\exp(-j1.77)$ A, $\bar{V}_o=19.6\exp(-j0.197)$ V (8-21) $Z_i=2.4(1-j2)$ (8-25) (a) $f_o=1299.5$ Hz, (b) $f_o=0.159$ Hz, (c) $f_o=159.15$ kHz (8-27) The network is resonant at all frequencies. (8-29) (1) $P_{ave}=642.8$ W, (2) $P_{ave}=98.3$ W, (3) $P_{ave}=250$ W, (4) $P_{ave}=0$ W

(8-31) $\omega_o = \dfrac{1}{\sqrt{LC}}\sqrt{\dfrac{R_L^2-L/C}{R_C^2-L/C}}$ (8-33) (a) $V_{rms}=0.577$ A, (b) $V_{rms}=A\sqrt{T/T_o}$, (c) $V_{rms}=0.707$ A (8-35) $V_{rms}=\sqrt{A^2+B^2}/2$

(8-37) (a) p.f.=1.0, (b) p.f.=0.507 lagging, (c) p.f.=0, (d) p.f.=0.781 leading, (e) p.f.=0.98 lagging, (f) p.f.=0.707 lagging (8-39) (a) p.f.=1.0, (b) p.f.=0.949 leading, (c) p.f.=0.985 lagging (8-41) (a) C=4.05 μF (8-43) $W_1=270.6+j156.3$, $W_2=62.5+j108.3$, $W_{total}=333.1+j264.6$ (8-45) (a) $R_L=25$ Ω. (b) $R_L=55.9$ Ω, (8-47) (a) R=50 Ω, X=0, G=0.02 mho, B=0, (b) R=75 Ω, X=-25 Ω, G=0.012 mho, B=0.004 mho, (c) R=33.53 Ω, X=5.88 Ω, G=0.029 mho, B=-0.0051 mho (8-51) $Z_1=Z_2=Z_3=1.67\exp(-j\pi/4)$

CHAPTER 9

(9-1) (a) $M(\omega)=1/\sqrt{\omega^2+25}$, $\phi(\omega)=-\tan^{-1}(\omega/5)$, (b) $M(\omega)=$

Answers to Selected Problems

$$\frac{1}{\sqrt{(\omega_n^2-\omega^2)^2+(2\zeta\omega\omega_n)^2}}, \quad \phi(\omega)=-\tan^{-1}\{2\zeta\omega\omega_n/(\omega_n^2-\omega^2)\}, \quad (c)$$

$$M(\omega)=\frac{1}{\omega\sqrt{(\omega^2+100)}}, \quad \phi(\omega)=-90°-\tan^{-1}(10/\omega), \quad (d) \ M(\omega)=$$

$$\frac{1}{|1-\omega^2|\sqrt{1+\omega^2}}, \quad \phi(\omega)=-\tan^{-1}(\omega) \quad (9\text{-}3) \quad M(4)=0.102, \ \phi(4)=-3.73°$$

(9-5) $G(s)=10^6\pi/(s+2000\pi)$ (9-7) (a) $M_p=M(0)=1$, $\omega_c=1.33$ rad/s, (b) $M_p=0.314$, $\omega_c=4.34, 3.35$ rad/s, (c) $M_p=1.15$ $\omega_c=1.17$ rad/s (9-9) (a) $y_{ss}(t)=0.018\cos(25t-116.57°)$, (b) $y_{ss}(t)=0.099\cos(5t-95.71°)$ (9-11) $M_p=134.6$, $\omega_p=5.0$ rad/s (9-13) (a) 27.96 db, (b) -60 db, (c) 58.7 db, (d) -32.04 db, (e) No db values for negative quantities. (9-17) (a) $T(\omega)=0$, (b) $T(\omega)=0$, (c) $T(\omega)=0$, (d) $T(\omega)=-a/(a^2+\omega^2)$, (e) $T(\omega)=a/(a^2+\omega^2)$, (f) $T(\omega)=5/(25+\omega^2)$, (g) $T(\omega)=5/(25+\omega^2)$, (h) $T(\omega)=-1/(1+\omega^2)+2/(4+\omega^2)$, (i) $T(\omega)=4(3+\omega^2)/(9+\omega^4)$.

CHAPTER 10

(10-1) (a) -42, (b) -5, (c) 50, (d) 1 (10-3) (a) $AB=\begin{bmatrix} -7 & -18 \\ 38 & 37 \end{bmatrix}$

(b) Not conformable, (c) $BA = \begin{bmatrix} 9 & 2 & 23 & -9 \\ 2 & 1 & 4 & 8 \\ 1 & 0 & 3 & -5 \\ 11 & 4 & 25 & 17 \end{bmatrix}$

(e) Not conformable, (f) Not conformable

(10-5) $A^T = \frac{1}{D}\begin{bmatrix} a_{22} & -a_{12} \\ -a_{21} & a_{11} \end{bmatrix}$ $D = a_{11}a_{22} - a_{21}a_{12}$

(10-7) (a) $A^{-1}=\text{Diag}[1/5, -1/2, -1/4]$, (b) $(I_3-A)^{-1}=$ $\text{Diag}[-1/4, 1/3, 1/5]$, (c) The matrix is singular and has no inverse, (d) $A^{-2}=\text{Diag}(1/25,1/4,1/16)$

(10-9) (a) $A^2 = \begin{bmatrix} -1 & 2 \\ -1 & -2 \end{bmatrix}$ (b) $A^3 = \begin{bmatrix} -3 & -2 \\ 1 & -2 \end{bmatrix}$ (c) $\exp(A) = \begin{bmatrix} 3/2 & 3 \\ -3/2 & 0 \end{bmatrix}$

(10-13) (a) $A = \begin{bmatrix} 0 & 1 \\ -5 & 0 \end{bmatrix}$ $\underline{b} = \begin{bmatrix} 0 \\ 1 \end{bmatrix}$ $\underline{x}(0) = \begin{bmatrix} 1 \\ 2 \end{bmatrix}$ $C = \begin{bmatrix} 1 & 0 \end{bmatrix}$, $D = 0$,

where $x_1 = y$ and $x_2 = \dot{y}$.

(b) $A = \begin{bmatrix} 0 & 1 & 0 \\ 0 & 0 & 1 \\ -6 & -5 & -2 \end{bmatrix}$ $\underline{b} = \begin{bmatrix} 0 \\ 0 \\ 20 \end{bmatrix}$ $f(t) = u(t)$, $C = \begin{bmatrix} 1 & 0 & 0 \end{bmatrix}$, $D = 0$, $\underline{x}(0) = \begin{bmatrix} 1 \\ 0 \\ 5 \end{bmatrix}$

where $x_1 = y$, $x_2 = \dot{y}$, and $x_3 = \ddot{y}$.

(c) $A = \begin{bmatrix} 0 & 1 & 0 & 0 \\ -5 & 0 & 0 & 0 \\ 0 & 0 & 0 & 0 \\ -2 & 0 & 0 & 0 \end{bmatrix}$ $\underline{b} = \begin{bmatrix} 0 \\ 10 \\ 0 \\ 0 \end{bmatrix}$ $f(t) = u(t)$, $C = \begin{bmatrix} 1 & 0 & 0 & 0 \\ 0 & 0 & 1 & 0 \end{bmatrix}$ $\underline{d} = \begin{bmatrix} 0 \\ 0 \end{bmatrix}$

where $x_1 = y$, $x_2 = \dot{y}$, $x_3 = g$, $x_4 = \dot{g}$, and $\underline{x}(0) = \underline{0}$.

(d) $A = \begin{bmatrix} 0 & 1 & 0 & 0 \\ -3 & 0 & 1 & 0 \\ 0 & 0 & 0 & 1 \\ -1 & 0 & 6 & 0 \end{bmatrix}$ $\underline{b} = \begin{bmatrix} 0 \\ 1 \\ 0 \\ 0 \end{bmatrix}$ $f(t) = u(t)$, $C = \begin{bmatrix} 1 & 0 & 0 & 0 \\ 0 & 0 & 1 & 0 \end{bmatrix}$ $\underline{d} = \begin{bmatrix} 0 \\ 0 \end{bmatrix}$

where $x_1 = g$, $x_2 = \dot{g}$, $x_3 = h$, $x_4 = \dot{h}$, and $\underline{x}(0) = \underline{0}$.

(10-16) None of the systems can be diagonalized as they all have complex eigenvalues.

(10-17) (a) $\underline{x}(t) = \begin{bmatrix} 1 - \exp(-t) \\ 1 \end{bmatrix}$ (b) $\underline{x}(t) = \begin{bmatrix} 5\exp(-2t) \\ 6\exp(-3t) \end{bmatrix}$

(c) $\underline{x}(t) = \begin{bmatrix} 0 \\ \exp(t) - 1 \end{bmatrix}$ (10-19) (a) $G(s) = 1/(s+1)$, (b) $G(s) = 0$,

(c) $G(s) = 2/(s-1)$

Answers to Selected Problems 663

CHAPTER 11

(11-1) (a) 1,1,1,1,1, (b) $1, \exp(-T/5), \exp(-2T/5), \exp(-3T/5)$, $\exp(-4T/5)$, (d) $3, 3+5T^2, 3+20T^2, 3+45T^2, 3+80T^2$ (11-3) $f(t)=$

$(Aj/\pi) \sum_{n=-\infty}^{\infty} \exp(jn\omega_o t)/n$ (11-6) (a) $F(j\omega) = \pi\{\delta(\omega-2)(5-j12)+$

$\delta(\omega+2)(5+j12)\}$, (b) $F(j\omega)=A/(b+j\omega)$, (c) $G(j\omega)=(1-\exp(j\omega T))/j\omega$,

(d) $Y(j\omega)=\{\exp(-2j\omega)(2j\omega+1)-1\}8/\omega^2$ (11-8) (a) $\omega_N = 6$ rad/s,

(b) $\omega_N = 80\pi$ rad/s, (c) $\omega_N = 1200\pi$ rad/s, (d) $\omega_N = 50$ rad/s

(11-10) 0.4798 (Zeroth harmonic not counted) (11-12) (a)

Yes, (b) $f_q = 40$ Hz (11-14) $\omega_c = 400\pi$ rad/s (11-18) R=0.0078125

(11-24) $F(j\omega) = \sum_{n=-\infty}^{\infty} C_n \delta(\omega-n\omega_o) 2\pi$

CHAPTER 12

(12-1) (a) $F(z) = \sum_{k=0}^{\infty} (kT)z^{-k}$, (b) $F(z) = \sum_{k=0}^{\infty} (kT)^2 z^{-k}$, (c) $F(z) =$

$\sum_{k=0}^{\infty} (kT)\exp(-akT)z^{-k}$, (d) $F(z) = \sum_{k=0}^{\infty} A\cos(\omega_o kT)z^{-k}$ (12-3) $Z\{kT\}=$

$Tz/(z-1)^2$ (12-5) (a) $Y(z)=z/(z-1)^2$ for $|z|>1$, (b) $Y(z)=$

$Az/(z-b)$ for $|z|>b$, (c) $Y(z)=bz/(z-b)^2$ for $|z|>b$, (d) $Y(z)=$

$\dfrac{z(z-\cos(\omega_o T))}{z^2 - 2z\cos(\omega_o T)+1}$ for $|z|>1$, (e) $Y(z)=z/(z-\exp(-T))$ for

$|z|>\exp(-T)$, (f) $Y(z) = \dfrac{z\{z\sin(\theta)+\sin(\omega_o T-\theta)\}}{z^2 - 2z\cos(\omega_o T)+1}$ for $|z|>1$ (12-7)

$F(z) = \frac{z}{z-0.5}\{1-(0.5z^{-1})^{26}\} + \frac{(0.25z^{-1})^{26}}{z-0.25}$ (12-9) The locus

is a circle of radius exp(-2.5) centered at the origin of the z-plane. (12-13) (a) y(k)=1.1sin(2k), (b) $y(k)=0.5(1)^k +$ 0.5cos(k/4), (c) $y(k)=2\{1-0.5^{(k-1)/2}\cos(\pi(k-1)/4)\}$
(12-21) (a) f(0)=0, f(1)=0, $f(k)=\delta(k-2)+1.125(0.125)^{k-2}$ for k≥2, (b) y(k)=0 for k=0,1,2,3 $y(k)=a^{k-4}$ for k≥4, (c) h(k)=0 for k=0,1,2,...25, $h(k)=(-0.5)^{k-26}$ for k=26,27,...
(12-23) (a), $y(k)=(5-2(0.4)^k)/3$, (b) $y(k)=2.5(0.45)^{k/2}(1+(-1)^k)$, (c) $q(k)=1.786(1)^k-1.286(0.3)^k+0.5(0.2)^k$, (d) $h(k)=(4(0.4)^k-(0.1)^k)/3$ (12-25) (a) y(-1)=2.222, (b) y(-1)=-4
(12-27) y(-2)=-12 (12-29) $y(k)=(k-1)(1)^k+(0.5)^k$ (12-31)
(a) G(z)=z/(z-0.4), (b) $G(z)=z^2/(z^2-0.45)$, (c) $G(z)=z^2/(z^2-0.5z+0.06)$, (d) $G(z)=z^2/(z^2-0.5z+0.04)$ (12-33) $G(z)=z^2/(z^2-0.25z+0.1)$, $\bar{f}(k)=f(k)-0.2\delta(k-1)$ (12-35) (a) $y(k)=(1)^k-(a)^{k+1}$, (b) $y(k)=(1-a)\{(-1)^k+(a)^{k+1}\}/(1+a)$ (12-37) $y(k)=2|R|(c^2+d^2)^{k/2}\cos\{(k)\tan^{-1}(d/c)+Arg(R)\}$ (12-39) (a) $G(z)=Kz^2/(z^2-1)$, (b) $G(z)=\frac{Kz^2(z-1)}{(z+1)(z-\exp(j\omega_o T))(z-\exp(-j\omega_o T))}$,

(c) $G(z) = \frac{K(z^2-1)}{(z-r\exp(j\omega_o T))(z-r\exp(-j\omega_o T))}$, (d) $G(z)=$

$\frac{K(z^2-1)(z^2+1)}{(z-r\exp(j\omega_o T))(z-r\exp(-j\omega_o T))(z-\exp(j\omega_o T)/r)(z-\exp(-j\omega_o T)/r)}$

CHAPTER 13

(13-1) $d^3y(i)/dt^3=\{y(i)-3y(i-1)+3y(i-2)-y(i-3)\}/T^3$,
$d^4y(i)/dt^4=\{y(i)-4y(i-1)+6y(i-2)-4y(i-3)+y(i-4)\}/T^4$ (13-3)

Answers to Selected Problems

$y(-1)=9.5$, $y(-2)=9.01$, $y(-3)=-8.528$ (13-7) (a) $y(k)=$
$y(k)=0.5\{1-0.75(0.5)^k\}$ $k\geq 1$, $y(0)=0$ (b) $q(k)=5(0.875)^k$
(13-9) Error$=b\{\exp(-akT)-(1-aT)^k\}$ (13-11)

$$\begin{bmatrix} x_1(k) \\ x_2(k) \end{bmatrix} = \begin{bmatrix} \exp(-T) & 0 \\ 0 & \exp(-2T) \end{bmatrix} \begin{bmatrix} x_1(k-1) \\ x_2(k-1) \end{bmatrix} + \begin{bmatrix} 2(1-\exp(-T)) \\ (1-\exp(-2T)/2 \end{bmatrix}(1)^k$$

(13-13) (a) $\exp(AT) = \begin{bmatrix} 1-T+T^2 & 0 \\ 0 & 1-2T+4T^2 \end{bmatrix}$

(b) $\exp(AT) = \begin{bmatrix} 1+T+T^2 & 0 \\ 2T-4T^2 & 1-3T+9T^2 \end{bmatrix}$

(c) $\exp(AT) = \begin{bmatrix} 1-T+T^2 & 0 & 0 \\ 0 & 1-5T+25T^2 & 0 \\ 0 & 0 & 1+2T+4T^2 \end{bmatrix}$

(d) $\exp(AT) = \begin{bmatrix} 1+2T+7T^2 & -T-6T^2 \\ -3T-18T^2 & 1+4T+19T^2 \end{bmatrix}$ (13-15)

$$\begin{bmatrix} x_1(k) \\ x_2(k) \end{bmatrix} = \begin{bmatrix} (1-T/2)/(1+T/2) & 0 \\ 0 & (1-T)/(1+T) \end{bmatrix} \begin{bmatrix} x_1(k-1) \\ x_2(k-1) \end{bmatrix} +$$

$$\begin{bmatrix} T/(1+T/2) & 0 \\ 0 & T/(1+T) \end{bmatrix} \{u(k-1)+u(k)\} \quad \text{for } k \geq 1$$

(13-17) $\underline{x}(k)=Z^{-1}\{(zI_n-I_n-AT)^{-1}TB\underline{F}(z)\}$

(13-19) $\underline{x}(k)=Z^{-1}\{(zI_n-(I_n+AT))^{-1}\{\underline{x}(0)-TBf(-1)+TBz^{-1}\underline{F}(z)\}\}$

(13-21) (a) $0<T<0.1$, (b) $0<T<0.2$ (13-23) $y(k)=2(0.60677)^k$

(13-25) $\underline{x}(0)=\begin{bmatrix}5\\4\end{bmatrix}$ $\underline{x}(1)=\begin{bmatrix}4.094\\2.6816\end{bmatrix}$ $\underline{x}(2)=\begin{bmatrix}3.350\\1.798\end{bmatrix}$ $\underline{x}(3)=\begin{bmatrix}2.743\\1.205\end{bmatrix}$

APPENDIX H

(H-1) $\underline{a}=0.3\underline{i}+0.2\underline{j}+0.9\underline{k}$ m/s^2 (H-3) (a) $J=mR^2$,
(b) $J=m(R^2/4+L^2/12)$, (c) $J=mL^2/12$ (H-9) $\ddot{y}_2+k_2y_2/m_2-k_2y_1/m_2=0$,

$\ddot{y}_1 + b\dot{y}_1/m_1 + y_1(k_1+k_2)/m_1 - k_2 y_2/m_1 = 0$, $y_2(0) = D$, All other initial conditions are zero.

(H-11) $\ddot{y}_1 + b_1\dot{y}_1/m_1 + k_1 y_1/m_1 - b_1\dot{y}_2/m_1 = -f(t)/m_1$,
$\ddot{y}_2 + (b_1+b_2)\dot{y}_2/m_2 + 2k_2 y_2/m_2 - b_1\dot{y}_1/m_2 - b_2\dot{y}_3/m_2 - 2k_2 y_3/m_2 = 0$,
$\ddot{y}_3 + 2b_2\dot{y}_3/m_3 + 2k_2 y_3/m_3 - 2b_2\dot{y}_2/m_3 - 2k_2 y_2/m_3 = 0$. All IC's are 0.

(H-13) $\ddot{y}_1 + b\dot{y}_1/m + k_1 y_1/m - b\dot{y}_2/m - k_1 y_2/m = W/m$,
$\dot{y}_2 + (k_1+k_2)y_2/b - \dot{y}_1 - k_1 y_1/b = 0$, $y_1(0) = -W(k_1+k_2)/k_1 k_2$, $\dot{y}_1(0) = 0$,
$y_2(0) = -W/k_2$ (H-15) (a) $\ddot{\theta}_1 + (b_1+b_2)\dot{\theta}_1/J_1 - b_2\dot{\theta}_2/J_1 = M(t)/J_1$,
$\ddot{\theta}_2 + b_2\dot{\theta}_2/J_2 + k_1\theta_2/J_2 - b_2\dot{\theta}_1/J_2 = 0$ All IC's are 0.
(b) $\ddot{\theta}_2 + b\dot{\theta}_2/J_2 + k_2 r^2\theta_2/J_2 = M(t)/J_2$, $\theta_2(0) = 0$, $\dot{\theta}_2(0) = 0$

(H-17) $J\dot{\omega} + b\omega = 0$, $\omega(0) = 0$

(H-19) $\ddot{\theta}_1 + k\theta_1/J_1 - k\theta_2/J_1 = M(t)/J_1$,
$\ddot{\theta}_2 + b\dot{\theta}_2/J_2 + k\theta_2/J_2 - k\theta_1/J_2 = 0$, all IC's are zero.

(H-21) (a) $\dot{y} - y/y_o^2 = f(t) - 2/y_o$, (b) $\ddot{y} + 3y_o^2 y = f(t) + 2y_o^2$,
(c) $\ddot{y} - y\sin(y_o) = g(t) - \cos(y_o) + y_o\sin(y_o)$,
(d) $\dot{y} + y\exp(y_o)(y_o+1) = f(t) + y_o^2\exp(y_o)$.

Index

A

Acceleration, angular, 620
Additivity, 45
Adjoint, 438
Admittance, 233
Aliasing, 492
Amplifier, operational, 114
Analog computation, 6, 321
Analog signal, 21
Analog-to-digital conversion, 496—501
Analog-to-digital converter
    multi-comparator, 496
    resolution, 499
    successive-approximation, 497
    time-interval, 500
Analysis, 2
Angular acceleration, 620
Anti-resonance, 357
Apparent power, 371

Array, Routh, 287
Asymptote, 414
Asymptotic exponential, 38
Auxiliary polynomial, 289

B

Bandlimited signal, 482
Bandwidth, 394
Binary signal, 21
Block diagram
    algebra of, 308
    definition, 307
    equivalent, 312
    feedback connection, 310
    for state-variable system, 447
    parallel connection, 309
    series connection, 308
    z-domain, 558

Bode plots, 409—423
    table of, 618
Branch, 61
Break point, 415

C

Capacitance
    definition, 66
    final-value model, 104
    in parallel, 88
    in series, 87
    initial-value model, 104
    phasor model, 348
    reference convention, 67
    s-domain, 227
Capacitor (see Capacitance)
Cascade programming, 555
Characteristic equation, 146
    z-domain, 595
Circuit
    definition, 61
    Norton equivalent, 97
    open, 101
    short, 101
    Thevenin equivalent, 95
Closed-loop system, 320
Co-factor, 438
Coefficients, Fourier, 41, 478
Combined inputs, 175
Combined response, 176, 273
Combined Signals, 38
Compensator, 320
Complex exponential, 32
Complex number, 607—612
    algebra, 609
    argument, 609
    conjugate, 611
    exponential form, 609
    modulus, 608
    polar form, 609
    rectangular form, 607
    trigonometric form, 609
Complex power, 371
Complex z-plane, 519
Components, passive
    electrical, 64
    impedance of, 229
    mechanical, 621—624
    phasor models, 348
    s-domain models, 229
Computation
    analog, 6
    digital, 5, 321
    hybrid, 7
Computer (see Computation)
Conditional stability, 284
Conductance, 367

Connection equations, 63, 450
Constant, 13
Continuous system, 1, 21
Continuous signal, 21
Convolution, 247—251
    discrete, 537—541
    integral, 250
    summation, 538
Coupled systems, 180, 445, 242—245
    mechanical, 629
    transfer function for, 303
Current
    definition, 56
    divider, 94
    Norton equivalent, 97
    source, 68
Cutoff frequency, 394

D

Damped sinusoid, 33, 39, 176—180, 270
    discrete, 551
Damper, viscous, 622, 624
Decade, 411
Decibel, 410
Decoupling transformation, 455
Delay distortion, 409
Dependent source, 70
Dependent variable, 12
Derivative operator
    rectangular, 579
    trapezoidal, 584
Design, 2
Determinant, 435
Dielectric, 67
Difference equation
    conversion to, 572
    definition, 531
    initial conditions, 535
    sequence-varying, 531
    solution, 532
    state-variable, 588
    z-transform of, 533
Differential equation
    definition, 15
    forcing function, 17
    initial conditions, 16
    Laplace transformed, 209
    nonlinear, 19
    order, 16
    simultaneous, 15
    time-varying, 19
    types of, 18
Digital computation, 5, 231
Digital signal, 21, 476
Digital-to-analog conversion, 493—495

Index

Digital-to-analog converter
    ladder-network, 495
    weighted-resistor, 494
Dimensions, 43
Direct programming, 554
Dirichlet conditions, 478
Discrete
    analysis, issues in, 472—476
    convolution, 537
    signal, 21, 476
    system, 530
    transfer function, 541—544
Discretization, 570
    rectangular, 578
    state-variable systems, 588
    Taylor's series method, 577
    trapezoidal, 583
    Tustin's approximation, 584
Distortion
    delay, 408
    magnitude, 407
    phase, 408
Dominant pole
    s-domain, 279
    z-domain, 551
Dynamic system, 1, 594

E

Eigenvalue, 456, 594
Eigenvector, 456
Electrical components, 64
Electrical units, 613
Electrical variables, 56
Envelope, 39
Equation
    characteristic, 46, 451, 595
    connection, 63, 450
    device, 450
    difference, 531
    differential, 15
    homogeneous, 145
    state-variable, 451
Equivalence transformation, 453—457
Equivalent systems, 453
Error signal, 321
Euler's identities, 33
Exponential function
    complex, 32
    decrement property, 31
    definition, 29
    Laplace transform of, 167
    matrix, 441
    slope property, 30
    time constant, 30
    properties, 30
Exponential order, 194
Extraneous elements, 90

F

Feedback
    connection, 310
    system, 317—322
Feed-forward programming
    s-domain, 325
    z-domain, 552
Filter
    characteristics, 395
    digital, 475
    high-pass, 339
Final-value theorem
    Laplace transforms, 207
    z-transforms, 528
Folding frequency, 490
Force-free response, 144, 461
Forced response
    definition, 144
    first-order system, 164
    second-order system, 170
    state-variable system, 462
Forcing function, 17
    discrete, 532
Fourier coefficients, 42, 478
Fourier series
    complex, 447
    exponential, 447
    trigonometric, 41
Fourier transform, 480
Frequency
    damped natural, 179
    folding, 490
    Nyquist, 484
    resonant, 355
    response functions, 392
    sampling, 484
    undamped natural, 179
Frequency response
    first-order, 396
    second-order, 398
    peak, 406
Function
    boxcar, 482
    delay, 409
    digital data, 476
    exponential, 29
    impulse, 34
    linear, 13
    piecewise-continuous, 194
    polynomial, 39
    ramp, 39
    rational, 211
    sampled, 512
    sinusoidal, 25
    step, 23

G

Gears, 635
Gravity, 630
Ground, 80

H

Half-power frequency, 395
Hard spring, 643
Homogeneity, 45
Homogeneous equation, 145, 596
Hybrid computation, 7

I

Ideal voltage source, 68
Impedance
    definition, 232
    input, 234
    parallel, 325
    phasor, 348
    s-domain, 232
    series, 236
Impulse function
    definition, 34
    discrete, 516
    integral property, 36
    Laplace transform of, 200
    sampling property, 35
Impulse response, 245
Independent variable, 12
Inductance
    definition, 65
    final-value model, 104
    impedance of, 228
    in parallel, 87
    in series, 87
    initial-value model, 348
    phasor model, 348
    s-domain, 227
Inductor (see Inductance)
Initial-value theorem
    Laplace transforms, 528
    z-transforms, 528
Inversion, matrix, 438

J

Junction, 59

K

Kirchhoff's laws
    current, 59
    voltage, 60

L

Lag, first-order, 268
Laplace transform
    definition, 193
    final-value theorem, 207
    initial-value theorem, 207
    inverse, 211—226
    inversion integral, 195
    linearity, 202
    of circuit components, 227
    of common functions, 196—200
    of derivatives, 204
    of differential equations, 209
    of source models, 228
    table of pairs, 615
    table of properties, 616
Linear
    differential equation, 13
    discrete system, 531
    state-variable system, 445
    spring, 621
    system, 13
Linearity
    definition, 13
    Laplace transforms, 202
    z-transforms, 527
Linearization, 640—644
Log-magnitude function, 410
Loop analysis, 72
Loop, 60

M

Magnitude function, 392
Massless points
    translational systems, 631
    rotational systems, 634
Mathematical model, 3
Matrix
    addition, 435
    adjoint, 438
    column, 433
    conformability, 435
    definition, 433
    diagonal, 433
    differentiation of, 440
    dimensions, 433
    exponential, 441
    function, 440
    identity, 433
    integration, 440
    inversion, 438
    multiplication, 435
    null, 433
    power of, 440
    row, 433
    singular, 438
    state transition, 458
    subtraction, 435
    transfer function, 459
    transposition, 434
    vector differential equation, 445
Maximum power transfer, 372

Index

Mechanical components
    linear spring, 621
    moment of inertia, 621
    rotary damper, 624
    rotary spring, 624
    torsion spring, 624
    viscous damper, 622
Mechanical systems, 619—652
Mechanical units, 653, 654
Mesh, 60
Mho, 234
Minimum phase system, 424
Moment of inertia, 621
Moment, 620
Momentum, linear, 619

N

Natural frequency
    damped, 197
    undamped, 197
Natural response
    first-order, 146
    second-order, 149
    state-variable systems, 461
Network
    equations, number of, 61
    extraneous elements in, 89
    initial-value, 99
    final-value, 103
    Laplace transformed, 233
    nonplanar, 71
    phasor, 348
    planar, 71
    topology, 61
Newton's second law, 619
Nodal analysis, 79
Node, 59
Nonlinear
    differential equation, 181
    discrete system, 531
    state-variable system, 445
    system, 19, 638—644
Nonplanar network, 71
Norton's theorem
    for impedances, 238
    for resistive networks, 97
Nyquist frequency, 484

O

Octave, 415
Ohm's law, 65
Open circuit, 101
Open-loop system, 319
Operational Amplifier
    applications, 117
    circuit model, 115
    frequency response, 116
    gain, 116
    integrator, 120
    isolation, 122
    linearity, 116
    resonator, 123
    summer, 119
    symbol, 115
    voltage follower, 121

P

Parallel
    capacitors, 87
    impedances, 325
    inductors, 87
    resistors, 86
    system blocks, 309
Parallel programming, 556
Parameter, 13
Parseval's theorem, 487
Partial fractions
    complex poles, 217
    imaginary poles, 216
    Laplace transforms, 211—216
    repeated poles, 214
    residue, 213
    simple poles, 213
    z-transforms, 523
Passive components, 64
Peak response, 400
Pendulum, 639
Periodic function, 26
Phase function, 392
Phasor
    definition, 347
    diagram, 352
    network, 348
    variables, 348
Planar network, 71
Pole-zero plot
    s-domain, 262
    z-domain, 547
Pole
    definition, 212
    dominant, 279, 551
    remote, 279
    s-plane, 263
    z-domain, 544
Polynomial
    auxiliary, 289
    characteristic, 456
    function, 39
    Laplace transform of, 199
Potentiometer, 94
Power
    apparent, 371
    average, 59, 358—361
    complex, 371
    factor, 366
    in network elements, 370
    in Fourier coefficients, 486
    in sampled signals, 486
    instantaneous, 58

maximum transfer of, 372—376
real, 371
triangle, 371
Power series
for inverse z-transforms, 525
Problem-oriented language, 475
Pulse transfer function, 542

R

Radius of convergence
Laplace transforms, 194
z-transforms, 519
Ramp function, 39
Rational function
definition, 211
improper, 212
Reactance, 366
Real power, 371
Rectangular integration
derivation, 578
first-order systems, 580
state equations, 591
Reference conventions, 67
Reference input, 320
Reference node, 80
Reference polarities, 61
Relative stability, 286
Remote pole, 279
Residue
definition, 213
s-plane determination, 274
Resistance
definition, 65
Norton, 97
parallel, 86
phasor model, 348
series, 86
s-domain model, 226
Thevenin, 95
Resistor (see Resistance)
Resonance
definition, 354, 401
frequency, 355
parallel, 356
series, 354
Response
forced, 144, 171
force-free, 144
natural, 144
overdamped, 171
transient, 176
RLC tank circuit, 153, 224, 255
RMS variables, 362
Root locus, 272, 321
Rotational mechanical systems, 632—638
Routh Array, 287
Routh criterion, 287
Runge-Kutta integration, 597—600

S

s-Domain
admittance, 233
capacitor, 227
impedance, 232
inductor, 227
resistor, 226
s-Plane, 262
Sampled-data signal, 21
Sampled time function, 51, 476
Sampling
frequency, 484
general process, 476—483
interval, 476
theorem, 483
Scalar system, 459
Sensitivity, 322—324
Sequence domain, 515
Series
capacitors, 87
impedances, 235
inductors, 87
resistors, 86
system blocks, 308
Short circuit, 101
Signal
bandlimited, 482
classification, 21
combined, 38
conversion, 492—502
definition, 12
discrete, 476
error, 321
exponential, 29
frequency content, 477
impulse, 34
ramp, 39
sinusoidal, 25
unit step, 23
Simulation, transfer function, 324—326
Singular matrix, 438
Sinusoidal function, 25
amplitude, 26
damped, 39, 176—180, 270
frequency, 26
period, 26
phase, 26
Sinusoidal steady state
transfer function in, 337
Slope property, 30
Slope-breakpoint approximation, 415
Source
combination, 89
conversion, 91
current, 68
dependent, 70
initial condition, 105
Norton, 97
s-domain, 228
Thevenin, 95

Index

Spectrum
  discrete magnitude, 478
  discrete phase, 478
  magnitude, 481
Spring
  linear, 621
  torsion, 624
Stability
  conditional, 284
  definition, 283
  discrete systems, 545, 594—597
  from poles, 284
  from transfer function, 305
  relative, 286
  Routh criterion, 287
  state-variable systems, 459
Stabilization, 285
State space, 448
State transition matrix, 458
State variable
  block diagram, 447
  definition, 443
  discrete systems, 588—594
Steady-state response
  discrete systems, 546
Step function, 23
Summing point, 310
Superposition, 13
Susceptance, 367
Switch notation, 170
System
  anticipatory, 538
  closed-loop, 320
  continuous, 1, 21
  coupled, 180, 445
  definition, 4, 443
  discrete, 21, 530
  dynamic, 1
  equivalent, 453
  feedback, 317—322
  linear, 13
  minimum-phase, 424
  nonlinear, 19, 638—644
  open-loop, 319
  rotational mechanical, 632—638
  scalar, 459
  state-variable, 444
  time-varying, 19
  translational mechanical, 627—632

T

Taylor's series
  discretization by, 577
  linearization by, 641
Thevenin's theorem
  impedances, 237
  resistive network, 95
Time-varying
  differential equation, 181
  state-variable system, 445
  system, 19

Torque, 620
Track store, 132
Trajectory, 448
Transfer function
  coupled systems, 303
  definition, 299
  discrete, 541—544
  experimental determination, 423
  matrix, 459
  pole-zero form, 300
  pulse, 542
  stability from, 305
  s-domain, 299
  z-domain, 542
Transform
  Fourier, 480
  Laplace, 193
  z-, 512
Transient response
  damping ratio, 177
  discrete systems, 545
  natural frequency, 179
Translational mechanical systems, 627—632
Trapezoidal integration
  derivation, 583
  first-order systems, 585
  state-variable equations, 592
Triangle, power, 371

U

Undetermined coefficients
  functional forms for, 169
  method of, 165
Units, 44
  electrical, 613
  mechanical, 653, 654
Unit prefixes, 614
Unit ramp function, 39
Unit step function, 23

V

Variable
  dependent, 12
  independent, 12
  state, 443
Vector
  column, 433
  input, 447
  output, 447
  row, 433
Vertex, 59
Viscous damper, 622
Voltage
  definition, 57
  divider, 93
  source, 68
  Thevenin, 95

W

Waveform, 24

Z

z-Plane, 519
    unit circle, 522
z-Transform
    closed form, 518
    convergence of, 519
    definition, 512
    divergence of, 519
    inverse, 523—526
    linearity, 527
    shifting theorems, 527
    table of pairs, 617
Zero
    s-domain, 212, 263
    z-domain, 544

Notes

Notes

Notes

Notes

Notes

Notes

Notes

Notes

Notes

Notes

Notes

Notes